Deep Learning and Computer Vision in Remote Sensing-II

Deep Learning and Computer Vision in Remote Sensing-II

Editors

Fahimeh Farahnakian
Jukka Heikkonen
Pouya Jafarzadeh

MDPI

Basel • Beijing • Wuhan • Barcelona • Belgrade • Novi Sad • Cluj • Manchester

Editors
Fahimeh Farahnakian
University of Turku
Turku, Finland

Jukka Heikkonen
University of Turku
Turku, Finland

Pouya Jafarzadeh
University of Turku
Turku, Finland

Editorial Office
MDPI
St. Alban-Anlage 66
4052 Basel, Switzerland

This is a reprint of articles from the Special Issue published online in the open access journal *Remote Sensing* (ISSN 2072-4292) (available at: https://www.mdpi.com/journal/remotesensing/special_issues/deep_learning_computer_rs_II).

For citation purposes, cite each article independently as indicated on the article page online and as indicated below:

Lastname, A.A.; Lastname, B.B. Article Title. *Journal Name* **Year**, *Volume Number*, Page Range.

ISBN 978-3-0365-9364-7 (Hbk)
ISBN 978-3-0365-9365-4 (PDF)
doi.org/10.3390/books978-3-0365-9365-4

Contents

About the Editors

Fahimeh Farahnakian

Fahimeh Farahnakian is currently an adjunct professor (docent) in the Algorithms and Computational Intelligence Research Lab, Department of Future Technologies, University of Turku, Finland.

Her research interests include the theory and algorithms of machine learning, computer vision and data analysis methods, and their applications in various fields. She has published +30 articles in journal and conference proceedings. She is a member of the IEEE and has also served on the program committees of numerous scientific conferences.

Jukka Heikkonen

Jukka Heikkonen is a full professor and head of the Algorithms and Computational Intelligence Research Lab, University of Turku, Finland. His research focuses on data analytics, machine learning, and autonomous systems. He has worked at top-level research laboratories and Centers of Excellence in Finland and international organizations (the European Commission and Japan), and has led many international and national research projects. He has authored more than 150 peer-reviewed scientific articles. He has served as an organizing/program committee member in numerous conferences and has acted as a guest editor in five Special Issues of scientific journals.

Pouya Jafarzadeh

Pouya Jafarzadeh received an MS degree in Technological Competence Management from the University of Applied Silence, Turku, Finland. He is currently working toward a PhD degree in the Algorithms and Computational Intelligence Research Lab, University of Turku, Finland. His research interests include artificial intelligence, machine learning, deep learning, computer vision, and data analysis. He is a frequent reviewer for research journals.

remote sensing

MDPI

Article

Sparse Signal Models for Data Augmentation in Deep Learning ATR

Tushar Agarwal *, Nithin Sugavanam and Emre Ertin

Department of Electrical and Computer Engineering, The Ohio State University, Columbus, OH 43210, USA; sugavanam.3@osu.edu (N.S.); ertin.1@osu.edu (E.E.)
* Correspondence: agarwal.270@osu.edu

Abstract: Automatic target recognition (ATR) algorithms are used to classify a given synthetic aperture radar (SAR) image into one of the known target classes by using the information gleaned from a set of training images that are available for each class. Recently, deep learning methods have been shown to achieve state-of-the-art classification accuracy if abundant training data are available, especially if they are sampled uniformly over the classes and in their poses. In this paper, we consider the ATR problem when a limited set of training images are available. We propose a data-augmentation approach to incorporate SAR domain knowledge and improve the generalization power of a data-intensive learning algorithm, such as a convolutional neural network (CNN). The proposed data-augmentation method employs a physics-inspired limited-persistence sparse modeling approach, which capitalizes on the commonly observed characteristics of wide-angle synthetic aperture radar (SAR) imagery. Specifically, we fit over-parametrized models of scattering to limited training data, and use the estimated models to synthesize new images at poses and sub-pixel translations that are not available in the given data in order to augment the limited training data. We exploit the sparsity of the scattering centers in the spatial domain and the smoothly varying structure of the scattering coefficients in the azimuthal domain to solve the ill-posed problem of the over-parametrized model fitting. The experimental results show that, for the training on the data-starved regions, the proposed method provides significant gains in the resulting ATR algorithm's generalization performance.

Keywords: machine learning; data augmentation; automatic target recognition; synthetic aperture radar

Citation: Agarwal, T.; Sugavanam, N.; Ertin, E. Sparse Signal Models for Data Augmentation in Deep Learning ATR. *Remote Sens.* **2023**, *15*, 4109. https://doi.org/10.3390/rs15164109

Academic Editors: Jukka Heikkonen, Fahimeh Farahnakian and Pouya Jafarzadeh

Received: 12 July 2023
Revised: 9 August 2023
Accepted: 11 August 2023
Published: 21 August 2023

1. Introduction

Synthetic aperture radar (SAR) sensors provide day and night high-resolution imaging capabilities that are robust to weather and other environmental factors. The SAR sensor consists of a moving radar platform with a collocated receiver and transmitter that traverses a wide aperture in the azimuth domain, acquiring coherent measurements of scene reflectivity. The returns for multiple pulses across the synthesized aperture are combined and coherently processed to produce high-resolution SAR imagery. A SAR imaging system achieves a high spatial resolution in both the radial direction, termed as range, as well as in the orthogonal direction, termed as cross-range. The range resolution is a function of the bandwidth of the signal used in illumination. The cross-range resolution is a function of the antenna aperture's size and the persistence of scattering centers [1]. A significant fraction of the energy in the back-scattered signal from the scene is due to a small set of dominant scattering centers that are resolved by the SAR sensor. The localization of back-scatter energy provides a distinct description of the targets of interest [2], such as in the case of man-made objects such as civilian and military vehicles. This sparsity structure has been utilized in [3,4] to design features like peak locations and edges that succinctly represent the scene. In the early works, these hand-crafted features were used in solving the target recognition problem in a statistical framework. Notably, the template-based methods exploited the geometric structure and variability of these features in the scattering centers in [5,6] to distinguish between the different target categories. The target signature of each of the scattering centers varied with the viewing angle of the sensor platform. Statistical methods can explicitly model and utilize this low-dimensional manifold structure of the scattering center descriptors [7,8] for improved decisions, as well as for integrating information across views [9,10].

However, ATR algorithms based on these hand-crafted features are limited to the information present in these descriptors, and they lack the generalization ability with respect to variability in clutter, pose, and noise. With the advent of data-driven algorithms such as artificial neural networks (ANN) [11], an appropriate feature set and a discriminating function can be jointly estimated using a unified objective function. Recent advances in techniques to incorporate the deep hierarchical structures used in ANN [12,13] has led to the widespread use of these methods to solve inference problems in a diverse set of application areas. Convolutional neural networks (CNN) in particular have been used as automatic feature extractors for image data. These methods have also been adopted in solving the ATR problem when using SAR images [14]. There have been several efforts in this direction, including the state-of-the-art ATR results in the MSTAR data set in [15]. These results establish that a CNN could be effective in radar image classification when provided with sufficient training data. However, this approach of designing ATR algorithms for new sensors that operate in different bands and elevations with limited training data from targets of interest is not feasible as the scattering behavior changes substantially as the wavelength of the operation changes. The major challenge is that neural networks usually require large data sets to have a good generalization performance. In general, labeled radar image data are not readily available in abundance unlike other image data sets. In this paper, we address the scarcity of training data and provide a general method that utilizes a model-based approach to capture and exploit the underlying scattering phenomenon to enrich a training data set.

Transfer learning is one of the most effective techniques through which to handle the availability of limited training data. Transfer learning uses the model parameters, which are estimated using a similar data set such as Image-net [16], as initialization for solving the problem of interest, and typically CNNs are used with little to no fine-tuning. There have been numerous experiments supporting the benefits of transfer learning, including two seminal papers [17,18]. However, radar images are significantly different from regular optical images. In particular, SAR works in the wavelength of 1 cm to 10 m, while visible light has a wavelength of the order of 1 nm. As a result, most surfaces in natural scenes are rough at visible wavelengths, leading to diffused reflections. In contrast, microwaves from radar transmitters undergo specular reflections. This difference in scattering behavior leads to substantially different images in SAR and optical imaging. Since specular reflections dominate the scattering phenomenon, the images are sensitive to instantaneous factors like the imaging device's orientation and background clutter. Therefore, readily available optical-imagery based deep neural network models like Alex-net and VGG16 [19] are not suitable for transferring knowledge to the SAR domain. In this paper, we pursue an alternative strategy for the data augmentation of limited data sets through using a principled approach that exploits the phenomenology of the RF backscatter data.

Next, we review the relevant research work and outline our contributions in Sections 1.1 and 1.2, respectively.

1.1. Related Work

Over-fitting is a modeling error that is common to data-driven machine learning methods when the learned classifier function is too closely aligned to the training data points and therefore fails to generalize to the data points outside the support of the training set. The over-fitting problem is exacerbated with smaller training sets. Several methods have been proposed to reduce over-fitting and to improve the generalization performance. Typically, the ill-posed problem of fitting an over-parametrized function to data is solved by using regularizers that impose structure and constraints in the solution space. The norm of the model parameters serves as a standard regularizing function. This keeps the parameter values small with a 2-norm ($||\cdot||_2$ called L_2 loss) space or sparse with a 1-norm ($||\cdot||_1$ called L_1 loss) space. Furthermore, the optimization algorithms, such as stochastic gradient descent and mirror descent, implicitly induce regularization [20,21]. Dropout, introduced by Srivastava et al. [22], is another popular method specifically for deep neural networks. The idea is to randomly switch off certain neurons in the network by multiplying a Bernoulli random variable with a predefined probability distribution. The overall model learned is an average of these sub-models, providing improved generalization performance. Batch normalization is another way through which to improve the generalization performance, and was proposed by Ioffe and Szegedy [23]. They proposed normalizing all the neuron values of the designated layers continuously while training them along with an adaptive mean and variance that would also be learned as part of the back-propagation training regime. Finally, the work by Neyshabur et al. [24] established the benefit of over-parameterizing in implicitly regularizing the optimization problem and in improving the generalization performance.

Transfer learning is another approach for improving the generalization performance in the cases where there is a limited availability of data. Pan and Yang [25] provided a comprehensive overview that illustrated the different applications and performance gains of transfer learning. For radar data, Huang et al. [26] suggested a promising approach along this direction by using a large corpus of SAR data to train feature extractors in an unsupervised manner. Huang et al. [27] recently extended this idea to high-resolution SAR data.

Much of the recent literature has gravitated towards using CNNs as the automatic feature extractors for the SAR ATR task, as mentioned earlier. One research direction has been to improve the classifier of CNNs by cascading CNN feature extraction with other machine learning algorithms, such as a large-margin softmax classifier in [28] and an ensemble learning-based classifier called the AdaBoost rotation forest in [29]. Other researchers have focused on improving ATR performance through multiple views of the same object as in [30], or by using multiple polarization information as in [31,32]. There is a great deal of potential in improving ATR performance, especially in challenging scenarios such as clutter. In such scenarios, polarimetric data and bistatic measurements serve as an important tool in improving the classification performance, especially in cases of low sample sizes in the training data. Our phase history model works with complex-valued data, and we previously extended the model to bistatic measurements in [33]. Another important research direction has been focused on reducing the space and computation requirements of CNN-based ATR. To achieve this, depthwise separable convolutions were used in [34]; in addition, Huffman coding and weight quantization were used in [35], as well as knowledge distillation in [36,37]. Few other approaches focus on learning a special type of features. Dong et al. [38] generated an augmented monogenic feature vector followed by a sparse representation-based classification. In [39], the authors used hand-designed features with supervised discriminative dictionary learning to perform SAR ATR. Song et al. used a sparse-representation-based classification (SRC) approach in [40]. In [41], Huang et al. designed a joint low rank and sparse dictionary to denoise the radar image while keeping the main texture of the targets. Yu et al. [42] proposed a combination of Gabor features and the features extracted by neural networks for better classification performance.

When there is a limited availability of SAR data, there exist several ANN-architecture-based approaches to improve generalization. Chen et al. [14] restricted the effective degrees of freedom of a network by using a fully convolutional network. Lin et al. [43] proposed a convolutional highway network to tackle the problem of limited data availability. In [44], the authors designed a specialized ResNet architecture that learns effectively even when the training data set is small. In addition to [26], the idea of semi-supervised learning has recently received much attention for improving SAR ATR performance in cases of limited data availability. Yue et al. [45] used a CNN to obtain the class probabilities of unlabeled data samples, which was followed by incorporating this knowledge into the classification loss of ATR via a novel linear discriminant analysis method. In [46], Wang et al. used the information in unlabeled SAR data to inform a deep classifier by using a self-consistent augmentation rule, a mixup-based mixture, and weighted loss. Recently, Chen et al. [47] used an unlabeled data-based consistency criterion, domain adaptation, and top-k loss to alleviate the requirement of labeled data.

Data augmentation is another example of a regularization strategy that reduces the generalization error while not affecting the training error [48,49], and is the main focus of this paper. The main idea is to use domain-specific transformations to augment the original training data set. J. Ding et al. [50] explored the effectiveness of the conventional transformations used for optical images, viz. translations, noise addition, and linear interpolation (for pose synthesis). They reported marginal improvements in classification performance on the MSTAR data set. Yan [51] used the original training images to generate noisy samples at different signal-to-noise ratios, multiresolution representations, and as partially occluded images. In [52], the authors proposed a generative adversarial network (GAN) to generate synthetic samples for the augmentation of SAR data, but they did not report any significant improvements in the error rate of the ATR task. Lewis et al. [53] explored multiple deep generative models for SAR data augmentation and recommend BicycleGAN after experimentation. In another effort that used a GAN, Gao et al. [54] used two jointly trained discriminators with a non-conventional architecture. They further used the trained generator to augment the base data set and reported significant improvements. Cui et al. [55] used a Wasserstein GAN, and Sun et al. [56] proposed an attribute-driven angular rotation generative network to produce synthetic samples for augmentation. Shi et al. [57] used a GAN to super-resolve samples for data augmentation. None of these deep generative methods used complex-valued imagery; therefore—unlike the proposed work here—they were unable to create imagery that was consistent with the frequency support of the imaging system. Cha et al. [58] used images from a SAR data simulator and refined them using a learned function from real images. Simple rotations of radar images were considered as a data augmentation method in [59]. In [15], Zhong et al. suggested key ideas for incorporating prior knowledge in training the model. They added samples that were flipped in the cross-range dimension with a reversed sign of the azimuthal angle. Such flip-augmentation exploits the symmetric nature of most objects in the MSTAR data set. They also added a loss that was auxillary to the primary objective of classification. The authors used the pose prediction (azimuthal angle) as the secondary objective of the network. They empirically showed that this helps by adding meaningful constraints to the network learning. Thus, the network was more informed about the auxiliary confounding factor, improving its generalization capability. Lv and Liu [60] proposed to extract attributed scattering centers (ASCs) through the sparse representation algorithm. The synthetic samples for data augmentation were then reconstructed by selecting a subset of these ASCs and by repeating the procedure.

1.2. Contributions

In this work, we introduce a novel data augmentation method for SAR domains, following a principled approach that exploits the phenomenology of the RF backscatter data over the azimuth and frequency domains. This paper is an extension of our previous work [61] with additional results that include a comparison to other existing techniques and an ablation study of the components of the proposed technique.

First, we introduce an approach for pose synthesis that models and exploits the limited persistence of the sparse set of scatterers over the azimuth domain. We assume that man-made objects comprise a small set of dominant scattering centers. Specifically, we first transform the image into the polar frequency domain to obtain the samples in the phase history domain. We then construct a model motivated by the scattering behavior of canonical reflectors in this phase history domain. The phase history model further decouples the point-spread function associated with the imaging setup. This model captures the phenomenology of the viewing-angle-dependent anisotropic scattering behavior of man made-objects, as well as provides realistic imagery at poses outside the training data set, with quality that far surpasses previous approaches such as linear interpolation in image domains [50].

Second, with modeling in the complex valued phase history domain, our algorithm can create realistic sub-pixel shift augmentations that capture the well-known scintillation effects in SAR imagery. These sub-pixel shifts are not possible in traditional image domains that use standard interpolators (linear, cubic, etc.), as the complex-valued interpolation kernels need to be appropriately designed by taking into account the azimuth and frequency windows of the sensor. We hypothesize that these two factors are essential for improving the network's knowledge about the SAR imaging systems' underlying physics.

Third, we focus on a state-of-the-art deep learning classifier for SAR ATR [15] using the MSTAR data set, as well as provide extensive simulation studies to illustrate the learning performance of different training data set sizes with un-augmented and augmented approaches to training. Our results show a significant boost in the generalization performance over both un-augmented and augmentation approaches with the previously suggested approaches. In particular, for the MSTAR data set when the training data set is reduced by a factor of 32, the proposed augmentation algorithm reduces the test error by more than 42% when compared to the baseline approach that includes image domain flips and integer pixel translations.

It is important to note that our data-augmentation-based strategy is generic and decoupled from the network architectures proposed in other works like [14,43]. Therefore, the proposed augmentation strategy may yield even further improvements in conjunction with the methods mentioned above. Our objective here is to demonstrate the benefits of the proposed data augmentation strategy. Hence, apart from data augmentation, we only use Zhong et al.'s [15] multi-task learning paradigm.

The rest of the paper is structured as follows. In Sections 3.1 and 3.2, we describe the data set and network architecture in detail. In Section 2.1, we provide an overview of our strategy and then describe the details of our pose-synthesis methodology in Section 2.2. Following those sections, we present the details of the experiments and corresponding results in Sections 3 and 4, respectively, which provide the empirical evidence for the effectiveness of the proposed data augmentation method. We then conclude with some possible directions for future research in Section 5.

2. Model-Based SAR Data Augmentation

An approach to ATR algorithm design is to train a parametric neural network classifier g, with parameters $w \in \mathbb{R}^{d_w}$, that predicts an estimate of output labels $Y \in \mathbb{R}^{d_Y}$ for an input $X \in \mathbb{C}^{d_X}$, i.e., $\hat{Y} = g(X; w)$, where d_X, d_w, and d_Y are dimensions of X, w, and Y, respectively. We consider a supervised learning setting, where a labeled training data set $\mathcal{D}_{train} = \{(X_u, Y_u)\}_{u=1}^{N_{train}}$ is used to estimate the classifier parameters w, where N_{train} is the total number of training samples. The training procedure is the minimization of an appropriate loss function $\mathcal{L} : (w, \mathcal{D}) \to \mathbb{R}$, which is achieved by using an iterative algorithm like the stochastic gradient descent. Therefore, the learned w^* is the solution of the following minimization problem $\mathcal{P}$:

$$w^* = \mathcal{P}(\mathcal{D}) = \arg\min_{w} \mathcal{L}(w, \mathcal{D}) \tag{1}$$

Data augmentation involves applying an appropriate transformation, such as $T\mathcal{D}_{in} \to \mathcal{D}_{out}$, to a data set (only using $\mathcal{D}_{train}$ for our purposes), and to then expand it to an augmented data set $T(\mathcal{D}_{train})$. We also use a validation data set, $\mathcal{D}_{val} = \{(X_u, Y_u)\}_{u=1}^{N_{val}}$, for cross-validation during training. Furthermore, we use a test data set $\mathcal{D}_{test} = \{(X_u, Y_u)\}_{u=1}^{N_{test}}$ for evaluating $g(X; w)$ in post-training. The evaluation can be conducted using a suitable metric $\mathcal{M} : (w, \mathcal{D}) \to \mathbb{R}$, which may be different from the $\mathcal{L}$ above. Our aim is to find T, such that the estimated parameters $w_{aug} = \mathcal{P}(T(\mathcal{D}_{train}))$ perform better than

$w_{train} = \mathcal{P}(\mathcal{D}_{train})$ in terms of the chosen metric, i.e., $\mathcal{M}(w_{aug}, \mathcal{D}_{test})$ is more desirable than $\mathcal{M}(w_{train}, \mathcal{D}_{test})$.

2.1. Exploiting SAR Phenomenology for Data Augmentation

Simple approaches to designing transformations T for data augmentation consider translation invariance and the symmetry of objects around its main axis in order to introduce discrete pixel shifts and flips along the cross-range dimension. Our augmentation strategy goes further and uses model-based transformations to improve the network's knowledge principally about two confounding factors, the pose as well as the scintillation effects that occur due to shifts in the range domain. Our method allows the synthesis of new poses in a close neighborhood of existing poses, based on a sparse modeling of the existing training data set that exploits the spatial sparsity and the scattering centers' limited persistence.

An overview of our approach is as follows: for every image in the training data set, we fit a sparse model in the phase history domain by exploiting SAR phenomenology. This model is henceforth referred as the PH model. We utilize the continuity of this PH model in the azimuth domain to extrapolate the phase history measurements and to synthesize new images in a close neighborhood of the original image. The PH model also allows for the introduction of arbitrary-valued sub-pixel shifts in both range and cross-range dimensions to images at both the original and synthesized poses. These fractional shifts provide information to the network regarding scintillation effects, which further improves its generalization capability. In the following section, we describe our modeling and pose synthesis strategy in full detail.

2.2. Modeling and Pose Synthesis Methodology

This section describes the pose synthesis methodology used for data augmentation when using the PH model. This work builds on our earlier work, which focused on modeling of the scattering behavior of targets in monostatic and bistatic setups [33,62–68]. We first constructed a model for each image in the training data set and locally extrapolated the measured images through using the model. We assumed that a SAR sensor that operated in the spotlight mode was used to create the images (as in the case of the MSTAR data set). The images were translated from the spatial domain to the Cartesian frequency domain via the steps described in [69]. Subsequently, we converted the frequency measurements to the polar coordinates to obtain the phase history measurements described in [70].

We considered a square patch on the ground of side lengths $L = 30$ m that were centered around the target. From the geometric theory of diffraction, we assume that a complex target can be decomposed into a sparse set of scattering centers. The scattering centers are then assumed to be K point targets, and are described through using $\{(x_k, y_k), h_k(\theta, \phi)\}_{k=1}^{K}$, where $(x_k, y_k) \in [-\frac{L}{2}, \frac{L}{2}] \times [-\frac{L}{2}, \frac{L}{2}]$ are the spatial coordinates of the point targets, θ is the azimuthal angle, ϕ is the angle of elevation of the radar platform, and $h_k(\theta, \phi)$ is the corresponding scattering coefficients that depend on the viewing angle. The samples of the received signal after the standard de-chirping procedure are given by

$$s(f_m; \theta, \phi) = \sum_{k=1}^{K} h_k(\theta, \phi) \exp\left(-j4\pi \frac{f_m \cos(\phi)}{c} (x_k \cos(\theta) + y_k \sin(\theta))\right), \tag{2}$$

where f_m is the illuminating frequencies such that $m \in [M]$, $M = \frac{2BL}{c}$; B is the bandwidth of the transmitted pulse; c is the speed of light; and the notation $[M]$ denotes the enumeration of natural numbers up till M. We estimated the function $h_k(\theta, \phi)\ \forall\ k \in [K]$ from the receiver samples.

Parametric models for standard reflectors, such as dihedral and trihedral reflectors, were studied in [71–73]. These models indicate that the reflectivity is a smooth function over the viewing angle, which is parameterized by the reflector's dimensions and orientation. Therefore, we exploited this smoothness to approximate this infinite-dimensional function through using interpolation strategies [74] with the available set of samples Θ in the angle domain. We denote the sampled returns from the scene by the matrix $\mathbf{S} = NUFFT(X) \in \mathbb{C}^{N_\theta \times M}$, where NUFFT represents the non-uniform Fourier transform. The elements of $\mathbf{S}$ are defined as follows:

$$s_{m,i} = n_{m,i} + \sum_{k=1}^{K} h_k(\theta_i, \phi) \exp\left(-j4\pi \frac{f_m \cos(\phi)}{c}(x_k \cos(\theta_i) + y_k \sin(\theta_i))\right). \quad (3)$$

where $n_{m,i}$ represents the measurement noise. In order to solve the estimation problem, we assume that the function h_k has a representation in the basis set denoted by the matrix $\mathbf{\Psi} \in \mathbb{C}^{N_\theta \times D}$ of a size D. For the MSTAR data set, the elevation angles we worked with are similar. We assumed that the variation in h_k with respect to ϕ was insignificant. This assumption lead to the following relation $h_k(\theta;\phi) = \sum_{v=1}^{D} c_{v,k}\psi_v(\theta) + \epsilon_P$. The estimated phase history matrix was now $\hat{\mathbf{S}}$, whose elements were given by

$$\hat{s}_{m,i} = \hat{n}_{m,i} + \sum_{k=1}^{K}\sum_{v=1}^{D} c_{v,k}\psi_v(\theta_i) \exp\left(-j4\pi \frac{f_m \cos(\phi)}{c}(x_k \cos(\theta_i) + y_k \sin(\theta_i))\right), \quad (4)$$

where $\hat{n}_{m,i}$ consists of the measurement noise and the approximation error. To estimate the coefficients $c_{v,k}$ from the noisy measurements in (4), we discretized the scene with a resolution of ΔR in the X, Y (range and cross-range, respectively) plane to obtain the $K = N_R^2$ grid points, where $N_R = \frac{2BL}{c}$ is the number of the range bins. Furthermore, we considered a smooth Gaussian function to perform the noisy interpolation. We partitioned the sub-aperture $2\Delta\theta$ into smaller intervals of equal length with a corresponding set containing the means of the intervals given by $\{\hat{\theta}_v\}_{v=1}^{D}$, where $D = 12$, and these were used as the centroids for the Gaussian interpolating functions. We assumed the width of the Gaussian function σ_G as a constant hyper-parameter, whose selection is described in Section 3.4. Hence, σ_G is the constant minimum persistence of the scattering center in the azimuth domain that we wish to detect. The radial basis functions used were

$$\psi_v(\theta) = \exp\left(-\left(\frac{\theta - \hat{\theta}_v}{2\sigma_G}\right)^2\right) \quad (5)$$

The elements of $\hat{\mathbf{S}}$ were obtained due to the scattering centers located at the discrete grid points, which are now given by

$$\hat{s}_{m,i} = \hat{n}_{m,i} + \sum_{k=1}^{N_R^2}\sum_{v=1}^{D} c_{v,k}\psi_v(\theta_i) \exp\left(-j4\pi \frac{f_m \cos(\phi)}{c}(x_k \cos(\theta_i) + y_k \sin(\theta_i))\right). \quad (6)$$

Here, the discrete grids for (x_k, y_k) and (θ_i, f_m) are now both known. Let the vectors containing all corresponding grid points for $x_k, y_k, \theta_i, and f_m$ be referred to as $\mathbf{x}, \mathbf{y}, \boldsymbol{\theta}, and\ \mathbf{f}$ respectively. The problem now is to find the coefficients $c_{v,k}$ that minimize the error between $\hat{\mathbf{S}}$ and $\mathbf{S}$. Let vector $\mathbf{c_k} = [c_{1,k} \cdots c_{D,k}]^T$. To recover the structured signal $\mathbf{h} = [h_1 \cdots h_{N_R^2}]$, which represents the scattering coefficient of a sparse scene that has a sparse representation in an underlying set of functions, we solve the following linear inverse problem via a sparse-group regularization on $\mathbf{c_k} \forall\, k \in [N_R^2]$.

$$\min_{\mathbf{C}}\left(\sum_{k=1}^{N_R^2} \lambda\|\mathbf{c_k}\|_2 + \|\mathbf{S} - \hat{\mathbf{S}}\|_F\right) \Longleftrightarrow \min_{\mathbf{C}} J(\mathbf{C}, \sigma_G) \quad (7)$$

where $\mathbf{C}$ refers to the matrix $[\mathbf{c_1} \cdots \mathbf{c_{N_R^2}}]$, σ_G is a constant hyper-parameter, and $\|\cdot\|_2, \|\cdot\|_F$ refer to the l^2, Frobenius norms, respectively.

The elements of the recovered model, $\mathbf{S}^*(\boldsymbol{\theta}; \mathbf{f})$ are now

$$s^*_{i,m} = \sum_{k=1}^{N_R^2} \sum_{v=1}^{D} c^*_{v,k} \Psi_v(\theta_i) \exp\left(-j4\pi \frac{f_m \cos(\phi)}{c} (x_k \cos(\theta_i) + y_k \sin(\theta_i)) \right) \quad (8)$$

where $c^*_{v,k}$ are the recovered coefficients. The phase history measurements were converted back to the image by using overlapping sub-apertures that spanned $2\Delta\theta = 3$ degrees in the azimuth domain, as shown in Figure 1. Here, $\Delta\theta$ is the angular span of the sub-aperture from the center azimuth. This azimuth span determines the cross-range resolution of the SAR image. The slant-plane cross-range resolution is given by

$$\Delta^{CR}_{slant} = \frac{\lambda_c}{4\sin(\Delta\Theta)},$$
$$\Delta\Theta = \sin^{-1}\left(\frac{\lambda_c}{4\Delta^{CR}_{ground}\cos(\phi)} \right).$$

Each image in the MSTAR data set contains a header that summarizes the imaging geometry information stored in the Phoenix format. The parameters, such as center-frequency f_c and angle of depression ϕ, are obtained from the header information stored in each file. We assume that the cross-range resolution given in the file is on the ground-plane. The cross-range resolution is given as $\Delta^{CR}_{ground} = 0.305$. We infer that $\Delta\Theta = 1.51$ degrees. We apply the same Taylor window with zero-padding and then translated it back to the Cartesian coordinates before applying the Fourier transform to generate the images to augment the data set.

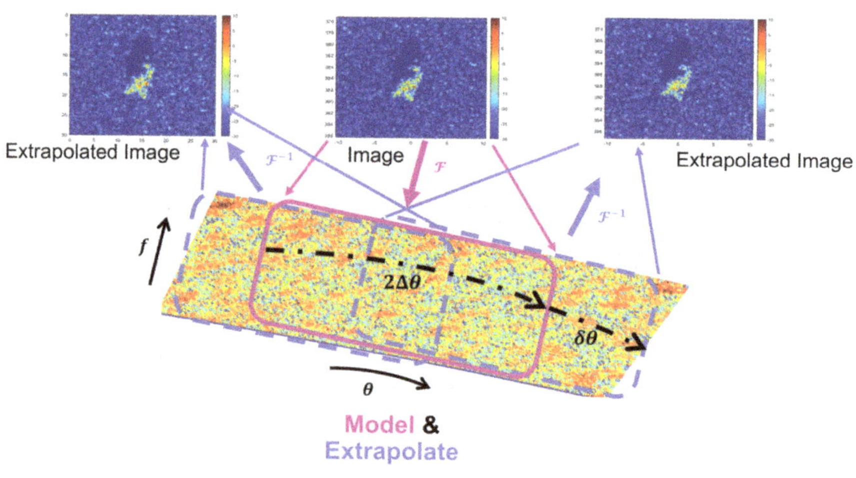

Figure 1. Pose synthesis using the phase history model. $\mathcal{F}$ denotes the Fourier transform operator. The phase history collected over an azimuth span of $2\Delta\theta = 3°$ was extrapolated by $\delta\theta$ via the model $\mathbf{S}^*(\boldsymbol{\theta}; \mathbf{f})$.

3. Experiments

We hypothesize that the underlying scattering mechanism is locally continuous or persistent in nearby look angles. We exploited this structure to generate realistic SAR images in the nearby look angles to augment the training data set. We evaluate that this hypothesis can be supported in the publicly available MSTAR data set, which has been typically used for evaluating algorithms.

3.1. MSTAR Data Set

One of the motivations for choosing the well-studied MSTAR data set for our experiments is its popularity, which enables us to compare our method with several other approaches in the literature. Additionally, imaging geometry parameters are provided in the MSTAR data set, which makes it straightforward to infer the frequency support of the images. Alternatively, for the data sets that do not provide the parameters, we can infer the frequency support of the image by applying Fourier transform to the complex-valued SAR image. The MSTAR data set consists of 10 classes, i.e., tanks (T62 and T72), armored vehicles (BRDM2, BMP2, BTR60, and BTR70), a rocket launcher (2S1), an air defense unit (ZSU234), a military truck (ZIL131), and a bulldozer (D7). We illustrate this observation in Figure 2.

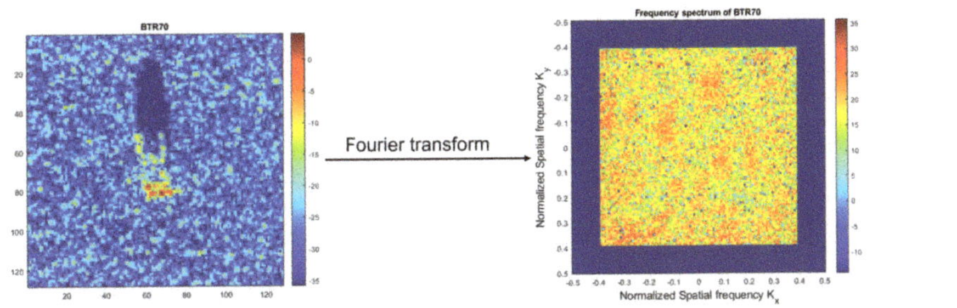

Figure 2. A 128 × 128 image chip from the MSTAR data set of a BTR-70 Tank and the corresponding spectral representation that was obtained by applying 2D Fourier transform to extract the K-space support.

We considered a image chip of BTR70 at the size of a 128 × 128 pixel-sized image, as well as the corresponding spectral content. The support exists on a central region of 100 × 100 frequency points. The corners of the square support can now be expressed in terms of the normalized spatial frequency domain, which is used to compute the K-space points. The phase history measurements at those K-space points were computed via the non-uniform Fourier transform [75,76] operator without a loss in generality. Therefore, we can estimate our model on this normalized domain by sampling in a wedge-shaped region. The radar platform used in constructing the MSTAR data set acquires the measurements through using $N_p = 100$ pulses over an aperture of 3 degrees. The phase history measurements obtained in the receiver were converted to images via the sub-aperture-based method described in [77]. The motion-compensation steps were followed by the application of a Taylor window to control the side-lobes. The measurements were zero-padded to obtain an over-sampled image via the Fourier transform. The complete MSTAR data set used in [15] was highly imbalanced. We replaced the data set used in [15] with a balanced subset, which is referred to as the standard operating conditions that were considered in [14,26]. We henceforth denote this subset as the SOC MSTAR data set.

Similar to the existing literature, we used the images at a depression angle of $\phi = 17°$ for training, while images at $\phi = 15°$ formed the test set. Similar to [15], we cropped the images to 64 × 64 with the objects in the center. Note that we cropped the images right before feeding it to the ANN. We performed the modeling and augmentation steps on the original images. Since our paper's objective is to investigate the effects of data augmentation, we worked with much smaller training data sets by artificially reducing the size of our data set to $\phi = 17°$. We exponentially sub-sampled by extracting only the $\mathcal{R}$ ratio of the samples from each class, where $\mathcal{R} \in \{2^{-5}, 2^{-4}, 2^{-3}, 2^{-2}, 2^{-1}, 2^{0}\}$. We ensured that the extracted images were uniformly distributed over the $[0, 2\pi]$ azimuthal angle domain for each sub-sampling ratio. This sub-sampling strategy is essential for ensuring that the learning algorithm obtains a complete view of the vehicle's scattering behavior. We further selected 15% of the uniformly distributed samples from this uniformly sub-sampled data as the validation set, we then utilized the remaining 85% as the new training set. The training data $\mathcal{D}_{train}$ include a flip augmentation along the cross-range domain [15], as well as real-time translations along both the range and cross-range domains. These translations (in no. of pixels) are randomly sampled from the set $\{-6, -4, -2, 0, 2, 4, 6\}$ at every epoch; as such, $\mathcal{D}_{train}$ will be henceforth referred to as just the baseline data. We also included the flip augmentation in the final validation set $\mathcal{D}_{val}$, and no augmentations were included in the final test-set $\mathcal{D}_{test}$. We formed our $T(\mathcal{D}_{train})$ by performing the proposed pose augmentation on each radar image in the training data set, as described in Section 2.2. Additionally, our net transformation T also includes sub-pixel level translations, as well as real-time pixel-level translations [50] in the range and cross-range domains. We used the sub-pixel shifts of the $\frac{1}{2}$ pixel, which corresponded to an approximately 0.15 m displacement in the Y-direction (range) as well as in the X-direction (cross-range) of the scene, where each pixel corresponds to 0.3 m in the range and cross-range domains.

3.2. Network Architecture

Our data augmentation algorithm was decoupled from the network architecture by realizing the ATR algorithm. For our experiments, we choose the simple CNN network architecture that was inspired by [15] and which is shown in Figure 3. We made such an architectural choice because, similar to Zhong and Ettinger [15], we wished to show that the classification performance of even simple CNN architectures can be improved through successful regularization through using domain-specific data augmentation for (in our case) SAR ATR. We modified the network and used batch-normalization layers after the ReLU activation in the convolutional layers. We deferred the use of dropout in the convolutional layers since batch normalization regularizes the optimization procedure [78]. After the last convolutional layer, we flattened out all the feature values and used a fully connected (FC) layer, which was followed by a dropout layer that was used to obtain the final set of features. These features were used to estimate class Y_1 of the input SAR images via training that was achieved by using the categorical cross-entropy loss function $\mathcal{L}_1$. We further modified the cosine loss, which is used for pose awareness in [15], to a pair of simpler losses by using the $Y_2 = \sin(\theta)$ and $Y_3 = \mathbb{1}_A(\theta)$ features, where θ is the azimuthal angle and $\mathbb{1}_A$ is the indicator function over set $A = [\frac{-\pi}{2}, \frac{\pi}{2}]$. The mean-squared-error loss $\mathcal{L}_2$ was used for training the network to estimate Y_2 and the binary cross-entropy loss $\mathcal{L}_3$ (which is used for training the network to estimate Y_3). These two features uniquely determined the azimuthal angle, and they remove the need for a cosine distance loss. In our experiments, while training the model, we found that this modification to the loss function resulted in improving the convergence of the optimization procedure. The loss function of $\mathcal{L}$ to find the network parameters is now

$$\begin{aligned}
\mathcal{L}(w, \mathcal{D}) &= \bar{\mathbb{E}}_{\mathcal{D}}[\mathcal{L}_1(w, X, Y_1) + \mathcal{L}_2(w, X, Y_2) + \mathcal{L}_3(w, X, Y_3)] \qquad (9) \\
\mathcal{L}_1(w, X, Y_1) &= -\sum_{p=1}^{10} Y_{1,p} \log(\hat{Y}_{1,p}(w, |X|)) \\
\mathcal{L}_2(w, X, Y_2) &= (Y_2 - \hat{Y}_2(w, |X|))^2 \\
\mathcal{L}_3(w, X, Y_3) &= -Y_3 \log(\hat{Y}_3(w, |X|)) - (1 - Y_3) \log(1 - \hat{Y}_3(w, |X|))
\end{aligned}$$

where $|.|$ denotes the absolute value, $X \in \mathbb{C}^{64\times 64}, Y_1 \in \{0,1\}^{10\times 1}, Y_2 \in [-1,1], and\ Y_3 \in \{0,1\}$ refer to the complex radar images, the one-hot vector of the 10 classes, $\sin(\theta)$, and $\mathbb{1}_A(\theta)$, respectively. $\bar{\mathbb{E}}_{\mathcal{D}}$ refers to the empirical mean over data set $\mathcal{D}$, and $Y_{1,p}$ is the pth component of the vector Y_1. All the quantities with ^(hat) are the corresponding estimates given by the ANN.

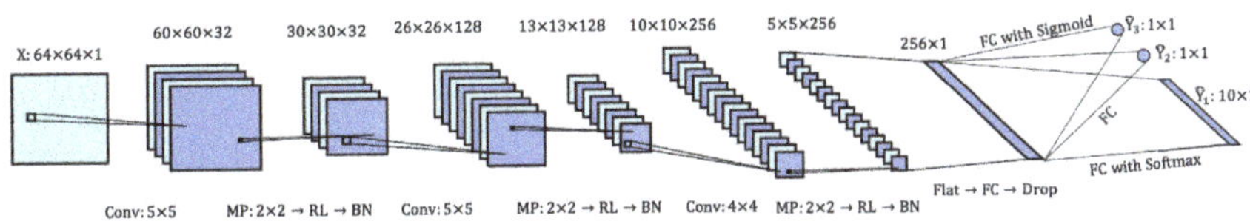

Figure 3. The neural network architecture. The abbreviations used are as follows. Conv is the convolutional layer followed by the kernel height × width. MP is the max pooling followed by the pooling size as the height × width. RL, BN, Flat, Drop, and FC are the ReLU, batch normalization, flattening, dropout and fully-connected layers, respectively. The sizes of the feature maps are mentioned at the top as height × width × channels.

3.3. Experimental Setup

The experiments were conducted using the network that is described in Section 3.2. This was used on the data sets described in Section 3.1. This model was trained on a local machine with a Titan Xp GPU. The Tensorflow (1.10) [79] library was used for its implementation through its Python API. We used the ReLU activation function everywhere except for in the final output layers of $\hat{Y}_1$, $\hat{Y}_2$, and $\hat{Y}_3$, where we instead used the Softmax, Linear, and Sigmoid activations, respectively.

An overview of the processing steps conducted to synthesize the radar images is as follows: Starting from the complex radar data, as described in Figure 4, we first transformed the image to a K-space by inverting the transformations applied to the MSTAR data in order to obtain the phase history representation. Through using the header information from the MSTAR data set, we determined the discrete grids for (x_k, y_k) and (θ_i, f_m). Next, we estimated the model coefficients by solving the optimization problem described in Equation (7). As a result, we obtained the $\mathbf{S}^*(\boldsymbol{\theta}; \mathbf{f})$ model that is given by Equation (8). This model was further used to synthesize new columns of phase history data (or to extend the $\boldsymbol{\theta}$ vector). Consequently, a synthesized image was produced, which was achieved by the procedure described in Section 2.2 and followed by a transformation of the phase history data to complex-valued image data. The complete MATLAB code that was used to perform our proposed augmentations on the MSTAR data set is available at https://github.com/SENSE-Lab-OSU/mstar_data_aug (accessed on 11 July 2023).

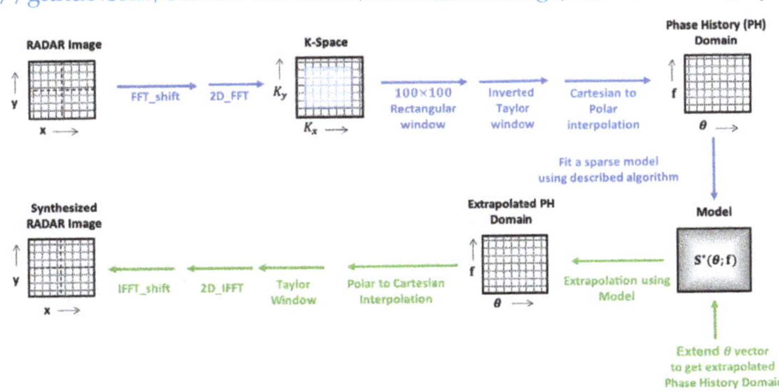

Figure 4. Overview of the Image Synthesizing Procedure. All boxes with grid lines represent the matrices of complex values across an $(x_k, y_k)\ \forall\ k \in [N_R^2]$ grid, where Cross-range and Range are labeled, as well as along an $(\theta_i, f_m)\ \forall\ i \in [N_\theta], m \in [M]$ grid, where $\boldsymbol{\theta}$ and $\mathbf{f}$ are also labeled. The blue arrows represent the pre-model fitting stage, and the green arrows represent the post-model fitting stage.

We used the magnitude of the complex-valued radar data as the input X, which is in agreement with the existing literature for training the network. We normalized all the input images to the unit norm to reduce some of the undesired effects that occur due to the Gaussian kernel during extrapolation. We also removed all of the synthetic images at poses that were already in the corresponding training set. Then, the optimization problem in Equation (1) was solved by using the off-the-shelf method, as well as by the Adam variant of the mini-batch stochastic gradient descent optimizer with a mini-batch size of 64. The training was carried out for many epochs (>400) while using the early-stopping criterion, and the model was saved for the best moving average validation performance metric. Although we care about accuracy (i.e., that the percentage of samples are classified correctly) as a performance metric, the $\mathcal{D}_{val}$ here becomes small, especially for small $\mathcal{R}$ values, thereby saturating the validation accuracy at 100% and thus yielding this metric as less useful. Instead, we then monitored the minimum classification loss $\mathcal{L}_1$ as the validation performance metric. We reported the percentage error (or misclassification), which was $100 -$ accuracy, as the test performance results.

3.4. Determining Hyper Parameters

The PH model for each image and the neural network model introduced a set of hyper parameters. We will now explain our choices for a subset of them and will mention some others. The neural network's hyper parameters were kept at the Tensorflow (1.10) library's default values unless specified.

The PH model has two main hyper parameters, the σ_G and $\delta\theta$. We determined, using a simple line-search, the optimum σ_G for every image by minimizing the following equation over all possible values of it.

$$\sigma_G^* = \arg\min_{\sigma_G}\left[\min_{\mathbf{C}} J(\mathbf{C}, \sigma_G)\right]$$

For determining the appropriate $\delta\theta$, we chose the heuristic approach for the grid search. We generated samples of up to 6° because the approximation error increases beyond that. We chose an appropriate $\delta\theta$ by running a grid search over a factor η, such that $\delta\theta = min\{6^\circ, \eta\sigma_G^*\}$. This was chosen because the amount of possible extrapolation per image depends on the corresponding kernel width σ_G^*. We ran the training on the smallest subset of the data set at a sub-sampling ratio of 2^{-5} for the purpose of searching over a grid of three values, i.e., $\eta = \{1, 2, 3\}$. We chose $\eta = 3$ as it gives the best validation performance. Although we experimented with $\eta > 3$, we found the results were comparable to when $\eta = 3$.

For the neural network model, we set the dropout rate for the last fully connected layer at 0.2.

4. Results

A scarcity of training data affects the performance of the resulting ATR classifier in two distinct ways: First, a small training data set interferes with the ability of learning how to extract informative features from the data. Second, given a set of features, limited training data results in suboptimal decision boundaries, thereby leading to a poor generalization performance. We hypothesize that data augmentation techniques primarily improve the former effect, i.e., it improves the test performance through an enhanced training of the CNN's convolutional layers that serve as the feature extractors. Our empirical results, which are presented below, support this observation. With an adequate feature set, the classifier can be trained even with small training data sets, and it will still generalize well. To disentangle the two effects, the convolutional layers of the network were trained with the augmented training data set, and the classifier layers (after and including the first FC layer) were trained using the corresponding non-augmented training data set.

For all the approaches, we generated the results for the 6 values of $\mathcal{R}$ that correspond to the different sub-sampling ratios of the original training data set. All the models have the same architecture as described in Section 3.2, and they use the same $\mathcal{D}_{test}$. The difference among them is the $\mathcal{D}_{train}$ and $\mathcal{D}_{val}$ that are used, as described in Section 3.1. For consistency in the results, we repeated the process described in Section 3.1 to obtain four different

$\mathcal{D}_{train}$ and $\mathcal{D}_{val}$ for each $\mathcal{R}$ (except for $\mathcal{R} = 2^{-1}$ and $\mathcal{R} = 2^0$, which is where only two and one such unique data sets were possible, respectively). Moreover, we reported the mean and standard deviation of the classification performance. The overall augmentation performance is summarized in Tables 1 and 2, and they are visualized in Figure 5. The bold numbers in these tables highlight the best performance in respective rows.

Table 1. The test errors that correspond to the analysis plot (Figure 5a).

Sub-Sampling Ratio ($\mathcal{R}$)	Baseline Data (B)	Adding Our Sub-Pixel Shifts (B + S)	Adding Our Poses and Sub-Pixel Shifts (B + S + P)	Full-Data Features (F)
2^0	0.50 $\{B_0\}$	0.58 $\{S_0\}$	**0.37** $\{SP_0\}$	0.50 $\{F_0\}$
2^{-1}	1.44 ± 0.33 $\{B_1\}$	1.03 ± 0.08 $\{S_1\}$	**0.54** ± 0.00 $\{SP_1\}$	0.72 ± 0.06 $\{F_1\}$
2^{-2}	4.21 ± 0.83 $\{B_2\}$	2.33 ± 0.33 $\{S_2\}$	1.00 ± 0.34 $\{SP_2\}$	**0.99** ± 0.17 $\{F_2\}$
2^{-3}	10.99 ± 0.73 $\{B_3\}$	5.68 ± 0.67 $\{S_3\}$	3.32 ± 1.31 $\{SP_3\}$	**1.22** ± 0.26 $\{F_3\}$
2^{-4}	18.78 ± 2.40 $\{B_4\}$	14.02 ± 0.28 $\{S_4\}$	7.33 ± 0.54 $\{SP_4\}$	**2.07** ± 0.21 $\{F_4\}$
2^{-5}	32.38 ± 2.93 $\{B_5\}$	29.98 ± 2.62 $\{S_5\}$	18.66 ± 3.22 $\{SP_5\}$	**4.55** ± 0.57 $\{F_5\}$

Table 2. Test Errors corresponding to the comparative plot (Figure 5b).

Sub-Sampling Ratio ($\mathcal{R}$)	Baseline Data (B)	Augmenting with Naively Rotated Poses (B + R)	Augmenting with Linearly Interpolated Poses (B + L)	Augmenting with Our Poses and Sub-Pixel Shifts (B + S + P)
2^0	0.50 $\{B_0\}$	0.62 $\{R_0\}$	0.50 $\{L_0\}$	**0.37** $\{SP_0\}$
2^{-1}	1.44 ± 0.33 $\{B_1\}$	1.22 ± 0.19 $\{R_1\}$	1.22 ± 0.14 $\{L_1\}$	**0.54** ± 0.00 $\{SP_1\}$
2^{-2}	4.21 ± 0.83 $\{B_2\}$	4.38 ± 0.72 $\{R_2\}$	2.56 ± 0.21 $\{L_2\}$	**1.00** ± 0.34 $\{SP_2\}$
2^{-3}	10.99 ± 0.73 $\{B_3\}$	10.01 ± 1.72 $\{R_3\}$	7.26 ± 2.56 $\{L_3\}$	**3.32** ± 1.31 $\{SP_3\}$
2^{-4}	18.78 ± 2.40 $\{B_4\}$	19.83 ± 2.21 $\{R_4\}$	12.99 ± 1.10 $\{L_4\}$	**7.33** ± 0.54 $\{SP_4\}$
2^{-5}	32.38 ± 2.93 $\{B_5\}$	32.05 ± 7.58 $\{R_5\}$	30.79 ± 3.04 $\{L_5\}$	**18.66** ± 3.22 $\{SP_5\}$

4.1. Ablation Study of the Proposed Approach

We performed an ablation study of the two proposed augmentations, i.e., the sub-pixel and pose augmentations, by incrementally adding them to the baseline data. We abbreviated the data sets as follows: the baseline data as B (which includes the image domain flips and integer pixel translations); the baseline data with proposed sub-pixel augmentations as B + S; and the baseline data with the proposed sub-pixel and pose augmentations as B + S + P. Moreover, to provide a lower bound on the test error at all the sub-sampling ratios of the data-augmentation approaches, we used a genie-aided approach (non-realizable in practice) by utilizing the full SOC data set to learn the CNN features, but we still used only the sub-sampled training data-set for training the fully connected classifier layers. This formed the test-error curve, which is referred to as F (for full data) in Figure 5a.

The full-data plot in Figure 5a (values are in Table 1) shows the importance of extracting good quality features, i.e., if we had access to all the poses, we would learn very good features. Having good features makes classification quite easy, and this is evident from the low test errors that are found even in cases of very low data availability when learning the classifier. The sub-sampling had little effect on the generalization performance for the genie-aided case. The baseline data plot showed a considerable degree of test error, especially in cases of low training-data availability. This test error was reduced in the B + S data plot, as well as further reduced in the B + S + P data plot, which shows the effectiveness of both our strategies in improving the quality of features extracted by the CNN. Note that the majority of the improvement comes from the pose augmentations. For $\mathcal{R} = 2^{-5}$, the proposed augmentations reduced the test error by more than 42% when compared to the baseline approach (which includes the image domain flips and integer pixel translations). For $\mathcal{R} > 2^{-2}$, the model that used both proposed augmentations provided an even better performance than the genie-aided features. This makes sense because our augmentation strategy is able to successfully fill in the pose information gaps in the complete SOC data.

However, there exists a considerable gap between the full data and the B + S + P data plots in the smaller data regimes of $\mathcal{R} < 2^{-2}$. So, there still exists room for further improvements in aiding the network-learning informative features for the data-starved regimes.

The confusion matrices for a sample data set at $\mathcal{R} = 2^{-4}$ are shown in Tables 3 and 4. These tables clearly show that the performance has considerably improved via the proposed augmentation of the training data in cases of low data availability. Not only that, but the performance also improved over all the classes except two. As there were four distinct sub-sampled data sets at $\mathcal{R} = 2^{-4}$, we picked the one that was a good representative of the average performance. The bold numbers in these tables highlight the best performing method for respective classes.

4.2. Comparison with Existing SAR-ATR Models

In comparing our test error with the full SOC MSTAR data set, it can be seen from Table 5 that our approach is on par with the existing approaches when using all of the data. We are interested in training the CNN models when data availability is extremely low, say $\leq$60 samples per class. To compare the results obtained from our approach to recent works in extremely low-data regimes, we utilized some results from [26,42]. We also conducted B + S + P experiments for 18% of data per class. These are also tabulated in Table 5. It is in this extreme sub-sampling regime where our approach outperforms all the other existing approaches. The proposed algorithm reduces the test error by more than 46% when compared to the next best approach of the CNN-TL-bypass [26]. We obtained the lowest test error even when using the smallest portion of the data. We reiterate that most of the tabulated approaches were decoupled from our data-augmentation approach. So, in principle, it may be possible to combine our data-augmentation strategy with the existing approaches in order to obtain even better results. The bold numbers in Table 5 highlight the best performing method in respective columns.

Table 3. Confusion matrix for the classifiers corresponding to $\mathcal{R} = 2^{-4}$ with no augmentation.

Class	2S1	BMP2	BRDM2	BTR60	BTR70	D7	T62	T72	ZIL131	ZSU234	Error (%)
2S1	196	0	1	0	3	1	40	5	18	10	28.467
BMP2	21	117	2	18	10	0	1	23	3	0	40.0
BRDM2	9	1	**256**	1	0	1	0	0	6	0	6.569
BTR60	2	2	4	161	10	3	2	4	4	3	17.436
BTR70	21	13	1	23	130	1	0	6	0	1	33.673
D7	0	0	0	0	0	**264**	1	0	7	2	3.65
T62	5	0	0	2	0	1	234	4	22	5	14.286
T72	5	2	0	3	0	1	16	164	5	0	16.327
ZIL131	1	0	0	0	0	34	4	0	234	1	14.599
ZSU234	0	0	0	0	0	17	11	0	29	217	20.803
					Overall						18.639

Table 4. Confusion matrix for the classifier corresponding to $\mathcal{R} = 2^{-4}$ with augmentation.

Class	2S1	BMP2	BRDM2	BTR60	BTR70	D7	T62	T72	ZIL131	ZSU234	Error (%)
2S1	**251**	0	0	1	0	1	9	8	4	0	8.394
BMP2	4	**169**	0	4	0	0	4	12	1	1	13.333
BRDM2	16	8	243	0	0	0	0	0	6	1	11.314
BTR60	2	1	4	**172**	6	1	3	1	1	4	11.795
BTR70	7	2	1	0	**184**	0	0	2	0	0	6.122
D7	0	0	0	0	0	263	0	0	0	11	4.015
T62	6	0	0	4	0	1	**257**	4	1	0	5.861
T72	1	0	0	0	0	0	10	**183**	0	2	6.633
ZIL131	6	0	0	0	0	8	6	1	**244**	9	10.949
ZSU234	0	0	0	0	0	0	2	0	0	**272**	0.73
					Overall						**7.711**

Table 5. Results from the various SAR-ATR efforts that used the SOC MSTAR data set.

Method	Error (%) Using 100% Data	Error (%) Using ≤20% Data
SVM (2016) [40]	13.27	47.75 (at 20%)
SRC (2016) [40]	10.24	36.35 (at 20%)
A-ConvNet (2016) [14]	0.87	35.90 (at 20%)
Ensemble DCHUN (2017) [43]	0.91	25.94 (at 20%)
CNN-TL-bypass (2017) [26]	0.91	2.85 (at 18%)
ResNet (2018) [44]	0.33	5.70 (at 20%)
DFFN (2019) [42]	**0.17**	7.71 (at 20%)
Our Method	0.37	**1.53** (at 18%)

Except for the works of [50,59], we did not find reproducible data-augmentation strategies that explicitly synthesize samples at new poses. As pointed out earlier, our approach can be used in conjunction with most of the other strategies outlined in Section 1.1. As such, we conducted a detailed comparison of our pose synthesis approach and sub-pixel level translations with the pose synthesis methods in [50,59]. We added real-time pixel level translations for all experiments. For the sake of completion, the simple rotations that are produced in [59] used the rotation matrix and were followed by appropriate cropping, and the linearly interpolated poses that were synthesized in [50] used the following equation

$$I_{\theta_c} = CR_{\theta_c}\left(\frac{|\theta_b - \theta_c|R_{\theta_a}(I_{\theta_a}) + |\theta_a - \theta_c|R_{\theta_b}(I_{\theta_b})}{|\theta_a - \theta_c| + |\theta_b - \theta_c|}\right) \tag{10}$$

where $R_\theta(I)$ denotes the rotation of the radar image I by θ degrees clockwise, $CR_\theta(I)$ denotes the same but counter-clockwise, I_θ denotes the radar image at the pose θ, θ_c is the desired new pose, and θ_a and θ_b are the poses closest to θ_c in the training data.

For a qualitative evaluation, we illustrated the images that were synthesized from the *T62* tank at $\theta_c = 57°$, and the corresponding ground-truth image (which is part of the Full SOC data) was used for comparison purposes only. We observed in Figure 6b the synthesized image at $\theta_c = 57°$ when using $\theta_a = 56°$ and $\theta_b = 85°$, and this was achieved by using a sub-sampled data set with $\mathcal{R} = 2^{-4}$. We note that the dominant scattering centers in the synthesized image were different compared to the ground-truth image. Next, we used the proposed model that was estimated for the azimuth angle of 56°. It is evident from Figure 6c that the synthesized image from our method captures all the dominant scattering centers present in the ground truth. Finally, we used the rotation operator to synthesize at $\theta_c = 57°$ when using $\theta = 56°$. We observed that, even when the rotation is small, the ground-truth image has a different scattering behavior that is not captured by rotation.

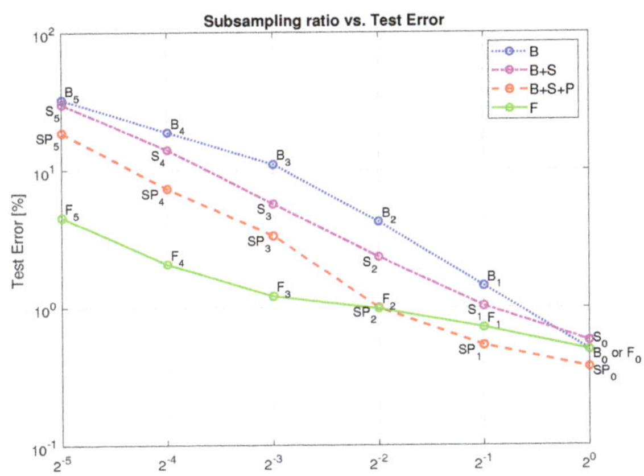

(a)

Figure 5. *Cont.*

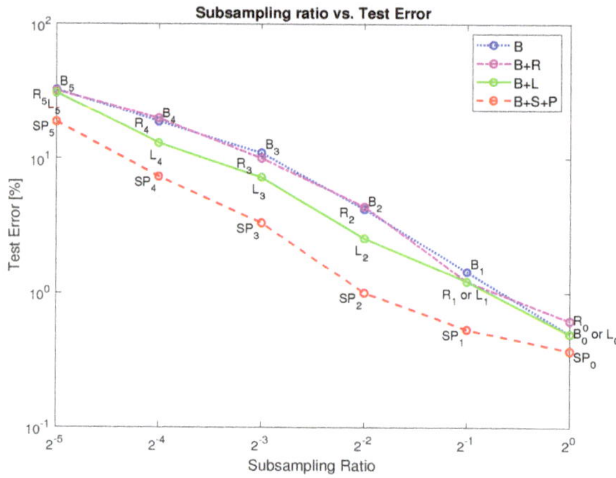

(**b**)

Figure 5. Quantitative evaluation of the proposed approach. (**a**) Ablation study. (**b**) Comparison with other augmentations.

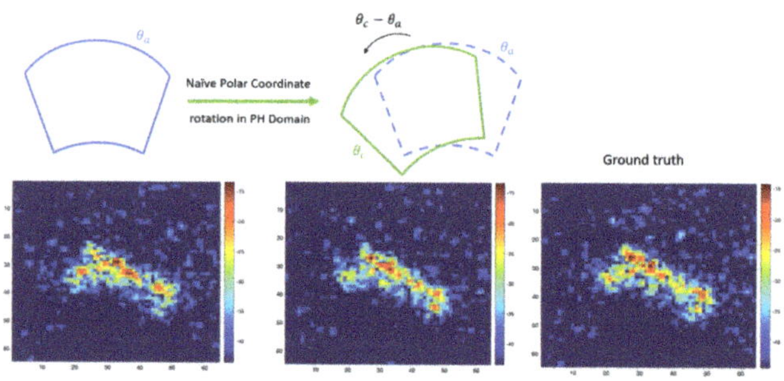

(**a**)

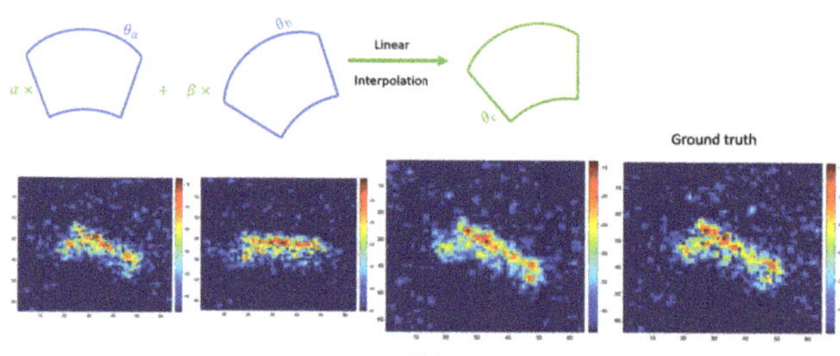

(**b**)

Figure 6. *Cont.*

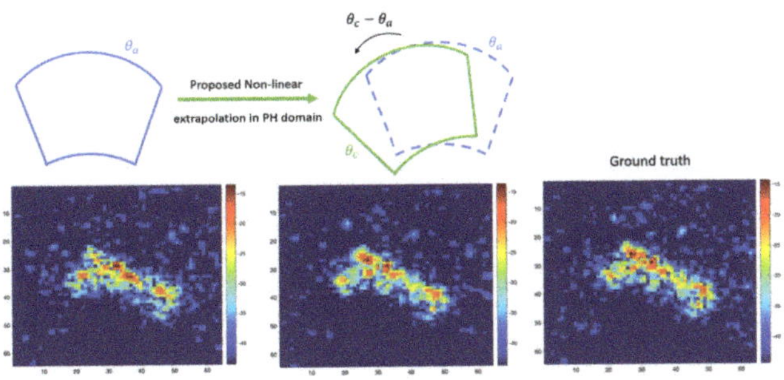

(c)

Figure 6. Comparison of the radar images synthesized for class T62 at the viewing angle of $\theta_c = 57°$ when using different augmentation strategies. (**a**) Radar image at a viewing angle of $\theta_c = 57°$, which was generated by rotating the closest available in the sub-sampled data set at $\theta_a = 56°$ (from [59]). (**b**) Linear interpolation strategy proposed in [50]. Here, α and β can be inferred from Equation (10), and the closest poses to $\theta_c = 57°$ in the sub-sampled data set were $\theta_a = 56°$ and $\theta_b = 85°$. (**c**) This radar image at the azimuth angle of $\theta_a = 56°$ was first approximated using the proposed set of basis functions in the frequency domain. Measurements from the unobserved viewing angles were synthesized via the model in the frequency domain, which was used to create the image at the viewing angle of $\theta_c = 57°$.

For the quantitative evaluation, we compared the ATR performance at all sub-sampling ratios similar to those conducted in the previous Section 4.1. We abbreviated the data sets as follows: the baseline data as B, the baseline data with simple rotations added (from [59]) as B + R, the baseline data with linearly interpolated poses (from [50]) as B + L, and the baseline data with proposed sub-pixel and pose augmentations as B + S + P.

The comparison of these is tabulated in Table 2, and it can also be seen in Figure 5b. It is evident that our approach is significantly better than both simple rotations and linearly interpolated poses for this CNN-based ATR task. For $\mathcal{R} = 2^{-5}$, the proposed augmentations reduced the test error by more than 39% when compared to the next best augmentation approach of [50].

5. Conclusions and Future Directions

In this paper, we proposed incorporating the domain knowledge of SAR phenomenology into a CNN by way of data augmentation. We presented a model-based approach to data augmentation for the purpose of training the neural network architecture to solve the ATR problem with limited labeled data. Through extensive simulation studies, we showed the effectiveness of the augmentation strategies by training a neural network with the augmented data set that was synthesized from the phase history models extracted from each available training image. Our results show that the proposed data augmentation strategy produced a significant improvement in the model's generalization performance when compared to the baseline performance over a wide range of sub-sampling ratios. As presented, the phase history approximation method is only valid in a local neighborhood of a given azimuth angle. Future work could focus on fitting a single global model to every class that is jointly derived from all the training images. Such a global model could produce a diverse set of SAR images over larger pose variations. Since, typically, target image chips are not perfectly registered and are aligned across different azimuth angles, the global model fit should incorporate unknown phase and spatial shifts for each image. As part of future research, we propose developing a network architecture to support a unified model that can account for these phase errors, and which can synthesize a larger data set to improve the classifier's performance further. We also aim to evaluate such augmentations on more challenging data sets such as the Military Ground Targets Database (MGTD) [80], the Synthetic and Measured Paired Labeled Experiment (SAMPLE) [81], and the Ship data set OpenSARShip (which was obtained from Sentinel-I imagery [82]). Additionally, since our data-augmentation method creates complex-valued synthetic data, it can potentially be used to regularize complex-valued neural networks [83,84], and it can be used to improve their performance for SAR ATR. Moreover, since the currently used Taylor windowing is sub-optimal, the problem in optimizing the window function that enhances ATR performance can be incorporated in the form of a multi-objective optimization—a topic that will be investigated in our future work.

Author Contributions: Conceptualization, T.A., N.S. and E.E.; methodology, T.A., N.S. and E.E.; software, T.A. and N.S.; validation, T.A. and N.S.; formal analysis, T.A. and N.S.; investigation, T.A.; resources, E.E.; data curation, N.S.; writing—original draft preparation, T.A.; writing—review and editing, N.S. and E.E.; visualization, T.A. and N.S.; supervision, E.E.; project administration, E.E.; funding acquisition, E.E. All authors have read and agreed to the published version of the manuscript.

Funding: This research was partially supported by the Army Research Office (grant W911NF-11-1-0391 and NSF Grant IIS-1231577).

Data Availability Statement: The publicly available MSTAR data set was used in this study. These data can be obtained from the US Air Force Research Laboratory (AFRL) website https://www.sdms.afrl.af.mil/index.php?collection=mstar (accessed on 11 July 2023). The data pre-processing conducted in this study is described in Section 3.1.

Conflicts of Interest: The authors declare no conflict of interest. The funders had no role in the design of the study; in the collection, analyses, or interpretation of data; in the writing of the manuscript; or in the decision to publish the results.

References

1. Moses, R.L.; Potter, L.C.; Cetin, M. Wide-angle SAR imaging. In Proceedings of the Algorithms for Synthetic Aperture Radar Imagery XI, Orlando, FL, USA, 12–16 April 2004 ; International Society for Optics and Photonics: Bellingham, WA, USA, 2004; Volume 5427, pp. 164–175.
2. Potter, L.C.; Moses, R.L. Attributed scattering centers for SAR ATR. *IEEE Trans. Image Process.* **1997**, *6*, 79–91. [CrossRef]
3. Çetin, M.; Stojanović, I.; Önhon, N.O.; Varshney, K.; Samadi, S.; Karl, W.C.; Willsky, A.S. Sparsity-Driven Synthetic Aperture Radar Imaging: Reconstruction, autofocusing, moving targets, and compressed sensing. *IEEE Signal Process. Mag.* **2014**, *31*, 27–40. [CrossRef]
4. Potter, L.C.; Ertin, E.; Parker, J.T.; Cetin, M. Sparsity and Compressed Sensing in Radar Imaging. *Proc. IEEE* **2010**, *98*, 1006–1020. [CrossRef]
5. Dungan, K.; Potter, L. Classifying transformation-variant attributed point patterns. *Pattern Recognit.* **2010**, *43*, 3805–3816. [CrossRef]
6. Dungan, K.E.; Potter, L.C. Classifying Vehicles in Wide-Angle Radar Using Pyramid Match Hashing. *IEEE J. Sel. Top. Signal Process.* **2011**, *5*, 577–591. [CrossRef]
7. Abdelrahman, T.; Ertin, E. Mixture of factor analyzers models of appearance manifolds for resolved SAR targets. In Proceedings of the Algorithms for Synthetic Aperture Radar Imagery XXII, Baltimore, MD, USA, 23 April 2015; International Society for Optics and Photonics: Bellingham, WA, USA, 2015; Volume 9475, p. 94750G.
8. Hegde, C.; Sankaranarayanan, A.C.; Yin, W.; Baraniuk, R.G. NuMax: A Convex Approach for Learning Near-Isometric Linear Embeddings. *IEEE Trans. Signal Process.* **2015**, *63*, 6109–6121. [CrossRef]
9. Teng, D.; Ertin, E. WALD-Kernel: A method for learning sequential detectors. In Proceedings of the 2016 IEEE Statistical Signal Processing Workshop (SSP), Palma de Mallorca, Spain, 26–29 June 2016; pp. 1–5. [CrossRef]
10. Cui, J.; Gudnason, J.; Brookes, M. Hidden Markov models for multi-perspective radar target recognition. In Proceedings of the 2008 IEEE Radar Conference, Rome, Italy, 26–30 May 2008; pp. 1–5. [CrossRef]
11. Lecun, Y.; Bottou, L.; Bengio, Y.; Haffner, P. Gradient-based learning applied to document recognition. *Proc. IEEE* **1998**, *86*, 2278–2324. [CrossRef]
12. Krizhevsky, A.; Sutskever, I.; Hinton, G.E. ImageNet Classification with Deep Convolutional Neural Networks. *Commun. ACM* **2017**, *60*, 84–90. [CrossRef]
13. Hinton, G.E.; Salakhutdinov, R.R. Reducing the Dimensionality of Data with Neural Networks. *Science* **2006**, *313*, 504–507.
14. Chen, S.; Wang, H.; Xu, F.; Jin, Y. Target Classification Using the Deep Convolutional Networks for SAR Images. *IEEE Trans. Geosci. Remote Sens.* **2016**, *54*, 4806–4817. [CrossRef]
15. Zhong, Y.; Ettinger, G. Enlightening deep neural networks with knowledge of confounding factors. In Proceedings of the IEEE International Conference on Computer Vision, Venice, Italy, 22–29 October 2017; pp. 1077–1086.
16. Russakovsky, O.; Deng, J.; Su, H.; Krause, J.; Satheesh, S.; Ma, S.; Huang, Z.; Karpathy, A.; Khosla, A.; Bernstein, M.; et al. ImageNet Large Scale Visual Recognition Challenge. *Int. J. Comput. Vis. (IJCV)* **2015**, *115*, 211–252. [CrossRef]

17. Yosinski, J.; Clune, J.; Bengio, Y.; Lipson, H. How transferable are features in deep neural networks? In Proceedings of the Advances in Neural Information Processing Systems, Montreal, QC, Canada, 8–13 December 2014; pp. 3320–3328.
18. Oquab, M.; Bottou, L.; Laptev, I.; Sivic, J. Learning and transferring mid-level image representations using convolutional neural networks. In Proceedings of the IEEE Conference on Computer Vision and Pattern Recognition, Columbus, OH, USA, 23–28 June 2014; pp. 1717–1724.
19. Simonyan, K.; Zisserman, A. Very Deep Convolutional Networks for Large-Scale Image Recognition. *arXiv* **2014**, arXiv:1409.1556.
20. Gunasekar, S.; Lee, J.; Soudry, D.; Srebro, N. Characterizing Implicit Bias in Terms of Optimization Geometry. In Proceedings of the 35th International Conference on Machine Learning, Stockholm, Sweden, 10–15 July 2018 ; Volume 80, pp. 1832–1841.
21. Ruhi, N.A.; Hassibi, B. Stochastic Gradient/Mirror Descent: Minimax Optimality and Implicit Regularization. In Proceedings of the 7th International Conference on Learning Representations, ICLR 2019, New Orleans, LA, USA, 6–9 May 2019.
22. Srivastava, N.; Hinton, G.; Krizhevsky, A.; Sutskever, I.; Salakhutdinov, R. Dropout: A simple way to prevent neural networks from overfitting. *J. Mach. Learn. Res.* **2014**, *15*, 1929–1958.
23. Ioffe, S.; Szegedy, C. Batch normalization: Accelerating deep network training by reducing internal covariate shift. *arXiv* **2015**, arXiv:1502.03167.
24. Neyshabur, B.; Li, Z.; Bhojanapalli, S.; LeCun, Y.; Srebro, N. The role of over-parametrization in generalization of neural networks. In Proceedings of the 7th International Conference on Learning Representations, ICLR 2019, New Orleans, LA, USA, 6–9 May 2019.
25. Pan, S.J.; Yang, Q. A survey on transfer learning. *IEEE Trans. Knowl. Data Eng.* **2009**, *22*, 1345–1359. [CrossRef]
26. Huang, Z.; Pan, Z.; Lei, B. Transfer learning with deep convolutional neural network for SAR target classification with limited labeled data. *Remote Sens.* **2017**, *9*, 907. [CrossRef]
27. Huang, Z.; Dumitru, C.O.; Pan, Z.; Lei, B.; Datcu, M. Classification of Large-Scale High-Resolution SAR Images with Deep Transfer Learning. *IEEE Geosci. Remote Sens. Lett.* **2021**, *18*, 107–111. [CrossRef]
28. Zhou, F.; Wang, L.; Bai, X.; Hui, Y. SAR ATR of Ground Vehicles Based on LM-BN-CNN. *IEEE Trans. Geosci. Remote Sens.* **2018**, *56*, 7282–7293. [CrossRef]
29. Zhang, F.; Wang, Y.; Ni, J.; Zhou, Y.; Hu, W. SAR Target Small Sample Recognition Based on CNN Cascaded Features and AdaBoost Rotation Forest. *IEEE Geosci. Remote Sens. Lett.* **2020**, *17*, 1008–1012. [CrossRef]
30. Pei, J.; Huang, Y.; Sun, Z.; Zhang, Y.; Yang, J.; Yeo, T.S. Multiview Synthetic Aperture Radar Automatic Target Recognition Optimization: Modeling and Implementation. *IEEE Trans. Geosci. Remote Sens.* **2018**, *56*, 6425–6439. [CrossRef]
31. Zhang, T.; Zhang, X. A polarization fusion network with geometric feature embedding for SAR ship classification. *Pattern Recognit.* **2022**, *123*, 108365. [CrossRef]
32. Shao, Z.; Zhang, T.; Ke, X. A Dual-Polarization Information-Guided Network for SAR Ship Classification. *Remote Sens.* **2023**, *15*, 2138. [CrossRef]
33. Sugavanam, N.; Ertin, E.; Burkholder, R. Compressing bistatic SAR target signatures with sparse-limited persistence scattering models. *IET Radar Sonar Navig.* **2019**, *13*, 1411–1420. [CrossRef]
34. Shao, J.; Qu, C.; Li, J.; Peng, S. A Lightweight Convolutional Neural Network Based on Visual Attention for SAR Image Target Classification. *Sensors* **2018**, *18*, 3039. [CrossRef] [PubMed]
35. Chen, H.; Zhang, F.; Tang, B.; Yin, Q.; Sun, X. Slim and Efficient Neural Network Design for Resource-Constrained SAR Target Recognition. *Remote Sens.* **2018**, *10*, 1618. [CrossRef]
36. Min, R.; Lan, H.; Cao, Z.; Cui, Z. A Gradually Distilled CNN for SAR Target Recognition. *IEEE Access* **2019**, *7*, 42190–42200. [CrossRef]
37. Zhang, F.; Liu, Y.; Zhou, Y.; Yin, Q.; Li, H.C. A Lossless Lightweight CNN Design for SAR Target Recognition. *Remote Sens. Lett.* **2020**, *11*, 485–494. [CrossRef]
38. Dong, G.; Wang, N.; Kuang, G. Sparse representation of monogenic signal: With application to target recognition in SAR images. *IEEE Signal Process. Lett.* **2014**, *21*, 952–956.
39. Song, S.; Xu, B.; Yang, J. SAR target recognition via supervised discriminative dictionary learning and sparse representation of the SAR-HOG feature. *Remote Sens.* **2016**, *8*, 683. [CrossRef]
40. Song, H.; Ji, K.; Zhang, Y.; Xing, X.; Zou, H. Sparse representation-based SAR image target classification on the 10-class MSTAR data set. *Appl. Sci.* **2016**, *6*, 26. [CrossRef]
41. Huang, Y.; Liao, G.; Zhang, Z.; Xiang, Y.; Li, J.; Nehorai, A. SAR automatic target recognition using joint low-rank and sparse multiview denoising. *IEEE Geosci. Remote Sens. Lett.* **2018**, *15*, 1570–1574. [CrossRef]
42. Yu, Q.; Hu, H.; Geng, X.; Jiang, Y.; An, J. High-Performance SAR Automatic Target Recognition Under Limited Data Condition Based on a Deep Feature Fusion Network. *IEEE Access* **2019**, *7*, 165646–165658. [CrossRef]
43. Lin, Z.; Ji, K.; Kang, M.; Leng, X.; Zou, H. Deep convolutional highway unit network for SAR target classification with limited labeled training data. *IEEE Geosci. Remote Sens. Lett.* **2017**, *14*, 1091–1095. [CrossRef]
44. Fu, Z.; Zhang, F.; Yin, Q.; Li, R.; Hu, W.; Li, W. Small sample learning optimization for ResNet based SAR target recognition. In Proceedings of the IGARSS 2018-2018 IEEE International Geoscience and Remote Sensing Symposium, Valencia, Spain, 22–27 July 2018; IEEE: Piscataway, NJ, USA, 2018; pp. 2330–2333.
45. Yue, Z.; Gao, F.; Xiong, Q.; Wang, J.; Huang, T.; Yang, E.; Zhou, H. A Novel Semi-Supervised Convolutional Neural Network Method for Synthetic Aperture Radar Image Recognition. *Cogn. Comput.* **2021**, *13*, 795–806. [CrossRef]

46. Wang, C.; Shi, J.; Zhou, Y.; Yang, X.; Zhou, Z.; Wei, S.; Zhang, X. Semisupervised Learning-Based SAR ATR via Self-Consistent Augmentation. *IEEE Trans. Geosci. Remote Sens.* **2021**, *59*, 4862–4873. [CrossRef]
47. Chen, K.; Pan, Z.; Huang, Z.; Hu, Y.; Ding, C. Learning from Reliable Unlabeled Samples for Semi-Supervised SAR ATR. *IEEE Geosci. Remote Sens. Lett.* **2022**, *19*, 4512205. [CrossRef]
48. Zhang, C.; Bengio, S.; Hardt, M.; Recht, B.; Vinyals, O. Understanding deep learning requires rethinking generalization. In Proceedings of the 5th International Conference on Learning Representations, ICLR 2017, Toulon, France, 24–26 April 2017.
49. Goodfellow, I.; Bengio, Y.; Courville, A. *Deep Learning*; MIT Press: Cambridge, MA, USA, 2016. Available online: http://www.deeplearningbook.org (accessed on 11 July 2023).
50. Ding, J.; Chen, B.; Liu, H.; Huang, M. Convolutional neural network with data augmentation for SAR target recognition. *IEEE Geosci. Remote Sens. Lett.* **2016**, *13*, 364–368. [CrossRef]
51. Yan, Y. Convolutional Neural Networks Based on Augmented Training Samples for Synthetic Aperture Radar Target Recognition. *J. Electron. Imaging* **2018**, *27*, 023024. [CrossRef]
52. Marmanis, D.; Yao, W.; Adam, F.; Datcu, M.; Reinartz, P.; Schindler, K.; Wegner, J.D.; Stilla, U. Artificial generation of big data for improving image classification: A generative adversarial network approach on SAR data. *arXiv* **2017**, arXiv:1711.02010.
53. Lewis, B.; DeGuchy, O.; Sebastian, J.; Kaminski, J. Realistic SAR Data Augmentation Using Machine Learning Techniques. In Proceedings of the Algorithms for Synthetic Aperture Radar Imagery XXVI, Baltimore, MD, USA, 18 April 2019; SPIE: Bellingham, WA, USA, 2019; Volume 10987, pp. 12–28. [CrossRef]
54. Gao, F.; Yang, Y.; Wang, J.; Sun, J.; Yang, E.; Zhou, H. A deep convolutional generative adversarial networks (DCGANs)-based semi-supervised method for object recognition in synthetic aperture radar (SAR) images. *Remote Sens.* **2018**, *10*, 846. [CrossRef]
55. Cui, Z.; Zhang, M.; Cao, Z.; Cao, C. Image Data Augmentation for SAR Sensor via Generative Adversarial Nets. *IEEE Access* **2019**, *7*, 42255–42268. [CrossRef]
56. Sun, Y.; Wang, Y.; Liu, H.; Wang, N.; Wang, J. SAR Target Recognition with Limited Training Data Based on Angular Rotation Generative Network. *IEEE Geosci. Remote Sens. Lett.* **2020**, *17*, 1928–1932. [CrossRef]
57. Shi, X.; Zhou, F.; Yang, S.; Zhang, Z.; Su, T. Automatic Target Recognition for Synthetic Aperture Radar Images Based on Super-Resolution Generative Adversarial Network and Deep Convolutional Neural Network. *Remote Sens.* **2019**, *11*, 135. [CrossRef]
58. Cha, M.; Majumdar, A.; Kung, H.; Barber, J. Improving SAR automatic target recognition using simulated images under deep residual refinements. In Proceedings of the 2018 IEEE International Conference on Acoustics, Speech and Signal Processing (ICASSP), Calgary, AB, Canada, 15–20 April 2018; IEEE: Piscataway, NJ, USA, 2018; pp. 2606–2610.
59. Zhai, Y.; Ma, H.; Liu, J.; Deng, W.; Shang, L.; Sun, B.; Jiang, Z.; Guan, H.; Zhi, Y.; Wu, X.; et al. SAR ATR with full-angle data augmentation and feature polymerisation. *J. Eng.* **2019**, *2019*, 6226–6230. [CrossRef]
60. Lv, J.; Liu, Y. Data Augmentation Based on Attributed Scattering Centers to Train Robust CNN for SAR ATR. *IEEE Access* **2019**, *7*, 25459–25473. [CrossRef]
61. Agarwal, T.; Sugavanam, N.; Ertin, E. Sparse signal models for data augmentation in deep learning ATR. In Proceedings of the 2020 IEEE Radar Conference (RadarConf20), Florence, Italy, 21–25 September 2020; IEEE: Piscataway, NJ, USA, 2020; pp. 1–6.
62. Sugavanam, N.; Ertin, E. Limited persistence models for SAR automatic target recognition. In Proceedings of the Algorithms for Synthetic Aperture Radar Imagery XXIV, Anaheim, CA, USA, 13 April 2017; International Society for Optics and Photonics: Bellingham, WA, USA, 2017; Volume 10201, p. 102010M.
63. Sugavanam, N.; Ertin, E.; Burkholder, R. Approximating Bistatic SAR Target Signatures with Sparse Limited Persistence Scattering Models. In Proceedings of the International Conference on Radar, Brisbane, Australia, 27–31 August 2018.
64. Sugavanam, N.; Ertin, E. Models of anisotropic scattering for 3D SAR reconstruction. In Proceedings of the 2022 IEEE Radar Conference (RadarConf22), New York, NY, USA, 21–25 March 2022; IEEE: Piscataway, NJ, USA, 2022; pp. 1–6.
65. Sugavanam, N.; Ertin, E. Interrupted SAR imaging with limited persistence scattering models. In Proceedings of the 2017 IEEE Radar Conference (RadarConf), Seattle, WA, USA, 8–12 May 2017; IEEE: Piscataway, NJ, USA, 2017; pp. 1770–1775.
66. Sugavanam, N.; Ertin, E. Recovery guarantees for MIMO radar using multi-frequency LFM waveform. In Proceedings of the 2016 IEEE Radar Conference (RadarConf), Philadelphia, PA, USA, 2–6 May 2016; IEEE: Piscataway, NJ, USA, 2016; pp. 1–6.
67. Sugavanam, N.; Ertin, E. Recovery Guarantees for High-Resolution Radar Sensing with Compressive Illumination. In *Compressive Sensing of Earth Observations*; CRC Press: Boca Raton, FL, USA, 2017; pp. 83–104.
68. Sugavanam, N.; Baskar, S.; Ertin, E. High Resolution MIMO Radar Sensing With Compressive Illuminations. *IEEE Trans. Signal Process.* **2022**, *70*, 1448–1463. [CrossRef]
69. Brito, A.E.; Chan, S.H.; Cabrera, S.D. SAR image formation using 2D reweighted minimum norm extrapolation. In Proceedings of the Algorithms for Synthetic Aperture Radar Imagery VI, Orlando, FL, USA, 5–9 April 1999; International Society for Optics and Photonics: Bellingham, WA, USA, 1999; Volume 3721, pp. 78–91.
70. Cetin, M. Feature-Enhanced Synthetic Aperture Radar Imaging. Ph.D. Thesis, Boston University, Boston, MA, USA, 2001.
71. Jackson, J.A.; Rigling, B.D.; Moses, R.L. Canonical Scattering Feature Models for 3D and Bistatic SAR. *IEEE Trans. Aerosp. Electron. Syst.* **2010**, *46*, 525–541. [CrossRef]
72. Sarabandi, K.; Chiu, T.C. Optimum corner reflectors for calibration of imaging radars. *IEEE Trans. Antennas Propag.* **1996**, *44*, 1348–1361. [CrossRef]

73. Potter, L.C.; Chiang, D.M.; Carriere, R.; Gerry, M.J. A GTD-based parametric model for radar scattering. *IEEE Trans. Antennas Propag.* **1995**, *43*, 1058–1067. [CrossRef]
74. Rauhut, H.; Ward, R. Interpolation via weighted L1 minimization. *Appl. Comput. Harmon. Anal.* **2016**, *40*, 321–351. [CrossRef]
75. Greengard, L.; Lee, J.Y. Accelerating the nonuniform fast Fourier transform. *SIAM Rev.* **2004**, *46*, 443–454. [CrossRef]
76. Duijndam, A.; Schonewille, M. Nonuniform fast Fourier transform. *Geophysics* **1999**, *64*, 539–551. [CrossRef]
77. Burns, B.L.; Cordaro, J.T. SAR image-formation algorithm that compensates for the spatially variant effects of antenna motion. In Proceedings of the Algorithms for Synthetic Aperture Radar Imagery, Orlando, FL, USA, 6–7 April 1994; International Society for Optics and Photonics: Bellingham, WA, USA, 1994; Volume 2230, pp. 14–24. [CrossRef]
78. Luo, P.; Wang, X.; Shao, W.; Peng, Z. Towards Understanding Regularization in Batch Normalization. *arXiv* **2018**, arXiv:1809.00846.
79. Abadi, M.; Barham, P.; Chen, J.; Chen, Z.; Davis, A.; Dean, J.; Devin, M.; Ghemawat, S.; Irving, G.; Isard, M.; et al. Tensorflow: A system for large-scale machine learning. In Proceedings of the 12th USENIX Symposium on Operating Systems Design and Implementation (OSDI 16), Berkeley, CA, USA, 2–4 November 2016; pp. 265–283.
80. Belloni, C.; Balleri, A.; Aouf, N.; Merlet, T.; Le Caillec, J.M. SAR image dataset of military ground targets with multiple poses for ATR. In Proceedings of the Target and Background Signatures III, Warsaw, Poland, 11–12 September 2017; SPIE: Bellingham, WA, USA, 2017; Volume 10432, pp. 218–225.
81. Lewis, B.; Scarnati, T.; Sudkamp, E.; Nehrbass, J.; Rosencrantz, S.; Zelnio, E. A SAR dataset for ATR development: The Synthetic and Measured Paired Labeled Experiment (SAMPLE). In Proceedings of the Algorithms for Synthetic Aperture Radar Imagery XXVI, Baltimore, MD, USA, 18 April 2019; SPIE: Bellingham, WA, USA, 2019; Volume 10987, pp. 39–54.
82. Huang, L.; Liu, B.; Li, B.; Guo, W.; Yu, W.; Zhang, Z.; Yu, W. OpenSARShip: A dataset dedicated to Sentinel-1 ship interpretation. *IEEE J. Sel. Top. Appl. Earth Obs. Remote Sens.* **2017**, *11*, 195–208. [CrossRef]
83. Sarroff, A.M. Complex Neural Networks for Audio. Ph.D. Thesis, Dartmouth College, Hanover, NH, USA, 2018.
84. Jojoa, M.; Garcia-Zapirain, B.; Percybrooks, W. A Fair Performance Comparison between Complex-Valued and Real-Valued Neural Networks for Disease Detection. *Diagnostics* **2022**, *12*, 1893. [CrossRef]

remote sensing

MDPI

Article

We Need to Communicate: Communicating Attention Network for Semantic Segmentation of High-Resolution Remote Sensing Images

Xichen Meng [1], Liqun Zhu [1], Yilong Han [1,2,3,*] and Hanchao Zhang [4,*]

1 College of Geodesy and Geomatics, Shandong University of Science and Technology, Qingdao 266590, China; 202001021017@sdust.edu.cn (X.M.); 202001021030@sdust.edu.cn (L.Z.)
2 Shandong Province Research Center of Engineering and Technology for "3S"—GPS, RS and GIS, Shandong University of Science and Technology, Qingdao 266590, China
3 Star-Rising Technologies Co., Ltd., Qingdao 266590, China
4 Chinese Academy of Surveying and Mapping, Beijing 100036, China
* Correspondence: hanyl@sdust.edu.cn (Y.H.); zhanghc@casm.ac.cn (H.Z.); Tel.: +86-0532-86057980 (Y.H.); +86-010-63880527 (H.Z.)

Abstract: Traditional models that employ CNNs as encoders do not sufficiently combine high-level features and low-level features. However, high-level features are rich in semantic information but lack spatial detail, while low-level features are the opposite. Therefore, the integrated utilization of multi-level features and the bridging of the gap between them is crucial to promote the accuracy of semantic segmentation. To address this issue, we presented communicating mutual attention (CMA) and communicating self-attention (CSA) modules to enhance the interaction and fusion of different levels of feature maps. On the one hand, CMA aggregates the global context information of high-level features into low-level features and embeds the spatial detail localization characteristics of low-level features in high-level features. On the other hand, the CSA module is deployed to integrate the spatially detailed representation of low-level features into the attention map of high-level features. We have experimented with the communicating attention network (CANet), a U-net-like network composed of multiple CMA and CSA modules, on the ISPRS Vaihingen and Potsdam datasets with mean F1-scores of 89.61% and 92.60%, respectively. The results demonstrate that CANet embodies superior performance in the semantic segmentation task of remote sensing of images.

Keywords: attention mechanism; remote sensing; semantic segmentation

Citation: Meng, X.; Zhu, L.; Han, Y.; Zhang, H. We Need to Communicate: Communicating Attention Network for Semantic Segmentation of High-Resolution Remote Sensing Images. *Remote Sens.* **2023**, *15*, 3619. https://doi.org/10.3390/rs15143619

Academic Editors: Jukka Heikkonen, Fahimeh Farahnakian and Pouya Jafarzadeh

Received: 4 June 2023
Revised: 5 July 2023
Accepted: 17 July 2023
Published: 20 July 2023

1. Introduction

As aerospace technology is constantly evolving, the spatial resolution, radiometric resolution, spectral resolution, and observation frequency of the sensors have been increased. As a result, both the quality and quantity of Earth Observation (EO) data have significantly improved. Remote sensing images find wide applications in various fields, including semantic segmentation [1–4], change detection [5,6], target detection [7,8], target extraction [9], and more. Semantic segmentation of remote sensing images in urban areas plays a crucial role in generating accurate land use maps [10]. Traditional methods applied to semantic segmentation based on edges [11] or regions [12], are known for their fast and efficient processing. However, different objects in urban areas may have the identical spectral signatures, such as grass and trees, roads and bridges. They are often difficult to distinguish by traditional methods, which leads to low accuracy in segmentation results. As a result, this is an obstacle to the widespread application of traditional segmentation methods. In recent years, there have been significant developments in deep learning techniques [13], and their high flexibility and accuracy have made them widely employed in semantic segmentation of remote sensing images. In addition, the ability of a convolutional neural network (CNN) to capture spatial features and the abstraction characteristics of object texture, edges, and shape [14] has led to a tremendous improvement in segmentation accuracy in the semantic segmentation task of remote sensing images [15]. By increasing in the number of layers in the CNN, the extracted features reach a higher level, characterized by a larger receptive field and richer semantic information [16]. However, relying solely on high-level features derived by the CNN for semantic segmentation while neglecting the geometric representation properties of low-level features results in a lack of spatial details, thus hindering the improvement of segmentation precision.

The distance from a satellite or UAV platform to an object on the Earth's surface is much greater than the distance from an ordinary camera to the photographed object. There should also be an awareness that both the distance and the viewpoint differ, with an overview for remotely sensed images and a sideview for close-range images. Therefore, traditional images and remote sensing images contain variations. They cover a very wide spatial area and have a larger scale [17]. Moreover, objects in remote sensing images are relatively smaller compared to conventional images, which makes it more difficult to access global semantic representational properties. On the one hand, high-level features have coarse image resolution and lack spatial detail characteristics, which reduces the capability of the model to segment fine objects. On the other hand, low-level features have finer resolution and smaller receptive fields, providing greater geometric representation capability suitable for handling small targets. The discrepancy between the different levels of features (high and low level) is shown in Figure 1. However, the shallow CNN layers that produce low-level features lack abstract semantic information and contextual content, which limits the accuracy of semantic segmentation. In other words, focusing only on high-level or low-level features is insufficient for the semantic segmentation task of high-resolution remote sensing in urban areas. Therefore, it is crucial to combine high and low-level features to reduce the gap between them.

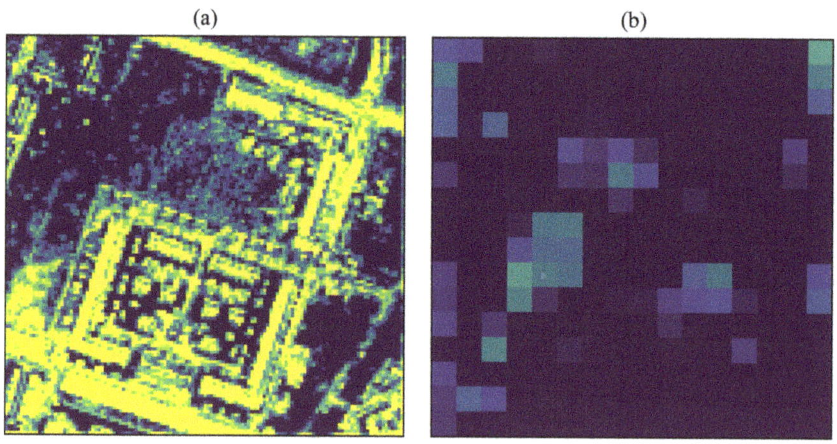

Figure 1. The gap between low-level features and high-level features. (**a**) is a visualization of low-level features. (**b**) is a visualization of high-level features.

Several models have recently been proposed to improve the accuracy of semantic segmentation models for high-resolution remote sensing images by reducing the gap between the high-level and low-level feature representation. LANet [18] embedded semantic information from high-level features into low-level features to strengthen the semantic reference of low-level features. However, this model does not embed spatial information from low-level features into high-level features, resulting in the spatial distribution and physical content of the latter remaining unchanged. Moreover, when extracting the contextual information of high-level features, the model barely utilizes local regions to generate channel attention maps, thus failing to exploit global long-range dependent semantic information.

Excessive emphasis on local regions leads to ambiguity in the pixel-level classification, while global contextual information can alleviate this problem [19,20].

We propose a novel communicating attention module to reduce the gap between the information representation capabilities of high-level and low-level features. The module is based on the discovery that high-level features are rich in contextual information and low-level features contain more spatially detailed content, while combining the two allows the model to have both long-range global modeling and segmentation capability for tiny objects. To address this challenge, we introduce two novel modules, namely CMA and CSA, to reduce the difference between high-level and low-level features. The CMA module consists of two branches, the high-level feature flow and the low-level feature flow, respectively. The two branches are combined into a loop to aggregate the global semantic information of high-level features into low-level features, while assigning the spatial information of low-level features to high-level features, which improves the long-range dependence and geometric representation of the model. The CSA module incorporates the spatial detail of low-level features when computing the attention map of high-level features, thereby avoiding excessive focus on global contextual information. This improves the perception of spatial details of high-level features and provides a trade-off between capturing global and local features. In this study, the model was tested using the ISPRS Potsdam and Vaihingen datasets to ensure the validity of the algorithm we propose.

To summarize, the contributions of this paper are as follows:

(1) To bridge the gap between high-level and low-level features in terms of spatial distribution and physical content, we introduce two attention modules, CMA and CSA. These modules enhance the model's ability to capture fine targets while maintaining the global semantic modeling capability.

(2) We propose the CANet model for semantic segmentation of high-resolution remote sensing images. The model improves the accuracy of semantic segmentation in urban areas by fusing output features of different scales and levels from the CNN in the encoding stage using the attention mechanism.

2. Related Work

2.1. Semantic Segmentation of Remote Sensing Images

Semantic segmentation of remote sensing images is a hot research theme in the field of computer vision and remote sensing technology, which aims to segment pixels or regions in remote sensing images into categories with specific semantic labels. This technique is of significant value in application scenarios such as land cover classification [21], urban planning [22], and environmental surveillance [23]. In recent years, remarkable progress has been achieved in semantic segmentation of remote sensing images with the development of deep learning techniques.

Deep learning methods, especially convolutional neural networks (CNN), have become the mainstream technique for semantic segmentation of remote sensing images. In this regard, a typical model is the full convolutional network (FCN) [24], which first applied CNNs to pixel-level image segmentation. Afterwards, many modified models based on FCNs came into being. For example, SegNet [25] employs an encoder–decoder structure to improve segmentation performance, while U-Net [26] achieves accurate edge segmentation by skip connections.

Some researchers have focused on fusing multiscale information to improve the performance of semantic segmentation in remote sensing images. For example, DeepLabV3+ [27] employed dilated convolution and pyramidal pooling modules to capture multi-scale information. In addition, HRNet [28] proposed a high-resolution network that integrates high-resolution features and multi-scale features to improve segmentation accuracy. DANet [29] enabled better contextual information acquisition through a self-attention mechanism that improved the performance of remote sensing image segmentation. Meanwhile, DANet also introduces spatial and channel attention to further optimize the segmentation results.

Spatial resolution and spectral resolution are often not available at the same time, so the number of bands in high-resolution remote sensing images is low, and the capture of spectral characteristics of features is deficient. The Gaussian-weighting spectral (GWS) feature and the area shape index (ASI) feature are based on adaptive areas to compensate for the lack of high spatial resolution images for land cover classification features [30].

2.2. Attention Mechanism

Traditional convolutional neural network models have limitations in handling complex remote sensing images for semantic segmentation. For example, there are certain barriers to fully obtaining the content of the image, resulting in low accuracy. To overcome these obstacles, the attention mechanism is introduced into models for semantic segmentation tasks of remote sensing images. By adopting the attention mechanism, the model adaptively adjusts the weights of different features to better capture the important features in remote sensing images and increase the accuracy and robustness of segmentation. In addition, the attention mechanism can also reduce the computational complexity and the efficiency of the algorithm. Therefore, the attention mechanism model becomes one of the essential techniques in the segmentation field of remote sensing image semantics and is extensively used in various remote sensing image segmentation tasks.

Attention mechanisms for the areas mentioned above are typically classified into two categories: spatial attention and channel attention.

The earliest spatial attention model is the Spatial Transformer Network (STNet) [31], which is a neural network module that can adaptively perform spatial transformations on the input. Since remote sensing images typically have large scale, high dimensionality and complex spatial structure, it is not possible for traditional image segmentation methods to process them effectively. In contrast, STN networks can preprocess the input images by adaptive spatial transformations to improve the segmentation accuracy. Although the STN network has many advantages in the field of semantic segmentation of remote sensing images, it also exhibits some limitations. First, STN networks need a large amount of training data to learn the transformation parameters, otherwise, the problem of overfitting may occur. Second, STN networks may not be able to be efficiently handle some complex transformations (e.g., nonlinear transformations).

Channel attention mechanisms are also widely applied in semantic segmentation of remote sensing images. The earliest channel attention model is the Squeeze-and-Excitation Network (SENet) [32]. Specifically, SENet achieves a better feature representation by calculating the significance of each channel and weighting the channels according to their importance. However, the limitation of SENet is that it requires a large amount of computational resources. Therefore, there may be some challenges in practical application.

Several modified models using channel attention mechanisms have been developed, including CBAM (Convolutional Block Attention Module) [33], ECA-Net (Efficient Channel Attention Network) [34], and SKNet (Selective Kernel Networks) [35]. Among them, CBAM is a module that can be embedded into existing deep learning models, and it contains both the channel attention mechanism and spatial attention mechanism. The channel attention mechanism is used to adjust the importance of each channel, and it employs the spatial attention mechanism to adapt the significance of the spatial location. This combination can improve the capabilities of the model and is more flexible than other attention mechanism models. ECA-Net is a lightweight model that strengthens the model's attention to important characteristics by introducing the ECA module. In addition, SKNet is a selective convolution-based model. It implements the channel attention mechanism by controlling the shape of each convolution kernel through selective weight vectors to enhance and augment the efficiency of the model.

MANet [36] utilizes multiple attention modules to model global long-range information and proposes linear complexity attention to solve the problem of operations that require huge memory and time consumption. UNetFormer [37] performs attention extraction for both global and local information, maintaining a balance between contextual and spatial detail information. CHGFNet [38] proposes a co-attention mechanism to fuse optical and SAR images, and thoroughly explores in depth the complementary relationship between the two images.

The attention mechanism plays an essential role in the semantic segmentation of remote sensing images. It can facilitate the model to capture the features of different regions and channels to promote the performance of the algorithm. In the future, with the continuous development of deep learning and remote sensing technology, the attention mechanism will also play a more crucial role in remote sensing image analysis.

2.3. Multi-Scale Feature Fusion

Multi-scale feature fusion is an extensively researched and employed technique in the field of semantic segmentation. Its primary objective is to integrate feature information from various scales and aims to enhance the accuracy of segmentation outcomes.

Some established researchers have utilized the traditional image pyramid approach, where the original image is scaled to different sizes, and the corresponding feature maps are retrieved and then integrated [39]. Other studies have adopted multiscale CNN models by introducing feature extraction modules at different scales directly in the network, for instance, the Pyramid Scene Parsing Network (PSPNet) [40] and DeepLabV3 [41].

Although multi-scale feature combination has been widely implemented in the field of semantic segmentation of remote sensing images, it still has some limitations and drawbacks. On the one hand, the conventional image pyramid method requires multiple scaling and feature abstraction of the original image, which is computationally intensive and prone to information loss and blurring. On the other hand, introducing multi-scale feature extraction modules directly into the network can reduce the computational effort and information loss, but tends to cause problems such as overfitting and a model complexity increase.

Some improvement approaches have been proposed for the issue of multi-scale feature fusion. For example, some studies have introduced attention mechanisms for adaptively selecting feature information at different scales [29,42]. Other studies have provided lightweight network structures based on deeply separable convolutions to reduce model complexity and computational complexity [43,44].

Building on the previous study, we propose the CMA and CSA modules for fusing high-level features with low-level features using the attention mechanism. Furthermore, we combine multiple attention modules to form the CANet model.

3. Materials and Methods

3.1. Approach Overview

To balance the ability to capture global contextual content and local spatial detail information, we propose the use of the CMA and CSA modules. Multiple CMA modules are embedded in CANet to reduce the gap between neighboring feature maps, while a CSA module is used to incorporate spatial geometric information when computing attention maps for high-level features. The redefined feature maps are then skip-connected to the decoder to output predicted images.

3.2. Communicating Mutual Attention (CMA)

The communicating mutual attention module includes two branches: high-level feature flow (HLF) and low-level feature flow (LLF), as shown in Figure 2. The high-resolution coarse features preserve more spatial and contextual characteristics, while the low-resolution fine features are rich in semantic characteristics. The balance between the two flows is essential for semantic segmentation. The LLF extracts the spatial and contextual information of the low-level features, which is transformed into an attention map, and then it is combined with the high-level characteristics. At the same time, in the high-level feature flow, attention extracted from the high-level features is aggregated with the low-level features. It can be seen that the high-level feature flow forms a closed loop with the low-level feature flow. Therefore, high-level features can communicate with low-level features in this closed loop.

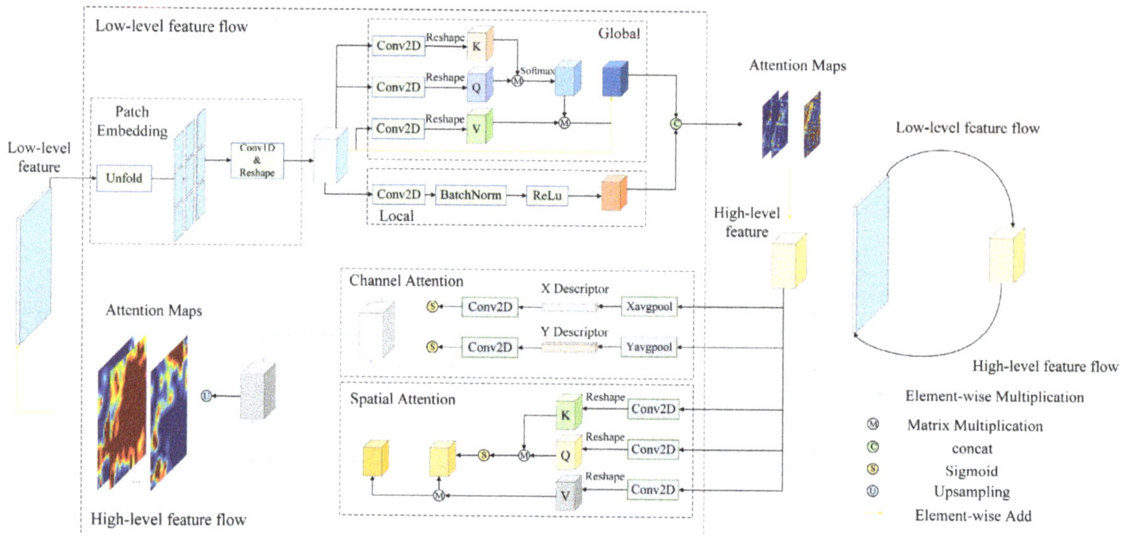

Figure 2. Illustration of Communicating Mutual Attention (CMA).

High-level features $\mathbf{X_h} \in \mathbb{R}^{B\times C_h\times H_h\times W_h}$ and low-level features $\mathbf{X_l} \in \mathbb{R}^{B\times C_l\times H_l\times W_l}$ are features output from different layers of the backbone network, so they have different shapes. In order to ensure that the features are the same shape, we have used patch embedding. To get the patches of low-level features, we first set $K = \frac{H_l}{H_h}$ and $S = \frac{H_l}{H_h}$, where K is the size of the kernel, S is the size of the stride. $\mathbf{X_l}$ is unfolded from $C \times H \times W$ to $CK^2 \times \left(\frac{H_l - K}{S} + 1\right)^2$, where $\left(\frac{H_l - K}{S} + 1\right)^2$ is equal to H_h. Then, $\mathbf{X_l}$ is reshaped to $CK^2 \times H_h \times H_h$. Eventually, the 1D convolution is used in $\mathbf{X_l}$ to ensure that $\mathbf{X_l}$ has the same number of channels as $\mathbf{X_h}$.

$$\mathbf{X_l} = \mathrm{conv}_{1\times1}(\mathrm{reshape}(\mathrm{unfold}(\mathbf{X_l}))) \tag{1}$$

Both exhaustive and detailed information is crucial for semantic segmentation because of the complexity of urban situations. This maintains a balance between global context and spatially detailed information [37]. Thus, we employ self-attention and 2D convolution to capture the global and local characteristics of low-level features. Features can be extracted from long-range information by self-attention, which can upgrade the global information of features. In order to achieve the above, we will first reshape $\mathbf{X_l} \in \mathbb{R}^{B\times C_l\times H_h\times W_h}$ into $\mathbf{X_l} \in \mathbb{R}^{B\times N\times C_l}$ and use the 1D convolution to get the query matrix $\mathbf{Q} \in \mathbb{R}^{B\times N\times C}$, $N = H_h \times W_h$, per column of the transformation result which is the 1D sequence of feature channels and each row is the value of a different channel at the same position in the feature map. Moreover, in the same way it is possible to obtain key matrix **K** and value matrix **V**.

$$\mathbf{Q} = \mathrm{reshape}(\mathrm{conv}_{1\times1}(\mathbf{X_l})) \tag{2}$$

$$\mathbf{K} = \mathrm{reshape}(\mathrm{conv}_{1\times1}(\mathbf{X_l})) \tag{3}$$

$$\mathbf{V} = \mathrm{reshape}(\mathrm{conv}_{1\times1}(\mathbf{X_l})) \tag{4}$$

We calculate the similarity of every row vector $\mathbf{q_i} \in \mathbb{R}^{1\times C}$ from $\mathbf{Q}$ to the corresponding elements of other row vectors $\mathbf{k_i} \in \mathbb{R}^{1\times C}$ from $\mathbf{K}$.

$$\mathbf{M_{similarity}} = \mathbf{QK^T} \tag{5}$$

By using the softmax function for each row of the similarity matrix, the normalized similarity weights of the elements at each position of the feature graph can be produced in relation to all other elements.

$$\mathbf{M_{normalized}} = \mathrm{softmax}\left(\mathbf{QK^T}\right) \tag{6}$$

The normalized similarity weight matrix is the matrix multiplied by $\mathbf{V}$ to generate a matrix of shape $\mathrm{N} \times \mathrm{C}$. Finally, the shape of this matrix is transformed to $\mathrm{C} \times \mathrm{H}_h \times \mathrm{W}_h$ to obtain the self-attention weight matrix.

$$\mathbf{A_{self}} = \mathrm{softmax}(\frac{\mathbf{QK^T}}{\sqrt{\mathrm{d}}})\mathbf{V} \tag{7}$$

where d is the channel number of the characteristic graph.

Features with contextual information are generated by adding self-attention weights to the original feature map.

$$\mathbf{A_{global}} = \mathbf{A_{self}} + \mathbf{X_l} \tag{8}$$

We deploy 2D convolution to obtain partial features. Finally, global features are aggregated with the local characteristics to generate the attention graph of the low-level feature stream.

$$\mathbf{A_{local}} = \mathrm{relu}(\mathrm{batchnorm}(\mathrm{conv}_{3\times3}(\mathbf{X_l}))) \tag{9}$$

$$\mathbf{A_{low\ to\ high}} = \mathrm{conv}_{1\times1}(\mathrm{cat}(\mathbf{A_{global}}, \mathbf{A_{local}})) \tag{10}$$

Using spatial or channel attention mechanism alone cannot fully extract contextual features. Therefore, we divided the high-level feature flow into spatial attention and channel attention in parallel to provide precise context features. Spatial attention is identical to self-attention in the LLF. Channel attention generally uses global pooling; however, global pooling cannot capture information about the position of an image in 2D [45]. In addition, global pooling causes the feature map to lose spatial detailed information. Thus, we employ two pooling kernels (H, 1) and (1, W) to pool $\mathbf{X_h}$ horizontally and vertically to generate feature descriptors $\mathbf{z_h}$ and $\mathbf{z_w}$.

$$\mathbf{z_h} = \mathrm{xavgpool}(\mathbf{X_h}) \tag{11}$$

$$\mathbf{z_w} = \mathrm{yavgpool}(\mathbf{X_h}) \tag{12}$$

The 2D convolution and sigmoid functions are used for $\mathbf{z_h}$ and $\mathbf{z_w}$ to obtain attention in the horizontal and vertical directions.

$$\mathbf{A_h} = \mathrm{sigmoid}(\mathrm{conv}_{3\times3}(\mathbf{z_h})) \tag{13}$$

$$\mathbf{A_w} = \mathrm{sigmoid}(\mathrm{conv}_{3\times3}(\mathbf{z_w})) \tag{14}$$

The spatial attention of $\mathbf{X_h}$ is fused with the channel attention and converted to the same shape as $\mathbf{X_l}$ by 1D convolution and upsampling.

$$\mathbf{A_{fusion}} = \mathrm{conv}_{1\times1}(\mathrm{cat}(\mathbf{A_{self}}, \mathbf{A_h A_w X_h})) \tag{15}$$

$$\mathbf{A_{high\ to\ low}} = \mathrm{upsample}(\mathrm{conv}_{1\times1}(\mathbf{A_{fusion}})) \tag{16}$$

The attention $\mathbf{A_{low\ to\ high}}$ and $\mathbf{A_{high\ to\ low}}$ extracted from the low-level feature stream and the high-level feature stream is added elementwise to the high-level and low-level features extracted by the encoder, respectively, for the purpose of feature communication. Hence, the rich spatially detailed information of the low-level features is aggregated into the high-level features. The semantic content is embedded in the low-level features.

$$\mathbf{X}_{\mathrm{l}} = \mathbf{X}_{\mathrm{l}} + \mathbf{A}_{\textbf{high to low}} \tag{17}$$

$$\mathbf{X}_{\mathrm{h}} = \mathbf{X}_{\mathrm{h}} + \mathbf{A}_{\textbf{low to high}} \tag{18}$$

3.3. *Communicating Self Attention (CSA)*

The attention maps for low-level and high-level features are typically computed separately without considering their interaction. However, the attention map of high-level features leads to a greater bias toward semantic information and misses the spatial detail representation of features, which is very unfavorable for semantic segmentation of urban areas that contain a variety of fine scenes. WiCoNet projects the contextual information of large area images into small local areas [46]. Consequently, we propose using the CSA module to improve the discrimination of tiny targets by introducing spatial characteristics of low-level features in the computation of attention maps of high-level features. Figure 3 shows the details of the CSA.

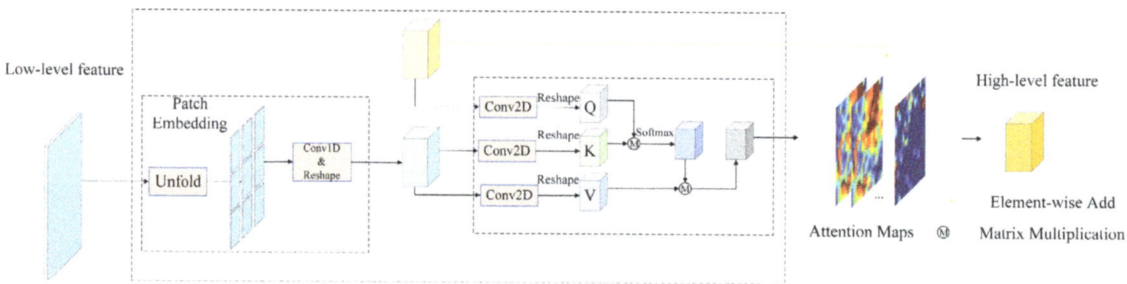

Figure 3. Proposed Communicating Self Attention (CSA).

We first utilized patch embedding to make the low-level characteristics the identical shape as the high-level features, similar to the CMA module. In contrast to the calculation of the self-attention graph in the CMA, the **Q** matrix in CSA is derived from high-level features, while the **K** and **V** matrices are generated by transforming low-level features. In the above operation, the weight of spatial information in the attention map can be strengthened, balancing the attention coefficients of high-level features on contextual information and spatial location information, which provides sufficient effort to exploit the low-level features.

$$\mathbf{Q} = \mathrm{reshape}(\mathrm{conv}_{1\times 1}(\mathbf{X}_{\mathrm{h}})) \tag{19}$$

$$\mathbf{K} = \mathrm{reshape}(\mathrm{conv}_{1\times 1}(\mathbf{X}_{\mathrm{l}})) \tag{20}$$

$$\mathbf{V} = \mathrm{reshape}(\mathrm{conv}_{1\times 1}(\mathbf{X}_{\mathrm{l}})) \tag{21}$$

The attention maps generated by the CSA module are added element-wise to the high-level features. The deployment of residual connections can mitigate network degradation.

$$\mathbf{X}_{\mathrm{h}} = \mathbf{X}_{\mathrm{h}} + \mathbf{A}_{\textbf{self}} \tag{22}$$

3.4. *Communicating Attention Network (CANet)*

Due to the fact that coarse-resolution high-level features lack rich spatial details and low-level features lack fine semantic content, both present complementary properties. Therefore, we propose the use of the CMA module to reduce the difference between the representation of high-level features and low-level features. The attention map provided by high-level features allows high-level features to focus further on contextual features and reduce the emphasis on spatial features, which is not conducive to improving the semantic segmentation accuracy of remote sensing images in urban areas. Hence, we designed the CSA module to incorporate the low-level features as auxiliary information added into the calculation of the attention graph of the high-level features.

As illustrated in Figure 4, CANet utilizes ResNet50 [16] pre-trained on ImageNet as the backbone network, and its four output feature maps $[\mathbf{X_1}, \mathbf{X_2}, \mathbf{X_3}, \mathbf{X_4}]$ correspond to the output results of the network from shallow to deep layers with low to high feature levels, respectively. The CMA and CSA modules contain dot-product attention, which is relatively computationally intensive. Therefore, we employed 1D convolution to reduce the number of channels of the output feature graph to one-fourth of the original to promote computational efficiency. We adopted three CMA modules and one CSA module to communicate information about the feature maps at different levels. The three CMA modules work on $\mathbf{X_1}$ and $\mathbf{X_2}$, $\mathbf{X_2}$ and $\mathbf{X_3}$, and $\mathbf{X_3}$ and $\mathbf{X_4}$, respectively, where the input feature maps of $\mathbf{X_2}$ and $\mathbf{X_3}$ are the original features output by ResNet50 when they are involved in the different modules. For the purpose of reducing the computational effort, we merely exploit $\mathbf{X_1}$ and $\mathbf{X_4}$ in the operations of the CSA module. The feature maps processed by the CMA and CSA modules are skip connected to the counterpart layers of the decoder. For the decoder part, the same structure is applied to each layer. First, feature maps with reduced numbers of channels can be produced by 1D convolution. Then, we use deconvolution to alter the shape of the feature graph. Finally, 1D convolution is utilized to convert the number of feature map channels to a specified number. Eventually, the output result of the last layer of the decoder is upsampled to a similar shape as the input image as the result of semantic segmentation.

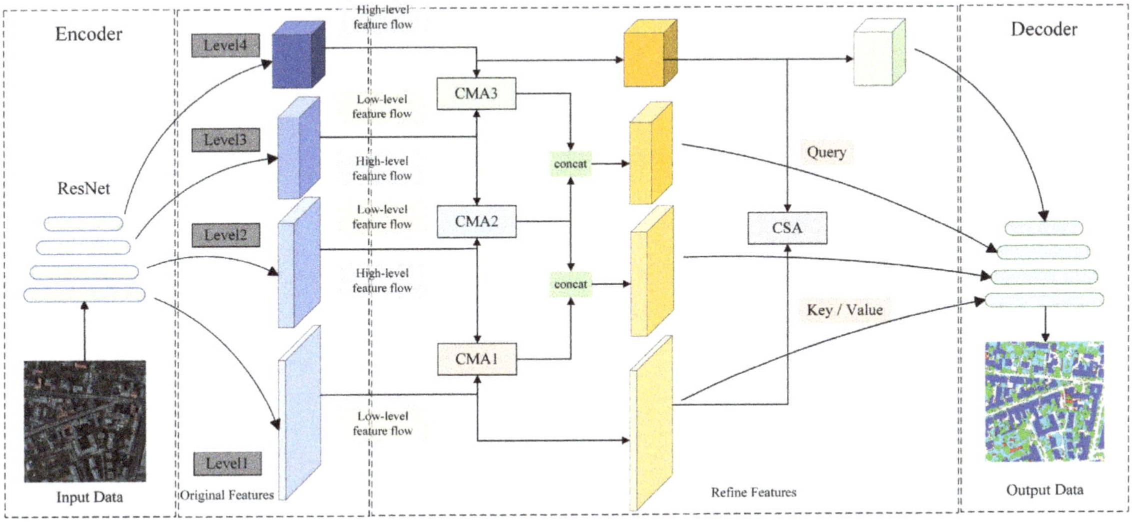

Figure 4. Structure of designed CANet.

The algorithm implementation flow of CANet is demonstrated in Algorithm 1.

Algorithm 1 The algorithm implementation process of CANet (Vaihingen dataset)

Input: **I** (NIR, R, G, DSM)
Output: **P** (Prediction results of semantic segmentation)
// Step1: Extracting multi-level feature maps from the encoder
  $[\mathbf{X_1}, \mathbf{X_2}, \mathbf{X_3}, \mathbf{X_4}] = ResNet50(\mathbf{I})$
// Step2: CMA and CSA performs characteristic aggregation of multi-level features
for *i* in {1, 2, 3} **do**
  $[\mathbf{X^i_{refine}}, \mathbf{X^{i+1}_{refine}}] = CMA_i(\mathbf{X_i}, \mathbf{X_{i+1}})$
end
$\mathbf{X^4_{new\ refine}} = CSA(\mathbf{X^1_{refine}}, \mathbf{X^4_{refine}})$
// Step3: Skip connection of refine features to decoders
for *i* in {1, 2, 3, 4} **do**
  $\mathbf{D_i} = \mathbf{D_i} + \mathbf{X^i_{refine}}$
end
// Step4: Output prediction results
$\mathbf{P} = Decoder(\mathbf{X^1_{refine}}, \mathbf{X^2_{refine}}, \mathbf{X^3_{refine}}, \mathbf{X^4_{refine}})$
end

3.5. Loss Function

In the training phase, we employ cross-entropy loss as a loss function to measure the difference between the prediction results of CANet and the ground truth data.

$$\text{Loss} = -\mathbf{y}\log(\mathbf{p}) - (1-\mathbf{y})\log(1-\mathbf{p}) \tag{23}$$

where **p** is the prediction result, and **y** is the ground truth data.

3.6. Dataset

For a more impartial test of the model's performance, we trained and tested it by employing the ISPRS Potsdam and Vaihingen datasets.

The Potsdam dataset provides 38 remotely sensed images of urban areas taken by UAV at high resolution. The images have a spatial resolution of 5 cm and a size of 6000 × 6000, which contain NIR, red, green, blue, DSM and normalized DSM images. In addition, each image corresponds to a surface truth image. The dataset contains six classes, which are impervious surfaces, building, low vegetation, tree, car, and clutter/background. We employed images numbered 2_13, 2_14, 3_13, 3_14, 4_13, 4_14, 4_15, 5_13, 5_14, 5_15, 6_13, 6_14, 6_15, and 7_13 for testing, images numbered 20_10 for validation, and the remaining 22 images for training.

The Vaihingen dataset consists of 33 remotely sensed images. Similar to the Potsdam dataset, this remote sensing image was also acquired by aerial photography from a UAV. Therefore, it has a high spatial resolution (i.e., 9 cm). Within this dataset, each image provides NIR, red, green, and DSM bands as well as surface truth data. The dataset comprises the identical classifiable categories as the Potsdam dataset. We utilized images numbered 2, 4, 6, 8, 10, 12, 14, 16, 20, 22, 24, 27, 29, 31, 33, 35, and 38 for testing, image number 30 for validation, and the remaining 15 images for training.

Figure 5 shows images of the entire Potsdam and Vaihingen datasets.

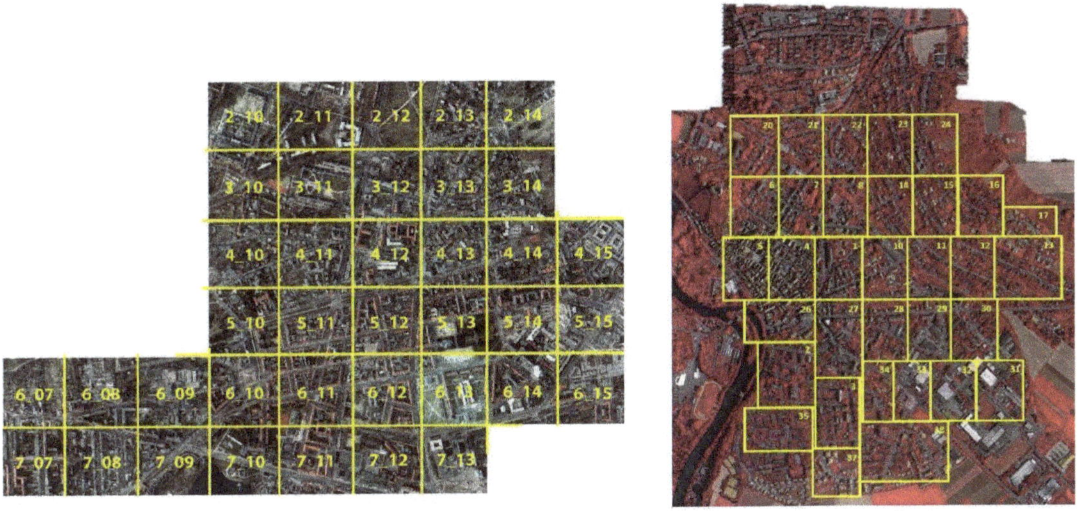

(a) (b)

Figure 5. Remote sensing images of the area where the dataset is located. (**a**) is an overview map of the Potsdam dataset set, (**b**) is an overview map of the Vaihingen dataset set. The different numbers are the IDs of the different images in the dataset.

3.7. Data Pre-Processing and Experimental Setting

To reduce the memory consumption and increase the diversity of the data, the original images are randomly cropped and randomly flipped, and the crop size is 512 × 512. Note that each original image is cropped by 1000 images to enhance the amount of trainable data. The images are re-cropped at each epoch. Simultaneously, the probability of random flipping in all four directions is 25%. Moreover, in the training and testing phases, the input data are all bands provided by the dataset. In the training phase, we set the initial learning rate to 0.0002 and selected the cosine annealing strategy for the decay of the learning rate. Meanwhile, Adam is utilized as the optimizer. All experiments of this study were accomplished by PyTorch on a NVIDIA Tesla V100 GPU with 16-GB RAM.

3.8. Accuracy Evaluation

We utilized three evaluation metrics to assess the capability of the model, namely, overall accuracy, F1-score, and mean intersection over union.

$$\mathrm{OA} = \frac{\sum_{k=1}^{N} \mathrm{TP_k}}{\sum_{k=1}^{N} \mathrm{TP_k} + \mathrm{FP_k} + \mathrm{TN_k} + \mathrm{FN_k}} \tag{24}$$

$$\mathrm{F1-score} = 2 \times \frac{\mathrm{precision} \times \mathrm{recall}}{\mathrm{precision} + \mathrm{recal}} \tag{25}$$

$$\mathrm{mIoU} = \frac{1}{N} \sum_{k=1}^{N} \frac{\mathrm{TP_k}}{\mathrm{TP_k} + \mathrm{FP_k} + \mathrm{TN_k}} \tag{26}$$

where $\mathrm{TN_k}$, $\mathrm{FN_k}$, $\mathrm{TP_k}$, $\mathrm{FP_k}$ are true negative, and false negatives, true positive, false positive, respectively, and k is the number of categories.

4. Results

4.1. Ablation Study

To evaluate the performance of the constructed CANet as well as the attention modules CMA and CSA, we performed an ablation study on the Vaihingen dataset. The training baseline for the study was an FCN [24] model with ResNet50 as the encoder. Table 1 demonstrates the comparison of the results of the ablation experiment executed on the Vaihingen dataset.

Table 1. Ablation experiments on the Vaihingen dataset.

Method	Imp. Surf.	Building	Low Veg.	Tree	Car	Mean F1	OA	mIoU
FCN [24]	89.05	91.54	79.01	87.10	84.22	86.18	87.0	75.97
FCN + CMA1	91.30	94.69	81.20	88.11	86.50	88.36	89.12	79.44
FCN + CMA2	92.21	94.96	82.29	88.51	86.75	88.95	89.79	80.38
FCN + CMA3	91.35	94.37	81.10	87.87	87.77	88.49	89.0	79.64
FCN + CMA123	91.67	94.58	83.31	88.94	**87.84**	89.27	89.83	80.83
FCN + CSA	90.82	94.44	80.56	87.95	86.48	88.05	88.80	78.96
CANet	**92.49**	**95.26**	**83.34**	**89.18**	87.75	**89.61**	**90.33**	**81.41**

Compared to FCN, Mean F1 improved by 1.87%, OA by 1.8% and mIoU by 2.99% with the addition of merely one CSA module. As can be shown from Table 1, CMA2 achieves better results with the application of barely a single CMA module. Among them, $\mathbf{X}_1$ and $\mathbf{X}_2$ are more inclined to contain spatial detail features, while the semantic information in $\mathbf{X}_3$ and $\mathbf{X}_4$ is richer, and the gap between the representational information of high-level features and low-level features is not large. Therefore, the effect of CMA on feature aggregation is not obvious. On the one hand, the comparison results of Figure 6a–c indicate that CANet can better segment the targets in the large-impervious surfaces and low vegetation categories. Meanwhile, for the low vegetation and tree categories with similar spectral characteristics, CANet can accurately distinguish them by aggregating the global information and capturing the small semantic feature variances between them. On the other hand, as illustrated in Figure 6d–e, CANet is very capable of identifying fine targets (e.g., cars) compared to FCN. In addition, CANet enhances 3.43%, 3.33% and 5.44% over FCN in Mean F1, OA and mIoU accuracy metrics, respectively. These results indicate that CANet achieves the bridging of the gap between high-level and low-level feature information characterization capabilities and physical content through the CMA and CSA modules, maintaining the ability to aggregate global contextual content while achieving high-precision extraction of tiny targets.

4.2. Quantitative Comparison of Various Modules

The communicating attention modules CMA and CSA narrow the gap between high-level features and low-level features by computing the attention map and embedding geometric detail features into high-level features to enable low-level features to aggregate semantic information. In other words, the CMA and CSA modules achieve multi-scale feature fusion through the attention mechanism. Therefore, we compare the mainstream attention mechanism modules (e.g., SE [32] and CBAM [33]) and multiscale feature fusion methods (e.g., PPM [40] and ASPP [27]) with the PAM and AEM modules proposed in LANet [18] in our experiments. The study results are illustrated by Table 2, where a single utilization of attention or multiscale feature fusion modules achieves only a limited improvement in accuracy compared to FCN [24]. It is worth noting that although the PAM and AEM module enables the semantic features of high-level features to be embedded in low-level features through the attention module, it does not fuse the spatial tiny information of low-level features into high-level features. Hence, both our proposed CMA and CSA modules attain better results than other attentional mechanisms and multi-scale feature fusion modules on both Vaihingen and Potsdam datasets.

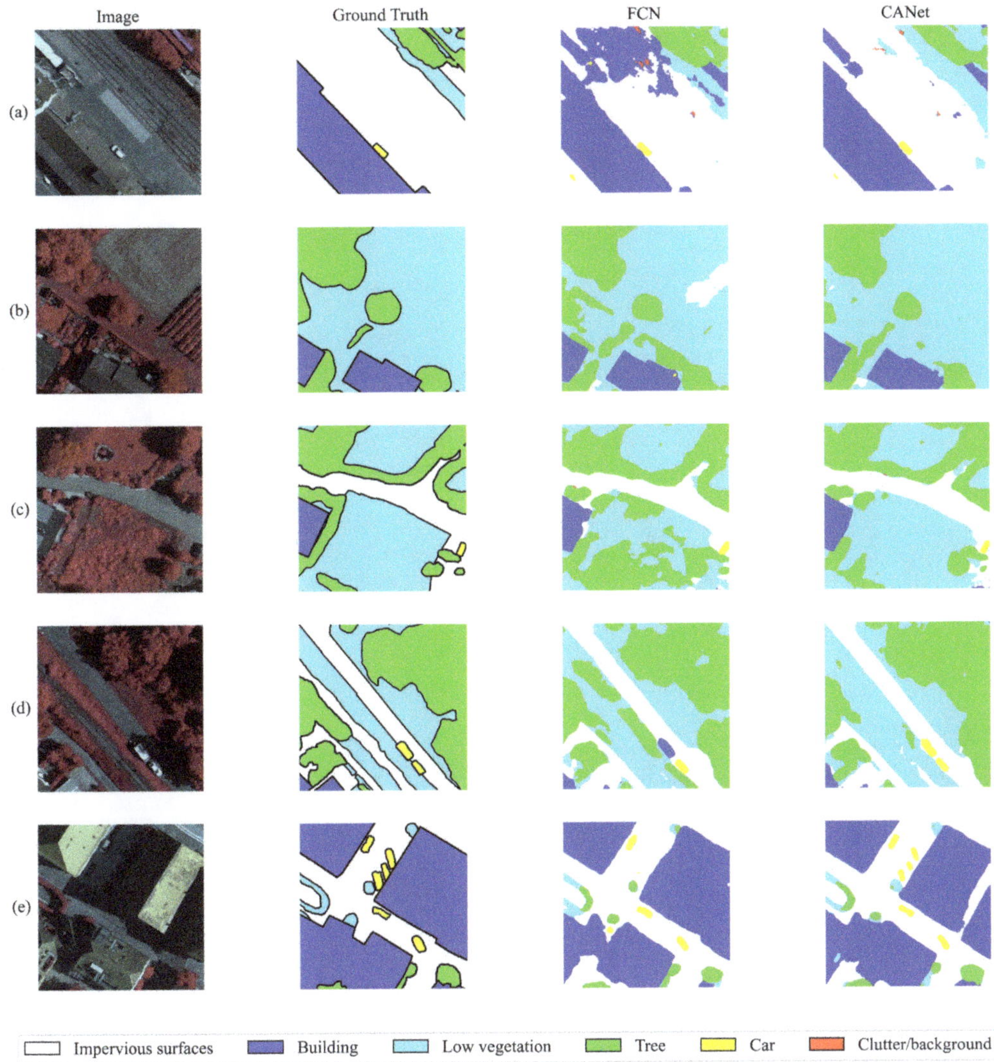

Figure 6. Comparison of ablation experimental results of FCN and CANet. (**a–e**) are the five regions in the Vaihingen dataset used to compare the original image, the ground truth image, the prediction results from FCN and CANet.

Table 2. Quantitative comparison results of multiple modules.

Dataset	Method	Mean F1	OA	mIoU
Vaihingen	FCN [24]	86.18	87.0	75.97
	FCN + SE [32]	87.23	89.71	77.89
	FCN + CBAM [33]	88.19	89.96	79.61
	FCN + PPM [40]	86.47	89.36	76.78
	FCN + ASPP [27]	86.77	89.12	77.12
	FCN + PAM and AEM [18]	88.09	89.83	-
	CANet	**89.61**	**90.33**	**81.41**

Table 2. *Cont.*

Dataset	Method	Mean F1	OA	mIoU
Potsdam	FCN [24]	88.05	88.02	81.41
	FCN + SE [32]	91.39	89.60	85.38
	FCN + CBAM [33]	91.73	89.89	85.65
	FCN + PPM [40]	89.98	90.14	81.99
	FCN + ASPP [27]	90.86	89.18	84.24
	FCN + PAM and AEM [18]	91.95	90.84	-
	CANet	**92.60**	**91.44**	**86.48**

4.3. Quantitative Comparison of Various Models

For a more impartial test of the model's performance, we quantitatively compared CANet with the current dominant semantic segmentation models. Both PSPNet [40] and DeepLabV3+ [27] models include a feature pyramid module to extract multi-scale features. MACUNet [47] and LANet [18] utilize a channel attention module to perform multi-scale feature interaction and fusion. The comparison results on the Vaihingen and Potsdam datasets can be seen in Tables 3 and 4. Compared to FCN [24], both PSPNet and DeepLabV3+ have improved F1-scores, but the acquisition of exhaustive information by enhancing the receptive field leads to a reduction in segmentation accuracy for small target objects (e.g., cars). MACUNet and LANet utilize only channel attention, which leads to a degradation in their ability to capture global semantic information. CANet maintains global contextual features while enhancing the geometric detail characterization capability, making it optimal on all three evaluation metrics (i.e., F1-score, OA, mIoU) for both datasets. It is worth noting that CANet improves the F1-score accuracy by 9.11% over the DeepLabV3+ in the car category on the Vaihingen dataset, which illustrates the strong segmentation capability of the proposed model for tiny objects.

Table 3. Quantitative comparison of model accuracy on the Vaihingen dataset.

Method	Imp.Surf.	Building	Low Veg.	Tree	Car	Mean F1	OA	mIoU
FCN [24]	89.05	91.54	79.01	87.10	84.22	86.18	87.0	75.97
PSPNet [40]	90.83	94.48	80.51	88.28	84.14	87.65	88.83	78.35
DeepLabV3+ [27]	90.41	94.05	80.27	88.31	78.64	86.34	88.50	76.44
MACUNet [47]	91.66	93.67	80.76	87.78	83.66	87.51	88.80	78.11
LANet [18]	92.41	94.90	82.89	88.92	81.31	88.09	89.83	-
CANet	**92.49**	**95.26**	**83.34**	**89.18**	**87.75**	**89.61**	**90.33**	**81.41**

Table 4. Quantitative comparison of model accuracy on the Potsdam dataset.

Method	Imp.Surf.	Building	Low Veg.	Tree	Car	Mean F1	OA	mIoU
FCN [24]	92.24	95.35	84.29	83.12	94.53	89.19	88.60	82.06
PSPNet [40]	90.80	95.17	85.76	86.99	91.14	89.97	88.82	81.94
DeepLabV3+ [27]	91.59	96.03	86.09	86.50	94.23	90.59	89.41	83.54
MACUNet [47]	92.64	97.00	86.30	87.49	95.14	91.71	90.36	84.97
LANet [18]	93.05	97.19	87.30	88.04	94.19	91.95	90.84	-
CANet	**93.91**	**97.22**	**87.59**	**88.23**	**96.07**	**92.60**	**91.44**	**86.48**

For the purpose of comparing the performance of each model more intuitively, we visualized their prediction results, in which the contrasting areas are highlighted with pink boxes. Figure 7. demonstrates the results for local areas in the Vaihingen dataset. Despite the successful segmentation of cars by each model, there is confusion between the impervious surfaces category and the car category for all models except CANet. In addition, CANet outperforms the other models in distinguishing between buildings and impervious surfaces. The local areas of the Potsdam dataset illustrated in Figure 8 are mainly low vegetation and tree categories, with very similar spectral characteristics. PSPNet and DeepLabV3+ enhance the representation of contextual information by increasing the receptive field and achieve better results, but the classification results are inferior for the more mixed location of low vegetation and tree. CANet maintains a balance between global contextual features and local geometric characterization ability and has better performance in classifying confusable categories with interleaved distribution.

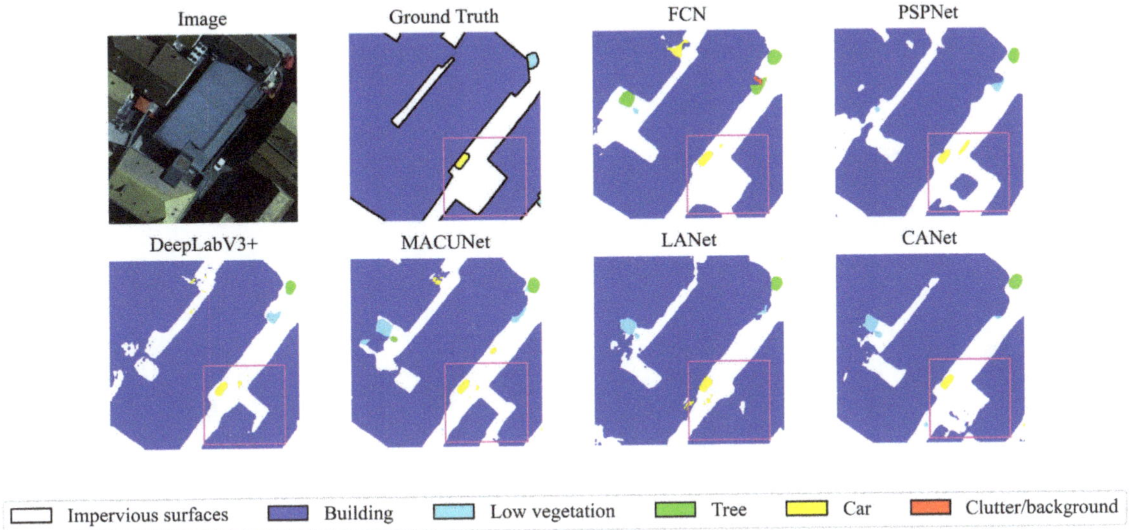

Figure 7. Model comparison visualization results on the Vaihingen datasets. The pink color selects the area where the contrast between the predictions of the different models is most pronounced.

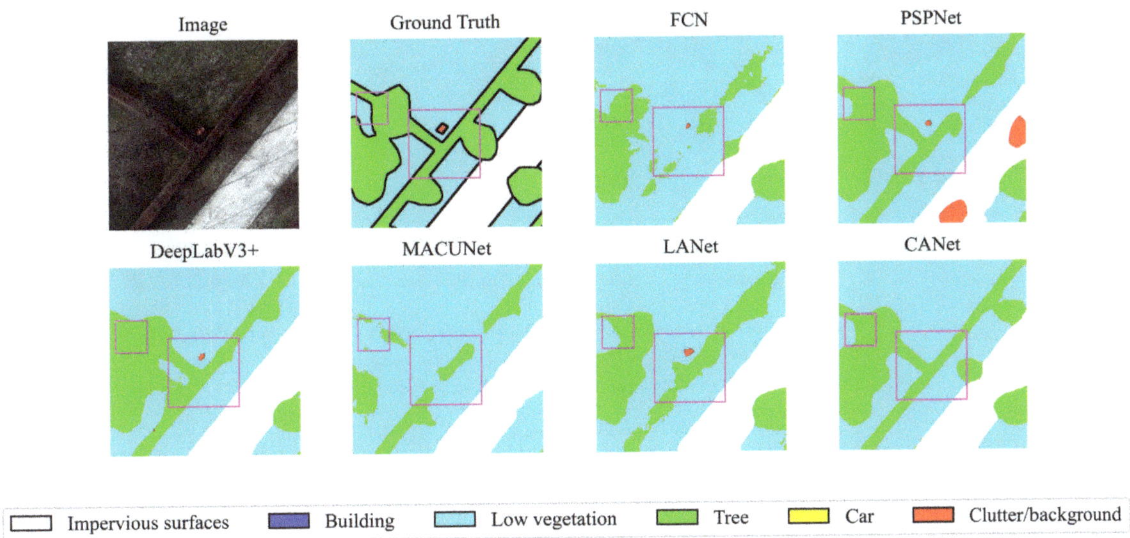

Figure 8. Model comparison visualization results on the Potsdam datasets. The pink color selects the area where the contrast between the predictions of the different models is most pronounced.

5. Discussion

5.1. Interaction of High-Level and Low-Level Features

In order to be capable of clearly illustrating the changes in the feature maps before and after the deployment of CMA and CSA, we visualized them. As demonstrated in Figure 9, before the utilization of CMA and CSA, the high-level features captured rich semantic information and the low-level features were more powerful in maintaining spatial details. However, there is a notable gap between the high-level features and the low-level features. Therefore, we implemented the interaction between high-level features and low-level features by reassigning and aggregating different features using CMA and CSA modules. By using this method, we can see that the gap between features of neighboring levels is reduced. Meanwhile, after involving the low-level features in the generation of the attention map of the high-level features through the CSA module, the high-level features are embedded with spatially detailed information while maintaining the global contextual contents. It is worth noting that after inserting the CMA and CSA modules into the model, we achieved a model that does not lose global semantic features but enhances the geometric representation of the feature maps.

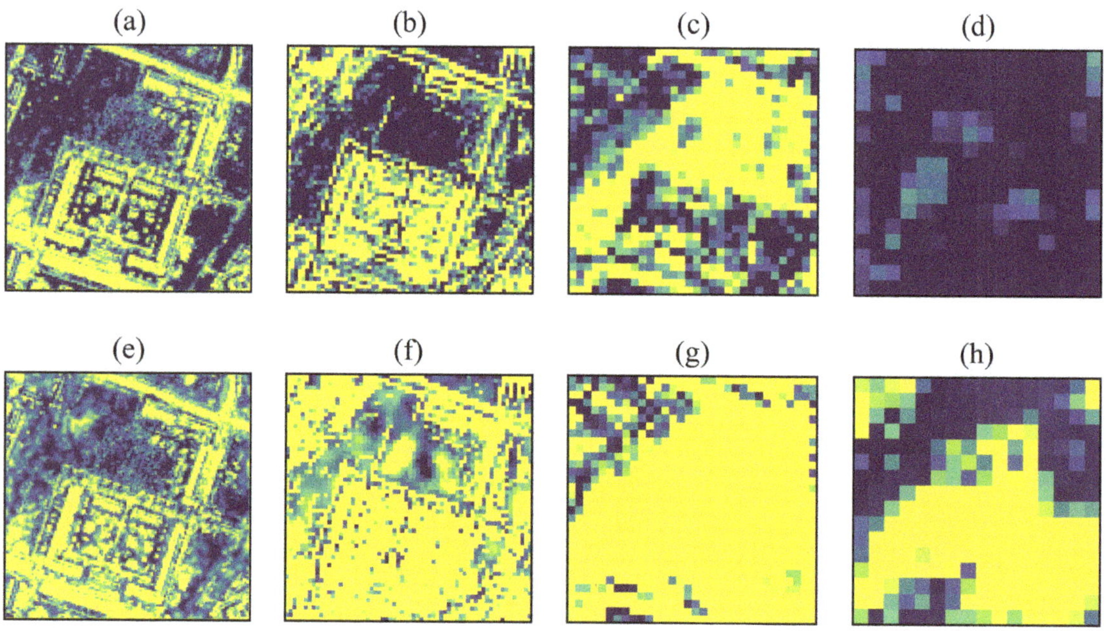

Figure 9. Comparison of multi-level feature maps before and after using the CMA and CSA modules. Where (**a**–**d**) are the feature maps of levels 1–4 before using the attention module. (**e**–**h**) are the feature maps of levels 1–4 after applying the attention module.

5.2. Number of High-Level and Low-Level Feature Cycles

As seen in Figure 2 (In Section 3.2), the high-level feature flow and the low-level feature flow in the CMA module are combined into one loop. In all the above experiments, when the high-level features and the low-level features pass through the CMA module, they do feature interaction merely once. Therefore, to explore the effect of the frequency of feature interactions on the semantic segmentation accuracy, we conducted experiments on the Vaihingen dataset with the number of exchanges set to two and three, respectively. As presented in Table 5, the OA, Mean F1 and mIoU precision metrics exhibit a decreasing trend as the number of interactions increases. On the one hand, the spatial details in the low-level features become ambiguous when the number of high-level and low-level feature interactions increases. On the other hand, the high-level features induce the noise of the low-level features, leading to the integrity of the contextual information being destroyed. Consequently, the best segmentation capability model can be attained only by setting a reasonable number of feature interactions.

Table 5. Comparison experiment of the number of interactions between high-level features and low-level features.

Cycle Number	Imp.Surf.	Building	Low Veg.	Tree	Car	Mean F1	OA	mIoU
1	**92.49**	**95.26**	**83.34**	**89.18**	**87.75**	**89.61**	**90.33**	**81.41**
2	92.41	95.02	82.82	88.91	87.12	89.26	90.07	80.86
3	91.75	94.57	82.54	88.27	86.04	88.64	89.55	79.85

5.3. Computational Complexity of the Algorithm

The time and space complexity of the model is crucial for the large-scale application and deployment of the algorithm. Therefore, it is necessary to analyze the computational complexity of the algorithm. The module that occupies the main computational time and memory in the CMA and CSA modules is the self-attention module, where the time complexity of computing the matrix product operation of $\mathbf{Q} \in \mathbb{R}^{B \times N \times C}$ and $\mathbf{K}^T \in \mathbb{R}^{B \times C \times N}$ is $O(CN^2)$, the time complexity of computing softmax$(\mathbf{QK}^T)$ is $O(N^2)$, and the time complexity of computing the matrix product operation of softmax$(\frac{\mathbf{QK}^T}{\sqrt{d}})$ and $\mathbf{V} \in \mathbb{R}^{B \times N \times C}$ is $O(CN^2)$. Therefore, the time complexity of the self-attention mechanism module is $O(CN^2)$. In addition, the space complexity of the self-attention mechanism module is $O(N^2)$.

6. Conclusions

Features at different levels have different data distributions and information contents. Therefore, the integration and aggregation of multi-scale features is essential to achieve accurate semantic segmentation. We propose a novel attention module and CNN-based neural network (CANet) for semantic segmentation of high-resolution remote sensing images of urban areas. To reduce the difference in feature characterization ability between high-level features and low-level features, we designed the CMA and CSA modules to enable the interaction of different levels of feature maps. We employed CMA to aggregate the spatially detailed information of low-level features into high-level features and embed the global semantic information of high-level features into low-level features. To maintain the balance between global contextual information and spatial characteristics, we utilized the CSA module to introduce the geometric features of the low-level features into the attention map computation of the high-level features. The model was tested in a series of ablation and comparison studies on the ISPRS Vaihingen and Potsdam datasets. The results (i.e., 89.61% and 92.60% mean F1-score) demonstrate the effectiveness of the method in the semantic segmentation task of high-resolution remote sensing images in urban areas. However, the temporal and spatial complexity of our model is high, which is challenging for large-scale deployment.

In the future, the work will inspire research on the fusion and interaction of features at different levels. Meanwhile, we will improve the model algorithm to reduce the computational complexity and we will continue to explore the relationship between multi-scale features and methods to aggregate global contextual information which further enhances the capability of the model.

Author Contributions: Conceptualization, Y.H.; Methodology, X.M. and L.Z.; Software, X.M.; Validation, L.Z.; Formal Analysis, H.Z.; Investigation, X.M. and L.Z.; Resources, H.Z.; Data Curation, X.M.; Writing—Original Draft Preparation, X.M. and L.Z.; Writing—Review and Editing, H.Z.; Visualization, L.Z.; Supervision, H.Z. and Y.H.; Project Administration, H.Z. and Y.H.; Funding Acquisition, H.Z. and Y.H. All authors have read and agreed to the published version of the manuscript.

Funding: This work was supported by the Natural Science Foundation of Shandong Province (grant number ZR2022QD141); National Natural Science Foundation of China (grant number 42271273);

National Natural Science Foundation of China (grant number 42201440); The Fundamental Research Funds for Chinese Academy of Surveying and Mapping (grant number AR2201); The China Post doctoral Science Foundation (grant number 2022M712936); and National Natural Science Foundation of China (grant number 42201278).

Data Availability Statement: Thanks to ISPRS for the training and testing datasets for remote sensing image semantic segmentation tasks, which can be found at the following URLs. Vaihingen: https://www2.isprs.org/commissions/comm2/wg4/benchmark/2d-sem-label-vaihingen/, accessed on 20 October 2020; Potsdam: https://www2.isprs.org/commissions/comm2/wg4/benchmark/2d-semlabel-potsdam/, accessed on 20 October 2020.

Acknowledgments: The authors would appreciate GPU support from the Chinese Academy of Surveying and Mapping.

Conflicts of Interest: The authors declare no conflict of interest.

References

1. Yuan, X.; Shi, J.; Gu, L. A review of deep learning methods for semantic segmentation of remote sensing imagery. *Expert Syst. Appl.* **2021**, *169*, 114417. [CrossRef]
2. Kemker, R.; Salvaggio, C.; Kanan, C. Algorithms for semantic segmentation of multispectral remote sensing imagery using deep learning. *ISPRS J. Photogramm. Remote Sens.* **2018**, *145*, 60–77. [CrossRef]
3. Diakogiannis, F.I.; Waldner, F.; Caccetta, P.; Wu, C. ResUNet-a: A deep learning framework for semantic segmentation of remotely sensed data. *ISPRS J. Photogramm. Remote Sens.* **2020**, *162*, 94–114. [CrossRef]
4. Wang, S.; Chen, W.; Xie, S.M.; Azzari, G.; Lobell, D.B. Weakly supervised deep learning for segmentation of remote sensing imagery. *Remote Sens.* **2020**, *12*, 207. [CrossRef]
5. Shafique, A.; Cao, G.; Khan, Z.; Asad, M.; Aslam, M. Deep learning-based change detection in remote sensing images: A review. *Remote Sens.* **2022**, *14*, 871. [CrossRef]
6. Chen, H.; Shi, Z. A spatial-temporal attention-based method and a new dataset for remote sensing image change detection. *Remote Sens.* **2020**, *12*, 1662. [CrossRef]
7. Deng, Z.; Sun, H.; Zhou, S.; Zhao, J.; Lei, L.; Zou, H. Multi-scale object detection in remote sensing imagery with convolutional neural networks. *ISPRS J. Photogramm. Remote Sens.* **2018**, *145*, 3–22. [CrossRef]
8. Carvalho, O.L.F.D.; de Carvalho Júnior, O.A.; Albuquerque, A.O.D.; Bem, P.P.D.; Silva, C.R.; Ferreira, P.H.G.; Moura, R.D.S.D.; Gomes, R.A.T.; Guimaraes, R.F.; Borges, D.L. Instance segmentation for large, multi-channel remote sensing imagery using mask-RCNN and a mosaicking approach. *Remote Sens.* **2020**, *13*, 39. [CrossRef]
9. Chen, M.; Wu, J.; Liu, L.; Zhao, W.; Tian, F.; Shen, Q.; Zhao, B.; Du, R. DR-Net: An improved network for building extraction from high resolution remote sensing image. *Remote Sens.* **2021**, *13*, 294. [CrossRef]
10. Maggiori, E.; Tarabalka, Y.; Charpiat, G.; Alliez, P. Convolutional neural networks for large-scale remote-sensing image classification. *IEEE Trans. Geosci. Remote Sens.* **2016**, *55*, 645–657. [CrossRef]
11. Ali, M.; Clausi, D. Using the Canny edge detector for feature extraction and enhancement of remote sensing images. In Proceedings of the IGARSS 2001. Scanning the Present and Resolving the Future. Proceedings. IEEE 2001 International Geoscience and Remote Sensing Symposium (Cat. No. 01CH37217), Sydney, NSW, Australia, 9–13 July 2001; pp. 2298–2300.
12. Wang, Z.; Jensen, J.R.; Im, J. An automatic region-based image segmentation algorithm for remote sensing applications. *Environ. Model. Softw.* **2010**, *25*, 1149–1165. [CrossRef]
13. Goodfellow, I.; Bengio, Y.; Courville, A. *Deep Learning*; MIT Press: Cambridge, MA, USA, 2016.
14. Kattenborn, T.; Leitloff, J.; Schiefer, F.; Hinz, S. Review on Convolutional Neural Networks (CNN) in vegetation remote sensing. *ISPRS J. Photogramm. Remote Sens.* **2021**, *173*, 24–49. [CrossRef]
15. Alam, M.; Wang, J.-F.; Guangpei, C.; Yunrong, L.; Chen, Y. Convolutional neural network for the semantic segmentation of remote sensing images. *Mob. Netw. Appl.* **2021**, *26*, 200–215. [CrossRef]
16. He, K.; Zhang, X.; Ren, S.; Sun, J. Deep residual learning for image recognition. In Proceedings of the IEEE Conference on Computer Vision and Pattern Recognition, Las Vegas, NV, USA, 27–30 June 2016; pp. 770–778.
17. Ghassemian, H. A review of remote sensing image fusion methods. *Inf. Fusion* **2016**, *32*, 75–89. [CrossRef]
18. Ding, L.; Tang, H.; Bruzzone, L. LANet: Local attention embedding to improve the semantic segmentation of remote sensing images. *IEEE Trans. Geosci. Remote Sens.* **2020**, *59*, 426–435. [CrossRef]
19. Yang, M.Y.; Kumaar, S.; Lyu, Y.; Nex, F. Real-time semantic segmentation with context aggregation network. *ISPRS J. Photogramm. Remote Sens.* **2021**, *178*, 124–134. [CrossRef]

20. Li, R.; Zheng, S.; Zhang, C.; Duan, C.; Wang, L.; Atkinson, P.M. ABCNet: Attentive bilateral contextual network for efficient semantic segmentation of Fine-Resolution remotely sensed imagery. *ISPRS J. Photogramm. Remote Sens.* **2021**, *181*, 84–98. [CrossRef]
21. Ju, J.; Gopal, S.; Kolaczyk, E.D. On the choice of spatial and categorical scale in remote sensing land cover classification. *Remote Sens. Environ.* **2005**, *96*, 62–77. [CrossRef]
22. Pham, H.M.; Yamaguchi, Y.; Bui, T.Q. A case study on the relation between city planning and urban growth using remote sensing and spatial metrics. *Landsc. Urban Plan.* **2011**, *100*, 223–230. [CrossRef]
23. Li, J.; Pei, Y.; Zhao, S.; Xiao, R.; Sang, X.; Zhang, C. A review of remote sensing for environmental monitoring in China. *Remote Sens.* **2020**, *12*, 1130. [CrossRef]
24. Long, J.; Shelhamer, E.; Darrell, T. Fully convolutional networks for semantic segmentation. In Proceedings of the IEEE Conference on Computer Vision and Pattern Recognition, Boston, MA, USA, 7–12 June 2015; pp. 3431–3440.
25. Badrinarayanan, V.; Kendall, A.; Cipolla, R. Segnet: A deep convolutional encoder-decoder architecture for image segmentation. *IEEE Trans. Pattern Anal. Mach. Intell.* **2017**, *39*, 2481–2495. [CrossRef]
26. Ronneberger, O.; Fischer, P.; Brox, T. U-net: Convolutional networks for biomedical image segmentation. In Proceedings of the Medical Image Computing and Computer-Assisted Intervention–MICCAI 2015: 18th International Conference, Munich, Germany, 5–9 October 2015; Proceedings, Part III 18. pp. 234–241.
27. Chen, L.-C.; Zhu, Y.; Papandreou, G.; Schroff, F.; Adam, H. Encoder-decoder with atrous separable convolution for semantic image segmentation. In Proceedings of the European Conference on Computer Vision (ECCV), Munich, Germany, 8–14 September 2018; pp. 801–818.
28. Wang, J.; Sun, K.; Cheng, T.; Jiang, B.; Deng, C.; Zhao, Y.; Liu, D.; Mu, Y.; Tan, M.; Wang, X. Deep high-resolution representation learning for visual recognition. *IEEE Trans. Pattern Anal. Mach. Intell.* **2020**, *43*, 3349–3364. [CrossRef] [PubMed]
29. Fu, J.; Liu, J.; Tian, H.; Li, Y.; Bao, Y.; Fang, Z.; Lu, H. Dual attention network for scene segmentation. In Proceedings of the IEEE/CVF Conference on Computer Vision and Pattern Recognition, Long Beach, CA, USA, 15–20 June 2019; pp. 3146–3154.
30. Lv, Z.; Zhang, P.; Sun, W.; Benediktsson, J.A.; Li, J.; Wang, W. Novel Adaptive Region Spectral-Spatial Features for Land Cover Classification with High Spatial Resolution Remotely Sensed Imagery. *IEEE Trans. Geosci. Remote Sens.* **2023**, *61*, 5609412. [CrossRef]
31. Jaderberg, M.; Simonyan, K.; Zisserman, A. Spatial transformer networks. In *Advances in Neural Information Processing Systems*; Curran Associates, Inc.: Red Hook, NY, USA, 2015; Volume 28.
32. Hu, J.; Shen, L.; Sun, G. Squeeze-and-excitation networks. In Proceedings of the IEEE Conference on Computer Vision and Pattern Recognition, Salt Lake City, UT, USA, 18–23 June 2018; pp. 7132–7141.
33. Woo, S.; Park, J.; Lee, J.-Y.; Kweon, I.S. Cbam: Convolutional block attention module. In Proceedings of the European Conference on Computer Vision (ECCV), Munich, Germany, 8–14 September 2018; pp. 3–19.
34. Wang, Q.; Wu, B.; Zhu, P.; Li, P.; Zuo, W.; Hu, Q. ECA-Net: Efficient channel attention for deep convolutional neural networks. In Proceedings of the IEEE/CVF Conference on Computer Vision and Pattern Recognition, Seattle, WA, USA, 13–19 June 2020; pp. 11534–11542.
35. Li, X.; Wang, W.; Hu, X.; Yang, J. Selective kernel networks. In Proceedings of the IEEE/CVF Conference on Computer Vision and Pattern Recognition, Long Beach, CA, USA, 15–20 June 2019; pp. 510–519.
36. Li, R.; Zheng, S.; Zhang, C.; Duan, C.; Su, J.; Wang, L.; Atkinson, P.M. Multiattention network for semantic segmentation of fine-resolution remote sensing images. *IEEE Trans. Geosci. Remote Sens.* **2021**, *60*, 1–13. [CrossRef]
37. Wang, L.; Li, R.; Zhang, C.; Fang, S.; Duan, C.; Meng, X.; Atkinson, P.M. UNetFormer: A UNet-like transformer for efficient semantic segmentation of remote sensing urban scene imagery. *ISPRS J. Photogramm. Remote Sens.* **2022**, *190*, 196–214. [CrossRef]
38. Li, X.; Lei, L.; Sun, Y.; Li, M.; Kuang, G. Collaborative attention-based heterogeneous gated fusion network for land cover classification. *IEEE Trans. Geosci. Remote Sens.* **2020**, *59*, 3829–3845. [CrossRef]
39. Lin, T.-Y.; Dollár, P.; Girshick, R.; He, K.; Hariharan, B.; Belongie, S. Feature pyramid networks for object detection. In Proceedings of the IEEE Conference on Computer Vision and Pattern Recognition, Honolulu, HI, USA, 21–26 July 2017; pp. 2117–2125.
40. Zhao, H.; Shi, J.; Qi, X.; Wang, X.; Jia, J. Pyramid scene parsing network. In Proceedings of the IEEE Conference on Computer Vision and Pattern Recognition, Honolulu, HI, USA, 21–26 July 2017; pp. 2881–2890.
41. Chen, L.-C.; Papandreou, G.; Schroff, F.; Adam, H. Rethinking atrous convolution for semantic image segmentation. *arXiv* **2017**, arXiv:1706.05587.
42. Zhang, T.; Qi, G.-J.; Xiao, B.; Wang, J. Interleaved group convolutions. In Proceedings of the IEEE International Conference on Computer Vision, Venice, Italy, 22–29 October 2017; pp. 4373–4382.
43. Howard, A.G.; Zhu, M.; Chen, B.; Kalenichenko, D.; Wang, W.; Weyand, T.; Andreetto, M.; Adam, H. Mobilenets: Efficient convolutional neural networks for mobile vision applications. *arXiv* **2017**, arXiv:1704.04861.
44. Sandler, M.; Howard, A.; Zhu, M.; Zhmoginov, A.; Chen, L.-C. Mobilenetv2: Inverted residuals and linear bottlenecks. In Proceedings of the IEEE Conference on Computer Vision and Pattern Recognition, Salt Lake City, UT, USA, 18–23 June 2018; pp. 4510–4520.
45. Hou, Q.; Zhou, D.; Feng, J. Coordinate attention for efficient mobile network design. In Proceedings of the IEEE/CVF Conference on Computer Vision and Pattern Recognition, Nashville, TN, USA, 20–25 June 2021; pp. 13713–13722.

46. Ding, L.; Lin, D.; Lin, S.; Zhang, J.; Cui, X.; Wang, Y.; Tang, H.; Bruzzone, L. Looking outside the window: Wide-context transformer for the semantic segmentation of high-resolution remote sensing images. *arXiv* **2021**, arXiv:2106.15754. [CrossRef]
47. Li, R.; Duan, C.; Zheng, S.; Zhang, C.; Atkinson, P.M. MACU-Net for semantic segmentation of fifine-resolution remotely sensed images. *IEEE Geosci. Remote Sens. Lett.* **2021**, *19*, 1–5.

remote sensing

MDPI

Article

TranSDet: Toward Effective Transfer Learning for Small-Object Detection

Xinkai Xu [1,2,3], Hailan Zhang [1], Yan Ma [2,3,*], Kang Liu [1], Hong Bao [2,3] and Xu Qian [1]

1 School of Artificial Intelligence, China University of Mining and Technology-Beijing, Beijing 100083, China
2 Beijing Key Laboratory of Information Service Engineering, Beijing Union University, Beijing 100101, China
3 College of Robotics, Beijing Union University, Beijing 100027, China
* Correspondence: jqrmayan@buu.edu.cn

Abstract: Small-object detection is a challenging task in computer vision due to the limited training samples and low-quality images. Transfer learning, which transfers the knowledge learned from a large dataset to a small dataset, is a popular method for improving performance on limited data. However, we empirically find that due to the dataset discrepancy, directly transferring the model trained on a general object dataset to small-object datasets obtains inferior performance. In this paper, we propose TranSDet, a novel approach for effective transfer learning for small-object detection. Our method adapts a model trained on a general dataset to a small-object-friendly model by augmenting the training images with diverse smaller resolutions. A dynamic resolution adaptation scheme is employed to ensure consistent performance on various sizes of objects using meta-learning. Additionally, the proposed method introduces two network components, an FPN with shifted feature aggregation and an anchor relation module, which are compatible with transfer learning and effectively improve small-object detection performance. Extensive experiments on the TT100K, BUUISE-MO-Lite, and COCO datasets demonstrate that TranSDet achieves significant improvements compared to existing methods. For example, on the TT100K dataset, TranSDet outperforms the state-of-the-art method by 8.0% in terms of the mean average precision (mAP) for small-object detection. On the BUUISE-MO-Lite dataset, TranSDet improves the detection accuracy of RetinaNet and YOLOv3 by 32.2% and 12.8%, respectively.

Keywords: object detection; transfer learning; dynamic resolution adaptation; small-object detection

Citation: Xu, X.; Zhang, H.; Ma, Y.; Liu, K.; Bao, H.; Qian, X. TranSDet: Toward Effective Transfer Learning for Small-Object Detection. *Remote Sens.* **2023**, *15*, 3525. https://doi.org/10.3390/rs15143525

Academic Editors: Fahimeh Farahnakian, Jukka Heikkonen and Pouya Jafarzadeh

Received: 2 June 2023
Revised: 27 June 2023
Accepted: 8 July 2023
Published: 12 July 2023

1. Introduction

With the application of automatic feature engineering in deep learning methods, significant progress has been made in object detection tasks in recent years [1–7], achieving accurate object recognition and localization in multiple scenarios. Existing cutting-edge object detection methods mainly focus on large objects, whereas small-object detection remains a challenging task due to the limited training samples and low-quality images. However, small-object detection is essential in many real-world applications of object detection. For example, in traffic scenes, the detection of traffic signs, small vehicles, and pedestrians is crucial for road safety [8–10], whereas in mining scenes, the detection of small objects such as miners and mining carts plays a positive role in enhancing safety and production efficiency [11–13]. Therefore, research on methods and technologies for small-object detection has significant academic and practical value.

In deep learning, the performance of methods is deeply affected by the size of the training dataset, and models trained on small datasets usually exhibit inferior performance. A solution to this problem is transfer learning [14–18], which aims to improve performance on small datasets by initializing the model with weights learned on a large dataset. Therefore, an intuitive idea for improving small-object detection performance is to adapt transfer learning to it. However, we empirically found that transferring the model from a large general object dataset (e.g., COCO [19]) to the target small-object dataset performs poorly. As shown in Figure 1, we transferred the Faster R-CNN model trained on COCO to the TT100K-Lite dataset, and the resulting detection performance AP_{50} was even worse than the AP_{50} of a traditional training strategy without transfer learning. A possible reason for this counter-intuitive failure is that the proportion of small objects in general object datasets is smaller than that of medium and large objects. Therefore, the learned weights are less effective for small objects. We present the distributions of small, medium, and large objects in popular object detection datasets in Table 1. Small objects [19] are defined as objects with sizes less than 32 square pixels, whereas large objects are defined as objects with sizes greater than 96 square pixels. In general object datasets, the proportion of small objects is low, whereas most objects in small-object datasets are small- and medium-sized. This discrepancy in data distribution between general and small-object datasets restricts the effectiveness of transfer learning.

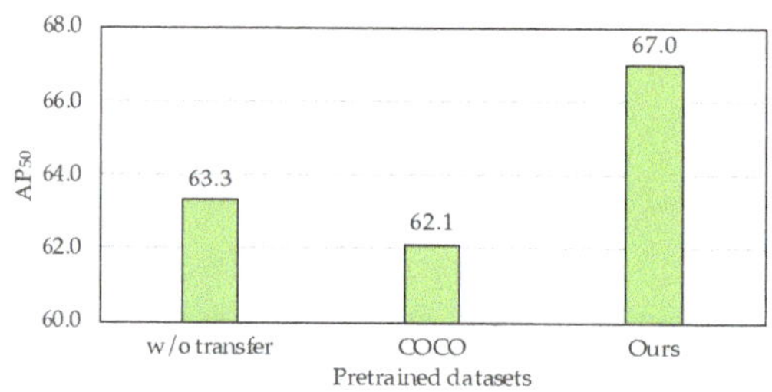

Figure 1. Comparison of object detection performance of Faster R-CNN using different pretrained models on the TT100K-Lite dataset (10% proportion). w/o transfer: the standard training strategy that uses an ILSVRC-pretrained backbone for initialization. COCO: transferring from the model trained on the COCO dataset. Ours: COCO model adapted with our proposed dynamic resolution adaptation scheme.

Table 1. Distributions of small, medium, and large objects in popular object detection training datasets.

	Dataset	Small Objects	Medium Objects	Large Objects
general object datasets	ILSVRC 2012 [20]	1.64%	11.54%	86.81%
	VOC 2007 [21]	11.20%	34.52%	54.28%
	VOC 2012 [21]	9.54%	27.59%	62.89%
	COCO 2017 [19]	31.13%	34.90%	33.97%
small object datasets	TT100K 2016 [22]	41.28%	51.66%	7.06%
	BUUISE-MO [11]	44.09%	33.79%	22.12%
	SODA-D [23]	48.20%	28.18%	23.62%
	Tiny Person [24]	85.80%	11.54%	2.66%

This paper presents TranSDet, a novel method for effective transfer learning for small-object detection, which aims to reduce the transfer discrepancy from a general object dataset to a small-object dataset by incorporating an additional dynamic resolution adaptation scheme. Specifically, we propose to adapt a model trained on a general dataset to a small-object-friendly model by augmenting the training images in the general dataset with diverse smaller resolutions. This gradually shifts the weights toward small objects without losing the discriminative information in the original model. The diverse resolutions are used to ensure consistent performance on various sizes of objects, and we introduce a meta-learning scheme to balance the learning of resolutions.

Another mainstream solution for improving small-object detection performance is to enhance the model architecture. For instance, FPN [25] connects features at different scales to enhance the semantic information of shallower features with deeper ones. Carafe [26] proposes learning image upsample kernels for better fusion of features. FPG [27] further improves the feature pyramid with fused multi-directional lateral connections, and MFR-CNN [28] combines global information with locally extracted multi-scale features to augment small-object features. However, most of the existing works require learning additional network modules, which inevitably disturb the pretrained features in transfer learning and make them incompatible with transfer learning. In this paper, we propose two components for small-object detection networks that are effective, efficient, and transfer learning-friendly: (1) SFA-FPN. We state that the traditional upsample operations in FPN perturb the features on shallower layers by interpolating pixels to their neighboring pixels, resulting in the imprecise recognition and localization of small objects. To address this problem, we propose an SAF-FPN module that uses shifted feature aggregation to shift the upsampled pixels to their correct positions. (2) Anchor relation module. We introduce an anchor relation module that captures the relationship between each object anchor with transformer blocks to enhance the anchor features. By combining both network components with the proposed dynamic resolution adaptation transfer learning, our TranSDet can achieve further improvements.

The contributions of this work can be summarized as follows:

(1) We propose a meta-learning-based dynamic resolution adaptation scheme for transfer learning that effectively improves the performance of transfer learning in small-object detection.
(2) We propose two network components, an SFA-FPN and an anchor relation module, which are compatible with transfer learning and effectively improve small-object detection performance.
(3) We conduct extensive experiments on the TT100K [22], BUUISE-MO-Lite [11], and COCO [19] datasets. The results demonstrate that our method, TranSDet, achieves significant improvements compared to existing methods. For example, on the TT100K-Lite dataset, TranSDet improves the detection accuracy of Faster R-CNN and RetinaNet by 8.0% and 22.7%, respectively. On the BUUISE-MO-Lite dataset, TranSDet improves the detection accuracy of RetinaNet and YOLOv3 by 32.2% and 12.8%, respectively, compared to the baseline models. These results suggest that TranSDet is an effective method for improving small-object detection accuracy using transfer learning.

2. Related Works

2.1. Small-Object Detection

In recent years, deep learning has gained increasing attention and has been successfully applied in many practical applications, leading to significant progress in object detection. However, the majority of prior efforts have been tuned for large-object detection, leaving limited experience and knowledge for small-object detection [29]. Detecting small objects in computer vision remains a challenging task [30]. Firstly, the features generated by basic CNNs lack the information needed for small-object detection. Secondly, small objects lack appearance information and have more location possibilities, requiring higher precision for accurate localization. Thirdly, context information is lacking, making it difficult to differentiate small objects from their surroundings. Fourthly, there is an imbalance of foreground and background training examples, and an insufficient number of positive training examples for small objects, making classification difficult. Even state-of-the-art networks exhibit significant performance gaps between the detection of small- and normal-sized objects. For example, DyHead [31] achieved only a 28.3% mean average precision (mAP) for small objects on the COCO test set, significantly lagging behind medium (50.3%) and large (57.5%) objects.

For small-object detection, low-level features of convolutional neural networks are often more effective than high-level features [32]. To fully utilize the semantic information from high-level features and the fine-grained features from low-level features, researchers have employed various methods to improve the detection accuracy of small objects. For example, SSD [3] increases the depth of the feature extraction network and uses a dense connection structure to improve small-object detection accuracy. FPN [25] connects feature maps of different scales from top to bottom to enhance the features at each scale. YOLOv3 [33] detects small, medium, and large objects on three independent feature maps of different scales. Carafe [26] proposes learning image upsample kernels instead of traditional interpolation operations, achieving better fusion quality. FPG [27] improves the feature pyramid with fused multi-directional lateral connections. PANet [34] enhances the entire feature hierarchy by using accurate localization signals in the lower layers via a bottom-up pathway. MFR-CNN [28] combines global information with locally extracted multi-scale features, improving detection accuracy for small objects and severely occluded objects in traffic scenes. SODNet [35] enhances small-object detection accuracy with a spatial parallel convolution module, split-fusion sub-module, and fast multi-scale fusion module, achieving high accuracy and real-time performance on multiple benchmark datasets. FE-CenterNet [36] enhances small-object detection accuracy with an attention mechanism, feature enhancements, and an anchor-free architecture, achieving a 7.2% higher AP metric with a 1.3 FPS decrease. SRODNet [8] improves vehicular detection accuracy by modifying the residual block in the super-resolution module and optimizing it jointly with YOLOv5. DetectFormer [9] incorporates a ClassDecoder and global information, with data augmentation and an attention mechanism in the backbone network, to enhance category sensitivity and real-time detection performance for traffic scenes. AMMFN [37] enhances small-object detection accuracy on remote sensing images with multi-scale feature fusion, attention mechanisms, and a normalized Wasserstein distance and generalized intersection-over-union location regression loss function. FusionPillars [38] utilizes the Set-Abstraction-Self (SAS) fusion module and the Pseudo-View-Cross (PVC) fusion module to fuse multisensor data for 3D object detection, resulting in enhanced detection precision and performance in detecting smaller objects.

There is another direct solution for reducing the resolution of the targets by increasing the size of input images, which enables the acquisition of high-resolution feature maps. MS-CNN [39] significantly improves the detection performance of small objects by adding an upsampling layer to the feature maps obtained by the deconvolutional layer. STDnet [40] employs a visual attention mechanism to select the most promising regions and discards the rest of the input image, thereby preserving high-resolution feature maps in deeper layers. Cascade R-CNN++ [41] employs an ensemble strategy, a modified loss function, and enhanced bounding box regression to effectively detect small objects in multi-resolution remote sensing images, outperforming previous methods. Unfortunately, these deep learning-based object detection algorithms strongly rely on massive training data, and they often perform poorly in situations with small samples.

2.2. Transfer Learning in Object Detection

Deep learning algorithms attempt to learn high-level features from massive amounts of data, which allows deep learning to go beyond traditional machine learning. However, collecting data is complex and expensive, especially in specific domains where it is very difficult to construct large-scale, high-quality datasets with annotations. Transfer learning relaxes the assumption that training data must be independent and identically distributed from test data, making it a better choice to use transfer learning to solve the problem of few-shot object detection.

Deep transfer learning is the study of how to utilize knowledge from other domains through deep neural networks. With the wide application of deep neural networks in various fields, a large number of deep transfer learning methods have been proposed. For example, Redmon et al. [42] jointly trained large classification and smaller detection datasets, allowing feature transfer between tasks to boost small-object detection accuracy. Wang et al. [43] revealed that fine-tuning only the final layer of the object detector on a balanced subset while keeping the rest of the model fixed, significantly improves detection accuracy. Liang et al. [44] introduced a transfer learning method utilizing residual thought and dilated convolutions. The method is initialized with large-scale datasets and aims to address issues such as low image resolution and partial occlusions. Wang et al. [45]

presented a deep transfer learning model using a modified ResNet-50 model and scale feature learners for bearing-fault diagnosis, resulting in a reliable and generalizable model. Loey et al. [46] proposed a hybrid deep transfer learning model using Resnet50 for feature extraction and ensemble algorithms, achieving up to 100% accuracy in face-mask detection on three datasets. Tang et al. [47] actively queried labels for the bounding boxes of source images using informativeness and transferability criteria to improve the target model with cost-effective supervision from source data. Sun et al. [48] utilized contrastive learning to train a proposal encoder, transferring knowledge from large datasets to few-shot target detection tasks, thereby enhancing model performance. Zhu et al. [49] enhanced few-shot target detection by introducing a semantic relation reasoning module. Kaul et al. [50] achieved comparable accuracy to traditional methods in few-shot scenarios by rapidly adapting to target categories and backgrounds through labeling, verification, and correction. Yan et al. [51] proposed an improved Faster R-CNN for tailings pond detection using a step-by-step transfer learning approach and increased inputs to four multispectral bands, resulting in more precise detection and a higher recall rate.

However, most existing methods focus on transfer learning between datasets of similar scales, with limited work considering the transfer of a model trained on large, general datasets to small-object, few-shot datasets. We observed that the aforementioned approaches underperformed in this context, leading us to propose a novel transfer learning method tailored for small-object detection.

3. Methodology

In this section, we formulate our proposed method TranSDet, which consists of two main components: (1) We propose a dynamic resolution adaptation scheme to better transfer the model trained on a normal dataset to a small-object detection dataset. (2) We propose a new small-object detection module to enhance small-object detection performance without affecting transfer learning efficacy.

3.1. Dynamic Resolution Adaptation Transfer Learning

We first review traditional transfer learning methods on object detection [43]. As illustrated in Figure 2, to improve object detection performance on a few-shot dataset, current transfer learning methods aim to first train a model on a large and general dataset (stage I), and then transfer the learned weights to the target few-shot dataset (stage III). By adopting this simple two-stage learning strategy, a model initialized with a pretrained model on a large dataset can obtain discriminative features on both foreground-background classification and object classification, thereby improving its generalization on few-shot datasets.

However, most methods are designed to transfer knowledge from a general dataset to another general few-shot dataset, whereas for a small-object few-shot dataset, we have empirically found it difficult to achieve significant transfer performance compared to a normal few-shot dataset, as shown in Figure 1. In this paper, we regard the task of solving the dataset gap in object sizes as an adaptation task, which encourages the model trained on general datasets to adapt to small-object detection datasets. As a result, we propose a meta-learning-based dynamic resolution adaptation transfer (DRAT) learning scheme to efficiently and effectively adapt the pretrained general model to a small-object-friendly model. As shown in Figure 2, we introduce an additional stage (stage II) to adjust the pretrained model using DRAT, and then transfer the adjusted model to the target dataset.

The pretrained model on a general dataset is trained with only a small proportion of small objects. To increase its generalization for small objects, we aim to resize the input images to a small resolution, so that the medium and large objects in the original resolution become small objects in the small resolution. Meanwhile, considering that the object sizes in our target dataset are not identical, we propose to fine-tune the pretrained model with multiple resolutions to allow it to adapt well to all object sizes.

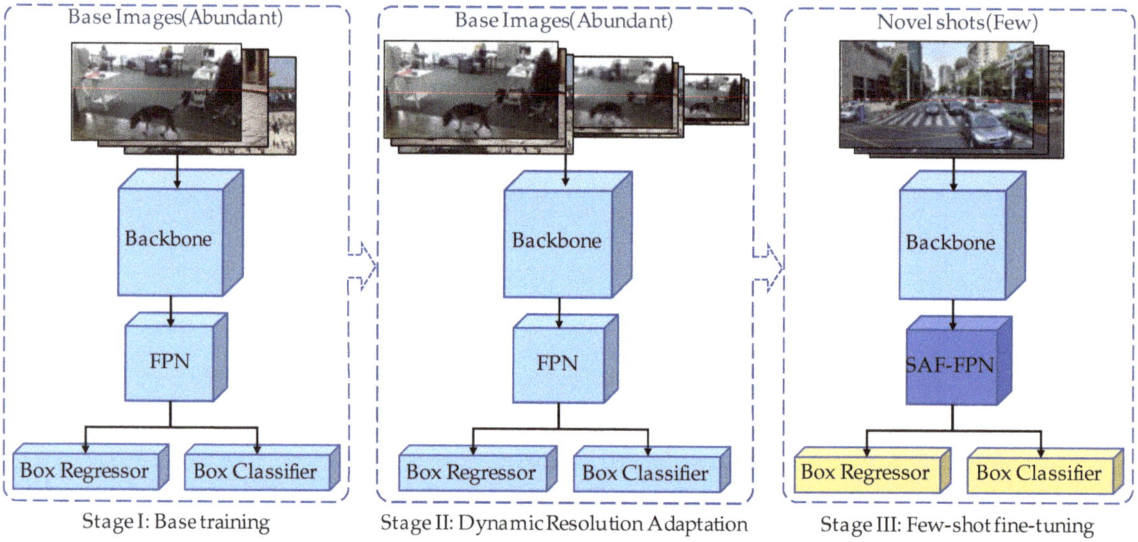

Figure 2. Framework of our proposed dynamic resolution adaptation transfer (DRAT) learning. Conventional transfer learning methods directly use the model pretrained on a base dataset to fine-tune the target few-shot dataset (stage I and stage III). We propose a dynamic resolution adaptation (stage II) to adapt the pretrained model to a small-object detection task and improve transfer learning performance.

Formally, given a set of resolutions $\mathcal{R}$ and a model $\mathcal{M}$ with pretrained weights θ_{pre} on a general dataset, our objective is to learn a model that generalizes well to all resolutions in $\mathcal{R}$ and adapts effectively to the new small-object dataset, i.e.,

$$\theta^* = \arg\min_{\theta} \mathop{\mathbb{E}}_{\mathcal{R}_i \in \mathcal{R}} \mathcal{L}(\mathcal{D}, \mathcal{R}_i, \mathcal{M}), \tag{1}$$

where $\mathcal{D}$ denotes the dataset and $\mathcal{L}$ is the loss function. To solve the above meta-learning problem, we adopt the widely-used MAML (Model-Agnostic Meta-Learning) model [52], which is model-agnostic and applicable to any model trained with gradient descent. The iterative learning strategy of MAML is summarized in Algorithm 1. At each training iteration, our DRAT performs (1) An inner update: for each resolution $\mathcal{R}_i$ in the resolution set $\mathcal{R}$, the parameters θ_i' are updated from the generic parameter θ by sampling and training on a mini-batch of images X_i at that resolution, i.e.,

$$\theta_i' = \theta - \alpha \nabla_\theta \mathcal{L}_{\mathcal{R}_i}(\mathcal{M}(\theta; X_i)), \tag{2}$$

where α denotes the step size for the inner update. (2) *An outer update*: the generic parameter θ is updated through gradient descent, where the meta-loss is the summation of losses across all meta-tasks (resolutions), i.e.,

$$\theta_{\text{new}} = \theta - \beta \nabla_\theta \sum_{\mathcal{R}_i \in \mathcal{R}} \mathcal{L}_{\mathcal{R}_i}(\mathcal{M}(\theta_i'; X_i)). \tag{3}$$

The updated generic parameter θ_{new} is then used for the next iteration. When the training is complete, the generic parameter becomes the final learned parameter of the model, and the model that adapts well to all resolutions is utilized for transferring to the small-object detection dataset. Note that for the consideration of training efficiency, we use the first-order approximation in MAML, which discards the production of the Hessian matrix and has a similar training speed to traditional training in detection.

Algorithm 1 Dynamic resolution adaptation meta-learning

Require: Adapting resolution set $\mathcal{R}$, model $\mathcal{M}$, dataset $\mathcal{D}$, training loss $\mathcal{L}$.
Require: Meta-learning step sizes of inner update and outer update α and β.
  Initialize model $\mathcal{M}$ with pretrained weights $\theta \leftarrow \theta_{pre}$;
  while training not complete **do**
    for all $\mathcal{R}_i \in \mathcal{R}$ **do**
      Sample a batch of images X_i and resize to $\mathcal{R}_i$;
      Evaluate $\nabla_\theta \mathcal{L}_{\mathcal{R}_i}(\mathcal{M}(\theta; X_i))$;
      Compute adapted weights using gradient descent:
      $\theta'_i = \theta - \alpha \nabla_\theta \mathcal{L}_{\mathcal{R}_i}(\mathcal{M}(\theta; X_i))$;
    end for
    Update $\theta \leftarrow \theta - \beta \nabla_\theta \sum_{\mathcal{R}_i \in \mathcal{R}} \mathcal{L}_{\mathcal{R}_i}(\mathcal{M}(\theta'_i; X_i))$;
  end while
  return Adapted model $\mathcal{M}$ with weights θ.

3.2. Enhanced Small-Object Detection for Transfer Learning

In order to enhance the performance of small-object detection modules, previous works usually add larger feature maps to the FPN [53], enhance the local information of shallower feature maps [54], and introduce new feature fusion strategies to the FPN [55]. Nevertheless, these methods usually require new network modules, which introduce new learnable parameters and significantly change the output features of the FPN. Note that for few-shot small-object detection, we transfer a pretrained model to a few-shot dataset, where the pretrained model is trained with a normal architecture designed for normal datasets. Directly injecting existing learnable and randomly initialized modules into the pretrained model would significantly change the semantic information present in the original features, leading to catastrophic forgetting of the pretrained knowledge.

In this paper, we propose a new FPN with a shifted feature aggregation (SFA-FPN) module and an anchor relation module to enhance small-object detection performance without disturbing the semantic information in the original model.

3.2.1. FPN with Shifted Feature Aggregation (SFA-FPN)

In a conventional FPN, the features at different scales are connected sequentially, and the feature map with a lower resolution is upsampled using linear interpolation to one with a higher resolution to perform a summation with its previous feature map. However, this naïve upsample interpolation can perturb the high-resolution feature map by assigning non-associated features to multiple pixels, as shown in Figure 3. As a result, the high-resolution feature maps, which are important for detecting small objects, are significantly polluted by low-resolution ones, leading to inferior performance.

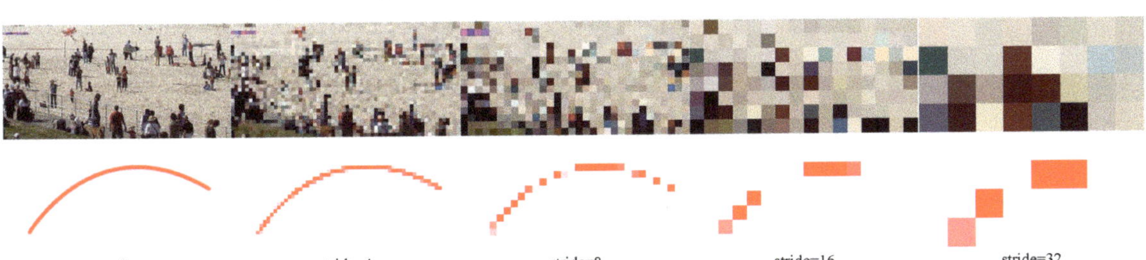

Figure 3. Comparisons of upsampling effects on different scales of images. We use a real detection image (**top**) and a simple curved line (**bottom**) for illustrations. The pixels are significantly polluted by their neighboring salient pixels when the stride is large.

In this paper, we aim to solve the above problem by learning to interpolate low-resolution features into high-resolution ones. Unlike previous methods that change the output features using learnable interpolation functions or additional convolutional layers, we propose a simple yet effective approach that involves shifting and aggregating pixels without changing their feature distributions. Specifically, as shown in Figure 4, for each pixel in an interpolated feature map $\boldsymbol{F} \in \mathbb{R}^{C\times 2H\times 2W}$ with a $2\times$ upsample ratio, where C, W, and H denote the channels, width, and height of the feature map, respectively, we first shift it along eight directions to obtain the shifted features. All nine features $\boldsymbol{F}^{(s)} \in \mathbb{R}^{9\times C\times 2H\times 2W}$ (including the original one) represent the possible true positions of the pixel. Then, we introduce a simple convolution module to predict the weights $\boldsymbol{W} \in \mathbb{R}^{9\times 2H\times 2W}$ for the nine feature maps by concatenating the feature map $\boldsymbol{F}$ with the feature map from its previous stage, which denotes the probabilities of the pixels being in the correct position. In other words, if a pixel in the interpolated feature map is not in the correct position, it should be discarded and has a weight of 0, whereas the pixel with the correct position should have a weight of 1. Multiplying the weights with the shifted feature maps and performing summation on the weighted feature maps yields our refined feature map $\boldsymbol{F}^{(r)} \in \mathbb{R}^{C\times 2H\times 2W}$, i.e.,

$$\boldsymbol{F}^{(r)} = \sum_{i=1}^{9} \boldsymbol{W}_i \odot \boldsymbol{F}_i^{(s)}, \tag{4}$$

where $\odot$ denotes the Hadamard product, and summation is performed element-wise in the first dimension (dimension C).

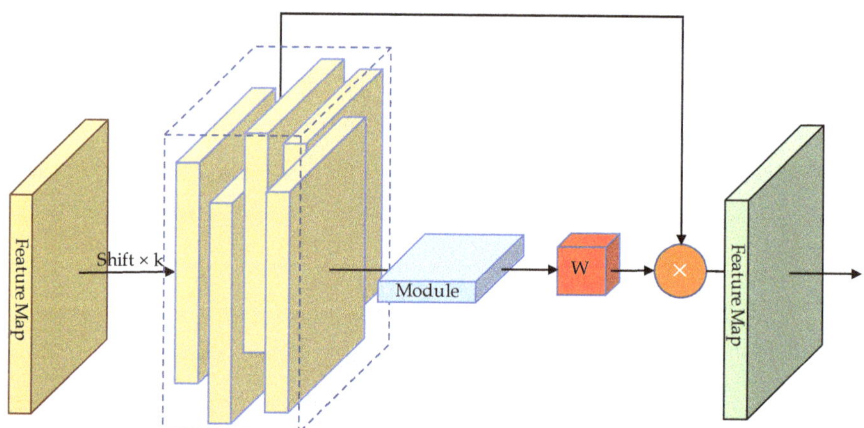

Figure 4. FPN with shifted feature aggregation.

After the shifted feature aggregation, we also adopt a channel attention module based on an SE module [56] to select and enhance the aggregated feature map $\boldsymbol{F}^{(r)}$. Specifically, as illustrated in Figure 5, the module first computes the global image features on channels by averaging across the spatial axes. Then, a squeeze-and-excitation structure with a reduction convolution, followed by an activation function and an expansion convolution, is introduced to extract the features and reduce the computational cost. Finally, we use a convolution layer to predict the attention weights of each channel and multiply these weights onto the refined feature map. This channel attention module adaptively assigns larger weights to valuable channels and smaller weights to noisy channels, helping us further reduce the disturbances caused by fusing two feature maps.

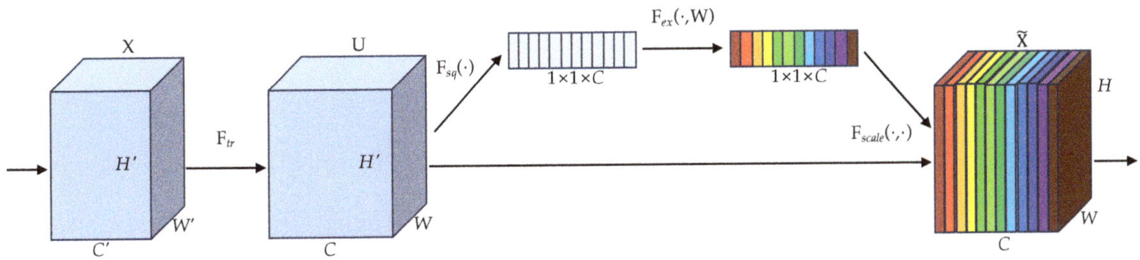

Figure 5. Channel attention module.

3.2.2. Anchor Relation Module

The small objects in images are usually blurred and low resolution, and thus the anchors of small objects are not as discriminating as the normal anchors of larger objects. In this paper, to enhance the region-of-interest (RoI) features of small objects, we propose an anchor relation module to model the relations of different anchors and improve the anchor features with the related anchors. The motivation behind this module is that for recognizing the object in one anchor, the content in other anchors can benefit the recognition through their relations, e.g., (1) similar objects with better image quality; (2) different parts of an object; (3) dependencies between objects (for instance, in mining scenarios, safety helmets co-occur with workers).

Our relation module leverages transformer blocks [57] to effectively model the relations between anchors. Specifically, given the input anchor features $\boldsymbol{A} \in \mathbb{R}^{N\times C_A}$ and their coordinates $\boldsymbol{P} \in \mathbb{R}^{N\times 4}$, where N and C_A denote the number of anchors in an image and the channels of an anchor, and each coordinate contains the positions (x, y) of the top-left and bottom-right points, we first bind the positional information to each anchor by adding the positional encoding to the anchor features. Following DETR [58], we generalize the positional encoding of Transformer to the 2D image scenario. For an anchor with normalized coordinates $(x_0, y_0, x_1, y_1) \in [0,1]^4$, its positional encoding is defined as $\boldsymbol{P} = [PE(x_0) : PE(y_0) : PE(x_1) : PE(y_1)]$, where [:] denotes concatenation and the function PE is formulated as

$$
\begin{aligned}
PE(a)_{2i} &= \sin(a/10{,}000^{2i/D}) \\
PE(a)_{2i+1} &= \cos(a/10{,}000^{2i/D})
\end{aligned}
\quad (5)
$$

where $D = C_A/4$ is the dimensions of PE. Then, the anchor features with positional encoding added are fed into a Transformer encoder to extract the relation features of the anchors. We use the output features of the encoder as the input features of the prediction heads (classification head and regression head) of the model.

The overall architecture of the anchor relation module is illustrated in Figure 6. Our encoder is stacked with L ($L = 2$ in our experiments) transformer blocks. With an input sequence of features $\boldsymbol{I} = \boldsymbol{A} + \boldsymbol{P}$, each transformer block computes it with a multi-head self-attention module and a multi-layer perception module. (1) Multi-head self-attention (MHSA): MHSA measures the similarities between each pair of features and then combines the features to obtain the output features based on the attention scores. Formally, with $\boldsymbol{I} \in \mathbb{R}^{N\times C_A}$, the output $\boldsymbol{O}_A$ of MHSA is computed as

$$
\boldsymbol{O}_A := Attention(\boldsymbol{I}_q, \boldsymbol{I}_k, \boldsymbol{I}_v) = softmax(\boldsymbol{I}_q \boldsymbol{I}_k^T)\boldsymbol{I}_v, \quad (6)
$$

where $\boldsymbol{I}_q, \boldsymbol{I}_k, \boldsymbol{I}_v$ are produced by the Q, K, V projections of $\boldsymbol{I}$, respectively. Then, $\boldsymbol{O}_A$ is fed into a multi-layer perception (also referred to as feed-forward network (FFN)), which consists of two fully-connected (FC) layers with an adjunct activation function, i.e.,

$$
\boldsymbol{O}_E = FC(Act(FC(\boldsymbol{O}_A))). \quad (7)
$$

The output of the last transformer block is then passed to the regression head and classification head of the detection model to generate the predictions.

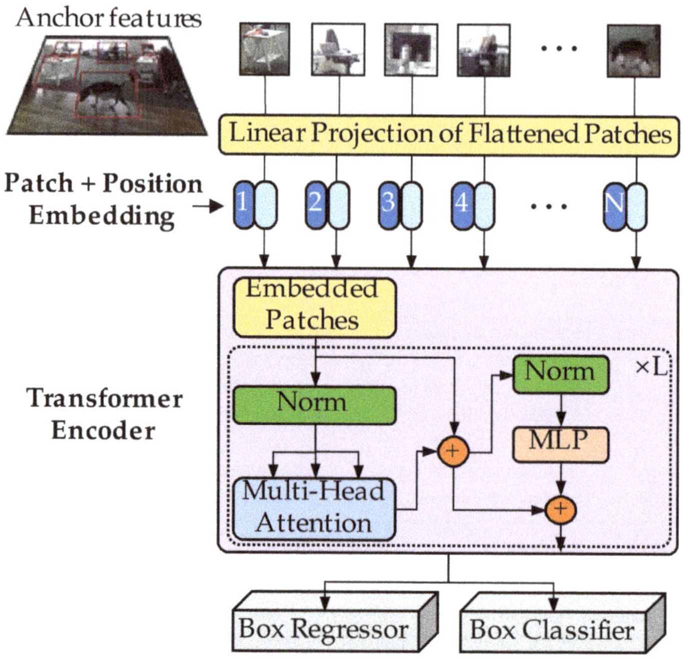

Figure 6. The architecture of the anchor relation module.

The overall network structure of TranSDet is illustrated in Figure 7. It contains shifted feature aggregation (SFA) and channel attention (CA) modules in the FPN and an anchor relation (AR) module before the prediction heads.

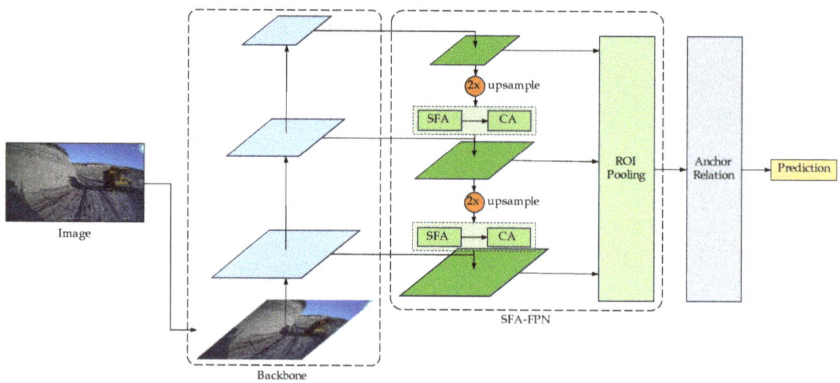

Figure 7. TranSDet network structure.

4. Experiments

4.1. Datasets and Evaluation Metrics

We chose the COCO dataset as the base dataset for this study, which is a widely used benchmark dataset in the field of object detection. Released by Microsoft in 2014, it contains over 330,000 well-labeled images of common objects from 80 different categories, with multiple objects in each image. Compared with VOC and ILSVRC (ImageNet), COCO contains smaller objects with a denser distribution, making it more similar to real-world scenarios. COCO has become the standard dataset for object detection.

For the novel datasets, we chose the TT100K [22] and BUUISE-MO datasets [11]. TT100K is a dataset for traffic sign detection and classification, whereas BUUISE-MO is a mining object detection dataset.

The evaluation metric used to assess the model detection precision in this paper was the average precision (AP). We used the AP_{50}, AP_S, AP_M, and AP_L to evaluate the detection capabilities of objects of different sizes on the TT100K and BUUISE-MO datasets. Among them, AP_{50} represents the average precision when the intersection over union (IoU) is greater than or equal to 0.5, and AP_S, AP_M, and AP_L represent the average precision of small, medium, and large objects, respectively. For the COCO dataset, we report the standard AP, AP_{50}, AP_{75}, AP_S, AP_M, and AP_L. We ran all the methods five times with different seeds and present the mean and the standard deviation as the final scores.

4.2. Models and Training Strategies

Our experiments were conducted on a computer equipped with an Intel Core i9-7900X CPU (3.3G) and an NVIDIA TITAN V GPU (12G). We used the MMDetection deep learning framework [59] with the default epoch value for the training process. We compared three object detection models: Faster R-CNN (two-stage), RetinaNet (one-stage), and YOLOv3 (efficient). Note that we only adopted SFA-FPN in RetinaNet and YOLOv3, since these networks do not contain RoI features needed for leveraging the anchor relation module. For the training strategies, we trained the models using an SGD optimizer with a momentum of 0.9 and a weight decay of 10^{-4}. A step learning rate schedule, which decayed the learning rate by a factor of 0.1 at the 8th and 11th epochs, was adopted with an initial value of 0.02. The training used standard data augmentations, including resizing, random flipping, and padding. The number of training epochs was set to 12 for the COCO dataset, whereas for the small TT100K-Lite and BUUISE-MO datasets, we trained the models for 36 epochs. The height resolutions of the input images on the COCO, TT100K-Lite, and BUUISE-MO datasets were resized to 800, 1440, and 1080, respectively, and the image widths were adjusted to maintain the original aspect ratios of the images.

4.3. Results on TT100K

The TT100K dataset contains 100,000 high-resolution (2400 $\times$ 2400) images and has been widely utilized for traffic sign detection and classification. With 30,000 instances of traffic signs, this dataset is an excellent benchmark for small-object detection, as it contains a considerable number of small objects. To address the imbalance in the number of instances across the different traffic sign classes, we excluded classes with fewer than 100 instances [22], resulting in 45 classes for our experiments.

We first conducted experiments to investigate the impact of the number of training images on the detection precision in small-object detection tasks. Faster R-CNN [1] is a highly accurate and scalable object detection algorithm that has gained attention for its performance in such tasks. We trained the Faster R-CNN model from scratch on the TT100K-Lite dataset, which is a subset of the original TT100K dataset containing only 45 categories. To sufficiently evaluate our performance on different numbers of training samples, we created four training sets with different proportions (100%, 10%, 5%, and 1%), where the proportions denote the randomly sampled proportions of the original training set. Our experimental results demonstrated that the number of training images had a significant impact on the detection performance of Faster R-CNN, as shown in Table 2. A limited number of training images led to a sharp decline in detection precision, indicating that the model's ability to detect objects was significantly reduced when the number of training images was limited. This decline in precision was particularly pronounced for small objects, as evidenced by the decreasing AP values as the proportion of training images decreased. These findings highlight the importance of transfer learning and dynamic resolution adaptation, as proposed in this paper, in enhancing object detection performance in few-shot scenarios. By leveraging transfer learning and dynamically adjusting the resolution of the input images during training, our proposed approach can effectively address the challenges of limited training samples and low-quality images, leading to significant improvements in small-object detection performance.

Table 2. Object detection performance of Faster R-CNN baseline on the TT100K-Lite dataset.

Proportion (%)	AP_{50}	AP_S	AP_M	AP_L
100	89.4 ± 0.90	75.4 ± 3.11	97.8 ± 0.78	91.7 ± 1.79
10	63.3 ± 0.53	48.2 ± 1.38	78.0 ± 1.22	79.4 ± 4.58
5	50.2 ± 0.98	40.4 ± 1.02	62.2 ± 1.35	59.1 ± 4.46
1	23.4 ± 1.44	18.5 ± 2.83	30.8 ± 0.94	43.3 ± 5.45

The aim of this paper was to address the above-mentioned problem by proposing an effective small-object detection method for transfer learning. Here, we conducted experiments to show the efficacy of our TranSDet model compared to various object detection methods, including Faster R-CNN [1], RetinaNet [2], and YOLOv3 [33]. RetinaNet [2] is a one-stage algorithm that employs a focal loss to address the class imbalance problem in object detection, whereas YOLOv3 [33] is an efficient detection algorithm that utilizes a fully convolutional neural network to detect objects. In comparative experiments, Faster R-CNN (FRCNN), RetinaNet, and YOLOv3 were used as typical algorithms to evaluate the performance of TranSDet. The experimental results in Table 3 show that TranSDet consistently outperformed the baseline algorithms on all three subsets of the TT100K-Lite dataset, with a significant improvement in detection accuracy for small-object detection tasks. Specifically, for the 10% subset, TranSDet achieved an 8% absolute improvement in detection accuracy for Faster R-CNN, a 22.7% improvement for RetinaNet, and a 6.6% improvement for YOLOv3, compared to their respective baselines. Similar improvements were observed for the 5% and 1% subsets, where TranSDet consistently outperformed the baseline algorithms across all three algorithms and achieved notable improvements in detection accuracy.

Furthermore, we conducted experiments on recently proposed and more advanced object detection models: RetinaNet with Swin Transformer [60], anchor-free RepPoints [61], and end-to-end Deformable DETR [62]. As summarized in Table 4, the advanced models also suffered due to the limited training samples in the TT100K-Lite small-object detection dataset, whereas our TranSDet achieved significant improvements over them, demonstrating the generality and effectiveness of our method. Specifically, on the 10% TT100K-Lite dataset, Deformable DETR only achieved an AP_{50} of 27.8 since its Transformer-based detection head required a large number of training samples to converge and avoid overfitting, whereas when using our proposed TranSDet model, performance was improved by 27.0.

It is worth noting that the performance of all the algorithms generally decreased as the dataset size decreased. However, we observed that TranSDet consistently exhibited performance improvements over the baselines, even when the proportion of training data was as low as 1%. These results suggest that TranSDet is effective in improving small-object detection and can robustly adapt to different dataset sizes.

Table 3. Performance comparison of our object detection model and classical models on the TT100K-Lite dataset with varying proportions of training data.

Pros. (%)	Method	AP_{50}	AP_S	AP_M	AP_L
10	FRCNN [1]	63.3 ± 0.53	48.2 ± 1.38	78.0 ± 1.22	79.4 ± 4.58
	FRCNN-TranSDet	71.3 ± 1.05	53.4 ± 1.59	85.0 ± 1.87	83.3 ± 3.97
	RetinaNet [2]	33.7 ± 2.34	33.3 ± 4.99	42.5 ± 2.66	55.7 ± 8.48
	RetinaNet-TranSDet	56.4 ± 0.98	45.9 ± 3.90	68.4 ± 1.89	68.2 ± 7.86
	YOLOv3 [33]	24.9 ± 0.98	15.6 ± 2.83	32.6 ± 2.34	37.0 ± 4.11
	YOLOv3-TranSDet	31.5 ± 0.32	19.6 ± 1.45	38.2 ± 2.05	41.7 ± 5.98
5	FRCNN [1]	50.2 ± 0.98	40.4 ± 1.02	62.2 ± 1.35	59.1 ± 4.46
	FRCNN-TranSDet	61.6 ± 0.72	44.8 ± 1.50	74.1 ± 1.19	70.1 ± 2.44
	RetinaNet [2]	17.1 ± 2.15	18.1 ± 1.72	21.5 ± 2.25	37.7 ± 8.56
	RetinaNet-TranSDet	44.1 ± 2.10	37.8 ± 0.59	56.7 ± 1.69	52.6 ± 4.09
	YOLOv3 [33]	14.6 ± 0.99	8.1 ± 1.90	19.9 ± 0.55	30.9 ± 4.36
	YOLOv3-TranSDet	22.0 ± 2.14	10.9 ± 5.71	30.5 ± 1.16	34.4 ± 3.20
1	FRCNN [1]	23.4 ± 1.44	18.5 ± 2.83	30.8 ± 0.94	43.3 ± 5.45
	FRCNN-TranSDet	30.9 ± 1.16	26.4 ± 1.35	38.7 ± 1.42	46.4 ± 4.28
	RetinaNet [2]	2.2 ± 0.35	1.8 ± 0.45	3.6 ± 0.55	16.2 ± 4.08
	RetinaNet-TranSDet	17.8 ± 0.89	17.6 ± 1.23	24.6 ± 2.32	37.3 ± 2.90
	YOLOv3 [33]	3.8 ± 0.14	1.9 ± 0.35	5.1 ± 1.06	16.5 ± 2.40
	YOLOv3-TranSDet	9.2 ± 0.92	3.9 ± 2.76	14.7 ± 0.97	18.6 ± 3.88

Table 4. Performance comparison of our object detection model and advanced models on the TT100K-Lite dataset with varying proportions of training data.

Pros. (%)	Method	AP_{50}	AP_S	AP_M	AP_L
10	RetinaNet (Swin) [60]	36.4 ± 1.73	32.3 ± 2.81	46.9 ± 1.24	55.1 ± 3.32
	RetinaNet-TranSDet (Swin)	61.4 ± 1.46	50.2 ± 2.59	73.9 ± 1.92	67.2 ± 4.81
	RepPoints [61]	42.3 ± 2.13	36.9 ± 3.17	50.5 ± 1.62	56.0 ± 2.91
	RepPoints-TranSDet	68.8 ± 1.53	51.3 ± 1.16	78.6 ± 1.63	69.2 ± 3.54
	Def. DETR [62]	27.8 ± 2.81	13.7 ± 3.19	38.3 ± 1.96	50.4 ± 5.32
	Def. DETR-TranSDet	54.8 ± 1.35	32.6 ± 1.63	61.6 ± 1.26	63.2 ± 4.63
5	RetinaNet (Swin) [60]	22.4 ± 1.65	19.2 ± 1.92	29.5 ± 1.57	47.8 ± 5.21
	RetinaNet-TranSDet (Swin)	46.9 ± 1.42	40.5 ± 2.51	58.1 ± 1.52	55.7 ± 4.21
	RepPoints [61]	31.0 ± 1.43	27.4 ± 2.71	38.7 ± 1.12	45.7 ± 4.83
	RepPoints-TranSDet	51.7 ± 2.16	40.1 ± 1.74	63.7 ± 1.76	58.2 ± 2.95
	Def. DETR [62]	15.2 ± 1.84	14.9 ± 1.05	25.7 ± 1.46	35.8 ± 3.69
	Def. DETR-TranSDet	43.7 ± 1.61	35.3 ± 1.31	56.9 ± 1.66	51.5 ± 5.03
1	RetinaNet (Swin) [60]	2.30 ± 0.52	1.70 ± 0.38	3.30 ± 1.15	18.1 ± 3.64
	RetinaNet-TranSDet (Swin)	19.5 ± 1.73	18.9 ± 1.56	28.2 ± 1.27	42.3 ± 4.29
	RepPoints [61]	3.30 ± 1.13	3.20 ± 0.94	4.51 ± 1.75	14.1 ± 2.89
	RepPoints-TranSDet	19.5 ± 2.04	18.2 ± 1.53	26.7 ± 1.39	40.2 ± 3.15
	Def. DETR [62]	1.7 ± 0.51	1.8 ± 0.37	3.28 ± 1.04	11.2 ± 2.46
	Def. DETR-TranSDet	16.2 ± 1.76	15.8 ± 1.53	24.1 ± 1.64	36.6 ± 2.78

4.4. Results on BUUISE-MO

The dataset used in this study was derived from the BUUISE-MO dataset, which is an open-pit-mine object detection dataset established by the team at the Beijing Information Service Engineering Key Laboratory of Beijing Union University [11]. The BUUISE-MO dataset comprises 9720 images with a resolution of 1920 × 1080, including 7220 training images and 2500 test images, as shown in Figure 8. The dataset has 15 categories, including truck, forklift, car, excavator, people, sign, etc. A total of 6041 large objects, 9230 medium objects, and 12,043 small objects are labeled, making the dataset suitable for small-object detection tasks.

Figure 8. Image samples in the BUUISE-MO-Lite dataset.

To create a few-shot dataset, we randomly selected 610 images from the training set and 500 images from the test set. We refer to this dataset as BUUISE-MO-Lite in this paper. Similar to the previous experiment, we trained Faster R-CNN, RetinaNet, and YOLOv3, and the results are shown in Table 5. The primary objective of this experiment was to demonstrate the ability of the TranSDet method to generalize well across different datasets and algorithms, given that the previous experiment (Table 3) had already demonstrated the effectiveness of TranSDet in improving small-object detection accuracy. The results showed that TranSDet has good generalization abilities, as it significantly improved the detection accuracy for all three algorithms on the BUUISE-MO-Lite dataset. The consistently better performance of TranSDet in small-object detection reaffirms its effectiveness, which can be beneficial in scenarios where large labeled datasets are not available. Notably, compared to the Faster R-CNN method, our method achieved more significant improvements on efficient one-stage detectors RetinaNet and YOLOv3. This demonstrates that our method also adapts well to detectors with limited FLOPs and parameters, and can help those detectors achieve competitive performance compared to the resource-heavy two-stage detector Faster R-CNN.

Table 5. Object detection performance on the proposed BUUISE-MO-Lite few-shot small-object detection dataset.

Method	AP_{50}	AP_S	AP_M	AP_L
FRCNN	57.6 ± 1.07	51.8 ± 0.55	66.6 ± 2.32	64.2 ± 1.88
FRCNN-TranSDet	61.6 ± 4.07	51.7 ± 1.96	68.2 ± 0.61	72.0 ± 7.76
RetinaNet	32.8 ± 1.53	28.4 ± 3.38	44.2 ± 1.26	46.7 ± 0.35
RetinaNet-TranSDet	65.0 ± 3.16	52.7 ± 2.25	65.9 ± 2.90	77.1 ± 2.87
YOLOv3	40.2 ± 4.19	16.1 ± 2.70	49.6 ± 7.28	55.4 ± 7.03
YOLOv3-TranSDet	53.0 ± 2.72	28.8 ± 6.84	59.1 ± 3.99	72.0 ± 4.45

4.5. Small-Object Results on COCO Dataset

We also conducted experiments on the large-scale general COCO dataset to validate the efficacy of our proposed small-object modules, namely the FPN with shifted feature aggregation (SFA-FPN) module and the anchor relation (AR) module. We trained Faster R-CNN with a ResNet-50 backbone using the standard 1× training schedule in MMDetection [59], and present the standard benchmark metrics of COCO in Table 6. Specifically, AP_S, AP_M, and AP_L are the AP_{50} on small, medium, and large objects, respectively. The results show that both the SFA-FPN and AR modules can improve detection performance on COCO, especially for small-object tasks. Our model achieved a 22.1% AP_S and significantly surpassed the baseline by 0.9%. This indicates that our proposed network modules not only benefits transfer learning in detection but also improves performance on the large-scale general COCO dataset without using transfer learning.

Table 6. Object detection performance on the COCO dataset.

Method	AP	AP_{50}	AP_{75}	AP_S	AP_M	AP_L
FRCNN	37.4	58.1	40.4	21.2	41.0	48.1
+ SFA-FPN	37.8 (+0.4)	58.3 (+0.2)	40.8 (+0.4)	21.8 (+0.6)	41.3 (+0.3)	48.2 (+0.1)
+ SFA-FPN + AR	38.0 (+0.6)	58.4 (+0.3)	41.0 (+0.6)	22.1 (+0.9)	41.4 (+0.4)	48.4 (+0.3)

4.6. Comparison with Transfer Learning Methods

We conducted experiments to compare our dynamic resolution adaptation (DRA) with existing transfer learning methods [43,63] designed for Faster R-CNN object detection. Specifically, a pretrained model fine-tuning method (PTD) [63] was considered as our baseline, which involved using a pretrained COCO model to initialize the target dataset. Frozen layer fine-tuning [43] was used to freeze the preceding layers to better preserve the semantic information from the source dataset.

The results presented in Table 7 demonstrate that TranSDet outperformed the other methods across various evaluation metrics. In particular, TranSDet achieved the highest AP scores at different IoU thresholds, including AP_{50}, AP_S, AP_M, and AP_L, indicating its superior performance in accurately detecting objects of different sizes. When comparing TranSDet with PTD, we found that TranSDet showed significant improvements in overall object detection performance, as measured by the AP_{50}. Additionally, TranSDet outperformed both PTD and FTD in terms of AP_S, AP_M, and AP_L, indicating its superiority in detecting small, medium, and large objects.

However, although the performance of PTD and TranSDet was reasonably stable across different proportions (percentage of samples in the dataset), FTD's performance decreased significantly, indicating its sensitivity to changes in object size distribution in the target dataset. This finding highlights the importance of carefully selecting an appropriate transfer learning method for different datasets based on the target distribution of objects. Note that our DRA only had to change the weights of the pretrained model in transfer learning, making it compatible with almost all the existing transfer learning methods. Therefore, one can easily implement DRA on more sophisticated transfer learning methods to improve small-object detection performance.

Table 7. Comparison of transfer learning methods on the Faster R-CNN model and the TT100K-Lite dataset.

Pros. (%)	Methods	AP_{50}	AP_S	AP_M	AP_L
10	PTD [63]	62.1 ± 1.79	46.6 ± 1.38	74.7 ± 1.79	86.2 ± 3.95
	FTD [43]	55.9 ± 0.71	42.9 ± 1.42	69.5 ± 1.25	72.5 ± 7.04
	TranSDet	71.3 ± 1.05	53.4 ± 1.59	85.0 ± 1.87	83.3 ± 3.97
5	PTD [63]	50.1 ± 0.75	38.4 ± 2.16	61.4 ± 1.49	61.9 ± 3.70
	FTD [43]	45.7 ± 1.26	35.1 ± 3.16	56.7 ± 5.63	58.7 ± 3.67
	TranSDet	61.6 ± 0.72	44.8 ± 1.50	74.1 ± 1.19	70.1 ± 2.44
1	PTD [63]	27.1 ± 2.30	22.1 ± 2.58	34.2 ± 1.46	48.7 ± 4.18
	FTD [43]	26.2 ± 1.12	21.5 ± 3.20	32.8 ± 1.36	41.7 ± 2.71
	TranSDet	30.9 ± 1.16	26.4 ± 1.35	38.7 ± 1.42	46.4 ± 4.28

4.7. Ablation Study

4.7.1. Ablation on the Proposed Modules

We first investigated the effects of our proposed dynamic resolution adaptation (DRA), SFA-FPN, and anchor relation (AR) modules. As shown in Table 8, our baseline Faster R-CNN with a ResNet-50 backbone achieved an AP_{50} of 50.2%, whereas adding DRA (second row) significantly improved the AP_{50} and AP_S by 5.7% and 3.9%, respectively. Meanwhile, adopting SFA-FPN (third row) achieved a 6.2% improvement in AP_{50} compared to the baseline, and leveraging AR achieved a further improvement of 2.8%. By combining all the proposed modules, our final method achieved the optimal performance of a 61.6% AP_{50} and a 44.8% AP_S and significantly outperformed our baseline by 11.4% and 4.4%, respectively.

Table 8. Ablation on the proposed modules on the TT100K-Lite dataset (5% proportion). We used Faster R-CNN as the baseline model. DRA: dynamic resolution adaptation. SFA-FPN: FPN with shifted feature aggregation. AR: anchor relation module.

DRA	SFA-FPN	AR	AP_{50}	AP_S
×	×	×	50.2 ± 0.98	40.4 ± 1.02
✓	×	×	55.9 ± 1.20	44.3 ± 1.49
×	✓	×	56.4 ± 1.81	43.5 ± 0.23
×	×	✓	59.2 ± 1.10	44.9 ± 2.15
×	✓	✓	59.2 ± 2.05	43.8 ± 2.44
✓	✓	✓	61.6 ± 0.72	44.8 ± 1.50

4.7.2. Comparison with Previous Small-Object Detection Methods

This experiment aimed to demonstrate the effectiveness of the TranSDet method for small-object detection. To evaluate its performance, comparison experiments were conducted using Faster R-CNN (FRCNN), Carafe [26], and FPG [27]. Both Carafe and FPG are designed for improving the small-object detection performance of the Faster R-CNN baseline. Carafe employs a learnable upsampling module to capture fine-grained information, whereas FPG enhances the feature pyramid structure to better capture object details at all scales. These modifications have been shown to significantly improve performance in small-object detection tasks on conventional datasets, indicating their efficacy in this context.

In Table 9, we present the comparison results when using the same Faster R-CNN model trained on the COCO dataset to initialize the model in the TT100K-Lite dataset but we replaced the neck with Carafe, FPG, and our approaches. The results show that TranSDet outperformed both Carafe and FPG in all evaluation metrics, with a significantly higher AP at all levels (AP_{50}, AP_S, AP_M, and AP_L). This demonstrates the superiority of the TranSDet structure in small-object detection tasks. Also, it is worth noting that using the existing Carafe and FPG modules to replace the original FPN module resulted in poorer object detection performance in transfer learning. This can be attributed to the random initialization of additional weights in these modules during transfer learning, which may have disrupted the learned semantic information and degraded accuracy. In conclusion, the experimental results provide strong evidence to support the effectiveness of transfer learning for small-object detection using TranSDet. The use of Carafe and FPG in the comparison experiments further demonstrates the superiority of TranSDet in small-object detection tasks.

Table 9. Comparison with previous small-object detection methods on the TT100K-Lite dataset (10% proportion).

Method	AP_{50}	AP_S	AP_M	AP_L
FRCNN	67.0 ± 1.15	47.4 ± 3.07	80.3 ± 2.24	81.0 ± 1.39
Carafe [26]	66.5 ± 1.27	51.2 ± 1.16	78.4 ± 1.14	80.4 ± 5.34
FPG [27]	35.1 ± 1.61	4.3 ± 2.04	50.0 ± 2.52	62.4 ± 1.61
TranSDet (FRCNN as baseline)	71.3 ± 1.05	53.4 ± 1.59	85.0 ± 1.87	83.3 ± 3.97

4.7.3. Effect of Transferring from a Small-Object Dataset

When comparing the small-object detection dataset to the large-scale general dataset, the latter usually contained more diverse samples. Therefore, the model trained on it is more suitable for transferring to other detection datasets. In order to clarify why we chose to adapt the model trained on a general dataset instead of directly transferring the model from a small-object dataset, we conducted experiments to transfer the Faster R-CNN model trained on the full TT100K small-object dataset to the BUUISE-MO-Lite dataset. As shown in Table 10, the model initialized with the TT100K pretrained weights obtained inferior results compared to the baseline, which only used the ILSVRC weights to initialize the backbone. This indicates that transferring the model from a small-object dataset with a specific scene would result in a loss of generality and discriminability of the model, leading to poorer performance. In contrast, our method adapted the model trained on the more diverse and general COCO dataset for initialization, resulting in significant improvements compared to the baselines.

Table 10. Comparison of transferring models trained on different datasets to the BUUISE-MO-Lite small-object dataset.

Pretrained Dataset	AP_{50}	AP_S	AP_M	AP_L
ILSVRC	59.4 ± 0.95	51.3 ± 1.07	70.5 ± 3.48	69.5 ± 3.87
TT100K	52.8 ± 0.49	48.1 ± 0.40	60.5 ± 1.33	59.3 ± 3.91
COCO	62.4 ± 1.45	51.0 ± 1.50	66.5 ± 1.66	71.7 ± 2.04
COCO-adapted (Ours)	65.0 ± 3.16	52.7 ± 2.25	65.9 ± 2.90	77.1 ± 2.87

4.7.4. Effects of Different Adaptation Resolutions in DRA

In this paper, we proposed DRA to adapt the pretrained normal model to the small-object detection task by fine-tuning it with smaller resolutions on the normal dataset. Here, we conducted experiments to show the effects of the different resolution choices in our DRA approach. In Figure 9, we report the performance of our method on Faster R-CNN and the TT100K-Lite dataset with different maximum resolutions. Here, the maximum resolution denotes the maximum value in our dynamic resolution set, i.e., 640 denotes a resolution set containing 640 and all the resolution choices smaller than it $\{640, 560, 480, 400, 320\}$. The results show that directly leveraging the original weights trained on COCO resulted in performance degradation at both 10% and 5% data. This indicates that the normally trained weights are not suitable for small-object detection tasks. In contrast, with our DRA, the performance was significantly improved at all resolutions, and the maximum resolution of 480 achieved the best performance.

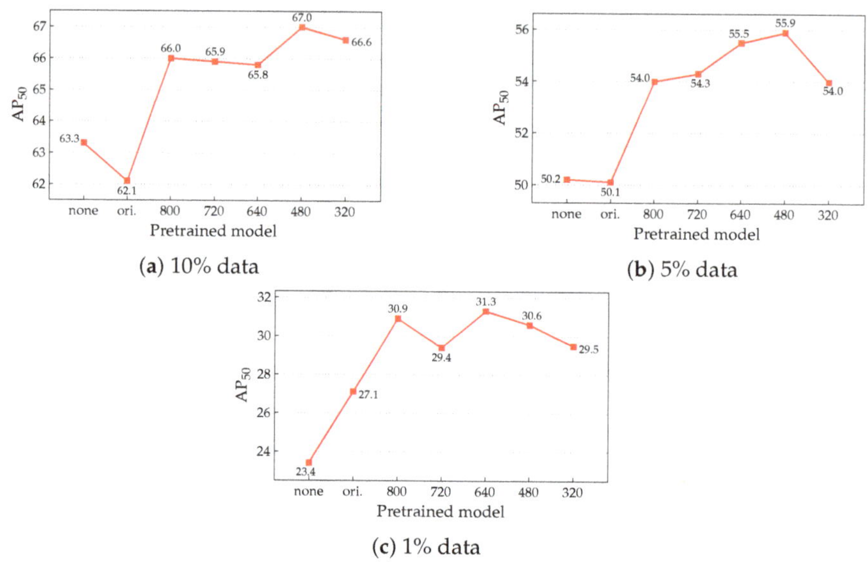

Figure 9. Comparison between different values of maximum resolutions in our dynamic resolution adaptation on TT100K-Lite. none: training on TT100K-Lite without transfer learning. ori.: transfer learning with the original pretrained weights on COCO.

4.8. Complexity Analysis

Furthermore, we analyzed the time complexity of our method. Compared to the original transfer learning method, TranSDet used an additional dynamic resolution adaptation stage to fine-tune the model learned on the source dataset, which increased the training cost while achieving significant improvements. Additionally, in our adaptation stage, we proposed using meta-learning for better learning of dynamic resolutions. But this had the same training complexity (same training time and number of iterations) since we used a first-order approximation in MAML.

To demonstrate the complexity of our proposed detection networks, we present the FLOPs, number of parameters, and inference speed in Table 11. Overall, TranSDet struck a balance between performance and computational complexity. It introduced a slight increase in the FLOPs and parameters compared to the baseline models, indicating reasonable optimization. Regarding the inference speed, the impact was marginal. Consequently, our method is efficient, leading to a significant improvement in performance with only a small computation overhead.

Table 11. Object detection performance on the TT100K-Lite dataset. We measured the inference speed using PyTorch on a single NVIDIA TITAN Xp GPU.

Method	FLOPs (GFLOPs)	Params. (M)	Inference Speed (Image/Second)
FRCNN	206.89	41.35	5.8
FRCNN-TranSDet	267.70	48.71	5.3
RetinaNet	223.83	37.02	6.5
RetinaNet-TranSDet	237.06	39.10	6.0
YOLOv3	194.79	61.76	7.3
YOLOv3-TranSDet	202.47	63.00	7.1

5. Conclusions

Our proposed method, TranSDet, offers a promising solution for small-object detection in deep learning. By leveraging transfer learning and augmenting training images with diverse smaller resolutions, TranSDet addresses the challenges of limited training samples and low-quality images. The dynamic resolution adaptation scheme ensures consistent performance on various object sizes, and the two network components, the FPN with shifted feature aggregation and the anchor relation module, can effectively improve small-object detection accuracy. Extensive experiments and ablation studies demonstrate the effectiveness and efficiency of TranSDet in improving small-object detection accuracy, where it outperformed existing state-of-the-art methods on various datasets. Our study provides valuable insights into designing transfer learning-based models for small-object detection and offers a promising solution for real-world applications, especially in scenarios where large labeled datasets are not available.

Furthermore, TranSDet is not limited to small-object recognition in computer vision. It can also be utilized for other tasks such as semantic segmentation, multi-label classification, and image classification. The transfer learning and adaptive resolution mechanisms employed in TranSDet can be extended to these tasks, enabling improved performance and generalization. This versatility makes TranSDet a valuable tool for a wide range of computer-vision applications beyond small-object detection.

Author Contributions: Conceptualization, X.X. and K.L.; methodology, X.X. and Y.M.; software, X.X.; validation, H.Z., Y.M. and K.L.; formal analysis, H.Z. and X.X.; investigation, H.Z.; resources, K.L.; data curation, H.Z.; writing—original draft preparation, X.X. and H.Z.; writing—review and editing, X.X. and H.B.; visualization, Y.M.; supervision, X.Q. and H.B.; project administration, X.Q. and K.L.; funding acquisition, H.B. All authors have read and agreed to the published version of the manuscript.

Funding: This research was supported in part by a key project of the National Nature Science Foundation of China (Grant No. 61932012); the Science and Technology Major Project of Shanxi Province, China (Grant No. 202101090301013); the Beijing Municipal Education Commission Science and Technology Program (Grant No. KM202111417007); and the Academic Research Projects of Beijing Union University (Grant No. ZK80202003).

Data Availability Statement: No new data were created or analyzed in this study. Data sharing is not applicable to this article.

Acknowledgments: The authors would like to thank the editors and anonymous reviewers for their valuable comments that greatly improved our manuscript.

Conflicts of Interest: The authors declare no conflicts of interest.

References

1. Ren, S.; He, K.; Girshick, R.; Sun, J. Faster R-CNN: Towards Real-Time Object Detection with Region Proposal Networks. In *IEEE Transactions on Pattern Analysis and Machine Intelligence*; IEEE: Piscataway, NJ, USA, 2017; Volume 39, pp. 1137–1149. [CrossRef]
2. Lin, T.Y.; Goyal, P.; Girshick, R.; He, K.; Dollar, P. Focal Loss for Dense Object Detection. In Proceedings of the IEEE International Conference on Computer Vision, Venice, Italy, 22–29 October 2017; pp. 2980–2988.
3. Liu, W.; Anguelov, D.; Erhan, D.; Szegedy, C.; Reed, S.; Fu, C.Y.; Berg, A.C. SSD: Single Shot MultiBox Detector. In *Computer Vision—ECCV 2016*; Springer: Cham, Switzerland, 2016; pp. 21–37. [CrossRef]
4. Shivappriya, S.N.; Priyadarsini, M.J.P.; Stateczny, A.; Puttamadappa, C.; Parameshachari, B.D. Cascade Object Detection and Remote Sensing Object Detection Method Based on Trainable Activation Function. *Remote Sens.* **2021**, *13*, 200. [CrossRef]
5. Fan, D.P.; Ji, G.P.; Cheng, M.M.; Shao, L. Concealed Object Detection. In *IEEE Transactions on Pattern Analysis and Machine Intelligence*; IEEE: Piscataway, NJ, USA, 2022; Volume 44, pp. 6024–6042. [CrossRef]
6. Nnadozie, E.C.; Iloanusi, O.N.; Ani, O.A.; Yu, K. Detecting Cassava Plants under Different Field Conditions Using UAV-Based RGB Images and Deep Learning Models. *Remote Sens.* **2023**, *15*, 2322. [CrossRef]
7. Wu, J.; Xu, W.; He, J.; Lan, M. YOLO for Penguin Detection and Counting Based on Remote Sensing Images. *Remote Sens.* **2023**, *15*, 2598. [CrossRef]
8. Musunuri, Y.R.; Kwon, O.S.; Kung, S.Y. SRODNet: Object Detection Network Based on Super Resolution for Autonomous Vehicles. *Remote Sens.* **2022**, *14*, 6270. [CrossRef]
9. Liang, T.; Bao, H.; Pan, W.; Fan, X.; Li, H. DetectFormer: Category-Assisted Transformer for Traffic Scene Object Detection. *Sensors* **2022**, *22*, 4833. [CrossRef]
10. Rasol, J.; Xu, Y.; Zhang, Z.; Zhang, F.; Feng, W.; Dong, L.; Hui, T.; Tao, C. An Adaptive Adversarial Patch-Generating Algorithm for Defending against the Intelligent Low, Slow, and Small Target. *Remote Sens.* **2023**, *15*, 1439. [CrossRef]

11. Xu, X.; Zhao, S.; Xu, C.; Wang, Z.; Zheng, Y.; Qian, X.; Bao, H. Intelligent Mining Road Object Detection Based on Multiscale Feature Fusion in Multi-UAV Networks. *Drones* **2023**, *7*, 250. [CrossRef]
12. Song, R.; Ai, Y.; Tian, B.; Chen, L.; Zhu, F.; Yao, F. MSFANet: A Light Weight Object Detector Based on Context Aggregation and Attention Mechanism for Autonomous Mining Truck. In *IEEE Transactions on Intelligent Vehicles*; IEEE: Piscataway, NJ, USA, 2023; Volume 8, pp. 2285–2295. [CrossRef]
13. Huang, L.; Zhang, X.; Yu, M.; Yang, S.; Cao, X.; Meng, J. FEGNet: A feature enhancement and guided network for infrared object detection in underground mines. *Proc. Inst. Mech. Eng. Part D J. Automob. Eng.* **2023**, 09544070231165627. [CrossRef]
14. Naz, S.; Ashraf, A.; Zaib, A. Transfer learning using freeze features for Alzheimer neurological disorder detection using ADNI dataset. *Multimed. Syst.* **2022**, *28*, 85–94. [CrossRef]
15. Chen, J.; Sun, J.; Li, Y.; Hou, C. Object detection in remote sensing images based on deep transfer learning. *Multimed. Tools Appl.* **2022**, *81*, 12093–12109. [CrossRef]
16. Neupane, B.; Horanont, T.; Aryal, J. Real-Time Vehicle Classification and Tracking Using a Transfer Learning-Improved Deep Learning Network. *Sensors* **2022**, *22*, 3813. [CrossRef] [PubMed]
17. Ghasemi Darehnaei, Z.; Shokouhifar, M.; Yazdanjouei, H.; Rastegar Fatemi, S.M.J. SI-EDTL: Swarm intelligence ensemble deep transfer learning for multiple vehicle detection in UAV images. *Concurr. Comput. Pract. Exp.* **2022**, *34*, e6726.
18. Narmadha, C.; Kavitha, T.; Poonguzhali, R.; Hamsadhwani, V.; Jegajothi, B. Robust Deep Transfer Learning Based Object Detection and Tracking Approach. *Intell. Autom. Soft Comput.* **2023**, *35*, 3613–3626. [CrossRef]
19. Lin, T.Y.; Maire, M.; Belongie, S.; Hays, J.; Perona, P.; Ramanan, D.; Dollár, P.; Zitnick, C.L. Microsoft COCO: Common Objects in Context. In *Computer Vision—ECCV 2014*; Springer: Cham, Switzerland, 2014; pp. 740–755. [CrossRef]
20. Russakovsky, O.; Deng, J.; Su, H.; Krause, J.; Satheesh, S.; Ma, S.; Huang, Z.; Karpathy, A.; Khosla, A.; Bernstein, M.; et al. ImageNet Large Scale Visual Recognition Challenge. *Int. J. Comput. Vis.* **2015**, *115*, 211–252. [CrossRef]
21. Everingham, M.; Eslami, S.M.A.; Van Gool, L.; Williams, C.K.I.; Winn, J.; Zisserman, A. The Pascal Visual Object Classes Challenge: A Retrospective. *Int. J. Comput. Vis.* **2015**, *111*, 98–136. [CrossRef]
22. Zhu, Z.; Liang, D.; Zhang, S.; Huang, X.; Li, B.; Hu, S. Traffic-Sign Detection and Classification in the Wild. In Proceedings of the IEEE Conference on Computer Vision and Pattern Recognition (CVPR), Las Vegas, NV, USA, 27–30 June 2016.
23. Cheng, G.; Yuan, X.; Yao, X.; Yan, K.; Zeng, Q.; Han, J. Towards large-scale small object detection: Survey and benchmarks. *arXiv* **2022**, arXiv:2207.14096.
24. Yu, X.; Gong, Y.; Jiang, N.; Ye, Q.; Han, Z. Scale Match for Tiny Person Detection. In Proceedings of the IEEE/CVF Winter Conference on Applications of Computer Vision (WACV), Snowmass Village, CO, USA, 1–5 March 2020.
25. Lin, T.Y.; Dollar, P.; Girshick, R.; He, K.; Hariharan, B.; Belongie, S. Feature Pyramid Networks for Object Detection. In Proceedings of the 2017 IEEE Conference on Computer Vision and Pattern Recognition (CVPR), Honolulu, HI, USA, 21–26 July 2017; IEEE: Piscataway, NJ, USA, 2017; pp. 936–944. [CrossRef]
26. Wang, J.; Chen, K.; Xu, R.; Liu, Z.; Loy, C.C.; Lin, D. Carafe: Content-aware reassembly of features. In Proceedings of the IEEE/CVF International Conference on Computer Vision, Seoul, Republic of Korea, 27 October–2 November 2019; pp. 3007–3016.
27. Chen, K.; Cao, Y.; Loy, C.C.; Lin, D.; Feichtenhofer, C. Feature pyramid grids. *arXiv* **2020**, arXiv:2004.03580.
28. Zhang, H.; Wang, K.; Tian, Y.; Gou, C.; Wang, F.Y. MFR-CNN: Incorporating Multi-Scale Features and Global Information for Traffic Object Detection. In *IEEE Transactions on Vehicular Technology*; IEEE: Piscataway, NJ, USA, 2018; Volume 67, pp. 8019–8030. [CrossRef]
29. Tong, K.; Wu, Y.; Zhou, F. Recent advances in small object detection based on deep learning: A review. *Image Vis. Comput.* **2020**, *97*, 103910. [CrossRef]
30. Liu, Y.; Sun, P.; Wergeles, N.; Shang, Y. A survey and performance evaluation of deep learning methods for small object detection. *Expert Syst. Appl.* **2021**, *172*, 114602. [CrossRef]
31. Dai, X.; Chen, Y.; Xiao, B.; Chen, D.; Liu, M.; Yuan, L.; Zhang, L. Dynamic Head: Unifying Object Detection Heads with Attentions. In Proceedings of the 2021 IEEE/CVF Conference on Computer Vision and Pattern Recognition (CVPR), Nashville, TN, USA, 20–25 June 2021; pp. 7369–7378.
32. Huang, J.; Shi, Y.; Gao, Y. Multi-Scale Faster-RCNN Algorithm for Small Object Detection. *J. Comput. Res. Dev.* **2019**, *56*, 319–327. [CrossRef]
33. Redmon, J.; Farhadi, A. Yolov3: An incremental improvement. *arXiv* **2018**, arXiv:1804.02767.
34. Liu, S.; Qi, L.; Qin, H.; Shi, J.; Jia, J. Path aggregation network for instance segmentation. In Proceedings of the IEEE Conference on Computer Vision and Pattern Recognition, Salt Lake City, UT, USA, 18–22 June 2018; pp. 8759–8768.
35. Qi, G.; Zhang, Y.; Wang, K.; Mazur, N.; Liu, Y.; Malaviya, D. Small Object Detection Method Based on Adaptive Spatial Parallel Convolution and Fast Multi-Scale Fusion. *Remote Sens.* **2022**, *14*, 420. [CrossRef]
36. Shi, T.; Gong, J.; Hu, J.; Zhi, X.; Zhang, W.; Zhang, Y.; Zhang, P.; Bao, G. Feature-Enhanced CenterNet for Small Object Detection in Remote Sensing Images. *Remote Sens.* **2022**, *14*, 5488. [CrossRef]
37. Qu, J.; Tang, Z.; Zhang, L.; Zhang, Y.; Zhang, Z. Remote Sensing Small Object Detection Network Based on Attention Mechanism and Multi-Scale Feature Fusion. *Remote Sens.* **2023**, *15*, 2728. [CrossRef]
38. Zhang, J.; Xu, D.; Li, Y.; Zhao, L.; Su, R. FusionPillars: A 3D Object Detection Network with Cross-Fusion and Self-Fusion. *Remote Sens.* **2023**, *15*, 2692. [CrossRef]

39. Cai, Z.; Fan, Q.; Feris, R.S.; Vasconcelos, N. A Unified Multi-scale Deep Convolutional Neural Network for Fast Object Detection. In *Computer Vision—ECCV 2016*; Springer: Cham, Switzerland, 2016; pp. 354–370. [CrossRef]
40. Bosquet, B.; Mucientes, M.; Brea, V.M. STDnet: Exploiting high resolution feature maps for small object detection. *Eng. Appl. Artif. Intell.* **2020**, *91*, 103615. [CrossRef]
41. Wu, B.; Shen, Y.; Guo, S.; Chen, J.; Sun, L.; Li, H.; Ao, Y. High Quality Object Detection for Multiresolution Remote Sensing Imagery Using Cascaded Multi-Stage Detectors. *Remote Sens.* **2022**, *14*, 2091. [CrossRef]
42. Redmon, J.; Farhadi, A. YOLO9000: Better, Faster, Stronger. In Proceedings of the 2017 IEEE Conference on Computer Vision and Pattern Recognition (CVPR), Honolulu, HI, USA, 21–26 July 2017 ; pp. 6517–6525. . [CrossRef]
43. Wang, X.; Huang, T.; Gonzalez, J.; Darrell, T.; Yu, F. Frustratingly Simple Few-Shot Object Detection. In Proceedings of the 37th International Conference on Machine Learning, Virtual, 13–18 July 2020; pp. 9919–9928.
44. Liang, G.; Zheng, L. A transfer learning method with deep residual network for pediatric pneumonia diagnosis. *Comput. Methods Programs Biomed.* **2020**, *187*, 104964. [CrossRef]
45. Wang, X.; Shen, C.; Xia, M.; Wang, D.; Zhu, J.; Zhu, Z. Multi-scale deep intra-class transfer learning for bearing fault diagnosis. *Reliab. Eng. Syst. Saf.* **2020**, *202*, 107050. [CrossRef]
46. Loey, M.; Manogaran, G.; Taha, M.H.N.; Khalifa, N.E.M. A hybrid deep transfer learning model with machine learning methods for face mask detection in the era of the COVID-19 pandemic. *Measurement* **2021**, *167*, 108288. [CrossRef]
47. Tang, Y.P.; Wei, X.S.; Zhao, B.; Huang, S.J. QBox: Partial Transfer Learning with Active Querying for Object Detection. In *IEEE Transactions on Neural Networks and Learning Systems*; IEEE: Piscataway, NJ, USA, 2021; pp. 1–13. [CrossRef]
48. Sun, B.; Li, B.; Cai, S.; Yuan, Y.; Zhang, C. FSCE: Few-Shot Object Detection via Contrastive Proposal Encoding. In Proceedings of the 2021 IEEE/CVF Conference on Computer Vision and Pattern Recognition (CVPR), Nashville, TN, USA, 20–25 June 2021; pp. 7348–7358.
49. Zhu, C.; Chen, F.; Ahmed, U.; Shen, Z.; Savvides, M. Semantic Relation Reasoning for Shot-Stable Few-Shot Object Detection. In Proceedings of the 2021 IEEE/CVF Conference on Computer Vision and Pattern Recognition (CVPR), Nashville, TN, USA, 20–25 June 2021; pp. 8778–8787.
50. Kaul, P.; Xie, W.; Zisserman, A. Label, Verify, Correct: A Simple Few Shot Object Detection Method. In Proceedings of the 2022 IEEE/CVF Conference on Computer Vision and Pattern Recognition (CVPR), New Orleans, LA, USA, 18–24 June 2022; pp. 14217–14227.
51. Yan, D.; Zhang, H.; Li, G.; Li, X.; Lei, H.; Lu, K.; Zhang, L.; Zhu, F. Improved Method to Detect the Tailings Ponds from Multispectral Remote Sensing Images Based on Faster R-CNN and Transfer Learning. *Remote Sens.* **2022**, *14*, 103. [CrossRef]
52. Finn, C.; Abbeel, P.; Levine, S. Model-Agnostic Meta-Learning for Fast Adaptation of Deep Networks. In Proceedings of the 34th International Conference on Machine Learning, Sydney, Australia, 6 –11 August 2017; pp. 1126–1135.
53. Deng, C.; Wang, M.; Liu, L.; Liu, Y.; Jiang, Y. Extended Feature Pyramid Network for Small Object Detection. *IEEE Trans. Multimed.* **2022**, *24*, 1968–1979. [CrossRef]
54. Xu, F.; Wang, H.; Peng, J.; Fu, X. Scale-aware feature pyramid architecture for marine object detection. *Neural. Comput. Appl.* **2021**, *33*, 3637–3653. [CrossRef]
55. Peng, F.; Miao, Z.; Li, F.; Li, Z. S-FPN: A shortcut feature pyramid network for sea cucumber detection in underwater images. *Expert Syst. Appl.* **2021**, *182*, 115306. .: 10.1016/j.eswa.2021.115306. [CrossRef]
56. Hu, J.; Shen, L.; Sun, G. Squeeze-and-excitation networks. In Proceedings of the IEEE Conference on Computer Vision and Pattern Recognition, Salt Lake City, UT, USA, 18–23 June 2018; pp. 7132–7141.
57. Vaswani, A.; Shazeer, N.; Parmar, N.; Uszkoreit, J.; Jones, L.; Gomez, A.N.; Kaiser, Ł.; Polosukhin, I. Attention is all you need. In *Advances in Neural Information Processing Systems*; Curran Associates, Inc.: NewYork, NY, USA, 2017 ; Volume 30.
58. Carion, N.; Massa, F.; Synnaeve, G.; Usunier, N.; Kirillov, A.; Zagoruyko, S. End-to-end object detection with transformers. In Proceedings of the Computer Vision–ECCV 2020: 16th European Conference, Glasgow, UK, 23–28 August 2020; Springer: Cham, Switzerland, 2020; pp. 213–229.
59. Chen, K.; Wang, J.; Pang, J.; Cao, Y.; Xiong, Y.; Li, X.; Sun, S.; Feng, W.; Liu, Z.; Xu, J.; et al. MMDetection: Open MMLab Detection Toolbox and Benchmark. *arXiv* **2019**, arXiv:1906.07155.
60. Liu, Z.; Lin, Y.; Cao, Y.; Hu, H.; Wei, Y.; Zhang, Z.; Lin, S.; Guo, B. Swin transformer: Hierarchical vision transformer using shifted windows. In Proceedings of the IEEE/CVF International Conference on Computer Vision, Montreal, BC, Canada, 11–17 October 2021; pp. 10012–10022.
61. Yang, Z.; Liu, S.; Hu, H.; Wang, L.; Lin, S. Reppoints: Point set representation for object detection. In Proceedings of the IEEE/CVF International Conference on Computer Vision, Seoul, Republic of Korea, 27 October– 2 November 2019; pp. 9657–9666.
62. Zhu, X.; Su, W.; Lu, L.; Li, B.; Wang, X.; Dai, J. Deformable {DETR}: Deformable Transformers for End-to-End Object Detection. In Proceedings of the International Conference on Learning Representations, Vienna, Austria, 3–7 May 2021.
63. Krizhevsky, A.; Sutskever, I.; Hinton, G.E. Imagenet classification with deep convolutional neural networks. *Commun. ACM* **2017**, *60*, 84–90. [CrossRef]

remote sensing MDPI

Article

AiTLAS: Artificial Intelligence Toolbox for Earth Observation

Ivica Dimitrovski [1,2], Ivan Kitanovski [1,2], Pančе Panov [1,3], Ana Kostovska [1,3], Nikola Simidjievski [1,3,4] and Dragi Kocev [1,3,*]

1 Bias Variance Labs, d.o.o., 1000 Ljubljana, Slovenia
2 Faculty of Computer Science and Engineering, University Ss Cyril and Methodius, 1000 Skopje, North Macedonia
3 Department of Knowledge Technologies, Jožef Stefan Institute, 1000 Ljubljana, Slovenia
4 Department of Computer Science and Technology, University of Cambridge, Cambridge CB3 0FD, UK
* Correspondence: dragi@bvlabs.ai

Abstract: We propose AiTLAS—an open-source, state-of-the-art toolbox for exploratory and predictive analysis of satellite imagery. It implements a range of deep-learning architectures and models tailored for the EO tasks illustrated in this case. The versatility and applicability of the toolbox are showcased in a variety of EO tasks, including image scene classification, semantic image segmentation, object detection, and crop type prediction. These use cases demonstrate the potential of the toolbox to support the complete data analysis pipeline starting from data preparation and understanding, through learning novel models or fine-tuning existing ones, using models for making predictions on unseen images, and up to analysis and understanding of the predictions and the predictive performance yielded by the models. AiTLAS brings the AI and EO communities together by facilitating the use of EO data in the AI community and accelerating the uptake of (advanced) machine-learning methods and approaches by EO experts. It achieves this by providing: (1) user-friendly, accessible, and interoperable resources for data analysis through easily configurable and readily usable pipelines; (2) standardized, verifiable, and reusable data handling, wrangling, and pre-processing approaches for constructing AI-ready data; (3) modular and configurable modeling approaches and (pre-trained) models; and (4) standardized and reproducible benchmark protocols including data and models.

Keywords: Earth observation; remote sensing; deep learning; semantic segmentation; object detection; land use and land cover classification

Citation: Dimitrovski, I.; Kitanovski, I.; Panov, P.; Kostovska, A.; Simidjievski, N.; Kocev, D. AiTLAS: Artificial Intelligence Toolbox for Earth Observation. *Remote Sens.* **2023**, *15*, 2343. https://doi.org/10.3390/rs15092343

Academic Editor: Gwanggil Jeon

Received: 8 March 2023
Revised: 15 April 2023
Accepted: 24 April 2023
Published: 28 April 2023

1. Introduction

Remotely gathered data are available from a wide range of sources using a wide range of data collection techniques. Satellites, airplanes, and Unmanned Aerial Vehicles (UAVs) are equipped with various sensors that gather huge amounts of remotely sensed images that provide comprehensive spatial and temporal coverage of the Earth [1,2]. On the other hand, the increase in data production is well matched by the rapidly growing development of Artificial Intelligence (AI), which probes various aspects of natural sciences, technology, and society. Recent trends in machine learning, and particularly in deep learning, have ushered a new era of image analysis and raised the predictive performance bar in many application domains, including remote sensing and Earth observation [3]. Remote sensing data have been used in various application areas, including land use and land cover analysis [4], forest mapping [5,6], monitoring of natural hazards and disasters [7,8], precision agriculture [9], assessing the weather and observing climate changes [10], and various environmental studies [11].

With the ever-growing availability of remote sensing data, there has been a significant research effort to prepare, label, and provide proper datasets that will support the development and evaluation of sophisticated machine-learning methods [12–15]. In the past years, several publicly available high-resolution remote sensing image datasets have been made available to support the research in a variety of remote sensing tasks, such as scene classification [4], semantic and instance segmentation [16], object detection [17], change detection, etc. Most of the annotated EO datasets are limited in scale and restricted in spatial coverage with a task-specific class distribution. Handling and pre-processing remote sensing image data can be very challenging due to their properties and heterogeneity, which greatly differ from conventional image data typically used in recent machine-learning pipelines. Namely, each type of sensor used for remote sensing has its own advantages (and disadvantages) conditioned by the geographical coverage, sensor resolution (spatial and temporal), and flight operations and specifics. For example, satellites are used for sensing at a global scale, and UAVs are typically used for sensing in small areas due to their flexibility and ease of operations in such conditions. Instead of 3-channel RGB imagery, the data in remote sensing are represented through different spectral, spatial, radiometric, and temporal resolutions:

- *Spectral resolution* defines the bandwidth and the sampling rate used to capture data. A high value for the spectral resolution means more narrow bands pertaining to small parts of the spectrum, and conversely, a low value means broader bands related to large parts of the spectrum. Spectral bands are groups of wavelengths, such as ultraviolet, visible, near-infrared, infrared, and microwave. Based on these, image sensors can be multi-spectral if they are able to cover tens of bands (e.g., Sentinel-2, which collects 12 bands) and hyper-spectral if they can collect thousands, such as Hyperion (part of the EO-1 satellite), which covers 220 spectral bands (0.4–2.5 μm) [18].
- *Spatial resolution* defines the size of the area on the Earth's surface represented by each pixel from an image. Spatial resolution relates to the level of detail captured in the image, with high resolutions (small pixel size) capturing more and low resolutions (large pixel size) capturing fewer details in an image. For example, most bands observed by the Moderate Resolution Imaging Spectroradiometer (MODIS) have a spatial resolution of 1 km, where each pixel represents a 1 km $\times$ 1 km area on the ground [19]. In contrast, images captured from UAVs or drones can have a very small spatial resolution of less than 1 cm [20].
- *Radiometric resolution* defines the number of discrete signals of given strengths that the sensor can record (also known as dynamic range). A large value of the dynamic range means that more details can be discerned in the recording, e.g., Landsat 7 records 8-bit images and can thus detect 256 unique gray values of the reflected energy [21]; similarly, Sentinel-2 has a 12-bit radiometric resolution (4095 gray values) [22]. In other words, a higher radiometric resolution allows for simultaneous observation of high and low-contrast objects in the scene. For example, a radiometric resolution is necessary to distinguish between subtle differences in ocean color when assessing water quality.
- *Temporal resolution* defines the frequency at which a given satellite revisits a given observation area. Polar-orbiting satellites have a temporal resolution that can vary from 1 day to 16 days (e.g., for Sentinel-2, this is ten days [22]). The temporal aspects of remote sensing are essential in monitoring and detecting changes in given observation areas (incl. land use change, mowing, and deforestation).

The increasing amount of available EO data is equally matched with the number of libraries and toolboxes designed to handle, process, and potentially provide a data analysis framework for such data. However, considering the complexity of EO data coupled with the diversity of EO tasks that can be addressed, many of the available libraries and toolboxes focus either on the data-specific processing aspect, have a very narrow application horizon, or include machine-learning approaches that are hardly accessible for domain experts. This relates to libraries such as eo-learn (open source) [23] and Up42 (commercial product) [24], which provide accessible means for EO data processing/feature extraction workflows from satellite imagery. These libraries focus mainly on the data acquisition, handling, and data pre-processing stages of the workflow and offer limited machine-learning capabilities. Similarly, Sentinels for Common Agriculture Policy (Sen4CAP), an open-source project based on the Sentinel Application Platform (SNAP), provides data pre-processing algorithms and workflows but only in the limited scope of agriculture monitoring relevant to the management of the CAP [25].

More recent libraries such as Orfeo ToolBox (OTB) [26] offer similar capabilities with respect to data pre-processing, data augmentation, and feature extraction pipelines, in addition to a catalog of more traditional machine-learning approaches for image data analysis. OTB further allows remote integration with deep-learning libraries such as TensorFlow [27]. However, this capability is part of an unofficial OTB module aimed primarily at AI users. On the other hand, CANDELA [28] is an end-to-end platform tailored for EO users, focusing on services that provide quick data access and exploratory data analysis. In addition to the core data handling capabilities, CANDELA allows for handcrafted application-centered data analysis blocks that employ data pre-processing and machine-learning methods but are specifically tailored for a particular EO task at hand (such as change detection). TorchGeo [29] is a recent library that builds on the PyTorch [30] deep-learning framework that includes more general methods for EO data analysis. Namely, it includes data loaders for standard benchmark datasets, methods for data handling, and data transformations, as well as a catalog of (pre-trained) vision models applicable to different tasks pertaining to EO applications.

An implicit but common theme among most of these libraries is the community barrier. Libraries that offer the most recent machine-learning approaches are tailored for data scientists and have a steep learning curve for domain experts, but libraries tailored for the remote sensing community are not easily applicable in modern machine-learning pipelines. While most of these attempts are a step in the right direction, there is still a gap related to the need for a common AI4EO framework that will provide (1) accessible and interoperability resources for data analysis (via configurable and readily usable pipelines); (2) standardized, verifiable, and reusable data handling, wrangling, and pre-processing approaches for constructing AI-ready data; (3) modular and configurable modeling approaches and (pre-trained) models; and (4) standardized and reproducible benchmark protocols (including data and models).

We present AiTLAS (http://aitlas.bvlabs.ai (accessed on 8 March 2023)), an open-source AI4EO toolbox that is designed based on the principles outlined above, thus facilitating the use of EO data in the AI community and, more importantly, accelerating the uptake of (advanced) machine-learning methods and approaches by EO experts. AiTLAS provides various resources, including customizable and easily usable data analysis pipelines; semantically annotated datasets, formalized to be used directly by AI methods; recent approaches for learning models de novo coupled with a model catalog consisting of large pre-trained vision models applicable to EO tasks; standardized frameworks for model benchmarking [3]; mechanisms for quantitative and qualitative model evaluation, etc.

In this paper, we provide extensive details of the many functionalities and capabilities of AiTLAS. We demonstrate its versatility for use in different EO tasks by exploiting the variety of EO data and AI methods made available for direct use within the toolbox. We first explain the design and implementation of AiTLAS and then discuss the supported EO tasks and data. Next, we explain the AI methods that are implemented within the toolbox. Furthermore, we showcase several use cases to illustrate the basic principles behind the toolbox, its modularity, and its flexibility. Lastly, we summarize the distinctive proprieties of AiTLAS and outline directions for its further development.

2. Materials and Methods

2.1. Design and Implementation of the AiTLAS Toolbox

The AiTLAS toolbox is designed such that leveraging recent (and sophisticated) deep-learning approaches over a variety of EO tasks (and data) is straightforward. On the one hand, it utilizes EO data resources in an AI-ready form; on the other hand, it provides a sufficient layer of abstraction for building and executing data analysis pipelines, thus facilitating better usability and accessibility of the underlying approaches—particularly useful for users with limited experience in machine learning, and in particular deep learning.

AiTLAS can be used both as an end-to-end standalone tool and as a *modular* library. Users can use and build on different toolbox components independently, be they related to the tasks, datasets, models, benchmarks, or complete pipelines. It is also *flexible* and *versatile*, facilitating the execution of a wide array of tasks on various domains and providing easy extension and adaptation to novel tasks and domains. Moreover, AiTLAS adheres to the principle *less is more*—it embeds the most common tasks and functionalities in easy-to-use interfaces that simplify the usage and adaptation of the toolbox with minimal modifications. Last but not least, AiTLAS is fully aligned with the principles of *open science*—its development is community-driven and open-source.

Figure 1 presents a high-level schematic diagram of the main modules and components of AiTLAS. It is designed around the concept of a workflow, where users need to define a specific task (`aitlas.tasks`), be it an exploratory analysis of a dataset or a predictive task of a different kind, such as image classification, object detection, image segmentation, etc. In turn, the instantiated task serves as an arbiter of the workflow and orchestrates the flow between the two central components of the toolbox—the datasets (`aitlas.datasets`) and the models (`aitlas.models`)—which relate to AI-ready formalized data and configurable model architectures, respectively. Programmatically, these modules are embedded within the core module `aitlas.base`, which contains all main abstract definitions related to every module, such as definitions of tasks, models, and datasets, but are also related to evaluations (`aitlas.metrics`), data transformations (`aitlas.transforms`), and various types of visualizations (`aitlas.visulizataions` and `aitlas.datasets.visulizataions`).

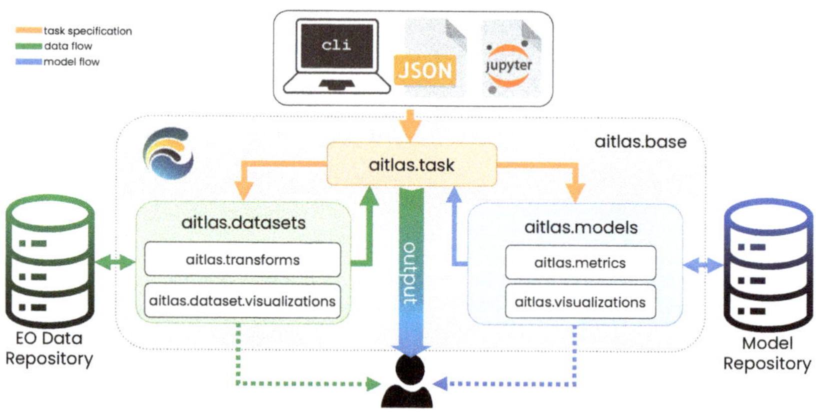

Figure 1. Diagram of the main modules and components in the AiTLAS toolbox.

More specifically, as a standalone application, the flow of AiTLAS begins with the user-specified definition of a task. These definitions can be provided at input via *command-line interface* (`CLI`) as a formatted `JSON` configuration file or more directly executed via Jupiter notebooks. This initiates the arbiter module `aitlas.tasks`. The `aitlas.tasks` act as a controller and component mediator during the entire workflow. To this end, AiTLAS can handle a variety of typical workflows, such as training and evaluating a model, data pre-processing and calculating statistics, extracting features, etc.; while the implemented tasks can be applied in many different scenarios, they can also serve as a blueprint for creating and instantiating new, more specific, tasks.

Typically, a *task* instantiates a specific *dataset* component as per the configuration and prepares it for processing. The *dataset* components, implemented in the `aitlas.datasets` module, encapsulate different operations for working with the underlying EO data, such as reading and writing from and to storage and preparing it for further processing. Each EO dataset has a separate and specific implementation of its *dataset* component since the different datasets have different formats, organizational structures, etc. (Tables 1–5 provide a list of currently supported datasets). Note that the datasets must be accessible for the machine that executes AiTLAS. Therefore, it is up to the user to download the datasets they work with, and organize their access, as AiTLAS only provides the means to access a dataset.

Once the data access is ready, AiTLAS offers various mechanisms for pre-processing, handling, and transforming raw EO data into an AI-ready format. These mechanisms are implemented in the `aitlas.transforms` module. More specifically, besides standard functions for handling image data, this module contains specific implementations of augmentations and transformations that can be applied to images, such as rotations, resizing, cropping, etc. These transformations can be configured to be applied to any raw image (regardless of a task), including target masks, as in the case of image segmentation. The processed (AI-ready) data is, in turn, used in the workflow. Moreover, for better reusability and reproducibility, AiTLAS also annotates and can store the processed data (with the accompanied meta-data) in an EO data repository such as the *AiTLAS semantic data catalog* (http://eodata.bvlabs.ai (accessed on 8 March 2023)), where users can further analyze and query the available data.

The *task* component also interacts with the *model* components within the `aitlas.models` module. The models in AiTLAS are based on the PyTorch framework [30]. The *model* component wraps the architecture of the deep-learning model and only exposes the operations for controlling their behavior, such as training the model, using it to perform predictions, saving it, or loading it from storage, etc. The `aitlas.models` module contains concrete implementations of the deep-learning models providing the means to work with individual model architectures (see Table 6 for a current list of implemented architectures). The concrete implementations are responsible for model instantiation and forwarding the input data. In the case of pre-trained models, the specific implementation can pull a remote or local version of the pre-trained model.

The models' performance is estimated using the *metrics* components implemented in the `aitlas.metrics` module. Depending on the task at hand (classification, segmentation), the module offers a variety of evaluation measures to assess the performance of the models, such as accuracy, F1 score, etc. Note that similar to storing processed data, AiTLAS also supports storing trained models. To this end, the AiTLAS model catalog [3] contains more than 500 trained models for EO image scene classification, trained and evaluated on 22 different EO datasets.

AiTLAS allows for certain aspects of the workflow to be illustrated, fostering better user interaction with the learning process and more interpretable outcomes. This is enabled via the *visualizations* components implemented in the `aitlas.visualization` and `aitlas.dataset.visualization` modules. These components provide the means to provide further inspection and analysis via visualizing datasets (samples and properties), tracking model performance, and visualizing model predictions.

AiTLAS is built in Python and uses a variety of other libraries related to different parts of the project. The core underlying library is PyTorch [30]—the AiTLAS model architecture extends PyTorch's model class. The extensions add the means for training, evaluating, predicting, resource utilization, saving, and loading the model from disk. The AiTLAS dataset management also extends the data module from PyTorch. As previously stated, AiTLAS can be used both as a standalone application and as a library embedded within other projects. The remaining dependencies are given in Table A1 in Appendix A, with further details of their scope of usage.

2.2. EO Data and Common Tasks

The AiTLAS toolbox can be applied for a variety of EO tasks (and datasets). For clarity, we present and discuss four common types of workflows pertaining to typical EO tasks: (1) image scene classification, (2) image semantic segmentation, (3) image object detection, and (4) crop type prediction using satellite time series data.

2.2.1. Image Scene Classification Tasks

The task of image scene classification refers to annotating images. In a typical scenario, working with large-scale EO images, this task addresses classifying smaller images (patches) extracted from a much larger remote sensing image. The extracted images can then be annotated based on the content using explicit semantic classes (e.g. forests, residential areas, rivers, etc.). Given an image as an input, the output would be single or multiple annotations with semantic labels, denoting land-use and/or land-cover (LULC) classes present in that image, as illustrated in Figure 2.

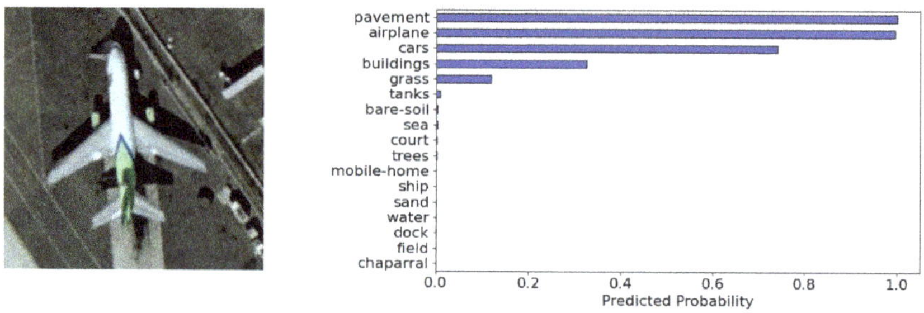

Figure 2. Remote sensing image scene classification: sample image patch provided on the left and the output (predicted LULC classes) shown on the right subfigure. The image is a sample from the UC Merced dataset from the MLC task [31].

Based on the number of semantic labels assigned to the images in the datasets, image scene classification tasks can be further divided into multi-class (MCC) and multi-label (MLC) classification. In the *multi-class classification* setting, each image is associated with a single class (label) from a set of predefined classes. The goal, in this case, is predicting one (and only one) class for each image in the dataset. In the *multi-label classification* setting, on the other hand, images are associated with multiple labels (from a predefined set) based on the information. The goal is then to predict the complete set of labels for each image in the dataset at hand [32]. To this end, AiTLAS offers 22 such datasets that can be readily used for a variety of MLC and MCC modeling tasks. These also serve as a blueprint for applying AiTLAS to other datasets pertaining to similar tasks. Tables 1 and 2 summarize the properties of the considered MCC and MLC datasets, respectively. The number of images across datasets can be quite diverse, ranging from datasets with ∼2 K images to datasets with ∼500 K images. This also holds for the number of labels per image, ranging from 2 to 60. Most of the datasets are comprised of aerial RGB images (with only a few comprised of satellite multi-spectral data) that are different in spatial resolution, size, and format. Finally, note that AiTLAS also provides standardized procedures for analyzing these datasets in terms of predefined splits (for training, validation, and testing), which will ensure reusable and reproducible experiments. A more detailed description of each dataset can be found in [3].

Table 1. Properties of the multi-class image scene classification (MCC) datasets available in the AiTLAS toolbox.

Name	Image Type	#Images	Image Size	Spatial Resolution	#Labels	Predefined Splits	Image Format
UC Merced [33]	Aerial RGB	2100	256 × 256	0.3 m	21	No	tif
WHU-RS19 [34]	Aerial RGB	1005	600 × 600	0.5 m	19	No	jpg
AID [35]	Aerial RGB	10,000	600 × 600	0.5 m–8 m	30	No	jpg
Eurosat [36]	Sat. Multispectral	27,000	64 × 64	10 m	10	No	jpg/tif
PatternNet [37]	Aerial RGB	30,400	256 × 256	0.06 m–4.69 m	38	No	jpg
Resisc45 [4]	Aerial RGB	31,500	256 × 256	0.2 m–30 m	45	No	jpg
RSI-CB256 [38]	Aerial RGB	24,747	256 × 256	0.3–3 m	35	No	tif
RSSCN7 [39]	Aerial RGB	2800	400 × 400	n/a	7	No	jpg
SAT6 [40]	RGB + NIR	405,000	28 × 28	1 m	6	Yes	mat
Siri-Whu [41]	Aerial RGB	2400	200 × 200	2 m	12	No	tif
CLRS [42]	Aerial RGB	15,000	256 × 256	0.26 m–8.85 m	25	No	tif
RSD46-WHU [43]	Aerial RGB	116,893	256 × 256	0.5 m–2 m	46	Yes	jpg
Optimal 31 [44]	Aerial RGB	1860	256 × 256	n/a	31	No	jpg
Brazilian Coffee Scenes (BSC) [45]	Aerial RGB	2876	64 × 64	10 m	2	No	jpg
SO2Sat [46]	Sat. Multispectral	400,673	32 × 32	10 m	17	Yes	h5

Table 2. Properties of the multi-label image scene classification (MLC) datasets available in the AiTLAS toolbox.

Name	Image Type	#Images	Image Size	Spatial Resolution	#Labels	#Labels per Image	Predefined Splits	Image Format
UC Merced (mlc) [31]	Aerial RGB	2100	256 × 256	0.3 m	17	3.3	No	tif
MLRSNet [47]	Aerial RGB	109,161	256 × 256	0.1 m–10 m	60	5.0	No	jpg
DFC15 [48]	Aerial RGB	3342	600 × 600	0.05 m	8	2.8	Yes	png
BigEarthNet 19 [13]	Sat. Multispectral	519,284	20 × 20 60 × 60 120 × 120	60 m 20 m 10 m	19	2.9	Yes	tif, json
BigEarthNet 43 [49]	Sat. Multispectral	519,284	20 × 20 60 × 60 120 × 120	60 m 20 m 10 m	43	3.0	Yes	tif, json
AID (mlc) [50]	Aerial RGB	3000	600 × 600	0.5 m–8 m	17	5.2	Yes	jpg
PlanetUAS [51]	Aerial RGB	40,479	256 × 256	3 m	17	2.9	No	jpg/tiff

2.2.2. Object Detection Tasks

Object detection is another common EO task that focuses on identifying and localizing objects present in an image. In the typical setting, this relates to annotating the identified objects with respect to different predefined classes and providing their location on the image (as a bounding box). Figure 3 illustrates an example of object detection, i.e., detecting ships at sea.

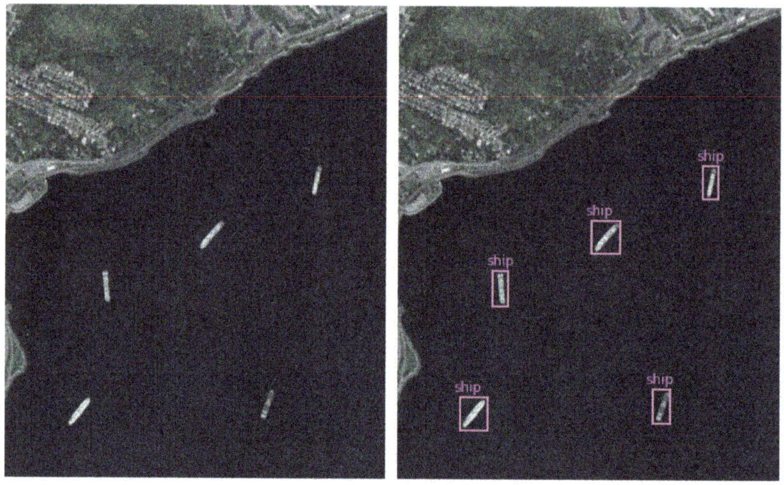

Figure 3. Remote sensing image object detection: sample image provided on the left and in the output image on the right, the objects are detected and localized with bounding boxes. The image is a sample from the HRRSD dataset [52].

Common instances of this task include the detection of specific objects such as buildings, vehicles, ships, and planes [53,54] from aerial image datasets. The AiTLAS toolbox readily supports such tasks and datasets, which follow the Pascal VOC and the COCO (Common Objects in Context) formatting guidelines for object annotations. To this end, it offers four such datasets (Table 3) for development and benchmarking object detection methods.

Table 3. Properties of the object detection datasets available in the AiTLAS toolbox.

Name	Image Type	#Images	#Instances	#Labels	Image Width	Spatial Resolution	Image Format
HRRSD [52]	Aerial RGB	21,761	55,740	13	152–10,569	0.15–1.2 m	jpeg
DIOR [54]	Aerial RGB	23,463	192,472	20	800	0.5–30 m	jpeg
NWPU VHR-10 [55]	Aerial RGB	800	3651	10	∼800	0.08–2 m	jpeg
SIMD [56]	Aerial RGB	5000	45,096	15	1024	0.15–0.35 m	jpeg

2.2.3. Image Semantic Segmentation Tasks

The tasks of image semantic segmentation aim at the fine-grained identification of objects in an image. In contrast to object detection, which aims at coarser localization of the detected objects, segmentation tasks focus on labeling each pixel of an image with a corresponding class of what the pixel represents. In the typical scenario, the input is an image, and the output is a mask (overlay) of categorized pixels based on a single semantic type present in the image. Figure 4 illustrates an example of image semantic segmentation. The more sophisticated extension of semantic segmentation tasks, referred to as *instance segmentation*, takes into account different semantic types and focuses on delineating multiple objects present in an image.

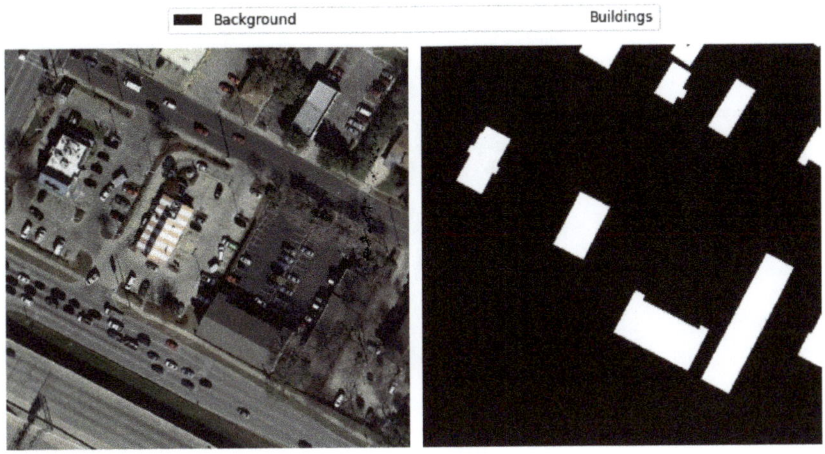

Figure 4. Remote sensing image semantic segmentation of buildings: sample image provided on the left and the output, which is the overlay mask of predictions on the right. The image is a sample from the Massachusetts Buildings dataset [57].

The AiTLAS toolbox offers several such datasets (summarized in Table 4) that can be readily used for EO image semantic segmentation tasks. All but one of the datasets are comprised of aerial RGB images (the remaining contains satellite multi-spectral data) with different spatial resolutions and sizes. The number of semantic labels in these datasets ranges from 2 to 5.

Table 4. Properties of the semantic segmentation datasets available in the AiTLAS toolbox.

Name	Image Type	#Images	Image Size	Spatial Resolution	#Labels	Image Format
LandCover.ai [58]	Aerial RGB	41	4200 × 4700 9000 × 9500	0.25–0.5 m	5	geo tif
Inria [59]	Aerial RGB	360	5000 × 5000	0.3 m	2	tif
AIRS [60]	Aerial RGB	1047	10,000 × 10,000	0.075 m	2	tif
Amazon Rainforest [61]	Aerial RGB	60	512 × 512	n/a	2	geo tif
Chactun [62]	Sat. Multispectral	2093	480 × 480	10 m	3	geo tiff
Massachusetts Roads [57]	Aerial RGB	1171	1500 × 1500	1 m	2	tiff
Massachusetts Buildings [57]	Aerial RGB	151	1500 × 1500	1 m	2	tiff

2.2.4. Crop Type Prediction Tasks

Crop type prediction is a semantic segmentation task that aims to map vegetation on the crops present in a given area. The main difference with the classical semantic segmentation task is that crop type prediction necessarily involves a temporal component. Namely, to properly train a model, it needs to be presented with data of the same area over different periods in time (preferably covering the whole growing season, e.g., the periods with longer daytime and more sunlight). The added complexity is that there is variability between different periods and locations (among different countries or even within the same country) as there is variability between different years. The key feature of any crop-type detection method is to utilize both the spatial and temporal data in multi-temporal satellite imagery. The input in this task is multi-temporal satellite imagery data for a specific geographic area. The output is a segmented mapping of the present crops in that geographic area—Figure 5 illustrates an example of this task.

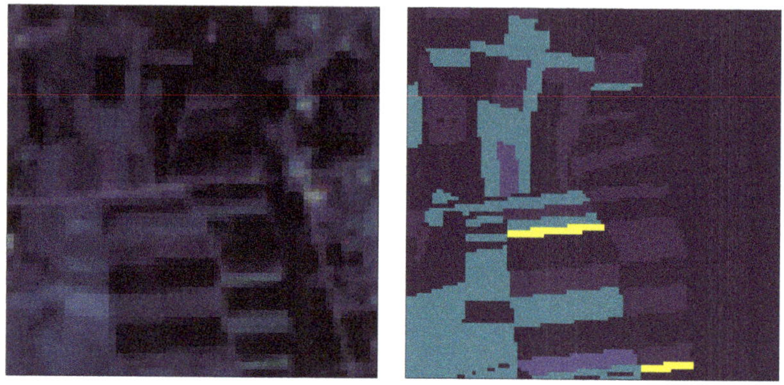

(a) (b)

Figure 5. Crop type prediction: The provided patch is represented by the image (**a**), and the output is the overlay mask of predictions (**b**). The image is a sample from the AiTLAS NLD dataset [63].

Datasets for crop type prediction are generally spatio-temporal, i.e., they contain time series data in addition to the EO image data. Table 5 presents several datasets available within the AiTLAS toolbox. On top of the recent Breizhcrops dataset [64], AiTLAS also presents novel Sentinel 2 imagery for three European countries (Denmark, the Netherlands, and Slovenia) across three years—2017, 2018, and 2019. These novel datasets are significant due to their size w.r.t. the geographic area they cover, the number of different parcels (i.e., polygons), and the number of distinct crop fields. The datasets are presented in detail in [63].

Table 5. Properties of the crop type prediction datasets available in the AiTLAS toolbox.

Dataset	# of Polygons	Area Covered	# of Crop Types
AiTLAS SLO [63]	800 k	5000 km^2	27
AiTLAS DNK [63]	580 k	26,000 km^2	27
AiTLAS NLD [63]	750 k	18,000 km^2	27
Breizhcrops [64]	580 k	27,200 km^2	9

2.3. Model Architectures

The AiTLAS toolbox contains a catalog of deep learning (DL) model architectures that support different EO tasks, including image classification, semantic segmentation, object detection, and crop-type prediction. To this end, AiTLAS implements 24 model architectures, listed in Table 6. For each model, we present the basis of its implementation, technical characteristics, and supported tasks. In the remainder, we discuss the different available architectures in AiTLAS, the basis of their implementation and technical characteristics, and the EO tasks to which they are applicable.

Table 6. Deep neural network architectures implemented in AiTLAS and their usability across the EO tasks.

Model	Im. Scene Class.	Seman. Segm.	Obj. Detection	Crop Type Pred.	Based on
	Supported Tasks				
AlexNet [65]	✓				[66]
CNN-RNN [67]	✓				[67]
ConvNeXt [68]	✓				[66]
DenseNet161 [69]	✓				[66]
EfficientNet [70]	✓				[66]
MLPMixer [71]	✓				[72]
ResNet152 [73]	✓				[66]
ResNet50 [73]	✓				[66]
Swin Transformer [74]	✓				[66]
VGG16 [75]	✓				[66]
Vision Transformer [76]	✓				[72]
DeepLabV3 [77]		✓			[66]
DeepLabV3+ [78]		✓			[79]
FCN [80]		✓			[66]
HRNet [81]		✓			[72]
UNet [82]		✓			[79]
RetinaNet [83]			✓		[66]
Faster R-CNN [84]			✓		[66]
InceptionTime [85]				✓	[86]
LSTM [87]				✓	[86]
MSResNet [88]				✓	[86]
OmniScaleCNN [89]				✓	[86]
StarRNN [90]				✓	[86]
TempCNN [91]				✓	[86]
Transformer for time series classification [92]				✓	[86]

Regarding **image scene classification tasks**, AiTLAS implements a variety of well-known DL models based on the traditional convolutional architectures, but also the more recent attention-based and mlp-based architectures. *Convolutional DL architectures* have contributed to many advances in computer vision. A convolutional neural network (CNN) typically consists of many (hidden) layers stacked together, designed to process (image) data in the form of multiple arrays. The distinctive component in these networks is the convolutional layers, which apply the convolution operation (passing the data through a kernel/filter) and forward the output to the next layer. This serves as a mechanism for constructing feature maps, with former layers typically learning low-level features (such as edges and contours) and subsequently increasing the complexity of the learned features with deeper layers in the network.

The convolutional layers are typically followed by pooling operations (serving as a downsampling mechanism) that aggregate the feature maps through local non-linear operations. In turn, these feature maps are fed to fully-connected layers which perform the ML task at hand—in this case, classification. All the layers in a network employ an activation function. In practice, the intermediate hidden layers employ a non-linear function such as rectified linear unit (ReLU) or Gaussian Error Linear Unit (GELU) as common choices. The choice of activation function in the final layer relates to the tasks at hand—typically, this is a sigmoid function in the case of classification. CNN architectures can also include different normalization and/or dropout operators embedded among the different layers, which can further improve the network's performance. CNNs have been extensively researched, with models applied in many contexts of remote sensing, and in particular EO image classification [93–96].

Recently, *attention-based network architectures* have shown state-of-the-art performance in various vision tasks, including tasks in EO domains. Very prominent in this aspect is the *Vision Transformers* (ViT) [76]—they are inspired by the popular NLP (natural language processing) transformer architecture [97], and leverage the attention mechanism for vision tasks. Much like the original transformer architecture that seeks to learn implicit relationships in sequences of word tokens via multi-head self-attention, ViTs focus on learning such relationships between image patches. Typically, ViTs employ a standard transformer encoder that takes a lower-dimensional (linear) representation of these image patches together with additional positional embedding from each, in turn feeding the encoder output to a standard MLP head. More recent and sophisticated attention network architectures such as the *Swin Transformers* (SwinT) [74,98] rely on additional visual inductive biases by introducing hierarchy, translation invariance, and locality in the attention mechanism. ViT and SwinT variants have shown excellent performance in practice on various vision tasks, including EO applications [3,99–101], particularly when pre-trained with large image datasets.

An attention mechanism can be obtained with different approaches, e.g., attending over channels and/or spatial information, and with convolutional architectures [102–104]. Another alternative is the *MLPMixer* [71]—it obtains its attention mechanism relying on the classical MLP architecture. Namely, similarly to a transformer architecture, an MLPMixer operates on image patches. It consists of two main components: a block of MLP layers for 'mixing' the spatial patch-level information on every channel and a block of MLP layers for 'mixing' the channel information of an image. This renders lightweight models, with performance on par with many much more sophisticated architectures [96,105,106].

The **semantic segmentation** tasks refer to pixel-wise classification [16]. In this scenario, segmentation models typically learn to extract meaningful features/representations from an image and use them to separate the image into multiple segments. In this context, convolutional architectures are frequently used for performing this task. As in the typical convolutional setting, the image is first passed through a series of layers (that learn image features). This process downsamples the image as it passes through a series of pooling layers. In turn, the image is upsampled/interpolated back to its original size (typically by using deconvolutional layers), but with some loss of information. The output is typically passed to a convolutional block with a sigmoid activation, which provides the resulting pixel-wise classification of the image. This relates to pixel-to-pixel fully convolutional networks (FCN) [80]. More sophisticated segmentation architectures typically combine existing architectures at different stages, such as for the model's downsampling (encoding), upsampling (decoding), and prediction blocks.

AiTLAS supports segmentation tasks and implements a variety of state-of-the-art architectures with a track record of successful applications for semantic image segmentation. This includes *UNet* [82], a robust, versatile, and accurate segmentation architecture [107]. UNet is a modification of FCN, consisting of encoder and decoder blocks (in a U shape): the encoder blocks relate the extracted features to the corresponding blocks of the decoder, with an additional shortcut connection in the decoder. This allows the model to capture more specific information from the image and retain more information by concatenating high-level features with low-level ones. AiTLAS also employs HRNet (High-Resolution Net), another fully convolutional network [81] with parallel architecture and multiple group convolutions. This allows for leveraging high-resolution images, which leads to better performance [108].

Common segmentation architectures employ a downsampling block, which broadens the receptive field (given the input) for the forthcoming filter(s), but at the cost of reduced spatial resolution. An alternative approach with the same effect but with the ability to preserve the spatial resolution is atrous (or dilated) convolutions. Here, the filter is upsampled along each spatial dimension by inserting zero values between two successive filters. The *DeepLab* segmentation architectures [109] employ this approach, incorporating both atrous convolutions and atrous spatial pyramid pooling (ASPP), leading to robust and accurate performance semantic segmentation of high-resolution images. To this end, AiTLAS implements DeepLabv3 [77] and DeepLabv3+ [78] architectures that have been successfully applied in a variety of EO application [110,111].

Another class of common tasks in EO domains is **object detection**, which aims at the localization and classification of objects present in an image. Typical deep-learning approaches that address object detection tasks can be divided into two groups: region proposal-based and regression-based [54]. Region proposal-based approaches tackle object detection in two stages. The first stage focuses on generating a series of candidate region proposals that may contain objects. The second stage classifies the candidate region proposals obtained from the first stage into object classes or backgrounds and further fine-tunes the coordinates of the bounding boxes. In contrast, *regression-based approaches* transform the problem to a multi-target regression task, focusing on directly predicting the coordinates of the (detection) bounding box.

In the domain of EO, object detection approaches are typically applied on aerial images, which can be challenging due to the significant variation in scale and viewpoint across the diverse set of object categories [53]. Most studies involving aerial images use region proposal-based methods to detect multi-class objects. To support these tasks, AiTLAS implements several state-of-the-art architectures such as the improved Faster R-CNN model with a ResNet-50-FPN backbone [84,112] and RetinaNet with ResNet-50-FPN backbone [83].

Finally, AiTLAS also provides approaches for addressing tasks of **crop type prediction**. This refers to a multi-dimensional time series classification task where the input is multi-spectral temporal data and the output is a discrete variable specifying the crop type; while these tasks have traditionally been tackled using standard machine-learning approaches (such as Random Forest [113]), more recent deep-learning approaches have shown better results [64]. These include approaches that build on convolutional, recurrence, and self-attention-based architectures.

The convolution-based models use a one-dimensional convolutional layer to extract features from a temporal local neighborhood by convolving the input time series with a filter bank learned by gradient descent. To this end, AiTLAS provides implementations of several such architectures that have been successfully used for crop type prediction and land cover mapping [91,114], including Temporal Convolutional Neural Network (TempCNN) [91] (TempCNN), Multi-Scale 1D Residual Network (MSResNet) [88], InceptionTime [85], and Omniscale Convolutional Neural Network (OmniscaleCNN) [89]. Recurrent Neural Network (RNN) models process a series of observations sequentially while maintaining a feature representation from the previous context. AiTLAS provides implementations for Long Short-Term Memory (LSTM) [115] models, which have been successfully used in remote sensing applications, especially for land cover mapping [116,117]. Finally, AiTLAS includes recent state-of-the-art attention-based transformer architectures based on [92], which can learn and use the most relevant parts of the input sequence via stacked self-attention layers for sequence-to-label classification.

3. Results and Discussion: Demonstrating the Potential of AiTLAS

In this section, we showcase the potential of the AiTLAS toolbox through a series of more detailed examples of its various functionalities and capabilities. We present five use cases that demonstrate the basic principles behind the toolbox, its modularity, and its flexibility. For each of the types of EO tasks that we highlighted earlier, we discuss every segment of the analysis pipeline. We start by loading a new dataset and performing exploratory data analysis, inspection, and pre-processing. Next, we demonstrate the use of a machine-learning model (provided in AiTLAS) for a given dataset as well as the evaluation of the model performance (through different evaluation measures, confusion matrices, and visualizations). Moreover, we show how users can utilize previously trained models for making predictions on unseen images and quantitatively and qualitatively analyze the obtained results. Finally, we provide recipes for including new machine-learning architectures into the AiTLAS toolbox.

3.1. Image Classification

We start with a showcase of image scene classification. Note that the presented examples (and the obtained results) can be easily reproduced via a Jupyter notebook presented and further discussed in Appendix C. Moreover, an extensive analysis, performed using AiTLAS, of more than 500 DL models across a variety of multi-class and multi-label image classification tasks is presented in [3].

3.1.1. Data Understanding and Preparation

Currently, the `aitlas.datasets` module includes 22 ready-to-use datasets for EO image scene classification. To use these datasets, one needs to set the location of the images and a csv file with the labels for each image. The loaded images can then be transformed (i.e., via data augmentation), inspected, and visualized (including summaries over the complete dataset). In the following, we show the process of adding a new dataset and illustrate the capabilities for exploratory analysis.

For this purpose, we use the CLRS dataset [42] as a running example without loss of generality of the `aitlas.datasets` module. The CLRS dataset (available at https://github.com/lehaifeng/CLRS (accessed on 8 March 2023)) is designed for the task named continual/lifelong learning for EO image scene classification [42]. It comprises 15,000 remote sensing images covering over 100 countries divided into 25 scene classes. Each class is associated with 600 images with a size of 256 × 256 pixels and spatial resolution in the range of 0.26 m to 8.85 m.

Given that the dataset is shared without predefined splits (for training/validation/testing), we can first create them using the split task within the `aitlas.tasks` module. The user needs to only specify the data location and the ratios (in percentages) for the desired splits (e.g., here, we use 60/20/20 splits). The split task will output three separate csv files for each split, where each row in the csv denotes the relative path of the image (including the sub-folder name and the file name) and its label. For reproducibility, we provide further details in Appendix B and already prepared dataset (available at https://github.com/biasvariancelabs/aitlas-arena (accessed on 8 March 2023)).

Next, through the class *MultiClassClassificationDataset* from the `aitlas.datasets` module, one can load a new dataset as shown in Listing 1. This class also implements additional data transformations, which can be applied if necessary.

Listing 1. Loading an MCC dataset for remote sensing image scene classification using the *MultiClassClassificationDataset* class from the AiTLAS toolbox.

```
dataset_config = {
"data_dir": "/datasets/CLRS",
"csv_file": "/datasets/CLRS/images.csv" }
dataset = MultiClassClassificationDataset(dataset_config)
```

Loading a completely new dataset (beyond the ones currently supported within AiTLAS) can be achieved by explicitly defining a new data loader class that inherits from the class *MultiClassClassificationDataset*. Within the definition of the class, one needs to manually set the list of the labels/classes (with labels matching the labels provided from the csv file). One can also provide additional meta-data (such as name and URL) together with additional methods for data manipulation. Listing 2 provides a recipe for this using the CLRS dataset.

Listing 2. Adding a new MCC dataset in the AiTLAS toolbox.

```
from .multiclass_classification import MultiClassClassificationDataset

LABELS = ["airport", "bare-land", "beach", "bridge", "commercial",
"desert", "farmland", "forest", "golf-course", "highway",
"industrial", "meadow", "mountain", "overpass", "park",
"parking", "playground", "port", "railway", "railway-station",
"residential", "river", "runway", "stadium", "storage-tank"]

class CLRSDataset(MultiClassClassificationDataset):

url = "https://github.com/lehaifeng/CLRS"
labels = LABELS
name = "CLRS dataset"

def __init__(self, config):
super().__init__(config)
```

Once the dataset is ready, one can use methods from the `aitlas.visulizataion` module for data visualization and inspection. For instance, one can easily plot images from the dataset (as shown in Figure 6) and analyze their properties in terms of data distributions. A more detailed discussion of data exploration capabilities is given in Appendix C.

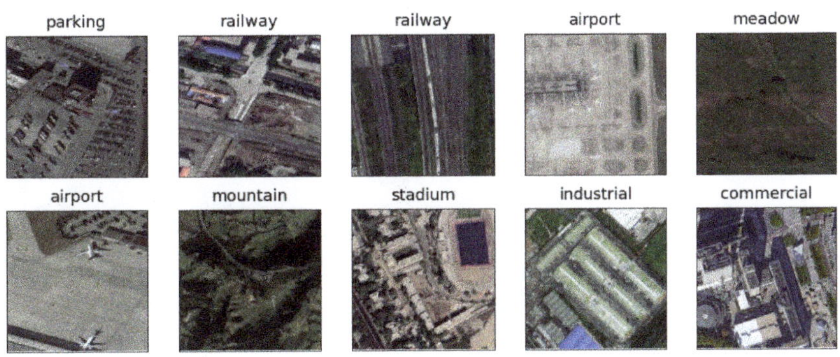

Figure 6. Example images with labels from the CLRS dataset.

3.1.2. Definition, Execution, and Analysis of a Machine Learning Pipeline

Given a dataset, we next focus on setting an approach that supports an image scene classification task. In this use case, we employ the Vision Transformer (ViT) model [76]. Specifically, we use a ViT with an input size of 224×224 and a patch resolution of 16×16 pixels. We showcase a pipeline with two variants: (i) a model "trained from scratch" using the CLRS dataset and (2) a pre-trained model on ImageNet-1K and then fine-tuned on the CLRS dataset. Note that the ViT model is trained/fine-tuned using the data splits we defined earlier.

We configure the model by setting several configuration parameters, i.e., the number of classes/labels, the learning rate, and the evaluation metrics. To use the pre-trained variant of the Vision Transformer (pre-trained on the ImageNet-1k dataset) in the configuration object, we also set the pre-trained parameter to *true*. We fine-tune the model on the CLRS dataset by calling the function `train_and_evaluate_model`. The code snippet for instantiating the model and running the training sequence is given in Listing 3.

Listing 3. Creating a model and executing model training.

```
epochs = 100
model_directory = "/experiments/CLRS"
model_config = {
"num_classes": 25,
"learning_rate": 0.0001,
"pretrained": True,
"metrics": ["accuracy", "precision", "recall", "f1_score"]}
model = VisionTransformer(model_config)
model.prepare()
model.train_and_evaluate_model(
train_dataset=train_dataset,
epochs=epochs,
model_directory=model_directory,
val_dataset=validation_dataset,
run_id='1',)
```

To evaluate the learned model, we load the dataset's test split, set the evaluation metrics list, and run the evaluation sequence on the test data (Listing 4). Note that the predictive performance of the models in this setting is typically assessed by *top-n accuracy* score (typically n is set to 1 or 5) [65]. This score calculates the number of correctly predicted labels among the n most probable labels the model outputs. Besides accuracy, AiTLAS supports additional prediction performance metrics such as Macro Precision, Weighted Precision, Macro Recall, Weighted Recall, Macro F1 score, and Weighted F1 score, etc.

Listing 4. Evaluating a trained model using images from the test split.

```
test_dataset_config = {
"batch_size": 128,
"shuffle": False,
"data_dir": "/dataset/CLRS",
"csv_file": "/dataset/CLRS/test.csv",
"transforms": ["aitlas.transforms.ResizeCenterCropToTensor"]}
test_dataset = CLRSDataset(test_dataset_config)
model_path = "best_checkpoint.pth.tar"
model.metrics = ["accuracy", "precision", "recall", "f1_score"]
model.running_metrics.reset()
model.evaluate(dataset=test_dataset, model_path=model_path)
model.running_metrics.get_scores(model.metrics)
```

Finally, once the model has been trained and evaluated, users can analyze their performance. In this particular example, the ViT model that was first pre-trained on ImageNet-1K achieved an accuracy of 93.20%, substantially outperforming the counterpart trained from scratch, which obtained an accuracy of 65.47%. The performance can also be investigated via confusion matrices (Figure 7), allowing for a more fine-grained analysis of the model performance on individual classes. Finally, using AiTLAS, users can further analyze the predictions made by the model's Grad-CAM [118] activation maps. For instance, Figure 8 presents a sample output obtained from the ViT model, highlighting where the model focused when making the correct (or incorrect) predictions. This capability allows for further validation and diagnosis of the models.

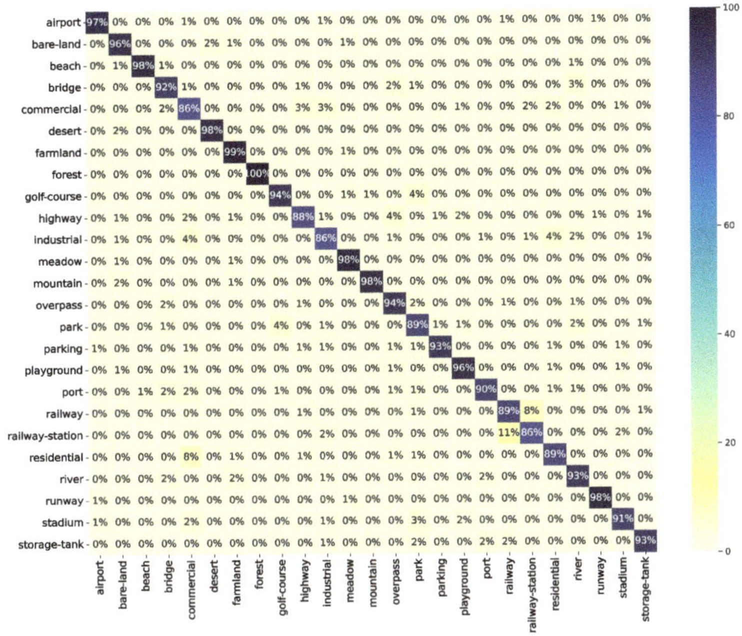

Figure 7. Confusion matrix obtained from a pre-trained Vision Transformer model applied on the CLRS dataset. The values denote percentages of correctly/incorrectly predicted labels.

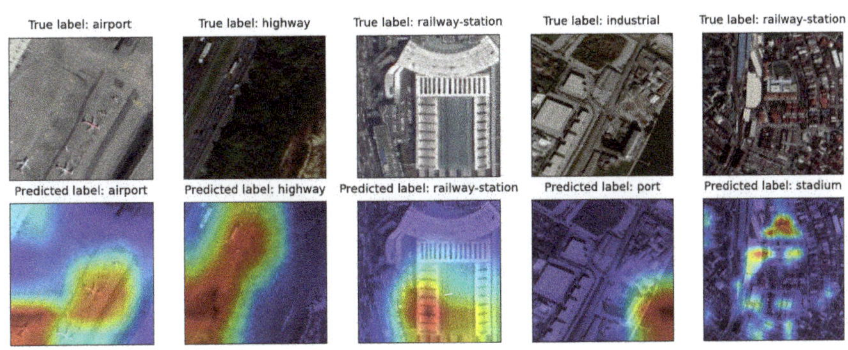

Figure 8. GradCAM visualizations for images, sampled from the CLRS dataset: (**top**) input images with their ground-truth label, (**bottom**) corresponding activation maps with predicted labels (from a ViT model).

A detailed discussion on the pipeline is presented in Appendix C, including templates for model learning, evaluation, and inspection, as well as guidelines on using the learned models for predicting tasks with unseen images.

3.2. Semantic Segmentation

In this section, we discuss the utility of the AiTLAS toolbox for semantic segmentation tasks. For this use case, we use the LandCover.ai (Land Cover from Aerial Imagery) dataset [119] and employ a DeepLabV3 model to perform the semantic segmentation task. The presented use case (together with the obtained results) can be easily reproduced using the Jupyter notebook presented and discussed in Appendix D.

3.2.1. Data Understanding and Preparation

LandCover.ai (Land Cover from Aerial Imagery) (available at https://landcover.ai.linuxpolska.com/download/landcover.ai.v1.zip (accessed on 8 March 2023)) is a dataset for automatic mapping of buildings, woodlands, water, and roads from aerial images [119]. It contains a selection of aerial images taken over the area of Poland. The images have a spatial resolution of 25 or 50 cm per pixel with three spectral bands (RGB bands). The original 41 images and their corresponding masks are split into 512 × 512 tiles. The tiles are then shuffled and organized into 70%/15%/15% of the tiles for training, validation, and testing, respectively.

AiTLAS implements data loaders for creating, loading, and preparing datasets for semantic segmentation. Specifically, the class *SemanticSegmentationDataset* from the `aitlas.datasets` module is the base class for creating a dataset, which loads images and the corresponding segmentation masks from a csv file. For example, in this case, this is performed within the instanced *LandCoverAiDataset* (presented in Listing 5), which sets the labels and their color mapping (used only for visualization as presented in Figure 9). A complete example of the data inspection capabilities is given in Appendix D.

Listing 5. Adding a new dataset for semantic segmentation in the AiTLAS toolbox.

```
class LandCoverAiDataset(SemanticSegmentationDataset):
url = "https://landcover.ai.linuxpolska.com/"
labels = ["Background", "Buildings", "Woodlands", "Water", "Road"]
color_mapping = [[255, 255, 0], [0, 0, 0], [0, 255, 0], [0, 0, 255], [200, 200,
    200]]
name = "Landcover AI"

def __init__(self, config):
super().__init__(config)
```

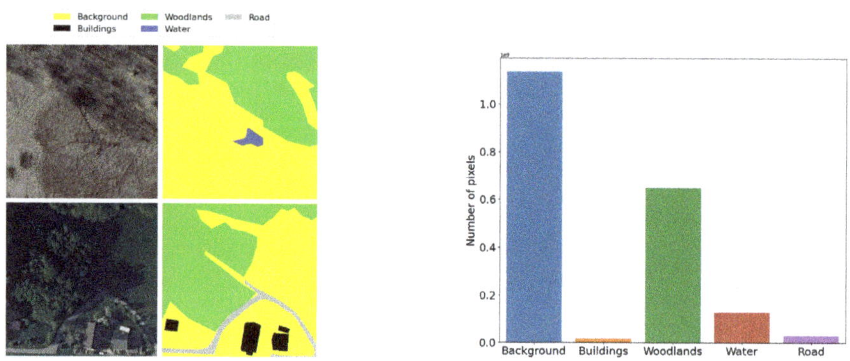

(**a**) (**b**)

Figure 9. Example images with masks from the LandCover.ai dataset. Each pixel is labeled with one of the following labels: background (yellow), buildings (black), woodlands (green), water (blue) and road (gray) (**a**) and pixel distribution within the labels (**b**).

3.2.2. Definition, Execution and Analysis of a Machine Learning Pipeline

Once the dataset has been appropriately set, we focus on setting, training, and evaluating the models. For this use case, we train two variants of DeepLabV3 models: (i) one "trained from scratch" using only the LandCover.ai dataset, and (2) a model with pre-trained weights on a subset of the COCO dataset [120] (using only the 20 categories also present in the Pascal VOC dataset [66]), subsequently fine-tuned on LandCover.ai dataset. Listing 6 presents the AiTLAS configuration for setting and executing the training procedure. We use mean intersection over union ($mIoU$) as an evaluation measure, which denotes the area of the overlap between the ground truth and predicted label divided by the total area, averaged across the different labels.

Listing 6. Creating an instance of a DeepLabv3 model and executing model training.

```
epochs = 100
model_directory = "/experiments/landcoverai"
model_config = {
"num_classes": 5,
"learning_rate": 0.0001,
"pretrained": True,
"threshold": 0.5,
"metrics": ["iou"] }

model = DeepLabV3(model_config)
model.prepare()
model.train_and_evaluate_model(
train_dataset=train_dataset,
val_dataset=validation_dataset,
epochs=epochs,
model_directory=model_directory,
run_id='1')
```

Once the training routine has finished, evaluating the model can be performed as presented in Listing 7.

Listing 7. Evaluating a trained DeepLabv3 model using images from the test split.

```
test_dataset_config = {
"batch_size": 4,
"shuffle": False,
"data_dir": "/dataset/landcoverai/images",
"csv_file": "/dataset/landcoverai/test.txt",
"transforms": ["aitlas.transforms.MinMaxNormTranspose"],
"target_transforms": ["aitlas.transforms.Transpose"]
}

test_dataset = LandCoverAiDataset(test_dataset_config)
model_path = "/experiments/landcoverai/best_checkpoint.pth.tar"
model.metrics = ["iou"]
model.running_metrics.reset()
model.evaluate(dataset=test_dataset, model_path=model_path)
model.running_metrics.get_scores(model.metrics)
```

The results presented in Table 7 summarize the segmentation performance of the trained models. In this example, both DeepLabv3 models resulted in a similar performance, with the pre-trained model having a (practically) insignificantly better performance than its counterpart trained from scratch. In general, the values of the IoU for the labels 'Road' and 'Building' are lower than the other labels since they are usually narrow (roads) and/or often small (buildings), thus typically challenging for segmentation models. Running additional experiments with different model architectures, setups, and datasets is straightforward within AiTLAS, requiring only simple modifications to the presented routines. Additional templates for this task, including additional visualizations and application of the model to external images, are given in Appendix D.

Table 7. Label-wise $IoU(\%)$ and $mIoU$ for the DeepLabv3 models trained on the LandCover.ai dataset.

Model/Label	Background	Buildings	Woodlands	Water	Road	mIoU	Training Time
Trained from scratch	93.813	80.304	91.952	94.877	69.190	86.027	4.5 h
Pre-trained on COCO	93.857	80.650	91.964	95.145	68.846	86.093	5 h

3.3. Object Detection

We next discuss using the AiTLAS toolbox for object detection. In this use case, we show using a Faster R-CNN model on the HRRSD dataset [52]. All of the necessary details of the developed resources are further given in Appendix E.

3.3.1. Data Understanding and Preparation

The AiTLAS toolbox supports the representation of the data for the object detection task through its `aitlas.datasets` module via the base classes *ObjectDetectionPascalDataset* and *ObjectDetectionCocoDataset*. They implement the two most widely used data representation formats for object detection: PascalVOC [121] and COCO [120], respectively. More specifically, the Pascal VOC format includes an XML file for each image in the dataset containing information about the bounding boxes of the objects present in the image, together with some additional metadata such as category/label, level of difficulty, and an indicator of whether the object is truncated (partially visible). On the other hand, the COCO annotation format stores the annotations in JSON files for the training, testing, and validation parts of the data. The JSON file contains a list of each object annotation from every image in the dataset. The annotation includes coordinates for the bounding box, the area of the bounding box, category/label, and an 'iscrowd' indicator for the number of objects in an image (which is 0 for single objects or 1 for a collection of objects).

The HRRSD (available at https://github.com/CrazyStoneonRoad/TGRS-HRRSD-Dataset (accessed on 8 March 2023)) [52,122] dataset contains 21,761 color images acquired from Google Earth with spatial resolution ranging from 0.15 to 1.2 m, and 4961 color images acquired from Baidu Maps with a spatial resolution ranging from 0.6 to 1.2 m. The dataset is divided into a training portion (5401 images), a validation portion (5417 images), and a test portion (10,943 images). An image from the dataset may contain several objects or just one and may contain objects from the 13 different categories/labels. The total number of object instances is 55,740.

Similarly to the use cases presented earlier, loading a dataset into AiTLAS involves instantiating *ObjectDetectionPascalDataset*, as shown in Listing 8. The class *ObjectDetectionPascalDataset* implements additional functionalities for further inspection of the loaded data. For example, one can visualize images from the dataset (as shown in Figure 10) coupled with the number of instances for each category within the dataset.

Listing 8. Loading HRRSD dataset using the class *ObjectDetectionPascalDataset* from the AiTLAS toolbox.

```
dataset_config = {
"image_dir": "/datasets/HRRSD/images",
"annotations_dir": "/datasets/HRRSD/annotations",
"imageset_file": "/datasets/HRRSD/train.txt", }

dataset = ObjectDetectionPascalDataset(dataset_config)
```

Figure 10. Example images with bounding boxes for the objects from the HRRSD dataset.

3.3.2. Definition, Execution and Analysis of a Machine Learning Pipeline

As mentioned earlier, in this use case, we use the Faster R-CNN model with a ResNet-50-FPN backbone [84]. Specifically, here we also train and evaluate two variants of the model: "trained from scratch" using the HRRSD dataset and a pre-trained model on the COCO dataset [66] and then fine-tuned on the HRRSD dataset. The code snippet for creating and training these models is shown in Listing 9. To evaluate the models, one needs to create a configuration object for the training split of the HRRSD data, load the data, set the path to the trained model, and run the evaluation process (as shown in Listing 10). As an evaluation measure, we use the mean average precision (mAP) as defined in the Pascal VOC Challenge [121], computed as the average precision value taken at recall values ranging from 0 to 1 (i.e., the area under the precision/recall curve) and then averaged over all classes. The performance of object detection models can also be evaluated using IoU between the predicted and ground-truth bounding boxes.

Listing 9. Creating an instance of a Faster R-CNN model and executing model training.

```
epochs = 100
model_directory = "/experiments/hrrsd"
model_config = {
"num_classes": 14,
"learning_rate": 0.0001,
"pretrained": True,
"threshold": 0.5,
"metrics": ["map"] }

model = FasterRCNN(model_config)
model.prepare()
model.train_and_evaluate_model(
train_dataset=train_dataset,
val_dataset=validation_dataset,
epochs=epochs,
model_directory=model_directory,
run_id='1'
)
```

Listing 10. Testing a trained model with images from the test split.

```
test_dataset_config = {
"batch_size": 4,
"shuffle": False,
"image_dir": "/datasets/HRRSD/images",
"annotations_dir": "/datasets/HRRSD/annotations",
"imageset_file": "/datasets/HRRSD/test.txt",
"joint_transforms": ["aitlas.transforms.ResizeToTensorV2"] }

test_dataset = ObjectDetectionPascalDataset(test_dataset_config)
model_path = "/experiments/hrrsd/best_checkpoint.pth.tar"
model.metrics = ["map"]
model.running_metrics.reset()
model.evaluate(dataset=test_dataset, model_path=model_path)
model.running_metrics.get_scores(model.metrics)
```

The results of this particular use case show that the pre-trained Faster R-CNN model leads to an mAP of 81.436%, outperforming the variant trained from scratch with an mAP of 77.412%. Further investigation shows that, in this case, the most challenging objects to detect are 'Crossroad' and 'T Junction' (due to the similarity of both objects). AiTLAS allows for further quantitative analysis of these results by examining the predicted outputs (images with the detected objects and the categories/labels for each object) as shown in Figure 11. Further details for this use case are presented in Appendix E.

Figure 11. Example images with the predicted bounding boxes and object labels ('ship', 'T junction' and 'airplane', respectively) using a Faster R-CNN model.

3.4. *Crop Type Prediction*

For our last use case, we show the capabilities of the AiTLAS toolbox on the task of crop type prediction. For this purpose, we train and evaluate an LSTM model applied to the AiTLAS NLD dataset [63]. Merdjanovska et al. [63] present an extensive analysis of this task, performed using the AiTLAS toolbox, comparing the performance of several state-of-the-art deep-learning architectures. All of the developed resources for this use case are also presented and discussed in Appendix F.

3.4.1. Data Understanding and Preparation

Typically, the data format for crop-type prediction tasks is different compared to data for the other EO tasks because of their temporal component that needs to be taken into consideration. The AiTLAS toolbox supports crop-type prediction datasets via the base class *CropsDataset* and the *EOPatchCrops* class (a wrapper for working with *EOPatches* [23]). The EOPatch format stores multi-temporal remotely sensed data of a single patch of the Earth's surface as constrained by the bounding box in a given coordinate system. The patch can be a rectangle, polygon, or pixel in space. The same object can also be used to store derived measures and indices from the patch, such as means, standard deviations, etc.

The AiTLAS NLD dataset [63] consists of Sentinel 2 data, resampled at 10-day intervals in the periods of March–November of 2017, 2018, and 2019 (resulting in 28 distinct dates for each year). We use images from 10 m and 20 m bands which contain eight spectral bands: B3, B4, B5, B6, B7, B8, B11, and B12. Additionally, the dataset contains three calculated indices: NDVI (normalized difference vegetation index), NDWI (normalized difference water index), and brightness (euclidean norm). Each polygon observation describes the temporal profile of a crop field and is associated with multivariate time series obtained by averaging the reflectance values at a crop-field level extracted from the Sentinel 2 data. In this way, each polygon is represented as a two-dimensional vector: the first dimension represents the (11) spectral bands, and the second one the (28) time steps. The crop type data categorization was created from the Land Parcel Identification System (LPIS). Each crop type label describes one crop field (or parcel), which in turn is identified with a polygon border. Figure 12 shows an example of the different field geometry present in the data.

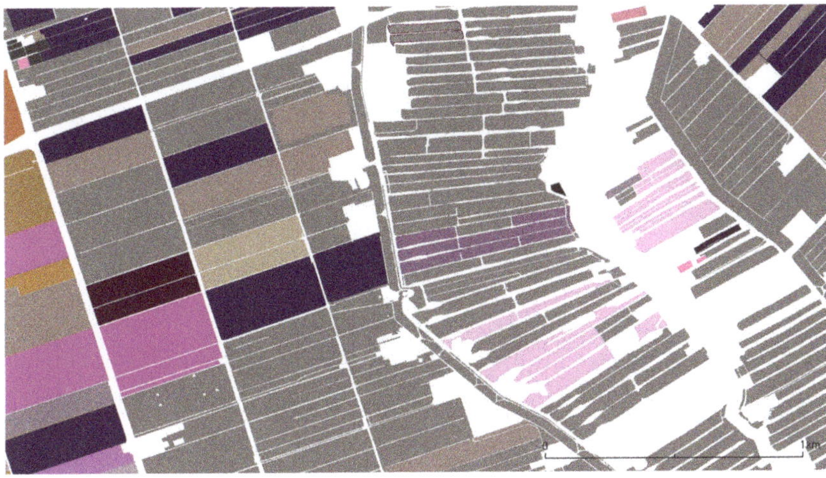

Figure 12. Example field geometries for the AiTLAS NLD dataset. The different colors of the polygons in the sample represent different crop types.

The loading configuration (shown on Listing 11) of such data includes mapping the path to the data, an index file (a separate csv file) which contains the class mappings of each polygon/patch, as well as the train/validation/test data splits (in terms of regions) for running the ML pipeline. Users can also specify other attributes for working with datasets, such as batch size, data shuffling, and the number of workers.

Listing 11. Loading the AiTLAS NLD dataset.

```
dataset_config = {
"root": "/home/user/data/CropTypeNetherlands/2019/",
"csv_file_path": "index.csv",
"batch_size": 128,
"shuffle": True,
"num_workers": 4,
"regions":["train", "test", "val",],
}
dataset = EOPatchCrops(dataset_config)
```

3.4.2. Definition and Execution of Machine Learning Tasks

AiTLAS casts the task of crop type prediction as a multi-class classification task, defined with the *BaseMulticlassClassifier* class, within the `aitlas.base` module. To illustrate the use of AiTLAS for crop type prediction, we use an LSTM model [87]. Specifically, we perform experiments on the AiTLAS NLD dataset independently for each of the three years. The goal is to examine the models' behavior and performance and how it varies across the years. For measuring the models' predictive performance, we use standard metrics such as accuracy, weighted F1 score, and Kappa coefficient. In the context of AiTLAS, initializing and creating the training routine is straightforward: one needs to set the model parameters in the configuration object, instantiate the model, and run the training sequence (as shown in Listing 12).

Listing 12. Creating an instance of an LSTM model and executing model training.

```
epochs = 100
model_directory = "./experiments/LSTM"
model_config = {
"input_dim":11,
"num_classes": 10,
"learning_rate": 0.001,
"dropout" : 0.2,
"weight_decay": 0.0001,
"metrics":["accuracy","f1_score", "kappa"] }
model = LSTM(model_config)
model.prepare()
model.train_and_evaluate_model(
train_dataset=train_dataset,
epochs=epochs,
model_directory=model_directory,
val_dataset=validation_dataset,
run_id='1',)
```

Once the model is trained, one can evaluate the model in a predictive setting with unseen data (Listing 13). In this example, the trained LSTM model shows consistent performance (across the three datasets/years) with accuracy in the range of ~84–85% and a weighted F1 score in the range ~82–84%. AiTLAS allows for more fine-grained per-label analysis of the model performance. For instance, in this example, the F1 score for *Temporary grasses and grazings* in 2017 is 45.23, while in 2018 raises to 59.34. Such insights can further help diagnose and improve the model's performance. Finally, AiTLAS also supports qualitative prediction analysis through visualizations of the predicted regions (and their labels). Further details of this use case, including complete results of the experiments, are given in Appendix F.

Listing 13. Evaluating a trained LSTM model.

```
labels = ["Permanent grassland", "Temporary grasses and grazings", "Green maize",
    "Potatoes (including seed potatoes)",
"Common winter wheat and spelt", "Sugar beet (excluding seed)", "Other farmland",
    "Onions",
"Flowers and ornamental plants (excluding nurseries)", "Spring barley",]

test_dataset_config = {
"batch_size": 32,
"shuffle": False,
"num_workers": 4,
"root": "/home/user/data/CropTypeNetherlands/2019/",
"csv_file_path": "index.csv",
"regions":["test", ],}

test_dataset = EOPatchCrops(test_dataset_config)
y_true, y_pred, y_prob = model.predict(dataset=test_dataset,)

eopatches_path = "/home/user/data/CropTypeNetherlands/2019/eopatches/"
patch = "eopatch_7495"
```

3.5. Adding a New Machine Learning Model in AiTLAS

In the use cases discussed previously, we showcased the capability of AiTLAS with the already available methods and model architectures. However, AiTLAS is modular and easily extensible to new approaches. Here we present a template for adding novel model architectures into the AiTLAS toolbox. This includes instantiating from one of the model base classes from the `aitlas.base` module, which corresponds to a particular EO task. For instance, adding a new model for image scene classification, such as *EfficientNetV2* as implemented in the PyTorch [66] model catalog. The model can be added by creating an inherited class from *BaseMulticlassClassifier*, which already implements all the requirements to train a deep-learning model successfully.

Namely, the implementation (e.g., Listing 14) includes initialization of the model variable from the base class and overriding the forward function. In the `init` function, the model variable is initialized using the `efficientnet_v2_m` function, which constructs the EfficientNetV2-M architecture introduced by Tan and Le [123]. Users can also include pre-trained model variants by setting the variable 'pretrained', which will lead to setting a pre-trained model from the ImageNet-1K dataset. The newly added *EfficientNetV2* model can then be used for remote sensing image scene classification in a similar manner as the use case presented in Section 3.1.

Listing 14. Adding EfficientNetV2 as a new model for remote sensing image scene classification in the AiTLAS toolbox.

```
import torchvision.models as models
import torch.nn as nn

from ..base import BaseMulticlassClassifier, BaseMultilabelClassifier

class EfficientNetV2(BaseMulticlassClassifier):
name = "EfficientNetV2"

def __init__(self, config):
super().__init__(config)
if self.config.pretrained:
self.model = models.efficientnet_v2_m(
weights=models.EfficientNet_V2_M_Weights.IMAGENET1K_V1,
progress=False
)
in_features = self.model.classifier[1].in_features
self.model.classifier[1] = nn.Linear(
in_features, self.config.num_classes
)
else:
self.model = models.efficientnet_v2_m(
weights=None, progress=False,
num_classes=self.config.num_classes
)

def forward(self, x):
return self.model(x)
```

4. Conclusions

We present AiTLAS (https://aitlas.bvlabs.ai (accessed on 8 March 2023)), an open-source, state-of-the-art toolbox for exploratory and predictive analysis of satellite imagery. AiTLAS is a versatile Python library applicable to a variety of different tasks from EO, such as image scene classification, image segmentation, object detection, and crop type prediction (time series classification) tasks. It provides the means for straightforward construction and execution of complete end-to-end EO pipelines catering to users' needs with different goals, domain backgrounds, and levels of expertise.

From an EO perspective, where users typically focus on a particular application, AiTLAS supports building complete data analysis pipelines starting from data preparation and understanding of the data, through leveraging state-of-the-art deep-learning architectures for various predictive tasks, to quantitative and qualitative analysis of the predicted outcomes. As such, the capabilities of AiTLAS expand significantly in comparison to other related libraries such as eo-learn [23], OTB [26], and CANDELA [28]. Namely, in addition to data handling, AiTLAS implements approaches for model learning, be they training models from scratch or employing/fine-tuning pre-trained models freely available via the AiTLAS model catalog. Moreover, AiTLAS provides an additional, more comprehensible, configuration layer for executing EO data-analysis pipelines that do not require significant familiarity with the underlining machine-learning technologies. We believe this capability substantially flattens the learning curve for constructing and executing novel EO data analysis pipelines. From an AI perspective, AiTLAS implements the necessary components for implementing novel methods for EO data analysis, both in terms of (novel) deep-learning architectures as well as approaches for data handling/transformation. Moreover, it further facilitates the development of new approaches by providing easy access to formalized AI-ready data (via the AiTLAS EO data catalog) and their evaluation via standardized and extensive benchmark framework [3].

The main motivation of AiTLAS is bringing together the AI and EO communities by providing extensible, easy-to-use, and, more importantly, open-source resources for EO data analysis. The design principles of AiTLAS, showcased in this work, build on this motivation by providing:

1. *User-friendly, accessible, and interoperable* resources for data analysis through easily configurable and readily usable pipelines. The resources can be easily adapted by adjusting the configuration (JSON) files to a specific task at hand.
2. *Standardized, verifiable, and reusable* data handling, wrangling, and pre-processing approaches for constructing AI-ready data. The AiTLAS datasets are readily available for use through the toolbox and accessible through its EO data catalog (http://eodata.bvlabs.ai) , which incorporates FAIR [124] ontology-based semantic (meta) data descriptions of AI-ready datasets;
3. *Modular and configurable* modeling approaches and (pre-trained) models. The implemented approaches can be easily adjusted to different setups and novel analysis pipelines. Moreover, AiTLAS includes the most extensive open-source catalog of pre-trained EO deep-learning models (currently with more than 500 models) that have been pre-trained on a variety of different datasets and are readily available for practical applications.
4. *Standardized and reproducible* benchmark protocols (for data and models) that are essential for developing trustworthy, reproducible, and reusable resources. AiTLAS provides the resources and the necessary mechanisms for reconciling these protocols across tasks, models (and model configurations), and processing details of the datasets being used.

The development of AiTLAS is an ongoing effort. We foresee its evolution in three primary directions. First, it will continue to keep pace with the growing body of resources developed by the AI4EO community, continuously extending the catalogs of available AI-ready EO datasets and novel mode architectures. Second, and more application focused, we will continue the development of templates for specific use cases and pipelines tailored for different domains (such as agriculture, urban planning, geology, archaeology, etc.). Third, we will focus on extending the AiTLAS capabilities with self-supervised learning (SSL) approaches as a response to various practical challenges akin to costly and tedious labeling processes of large amounts of unlabeled data [125]. We believe that such extensions will bring many practical benefits in different downstream applications [99,126–128], which will undoubtedly further increase the utility of AiTLAS.

Author Contributions: Conceptualization, I.D., N.S. and D.K.; methodology, I.D., I.K. N.S. and D.K.; software, I.D. and I.K.; validation, I.D., I.K. and N.S.; formal analysis, I.D., I.K., N.S. and D.K.; investigation, I.D., I.K., N.S. and D.K.; resources, I.D., I.K., N.S.; data curation, A.K. and P.P. ; writing—original draft preparation, I.D., I.K. and D.K.; writing—review and editing, A.K., P.P., N.S. ; visualization, I.D. and N.S.; supervision, D.K.; project administration, D.K.; funding acquisition, N.S., P.P. and D.K. All authors have read and agreed to the published version of the manuscript.

Funding: The AiTLAS toolbox is developed within the grant from the European Space Agency (ESRIN): AiTLAS–Artificial Intelligence toolbox for Earth Observation (ESA RFP/3-16371/19/I-NB) awarded to Bias Variance Labs, d.o.o.

Data Availability Statement: AiTLAS toolbox is available at https://aitlas.bvlabs.ai (accessed on 8 March 2023), together with the accompanying results and the Jupyter notebooks from the use cases. The datasets used in the use cases are described and available at the AiTLAS EO data repository http://eodata.bvlabs.ai (accessed on 8 March 2023). Further benchmark experiments, including pre-trained models for image scene classification, are available at https://github.com/biasvariancelabs/aitlas-arena (accessed on 8 March 2023).

Acknowledgments: We thank Žiga Kokalj, Sveinung Loekken, Sara Aparicio, Bertrand Le Saux, Elena Merdjanovska, Stefan Popov, Stefan Kalabakov and Sašo Džeroski—for their useful feedback and support during the development of AiTLAS. We also thank Sofija Dimitrovska for her thoughtful feedback on the manuscript.

Conflicts of Interest: The authors declare no conflicts of interest.

Appendix A. Third-Party Dependencies

Table A1. Third-party libraries used by AiTLAS described with their scope of usage and purpose within our toolbox.

Library	Scope of Usage	Purpose
PyTorch Vision [66]	`aitlas.models`	Pre-built deep-learning model architectures.
PyTorch Image Models [72]	`aitlas.models`	Pre-built deep-learning model architectures.
Segmentation Models PyTorch [79]	`aitlas.models`	Pre-built deep-learning model architectures specifically for segmentation tasks.
Albumentations [129]	`aitlas.transforms`	Applying transormations and augmentations.
NumPy [130]	`aitlas.base`	General scientific computing
Scikit-learn [131]	`aitlas.base`	Used for metric calculations
Scikit-multilearn [132]	`aitlas.tasks`	Used for stratified dataset splitting.
Seaborn [133]	`aitlas.visualizations`	For visualizations.
Matplotlib [133]	`aitlas.visualizations`	For visualizations
TensorBoard [27]	`aitlas.base`	Enable logging the train/validation loss during model training as well as any other supported metrics. This then allows for those statistics to be visualized in the *TensorBoard* UI.
zipp [134]	`aitlas.utils`	Enables working with zip files.
dill [135]	`aitlas.utils`	Extends Python's pickle module for serializing and de-serializing Python objects to the majority of the built-in data types.
lmdb [136]	`aitlas.datasets`	Enables working with LMDB data.
tifffile [137]	`aitlas.utils`	Reads TIFF files from any storage.
h5py [138]	`aitlas.datasets`	Provides an interface to the HDF5 binary data format.
click [139]	`aitlas.base`	Enables creating command line interfaces.
munch [140]	`aitlas.base`	Provides attribute-style access to objects.
marshmallow [141]	`aitlas.base`	It is an ORM/ODM/framework-agnostic library for converting complex data types to and from native Python data types.
Pytoch Metrics [142]	`aitlas.metrics`	Utility library used for computing performance metrics.

Appendix B. Split Task within the AiTLAS Toolbox for Creating Train, Validation and Test Splits

When constructing a data analysis pipeline, it is common to split the available data into separate sets/splits that can be used for model training, validation, and testing. While the training split is used for training the ML model, the validation split is used for estimating the optimal hyper-parameters of the model. The test split is then used to evaluate the model's performance on new, unseen data. In the context of remote sensing image scene classification, the AiTLAS toolbox provides an interface for performing the splitting task to generate these data splits. The task is very convenient as most of the datasets are published without these splits, and it is up to the practitioner to define the splits.

The split task in the AiTLAS toolbox supports sampling strategies for generating the splits: random and stratified. The latter method ensures that the distribution of the target/class variable(s) is the same among the different splits [143]. The configuration file that can be used to run the AiTLAS toolbox and obtain the required splits for a given dataset is provided in Listing A1.

Listing A1. Configuration file for creating train, validation and test splits.

```
{
"task": {
"classname": "aitlas.tasks.StratifiedSplitTask",
"config": {
"split": {
"train": {
"ratio": 60,
"file": "./data/CLRS/train.csv"
},
"val": {
"ratio": 20,
"file": "./data/CLRS/val.csv"
},
"test": {
"ratio": 20,
"file": "./data/CLRS/test.csv"
}
},
"data_dir": "./data/CLRS"
}
}
}
```

An example csv file generated using the split task is given in Listing A2.

Listing A2. Example csv file with the required format by the AiTLAS toolbox for remote sensing image scene classification datasets.

```
airport/airport_321_Level1_0.53m.tif,airport
storage-tank/storage-tank_489_Level1_0.50m.tif,storage-tank
golf-course/golf-course_374_Level3_2.05m.tif,golf-course
parking/parking_561_Level1_0.39m.tif,parking
residential/residential_77_Level1_0.49m.tif,residential
meadow/meadow_5_Level3_1.37m.tif,meadow
mountain/mountain_538_Level3_7.60m.tif,mountain
mountain/mountain_180_Level2_3.66m.tif,mountain
```

In the configuration file, the split task has to be set. In Listing A1, in order to perform a stratified splitting of the data, we have chosen the `aitlas.tasks.StratifiedSplitTask`. Then, we include the path to the folder in which the images are stored. Additionally, for the MLC dataset, a csv file containing one-hot encoded classes or labels must be provided alongside the images. However, for the MCC dataset, a csv file is not necessary since the images for different classes or labels are located in separate sub-folders within the main or root folder. We use ratios to generate the splits required for training, validation, and testing, used to determine the proportion of images in each split. Based on these ratios, a split-task automatically generates a csv file that contains the image names for each split. Additional configuration files for different datasets can be found in the AiTLAS repository (https://github.com/biasvariancelabs/aitlas/blob/master/examples/ (accessed on 8 March 2023)).

Appendix C. Remote Sensing Image Scene Classification

A Jupyter Notebook for demonstrating remote sensing image scene classification is available in the AiTLAS repository (https://github.com/biasvariancelabs/aitlas/blob/master/examples/multiclass_classification_example_clrs.ipynb (accessed on 8 March 2023)). The notebook contains a step-by-step sample code for running an image scene classification task. More specifically, we demonstrate the standard steps on how to load and split the data and examine its label distribution. It also showcases the steps for training and evaluation of the model and visualizing the prediction.

We use the CLRS multi-class dataset, which consists of 25 land cover classes. To obtain the dataset, it first has to be downloaded from the repository (https://github.com/lehaifeng/CLRS (accessed on 8 March 2023) and unzipped, after which we obtain the data organized in subfolders containing images for each label/class. However, since the

CLRS dataset does not come with predefined train, validation, and test splits, the split task defined within the AiTLAS toolbox is used to generate the splits. More details about the split task can be found in Appendix B.

Next, we create an instance of the CLRSDataset class from the toolbox. To create the instance, we provide the folder with the images and a csv file with a list of images and labels. Additionally, we set the batch size for the data loader, the shuffle parameter to reshuffle the data at each epoch, and we set the number of workers/sub-processes to load the data. Listing A3 provides a code snippet illustrating how to create the instance. The created instance loads the image data from the dataset and offers additional functionalities for inspection and visualization.

Listing A3. Load a dataset using the AiTLAS toolbox, inspect images and calculate the class distribution.

```
dataset_config = {
"data_dir": "/datasets/CLRS",
"csv_file": "/datasets/CLRS/train.csv",
"batch_size": 128,
"shuffle": True,
"num_workers": 4
}
dataset = CLRSDataset(dataset_config)
fig = dataset.show_image(340)
dataset.data_distribution_table()
```

We can use the `show_image` function to display images from the dataset. To inspect the label distribution in the dataset, we can use the `data_distribution_table` function. The class distribution of the train, test, and validation splits can be calculated and presented as shown in Figure A1.

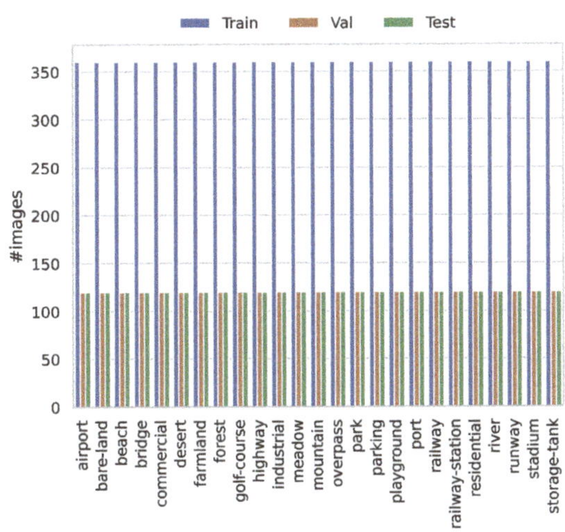

Figure A1. Class distribution for the CLRS dataset across the training, validation, and testing splits of the data.

In the next step, we continue specifying the model learning task by providing the training and validation data and a deep-learning model. The code snippet for creating train and validation datasets is given in Listing A4. Additionally, Listing A4 shows an example of specifying the data transformation parameter. Transformations are used to process the data to make it suitable for training. They can also augment the data to represent a more comprehensive set of possible data points, resulting in better performance and model generalization to address overfitting. Here we give an example of applying the *ResizeRan-*

domCropFlipHVToTensor on the training data, which first resizes all the images to 256 × 256, selects a random crop of size 224 × 224, and then applies random horizontal and/or vertical flips. On the test data, we apply the *ResizeCenterCropToTensor* transformation, which resizes the images to 256 × 256 and then applies a central crop of size 224 × 224.

Listing A4. Load the training and validation dataset.

```
train_dataset_config = {
"batch_size": 128,
"shuffle": True,
"data_dir": "/datasets/CLRS",
"csv_file": "/datasets/CLRS/train.csv",
}

train_dataset = CLRSDataset(train_dataset_config)
train_dataset.transform = ResizeRandomCropFlipHVToTensor()

validation_dataset_config = {
"batch_size": 128,
"shuffle": False,
"data_dir": "/datasets/CLRS",
"csv_file": "/datasets/CLRS/val.csv",
"transforms": ["aitlas.transforms.ResizeCenterCropToTensor"]
}

validation_dataset = CLRSDataset(validation_dataset_config)
```

For learning, we use the Vision Transformer model, available in the toolbox. We configure the model by setting several configuration parameters, i.e., the number of classes/labels, the learning rate, and the evaluation metrics. To use the pre-trained variant of the Vision Transformer (pre-trained on the ImageNet-1k dataset) in the configuration object, we also set the pre-trained parameter to *true*. We fine-tune the model on the CLRS dataset by calling the function `train_and_evaluate_model`. The code snippet for instantiating the model and running the training sequence is given in Listing A5.

Listing A5. Creating a model and start of model training.

```
epochs = 100
model_directory = "/experiments/CLRS"
model_config = {
"num_classes": 25,
"learning_rate": 0.0001,
"pretrained": True,
"metrics": ["accuracy", "precision", "recall", "f1_score"]
}
model = VisionTransformer(model_config)
model.prepare()
model.train_and_evaluate_model(
train_dataset=train_dataset,
epochs=epochs,
model_directory=model_directory,
val_dataset=validation_dataset,
run_id='1',
)
```

We use *ReduceLROnPlateau* as a learning scheduler—it reduces the learning rate when the loss has stopped improving. Namely, models often benefit from reducing the learning rate by a factor once learning stagnates: ReduceLROnPlateau tracks the values of the loss measure, reducing the learning rate by a given factor when there is no improvement for a certain number of epochs (denoted as 'patience'). In our experiments, we track the value of the validation loss with patience set to 5 and a reduction factor set to 0.1 (the new learning rate will thus be $lr * factor$). The maximum number of epochs is set to 100. We also apply early stopping criteria if no improvements in the validation loss are observed over 10 epochs. The best checkpoint/model found (with the lowest validation loss) is saved and then applied to the test part to obtain the final assessment of the predictive performance.

To evaluate the learned model, we load the test split of the dataset, set the list of evaluation metrics, and run the evaluation sequence on the test data (Listing A6). The predictive performance of the models for MCC is typically assessed by reporting the *top-n accuracy* score (typically n is set to 1 or 5) [65]. This score is calculated by checking whether the correct label is placed among the n most probable labels outputted by the model. In this use case, we report *top-1 accuracy*, denoted as 'Accuracy', Macro Precision, Weighted Precision, Macro Recall, Weighted Recall, Macro F1 score, and Weighted F1 score. Note that since for MCC tasks, the micro-averaged measures such as F1 score, Micro Precision, and Micro Recall have values equal to accuracy, we do not report them separately.

Listing A6. Testing the model using the images from the test split.

```
test_dataset_config = {
"batch_size": 128,
"shuffle": False,
"data_dir": "/dataset/CLRS",
"csv_file": "/dataset/CLRS/test.csv",
"transforms": ["aitlas.transforms.ResizeCenterCropToTensor"]
}

test_dataset = CLRSDataset(test_dataset_config)
model_path = "best_checkpoint.pth.tar"
model.metrics = ["accuracy", "precision", "recall", "f1_score"]
model.running_metrics.reset()
model.evaluate(dataset=test_dataset, model_path=model_path)
model.running_metrics.get_scores(model.metrics)
```

The results from the ViT models trained from scratch, pre-trained on ImageNet-1K, and then fine-tuned on the specific dataset are given in Table A2. From the presented results, it is evident that leveraging pre-trained models can lead to significant performance improvements on image classification tasks [144], and in particular on tasks in EO domains [145]. The results also show the average training time per epoch, the total training time, and the epoch in which the lowest value for the validation loss has been obtained.

Table A2. Detailed results for the ViT models trained on the CLRS dataset.

Model/Metric	Accuracy	Macro Precision	Weighted Precision	Macro Recall	Weighted Recall	Macro F1 score	Weighted F1 Score	Avg. Time/Epoch (s)	Total Time (s)	Best Epoch
Trained from scratch	65.47	66.41	66.41	65.47	65.47	65.49	65.49	24.96	1173	32
Pre-trained on ImageNet-1K	93.20	93.29	93.29	93.20	93.20	93.22	93.22	25.32	785	21

The performance of the ViT model for the individual classes from the CLRS dataset can be analyzed from the results presented in Table A3 and Figure 7. Table A3 gives the precision, recall, and F1 score at a class level for the pre-trained ViT model on the CLRS dataset, and Figure 7 shows the confusion matrix. We can note that the F1 score of most classes is over 90%. The confusion matrix shows the effect of class similarity for the classes 'railway' and 'railway-station' to the overall performance—the model has difficulty discerning between these two classes.

Table A3. Per class results for the pre-trained Vision Transformer model on the CLRS dataset.

Label	Precision	Recall	F1 Score
airport	97.48	96.67	97.07
bare-land	92.00	95.83	93.88
beach	99.15	97.50	98.32
bridge	90.91	91.67	91.29
commercial	79.84	85.83	82.73
desert	97.50	97.50	97.50
farmland	93.70	99.17	96.36
forest	100.00	100.00	100.00
golf-course	94.96	94.17	94.56
highway	92.11	87.50	89.74
industrial	88.79	85.83	87.29
meadow	96.72	98.33	97.52
mountain	99.15	97.50	98.32
overpass	89.68	94.17	91.87
park	85.60	89.17	87.35
parking	98.25	93.33	95.73
playground	95.04	95.83	95.44
port	94.74	90.00	92.31
railway	86.29	89.17	87.70
railway-station	88.79	85.83	87.29
residential	90.68	89.17	89.92
river	90.32	93.33	91.80
runway	98.33	98.33	98.33
stadium	95.61	90.83	93.16
storage-tank	96.55	93.33	94.92

Additionally, the learned model can be used for predicting the labels of unseen images using the `predict_image` function. The function takes the image, the labels, and the transformation instance as arguments and returns the predicted class and the confidence score of the prediction for the given image. The output of this function is shown in Figure A2. The code snippet to obtain the prediction for a given image is shown in Listing A7.

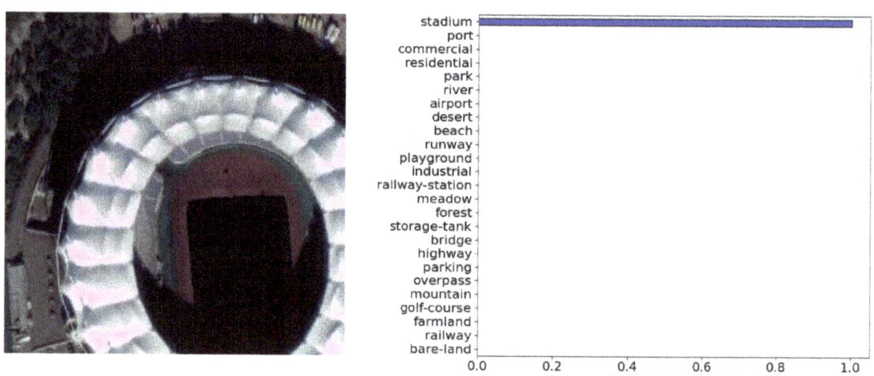

Figure A2. Example image with the predicted class and probability using the Vision Transformer model.

Listing A7. Getting predictions for images from external source.

```
labels = ["airport", "bare-land", "beach", "bridge", "commercial", "desert", "
    farmland", "forest", "golf-course", "highway", "industrial", "meadow", "
    mountain", "overpass", "park", "parking", "playground", "port", "railway", "
    railway-station", "residential", "river", "runway", "stadium", "storage-tank"
    ]
transform = ResizeCenterCropToTensor()
image = image_loader('/images/image1.tif')
fig = model.predict_image(image, labels, transform)
```

Appendix D. Semantic Segmentation of Remote Sensing Images

A Jupyter Notebook for demonstrating the task of semantic segmentation of remote sensing images is available in the AiTLAS repository (https://github.com/biasvariancelabs/aitlas/blob/master/examples/semantic_segmentation_example_landcover_ai.ipynb (accessed on 8 March 2023)). The notebook provides the code needed for loading a semantic segmentation dataset and examining the distribution of the pixels across the semantic labels/classes. The notebook also provides instructions on how to train and evaluate the DeepLabv3 model and how to use the learned model for running predictions on unseen data.

In this example, we use the LandCover.ai image classification dataset. To obtain the dataset, we download it from the repository (https://landcover.ai.linuxpolska.com/download/landcover.ai.v1.zip (accessed on 8 March 2023)) and unzip it, after which we obtain folders with images and masks, as well as separate files containing information about the train, validation, and test splits.

Next, we create an instance of the class *LandCoverAiDataset* available in the toolbox. To create the instance, we provide the folder with the images and masks and a file that contains the list of images to be loaded. The class *LandCoverAiDataset* has all the functionalities to load the images and masks for processing and training. Additionally, in the data configuration object, we set the batch size, the shuffle parameter, and the number of workers for loading the data (see Listing A8).

Listing A8. Load the LandCoverAiDataset dataset from the AiTLAS toolbox, inspect images and masks, and calculate the pixel distribution across the labels.

```
dataset_config = {
"data_dir": "/dataset/landcoverai/images",
"csv_file": "/dataset/landcoverai/train.txt",
"batch_size": 128,
"shuffle": True,
"num_workers": 4
}
dataset = LandCoverAiDataset(dataset_config)
dataset.show_image(2000);
dataset.data_distribution_table()
dataset.data_distribution_barchart();
```

We use the `show_image` function to display images from the dataset. In the displayed image, the different semantic regions in the mask are color coded with the colors from the class definition of the dataset. To inspect the distribution of the pixels across the labels in the dataset, we use the `data_distribution_table` and/or `data_distribution_barchart` function.

We continue with the model learning task by creating the training and validation datasets and training the deep-learning model. The code snippet for creating train and validation datasets is given in Listing A9. Additionally, Listing A9 shows an example of specifying the data transformation parameter. In the AiTLAS toolbox, three different transformations are available: transformation on the images, transformations on the targets (i.e., labels, masks, or bounding boxes), and joint transformations that are simultaneously applied on the input images and the targets. For example, in the case of semantic segmentation, if the input image is horizontally flipped, the mask should also be flipped. During training, in this use case, we perform *data augmentation* by random horizontal and/or vertical flips on the input images and the masks. For the input images, min-max normalization is applied, and the targets/masks are transposed. During evaluation/testing, we do not use joint transformations.

Listing A9. Load the training and validation dataset for semantic segmentation.

```
train_dataset_config = {
"batch_size": 16,
"shuffle": True,
"csv_file": "/dataset/landcoverai/train.txt",
"data_dir": "/dataset/landcoverai/images",
"joint_transforms": ["aitlas.transforms.FlipHVRandomRotate"],
"transforms": ["aitlas.transforms.MinMaxNormTranspose"],
"target_transforms": ["aitlas.transforms.Transpose"]
}
train_dataset = LandCoverAiDataset(train_dataset_config)

validation_dataset_config = {
"batch_size": 16,
"shuffle": False,
"csv_file": "../dataset/landcoverai/val.txt",
"data_dir": "../dataset/landcoverai/images",
"transforms": ["aitlas.transforms.MinMaxNormTranspose"],
"target_transforms": ["aitlas.transforms.Transpose"]
}
validation_dataset = LandCoverAiDataset(validation_dataset_config)
```

We use the DeepLabv3 model, which is available in the AiTLAS toolbox. We configure the model by setting several configuration parameters, i.e., the number of classes/labels, the learning rate, the pretraining mode, and the evaluation metrics. The code snippet for instantiating the model and running the training sequence is given in Listing A10.

Listing A10. Creating a DeepLabv3 model and starting the training.

```
epochs = 100
model_directory = "/experiments/landcoverai"
model_config = {
"num_classes": 5,
"learning_rate": 0.0001,
"pretrained": True,
"threshold": 0.5,
"metrics": ["iou"]
}

model = DeepLabV3(model_config)
model.prepare()
model.train_and_evaluate_model(
train_dataset=train_dataset,
val_dataset=validation_dataset,
epochs=epochs,
model_directory=model_directory,
run_id='1'
)
```

We train two variants of DeepLabV3: (i) model "trained from scratch" using only the dataset at hand and initialized with random weights at the start of the training procedure, and (2) model with pre-trained weights on a subset of the COCO dataset, using only the 20 categories that are also present in the Pascal VOC dataset [66] and then fine-tuned on the dataset at hand. The DeepLabv3 model is trained or fine-tuned on the train set of images, including *data augmentation* operations such as random horizontal and/or vertical flips and random rotations. Using the validation set of images, we perform a hyper-parameter search over different values for the learning rate: 0.01, 0.001, and 0.0001. We use fixed values for some of the hyper-parameters: batch size is set to 16, *Adam optimizer* [146] without weight decay, and *ReduceLROnPlateau* as a learning scheduler which reduces the learning rate when the loss has stopped improving (same as for the image scene classification task as discussed in Section 3.1.2). Furthermore, to prevent over-fitting, we perform early stopping on the validation set—the model with the lowest validation loss is saved and then applied on the original test set to obtain the estimate of the model's predictive performance. Finally, we use mean intersection over union ($mIoU$) as an evaluation measure: $mIoU$ is a widely used measure for semantic segmentation and is calculated as the average of intersection over union (IoU) across all labels. Note that IoU is defined as the area of the overlap between the ground truth and predicted label divided by the area of their union (a value of 0 for IoU means no overlap, while a value of 1 means complete overlap).

The train and validation learning curves are shown in Figure A3. After the 50 epoch, the value of the loss function is very stable, and the model starts to over-fit. The stop criteria prevent over-fitting. At this point, we save the model from the 49th epoch (with the lowest validation loss) and evaluate its performance on the test set.

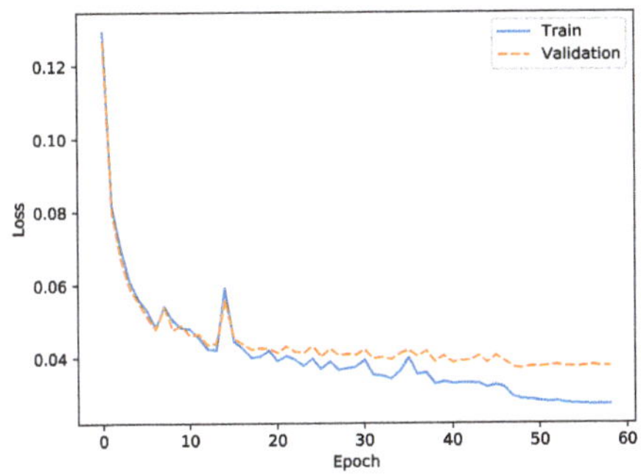

Figure A3. Train and validation learning curves for the DeepLabv3 model pre-trained on COCO and fine-tuned on the LandCover.ai dataset.

To evaluate the model, we load the test split of the dataset, set the path to the learned model, and set the list of evaluation metrics. In this example, we use IoU as an evaluation measure. We then proceed with running the evaluation sequence on the test data (see Listing A11).

Listing A11. Testing the model using the images from the test split.

```
test_dataset_config = {
"batch_size": 4,
"shuffle": False,
"data_dir": "/dataset/landcoverai/images",
"csv_file": "/dataset/landcoverai/test.txt",
"transforms": ["aitlas.transforms.MinMaxNormTranspose"],
"target_transforms": ["aitlas.transforms.Transpose"]
}

test_dataset = LandCoverAiDataset(test_dataset_config)
model_path = "/experiments/landcoverai/best_checkpoint.pth.tar"
model.metrics = ["iou"]
model.running_metrics.reset()
model.evaluate(dataset=test_dataset, model_path=model_path)
model.running_metrics.get_scores(model.metrics)
```

The results summarizing the segmentation performance of the DeepLabV3 model on the LandCover.ai dataset are given in Table A4. The DeepLabv3 model trained from scratch yields a $mIoU$ score of 86.027%, while the pre-trained one yields a score of 86.093%. Hence, there is no clear benefit of using the pre-trained weights for this particular dataset because the results in the different setups are very similar. The values of the IoU for the labels 'Road' and 'Building' are lower compared to the other labels. These labels are more challenging in the context of semantic segmentation because they are usually narrow, in the case of roads, or often small, in the case of buildings. Additionally, they are sometimes obscured by other objects, such as trees.

Table A4. Label-wise $IoU(\%)$ and $mIoU$ for the DeepLabv3 models trained on the LandCover.ai dataset.

Model/Label	Background	Buildings	Woodlands	Water	Road	mIoU	Avg. Time/Epoch (s)	Total Time (s)	Best Epoch
Trained from scratch	93.813	80.304	91.952	94.877	69.190	86.027	299.241	16159	44
Pre-trained on COCO	93.857	80.650	91.964	95.145	68.846	86.093	300.58	17734	49

Additionally, the learned model can be used for predicting the segmentation masks of unseen images by using the `predict_masks` function. The function takes the image, the labels, and the transformation instance as arguments and returns separate masks for each semantic label for the given image. An example output of this function is shown in Figure A4. The code snippet to obtain the prediction for a given image is shown in Listing A12.

Listing A12. Getting predictions for images from external source.

```
labels = ["Background", "Buildings", "Woodlands", "Water", "Road"]
transform = MinMaxNormTranspose()
model_path = "/experiments/landcoverai/best_checkpoint.pth.tar"
model.load_model(model_path)
image = image_loader('/images/image1.png')
fig = model.predict_masks(image, labels, transform)
```

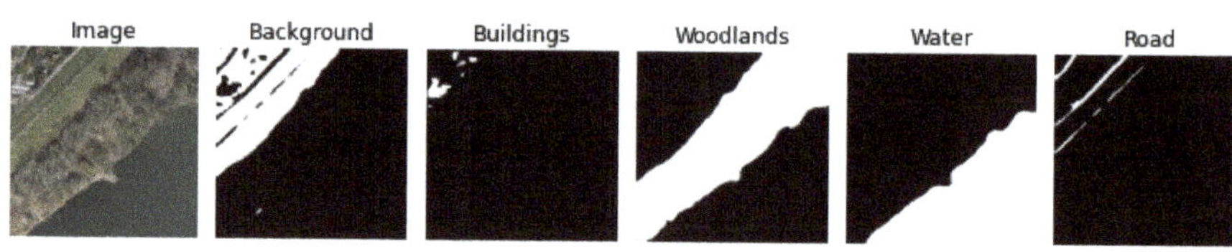

Figure A4. Example image with the predicted masks for each semantic label using the DeepLabv3 model.

Appendix E. Remote Sensing Image Object Detection

An example Jupyter Notebook for object detection in remote sensing images is available at the AiTLAS toolbox (https://github.com/biasvariancelabs/aitlas/blob/master/examples/object_detection_example_hrrsd.ipynb (accessed on 8 March 2023)). The notebook provides step-by-step code on how to load an object detection dataset and inspect the number of object instances for each object category. The notebook also demonstrates how to train and validate the Faster R-CNN model on the predefined test split, as well as how to use the learned model for inference on unseen data.

To prepare the HRRSD object detection dataset, we need to download the data from the repository (https://github.com/CrazyStoneonRoad/TGRS-HRRSD-Dataset (accessed on 8 March 2023)) and unzip it. This generates a folder with images in .jpg format, annotations in .xml format, and separate files containing information about the train, validation, and test splits. The Pascal VOC object annotation format is used in this example. Thus, we create an instance of the *ObjectDetectionPascalDataset* class defined within the AiTLAS toolbox. To instantiate the class, we specify the path to the folders containing the images and annotations and the file containing the list of images. We also set the batch size for the data loader during training, the shuffle parameter, and the number of workers to specify the number of sub-processes used for data loading (see Listing A13).

Listing A13. Load a dataset using the AiTLAS toolbox, inspect images and calculate the number of object instances for each category/label.

```
dataset_config = {
"image_dir": "/datasets/HRRSD/images",
"annotations_dir": "/datasets/HRRSD/annotations",
"imageset_file": "/datasets/HRRSD/train.txt"
"batch_size": 16,
"shuffle": True,
"num_workers": 4
}
dataset = ObjectDetectionPascalDataset(dataset_config)
dataset.show_image(2458);
dataset.show_batch(15);
dataset.data_distribution_table()
dataset.data_distribution_barchart();
```

We use the `show_image` function to display images from the dataset. The function displays the image with the bounding boxes and labels for each object present in the image. The function `show_batch` displays a batch of images with the bounding boxes of the objects in the image. The statistics for the number of object instances for each category/label in the dataset can be calculated using the functions `data_distribution_table` or `data_distribution_barchart`.

In the next step, we continue with the model learning part. First, we create the configuration objects for the training and validation data and specify the data transformation method we wish to be applied. For this use case, we apply the joint transformation *ResizeToTensorV2* to resize the images to a resolution of 480 × 480 pixels and convert them to tensors. This transformation also resizes the bounding boxes for the object to fit the new resolution of the images. After defining the data configuration objects, we load the data (see Listing A14).

Listing A14. Load the training and validation dataset.

```
train_dataset_config = {
"image_dir": "/datasets/HRRSD/images",
"annotations_dir": "/datasets/HRRSD/annotations",
"imageset_file": "/datasets/HRRSD/train.txt",
"joint_transforms": ["aitlas.transforms.ResizeToTensorV2"]
"batch_size": 16,
"shuffle": True,
"num_workers": 4
}
train_dataset = ObjectDetectionPascalDataset(train_dataset_config)

validation_dataset_config = {
"image_dir": "/datasets/HRRSD/images",
"annotations_dir": "/datasets/HRRSD/annotations",
"imageset_file": "/datasets/HRRSD/val.txt",
"joint_transforms": ["aitlas.transforms.ResizeToTensorV2"]
"batch_size": 16,
"shuffle": True,
"num_workers": 4
}
validation_dataset = ObjectDetectionPascalDataset(validation_dataset_config)
```

We train two variants of the model: (i) model "trained from scratch" using only the dataset at hand and initialized with random weights at the start of the training procedure, and (2) model with pre-trained weights on the COCO dataset [66] and then fine-tuned on the dataset at hand. The Faster R-CNN model is trained or fine-tuned using the training part of the images, with parameters selection/search performed using the validation part. Namely, we search over different values for the learning rate: 0.01, 0.001, and 0.0001. We use fixed values for some of the hyper-parameters: batch size is set to 16, *Adam optimizer* [146] without weight decay, and *ReduceLROnPlateau* as a learning scheduler which reduces the learning rate when the loss has stopped improving.

Furthermore, to prevent over-fitting, we perform early stopping on the validation set—the model with the best value for the evaluation measure is saved and then applied to the original test set to obtain the estimate of the model's predictive performance. Finally, as an evaluation measure, we use mean Average Precision (mAP) as defined in the Pascal VOC Challenge [121]. Average Precision (AP) is computed as the average precision value taken at recall values ranging from 0 to 1 (i.e., the area under the precision/recall curve). mAP is the average of AP over all classes. Next, IoU is crucial in determining true positives and false positives, and its threshold is set to 0.5. More details on the evaluation measures for object detection are provided in [121,142].

Next, we use the Faster R-CNN model, which is implemented in the AiTLAS toolbox. To configure the model, create a model configuration object and set several configuration parameters, i.e., the number of classes/labels, the learning rate, the pretraining mode, and the evaluation metrics. The code snippet for instantiating the model and running the training sequence is given in Listing A15.

Listing A15. Creating a Faster R-CNN model and start of model training.

```
epochs = 100
model_directory = "/experiments/hrrsd"
model_config = {
"num_classes": 14,
"learning_rate": 0.0001,
"pretrained": True,
"threshold": 0.5,
"metrics": ["map"]
}
model = FasterRCNN(model_config)
model.prepare()
model.train_and_evaluate_model(
train_dataset=train_dataset,
val_dataset=validation_dataset,
epochs=epochs,
model_directory=model_directory,
run_id='1'
)
```

Table A5 summarizes the results of the object detection task. The Faster R-CNN model trained from scratch yields 77.412% of mAP, while the pre-trained model yields 81.436% of mAP, as estimated using the test dataset. The results show the clear benefit of using the pre-trained weights for this particular dataset. The most challenging objects to detect are 'Crossroad' and 'T Junction' (due to the similarity of both objects).

Table A5. Mean average precision (mAP%) of the Faster R-CNN models trained from scratch and pre-trained for the HRRSD dataset.

Label	Faster R-CNN (Pretrained)	Faster R-CNN
Airplane	96.86	94.71
Baseball Diamond	79.75	80.13
Basketball Court	59.25	44.96
Bridge	82.22	78.55
Crossroad	77.06	71.78
Ground Track Field	95.62	92.84
Harbor	89.00	88.61
Parking Lot	53.80	51.22
Ship	86.61	78.50
Storage Tank	93.56	89.67
T Junction	66.83	69.12
Tennis Court	87.97	79.31
Vehicle	90.14	86.96
Mean AP	81.436	77.412
Avg. time / epoch (s)	221.96	244.63
Total time (s)	5993	10030
Best epoch	17	31

To evaluate the model, we create a configuration object for the training split of the HRRSD data, load the data, set the path to the trained model, and run the evaluation process (see Listing A16).

Listing A16. Testing the model using the images from the test split.

```
test_dataset_config = {
"batch_size": 4,
"shuffle": False,
"image_dir": "/datasets/HRRSD/images",
"annotations_dir": "/datasets/HRRSD/annotations",
"imageset_file": "/datasets/HRRSD/test.txt",
"joint_transforms": ["aitlas.transforms.ResizeToTensorV2"]
}

test_dataset = ObjectDetectionPascalDataset(test_dataset_config)
model_path = "/experiments/hrrsd/best_checkpoint.pth.tar"
model.metrics = ["map"]
model.running_metrics.reset()
model.evaluate(dataset=test_dataset, model_path=model_path)
model.running_metrics.get_scores(model.metrics)
```

Additionally, the model can be used for predicting the bounding boxes and labels for the objects in new, unseen images from an external source by using the `predict_objects` function. The function takes the image, the labels, and the transformation instance as arguments and returns an image with the bounding boxes and labels for the detected object in the image. Figure A5 shows an example output of this function. The code snippet to obtain the prediction for a given image is shown in Listing A17.

Listing A17. Getting bounding boxes and labels for images from external source.

```
labels = [None, 'T junction', 'airplane', 'baseball diamond', 'basketball court',
'bridge', 'crossroad', 'ground track field', 'harbor', 'parking lot',
'ship', 'storage tank', 'tennis court', 'vehicle']
transform = Resize()
model.load_model(model_path)
image = image_loader('../data/HRRSD/JPEGImages/00042.jpg')
fig = model.detect_objects(image, labels, transform)
```

Figure A5. Example image with the predicted bounding boxes and labels for the objects using the Faster R-CNN model.

Appendix F. Crop Type Prediction Using Satellite Time Series Data

We have added a Jupyter Notebook (https://github.com/biasvariancelabs/aitlas/blob/master/examples/crop_type_prediction_example_netherlands.ipynb (accessed on 8

March 2023)) in the AiTLAS toolbox that contains a step-by-step sample code for running through a crop type prediction task. As previously, we will demonstrate the standard steps to load and inspect the data, load, configure, train, and evaluate the models, and visualize predictions.

We use the AiTLAS NLD dataset for the year 2019. First, we set the configuration for the dataset and initiate with the `EOPatchCrops` class, which is a wrapper for working with `EOPatches`. We need to set the path to the root folder containing the patches. The patches are also folders containing detailed EO data. We need to specify the index, which is a `csv` file containing the class mappings for the polygons and patches they belong to, as well as the split (train, validation, or test). We also specify the standard PyTorch attributes for working with datasets, such as batch size, shuffle, and number of workers. The `regions` configurations specify which splits we should load from the index. Sample code is shown in Listing A18.

Listing A18. Load the dataset.

```
dataset_config = {
"root": "/home/user/data/CropTypeNetherlands/2019/",
"csv_file_path": "index.csv",
"batch_size": 128,
"shuffle": True,
"num_workers": 4,
"regions":["train", "test", "val",],
}
dataset = EOPatchCrops(dataset_config)
```

To display the time series values of the bands of a specific polygon, we can use the function `show_timeseries(index)`. Furthermore, to show sample data from the underlying index, we can use the `show_samples()` function (Listing A19).

Listing A19. Inspect the dataset.

```
fig = dataset.show_timeseries(0)
dataset.show_samples()
```

An example of the data representation for a selected polygon labeled as permanent grassland is illustrated in Figure A6.

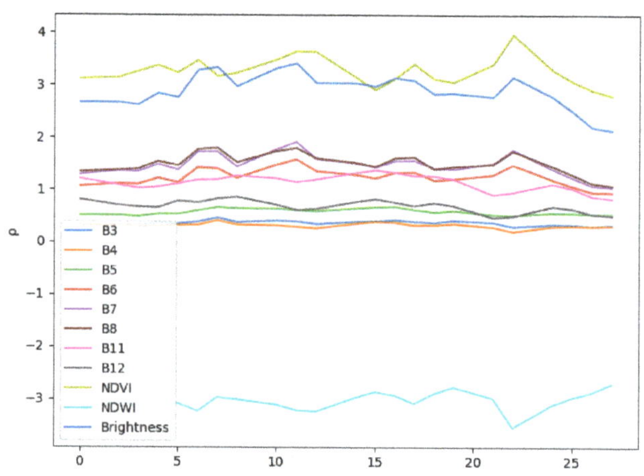

Figure A6. Examples of the input time series of reflectances ρ for the spectral bands of the Sentinel 2 satellite and calculated indices for the crop type *Permanent grassland*.

Once, we have a sense of the data, we load the train and validation splits of the dataset (Listing A20).

Listing A20. Load train and validation the dataset.

```
train_dataset_config = {
"root": "/home/user/data/CropTypeNetherlands/2019/",
"csv_file_path": "index.csv",
"batch_size": 128,
"shuffle": True,
"num_workers": 4,
"regions":["train", ],
}

train_dataset = EOPatchCrops(train_dataset_config)

validation_dataset_config = {
"batch_size": 32,
"shuffle": False,
"num_workers": 4,
"root": "/home/user/data/CropTypeNetherlands/2019/",
"csv_file_path": "index.csv",
"regions":["val", ],
}

validation_dataset = EOPatchCrops(validation_dataset_config)
```

In the next step, we need to initialize and create the model. For this task, we will use the LSTM model supported by the toolbox. We set the model parameters in the configuration object, instantiate the model, and run the training sequence (Listing A21).

Listing A21. Train the model.

```
epochs = 100
model_directory = "./experiments/LSTM"
model_config = {
"input_dim":11,
"num_classes": 10,
"learning_rate": 0.001,
"dropout" : 0.2,
"weight_decay": 0.0001,
"metrics":["accuracy","f1_score", "kappa"]
}
model = LSTM(model_config)
model.prepare()
model.train_and_evaluate_model(
train_dataset=train_dataset,
epochs=epochs,
model_directory=model_directory,
val_dataset=validation_dataset,
run_id='1',
)
```

We can use the model to run predictions and visualize them. We can load the test split of the dataset. Then, use the model to run predictions on that dataset (Listing A22). Finally, we can pick a sample patch and visualize its predicted polygons (Figure A7).

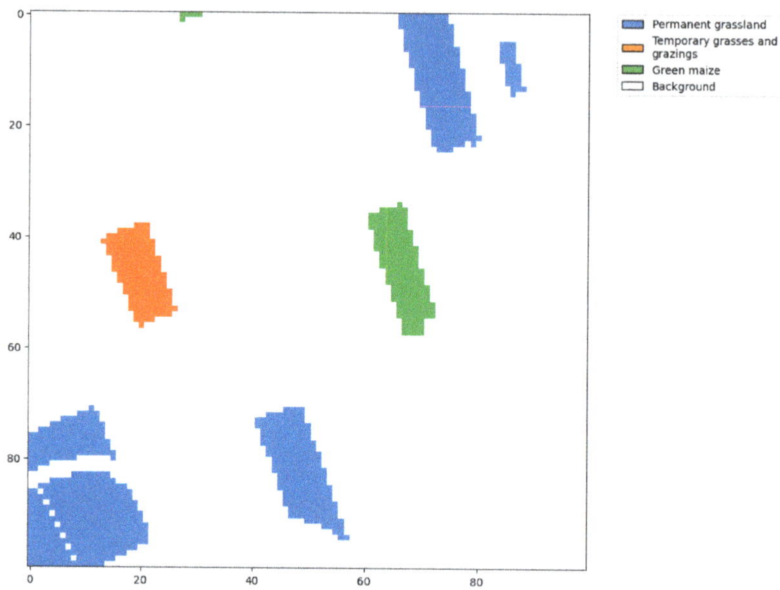

Figure A7. Example patch with the predicted polygons using the LSTM model.

Listing A22. Test the model.

```
labels = ["Permanent grassland", "Temporary grasses and grazings", "Green maize",
    "Potatoes (including seed potatoes)",
"Common winter wheat and spelt", "Sugar beet (excluding seed)", "Other farmland",
    "Onions",
"Flowers and ornamental plants (excluding nurseries)", "Spring barley",]

test_dataset_config = {
"batch_size": 32,
"shuffle": False,
"num_workers": 4,
"root": "/home/user/data/CropTypeNetherlands/2019/",
"csv_file_path": "index.csv",
"regions":["test", ],
}

test_dataset = EOPatchCrops(test_dataset_config)

y_true, y_pred, y_prob = model.predict(dataset=test_dataset,)

eopatches_path = "/home/user/data/CropTypeNetherlands/2019/eopatches/"
patch = "eopatch_7495"

fig = display_eopatch_predictions(
eopatches_path,
patch,
y_pred,
test_dataset.index,
y_true,
test_dataset.mapping,
)
```

The results summarizing the predictive performance on the AiTLAS NLD dataset across the different years are presented in Table A6. Using LSTM, we obtain a weighted F1 score in the range of ∼82–84% for the different years.

Table A6. Results for the crop type prediction on the AiTLAS NLD dataset using the LSTM model.

Dataset Year	Accuracy	Weighted F1 Score	Kappa
2017	84.28	82.48	77.22
2018	84.49	84.32	78.79
2019	85.25	84.10	79.55

We also present the per class F1 scores across all years in Table A7. We can note that the per-class performance is generally consistent between different years. The F1 score does not change significantly for any class across the different years. The only exception is the *Temporary grasses and grazings* class, where for 2017, the F1 score is 45.23, while for 2018, it is 59.34, which is around a 14% difference. The performance for this class is also the lowest compared to the other classes. The model performs best on the classes *Green maize, Common winter wheat and spelt, Potatoes (including seed potatoes),* and *Sugar beet (excluding seed).* For these classes, the F1 score is consistent across the different years with score ~95%. We provide the resources and explanations on the execution of this use case, including additional visualizations and application of the model to external images in Appendix F.

Table A7. Per class F1 score for the crop type prediction on the AiTLAS NLD dataset for every year with LSTM

Crop Type	2017	2018	2019
Permanent grassland	86.72	85.78	86.82
Temporary grasses and grazings	45.23	59.34	53.66
Green maize	96.10	95.09	96.14
Potatoes (including seed potatoes)	95.37	94.84	95.45
Common winter wheat and spelt	94.86	95.75	96.28
Sugar beet (excluding seed)	95.62	92.36	95.28
Other farmland	62.45	61.81	60.05
Onions	89.76	93.68	92.51
Flowers and ornamental plants (excluding nurseries)	83.18	78.30	79.67
Spring barley	92.30	91.12	90.80

References

1. Christopherson, J.; Chandra, S.N.R.; Quanbeck, J.Q. *2019 Joint Agency Commercial Imagery Evaluation—Land Remote Sensing Satellite Compendium;* Technical Report; US Geological Survey: Reston, VA, USA, 2019.
2. Tupin, F.; Inglada, J.; Nicolas, J.M. *Remote Sensing Imagery;* John Wiley & Sons: Hoboken, NJ, USA, 2014.
3. Dimitrovski, I.; Kitanovski, I.; Kocev, D.; Simidjievski, N. Current trends in deep learning for Earth Observation: An open-source benchmark arena for image classification. *ISPRS J. Photogramm. Remote Sens.* **2023**, *197*, 18–35. [CrossRef]
4. Cheng, G.; Han, J.; Lu, X. Remote Sensing Image Scene Classification: Benchmark and State of the Art. *Proc. IEEE* **2017**, *105*, 1865–1883. [CrossRef]
5. Tang, L.; Shao, G. Drone remote sensing for forestry research and practices. *J. For. Res.* **2015**, *26*, 791–797. [CrossRef]
6. Roy, P.; Ranganath, B.; Diwakar, P.; Vohra, T.; Bhan, S.; Singh, I.; Pandian, V. Tropical forest typo mapping and monitoring using remote sensing. *Remote Sens.* **1991**, *12*, 2205–2225. [CrossRef]
7. Sunar, F.; Özkan, C. Forest fire analysis with remote sensing data. *Int. J. Remote Sens.* **2001**, *22*, 2265–2277. [CrossRef]
8. Poursanidis, D.; Chrysoulakis, N. Remote Sensing, natural hazards and the contribution of ESA Sentinels missions. *Remote Sens. Appl. Soc. Environ.* **2017**, *6*, 25–38. [CrossRef]
9. Sishodia, R.P.; Ray, R.L.; Singh, S.K. Applications of remote sensing in precision agriculture: A review. *Remote Sens.* **2020**, *12*, 3136. [CrossRef]
10. Cox, H.; Kelly, K.; Yetter, L. Using remote sensing and geospatial technology for climate change education. *J. Geosci. Educ.* **2014**, *62*, 609–620. [CrossRef]
11. Collis, R.T.; Creasey, D.; Grasty, R.; Hartl, P.; deLoor, G.; Russel, P.; Salerno, A.; Schaper, P. *Remote Sensing for Environmental Sciences;* Springer Science & Business Media: Berlin, Germany, 2012; Volume 18.
12. Christie, G.; Fendley, N.; Wilson, J.; Mukherjee, R. Functional Map of the World. In Proceedings of the 2018 IEEE/CVF Conference on Computer Vision and Pattern Recognition, Salt Lake City, UT, USA, 18–23 June 2018; pp. 6172–6180. [CrossRef]

13. Sumbul, G.; de Wall, A.; Kreuziger, T.; Marcelino, F.; Costa, H.; Benevides, P.; Caetano, M.; Demir, B.; Markl, V. BigEarthNet-MM: A Large-Scale, Multimodal, Multilabel Benchmark Archive for Remote Sensing Image Classification and Retrieval [Software and Data Sets]. *IEEE Geosci. Remote Sens. Mag.* **2021**, *9*, 174–180. [CrossRef]
14. Long, Y.; Xia, G.S.; Li, S.; Yang, W.; Yang, M.Y.; Zhu, X.X.; Zhang, L.; Li, D. On Creating Benchmark Dataset for Aerial Image Interpretation: Reviews, Guidances and Million-AID. *IEEE J. Sel. Top. Appl. Earth Obs. Remote Sens.* **2021**, *14*, 4205–4230. [CrossRef]
15. Bastani, F.; Wolters, P.; Gupta, R.; Ferdinando, J.; Kembhavi, A. Satlas: A Large-Scale, Multi-Task Dataset for Remote Sensing Image Understanding. *arXiv* **2022**, arXiv:2211.15660. https://doi.org/10.48550/ARXIV.2211.15660.
16. Neupane, B.; Horanont, T.; Aryal, J. Deep Learning-Based Semantic Segmentation of Urban Features in Satellite Images: A Review and Meta-Analysis. *Remote Sens.* **2021**, *13*, 808. [CrossRef]
17. Xia, G.S.; Bai, X.; Ding, J.; Zhu, Z.; Belongie, S.; Luo, J.; Datcu, M.; Pelillo, M.; Zhang, L. DOTA: A Large-Scale Dataset for Object Detection in Aerial Images. In Proceedings of the IEEE Conference on Computer Vision and Pattern Recognition (CVPR), Salt Lake City, UT, USA, 18–23 June 2018.
18. Pearlman, J.; Barry, P.; Segal, C.; Shepanski, J.; Beiso, D.; Carman, S. Hyperion, a space-based imaging spectrometer. *IEEE Trans. Geosci. Remote Sens.* **2003**, *41*, 1160–1173. [CrossRef]
19. King, M.; Herring, D. SATELLITES | Research (Atmospheric Science). In *Encyclopedia of Atmospheric Sciences*; Holton, J.R., Ed.; Academic Press: Oxford, UK, 2003; pp. 2038–2047. . [CrossRef]
20. Osco, L.P.; Marcato Junior, J.; Marques Ramos, A.P.; de Castro Jorge, L.A.; Fatholahi, S.N.; de Andrade Silva, J.; Matsubara, E.T.; Pistori, H.; Gonçalves, W.N.; Li, J. A review on deep learning in UAV remote sensing. *Int. J. Appl. Earth Obs. Geoinf.* **2021**, *102*, 102456. [CrossRef]
21. Roy, D.; Wulder, M.; Loveland, T.; C.E., W.; Allen, R.; Anderson, M.; Helder, D.; Irons, J.; Johnson, D.; Kennedy, R.; et al. Landsat-8: Science and product vision for terrestrial global change research. *Remote Sens. Environ.* **2014**, *145*, 154–172. [CrossRef]
22. Drusch, M.; Del Bello, U.; Carlier, S.; Colin, O.; Fernandez, V.; Gascon, F.; Hoersch, B.; Isola, C.; Laberinti, P.; Martimort, P.; et al. Sentinel-2: ESA's Optical High-Resolution Mission for GMES Operational Services. *Remote Sens. Environ.* **2012**, *120*, 25–36.
23. Sentinel Hub: Cloud API For Satellite Imagery. 2023. Available online: https://www.sentinel-hub.com/ (accessed on 8 March 2023).
24. UP42: Simplified Access to Geospatial Data and Processing. 2023. Available online: https://up42.com/ (accessed on 8 March 2023).
25. De Vroey, M.; Radoux, J.; Zavagli, M.; De Vendictis, L.; Heymans, D.; Bontemps, S.; Defourny, P. Performance Assessment of the Sen4CAP Mowing Detection Algorithm on a Large Reference Data Set of Managed Grasslands. In Proceedings of the 2021 IEEE International Geoscience and Remote Sensing Symposium IGARSS, Brussels, Belgium, 11–16 July 2021; pp. 743–746. [CrossRef]
26. Grizonnet, M.; Michel, J.; Poughon, V.; Inglada, J.; Savinaud, M.; Cresson, R. Orfeo ToolBox: open source processing of remote sensing images. *Open Geospat. Data Softw. Stand.* **2017**, *2*, 15. [CrossRef]
27. Abadi, M.; Agarwal, A.; Barham, P.; Brevdo, E.; Chen, Z.; Citro, C.; Corrado, G.S.; Davis, A.; Dean, J.; Devin, M.; et al. TensorFlow: Large-Scale Machine Learning on Heterogeneous Systems. *arXiv* **2015**, arXiv:1603.04467.
28. Rolland, J.F.o.; Castel, F.; Haugommard, A.; Aubrun, M.; Yao, W.; Corneliu.; Dumitru, O.; Datcu, M.; Bylicki, M.; Tran, B.H.; et al. Candela: A Cloud Platform for Copernicus Earth Observation Data Analytics. In Proceedings of the IGARSS 2020—2020 IEEE International Geoscience and Remote Sensing Symposium, Waikoloa, HI, USA, 26 September–2 October 2020; pp. 3104–3107. [CrossRef]
29. Stewart, A.J.; Robinson, C.; Corley, I.A.; Ortiz, A.; Lavista Ferres, J.M.; Banerjee, A. TorchGeo: Deep Learning with Geospatial Data. In Proceedings of the SIGSPATIAL '22: 30th International Conference on Advances in Geographic Information Systems; Association for Computing Machinery: Seattle, WA, USA, 2022; pp. 1–12. [CrossRef]
30. Paszke, A.; Gross, S.; Massa, F.; Lerer, A.; Bradbury, J.; Chanan, G.; Killeen, T.; Lin, Z.; Gimelshein, N.; Antiga, L.; et al. PyTorch: An Imperative Style, High-Performance Deep Learning Library. In *Advances in Neural Information Processing Systems 32*; Curran Associates, Inc.: Red Hook, NY, USA, 2019; pp. 8024–8035.
31. Chaudhuri, B.; Demir, B.; Chaudhuri, S.; Bruzzone, L. Multilabel Remote Sensing Image Retrieval Using a Semisupervised Graph-Theoretic Method. *IEEE Trans. Geosci. Remote Sens.* **2018**, *56*, 1144–1158. [CrossRef]
32. Tsoumakas, G.; Katakis, I. Multi-Label Classification: An Overview. *Int. J. Data Warehous. Min.* **2009**, *3*, 1–13. [CrossRef]
33. Yang, Y.; Newsam, S. Bag-of-Visual-Words and Spatial Extensions for Land-Use Classification. In Proceedings of the GIS '10: 18th SIGSPATIAL International Conference on Advances in Geographic Information Systems, San Jose, CA, USA, 2–5 November 2010; Association for Computing Machinery: New York, NY, USA, 2010; pp. 270–279.
34. Xia, G.S.; Yang, W.; Delon, J.; Gousseau, Y.; Sun, H.; Maître, H. Structural High-resolution Satellite Image Indexing. *Int. Arch. Photogramm. Remote Sens. Spat. Inf. Sci. ISPRS Arch.* **2010**, *38*, 1–6.
35. Xia, G.S.; Hu, J.; Hu, F.; Shi, B.; Bai, X.; Zhong, Y.; Zhang, L.; Lu, X. AID: A Benchmark Data Set for Performance Evaluation of Aerial Scene Classification. *IEEE Trans. Geosci. Remote Sens.* **2017**, *55*, 3965–3981. [CrossRef]
36. Helber, P.; Bischke, B.; Dengel, A.; Borth, D. Eurosat: A novel dataset and deep learning benchmark for land use and land cover classification. *IEEE J. Sel. Top. Appl. Earth Obs. Remote Sens.* **2019**, *12*, 2217–2226. [CrossRef]
37. Zhou, W.; Newsam, S.; Li, C.; Shao, Z. PatternNet: A benchmark dataset for performance evaluation of remote sensing image retrieval. *ISPRS J. Photogramm. Remote Sens.* **2018**, *145*, 197–209. [CrossRef]
38. Li, H.; Dou, X.; Tao, C.; Wu, Z.; Chen, J.; Peng, J.; Deng, M.; Zhao, L. RSI-CB: A Large-Scale Remote Sensing Image Classification Benchmark Using Crowdsourced Data. *Sensors* **2020**, *20*, 1594. [CrossRef] [PubMed]
39. Zou, Q.; Ni, L.; Zhang, T.; Wang, Q. Deep Learning Based Feature Selection for Remote Sensing Scene Classification. *IEEE Geosci. Remote Sens. Lett.* **2015**, *12*, 2321–2325. [CrossRef]

40. Basu, S.; Ganguly, S.; Mukhopadhyay, S.; DiBiano, R.; Karki, M.; Nemani, R. DeepSat: A Learning Framework for Satellite Imagery. In *SIGSPATIAL '15: Proceedings of the 23rd SIGSPATIAL International Conference on Advances in Geographic Information Systems*; Association for Computing Machinery: New York, NY, USA, 2015.
41. Zhu, Q.; Zhong, Y.; Zhao, B.; Xia, G.S.; Zhang, L. Bag-of-Visual-Words Scene Classifier with Local and Global Features for High Spatial Resolution Remote Sensing Imagery. *IEEE Geosci. Remote Sens. Lett.* **2016**, *13*, 747–751. [CrossRef]
42. Li, H.; Jiang, H.; Gu, X.; Peng, J.; Li, W.; Hong, L.; Tao, C. CLRS: Continual Learning Benchmark for Remote Sensing Image Scene Classification. *Sensors* **2020**, *20*, 1226. [CrossRef]
43. Long, Y.; Gong, Y.; Xiao, Z.; Liu, Q. Accurate Object Localization in Remote Sensing Images Based on Convolutional Neural Networks. *IEEE Trans. Geosci. Remote Sens.* **2017**, *55*, 2486–2498. [CrossRef]
44. Wang, Q.; Liu, S.; Chanussot, J.; Li, X. Scene Classification With Recurrent Attention of VHR Remote Sensing Images. *IEEE Trans. Geosci. Remote Sens.* **2019**, *57*, 1155–1167. [CrossRef]
45. Penatti, O.A.; Nogueira, K.; Dos Santos, J.A. Do deep features generalize from everyday objects to remote sensing and aerial scenes domains? In Proceedings of the IEEE Conference on Computer Vision and Pattern Recognition Workshops, Boston, MA, USA, 7–12 June 2015; pp. 44–51.
46. Zhu, X.X.; Hu, J.; Qiu, C.; Shi, Y.; Kang, J.; Mou, L.; Bagheri, H.; Haberle, M.; Hua, Y.; Huang, R.; et al. So2Sat LCZ42: A Benchmark Data Set for the Classification of Global Local Climate Zones [Software and Data Sets]. *IEEE Geosci. Remote Sens. Mag.* **2020**, *8*, 76–89. [CrossRef]
47. Qi, X.; Zhu, P.; Wang, Y.; Zhang, L.; Peng, J.; Wu, M.; Chen, J.; Zhao, X.; Zang, N.; Mathiopoulos, P.T. MLRSNet: A multi-label high spatial resolution remote sensing dataset for semantic scene understanding. *ISPRS J. Photogramm. Remote Sens.* **2020**, *169*, 337–350. [CrossRef]
48. Hua, Y.; Mou, L.; Zhu, X.X. Recurrently exploring class-wise attention in a hybrid convolutional and bidirectional LSTM network for multi-label aerial image classification. *ISPRS J. Photogramm. Remote Sens.* **2019**, *149*, 188–199. [CrossRef]
49. Sumbul, G.; Charfuelan, M.; Demir, B.; Markl, V. Bigearthnet: A Large-Scale Benchmark Archive for Remote Sensing Image Understanding. In Proceedings of the IGARSS 2019—2019 IEEE International Geoscience and Remote Sensing Symposium Yokohama, Japan, 28 July–2 August 2019; pp. 5901–5904.
50. Hua, Y.; Mou, L.; Zhu, X.X. Relation Network for Multilabel Aerial Image Classification. *IEEE Trans. Geosci. Remote Sens.* **2020**, *58*, 4558–4572. [CrossRef]
51. Kaggle. Planet: Understanding the Amazon from Space. 2022. Available online: https://www.kaggle.com/c/planet-understanding-the-amazon-from-space (accessed on 8 March 2023).
52. Zhang, Y.; Yuan, Y.; Feng, Y.; Lu, X. Hierarchical and Robust Convolutional Neural Network for Very High-Resolution Remote Sensing Object Detection. *IEEE Trans. Geosci. Remote Sens.* **2019**, *57*, 5535–5548. [CrossRef]
53. Ding, J.; Xue, N.; Xia, G.S.; Bai, X.; Yang, W.; Yang, M.Y.; Belongie, S.; Luo, J.; Datcu, M.; Pelillo, M.; et al. Object Detection in Aerial Images: A Large-Scale Benchmark and Challenges. *IEEE Trans. Pattern Anal. Mach. Intell.* **2022**, *44*, 7778–7796. [CrossRef]
54. Li, K.; Wan, G.; Cheng, G.; Meng, L.; Han, J. Object detection in optical remote sensing images: A survey and a new benchmark. *ISPRS J. Photogramm. Remote Sens.* **2020**, *159*, 296–307. . [CrossRef]
55. Cheng, G.; Han, J.; Zhou, P.; Guo, L. Multi-class geospatial object detection and geographic image classification based on collection of part detectors. *ISPRS J. Photogramm. Remote Sens.* **2014**, *98*, 119–132. [CrossRef]
56. Haroon, M.; Shahzad, M.; Fraz, M.M. Multisized object detection using spaceborne optical imagery. *IEEE J. Sel. Top. Appl. Earth Obs. Remote Sens.* **2020**, *13*, 3032–3046. [CrossRef]
57. Mnih, V. Machine Learning for Aerial Image Labeling. Ph.D. Thesis, University of Toronto, Toronto, ON, Canada, 2013.
58. Boguszewski, A.; Batorski, D.; Ziemba-Jankowska, N.; Dziedzic, T.; Zambrzycka, A. LandCover.ai: Dataset for Automatic Mapping of Buildings, Woodlands, Water and Roads from Aerial Imagery. In Proceedings of the IEEE/CVF Conference on Computer Vision and Pattern Recognition (CVPR) Workshops, Nashville, TN, USA, 19–25 June 2021; pp. 1102–1110.
59. Maggiori, E.; Tarabalka, Y.; Charpiat, G.; Alliez, P. Can Semantic Labeling Methods Generalize to Any City? The Inria Aerial Image Labeling Benchmark. In Proceedings of the IEEE International Geoscience and Remote Sensing Symposium (IGARSS), Fort Worth, TX, USA, 23–28 July 2017.
60. Chen, Q.; Wang, L.; Wu, Y.; Wu, G.; Guo, Z.; Waslander, S.L. Aerial imagery for roof segmentation: A large-scale dataset towards automatic mapping of buildings. *ISPRS J. Photogramm. Remote Sens.* **2019**, *147*, 42–55. [CrossRef]
61. Bragagnolo, L.; da Silva, R.V.; Grzybowski, J.M.V. Amazon Rainforest Dataset for Semantic Segmentation. 2019. Available online: https://zenodo.org/record/3233081#.ZENXKc5ByUk (accessed on 8 March 2023).
62. Kocev, D.; Simidjievski, N.; Kostovska, A.; Dimitrovski, I.; Kokalj, Z. Discover the Mysteries of the Maya: Selected Contributions from the Machine Learning Challenge: The Discovery Challenge Workshop at ECML PKDD 2021. *arXiv* **2022**, arXiv:2208.03163. https://doi.org/10.48550/arXiv.2208.03163.
63. Merdjanovska, E.; Kitanovski, I.; Kokalj, Ž.; Dimitrovski, I.; Kocev, D. Crop Type Prediction Across Countries and Years: Slovenia, Denmark and the Netherlands. In Proceedings of the IGARSS 2022—2022 IEEE International Geoscience and Remote Sensing Symposium, Kuala Lumpur, Malaysia, 17–22 July 2022; pp. 5945–5948.
64. Rußwurm, M.; Lefèvre, S.; Körner, M. Breizhcrops: A satellite time series dataset for crop type identification. In Proceedings of the International Conference on Machine Learning Time Series Workshop, Long Beach, CA, USA, 9–15 June 2019; Volume 3.

65. Krizhevsky, A.; Sutskever, I.; Hinton, G.E. Imagenet classification with deep convolutional neural networks. *Adv. Neural Inf. Process. Syst.* **2012**, *25*, 1097–1105. [CrossRef]
66. Marcel, S.; Rodriguez, Y. Torchvision the machine-vision package of torch. In Proceedings of the 18th ACM International Conference on Multimedia, Firenze, Italy, 25–29 October 2010; pp. 1485–1488.
67. Wang, J.; Yang, Y.; Mao, J.; Huang, Z.; Huang, C.; Xu, W. CNN-RNN: A Unified Framework for Multi-label Image Classification. In Proceedings of the 2016 IEEE Conference on Computer Vision and Pattern Recognition (CVPR), Las Vegas, NV, USA, 27–30 June 2016; IEEE Computer Society: Los Alamitos, CA, USA, 2016; pp. 2285–2294.
68. Liu, Z.; Mao, H.; Wu, C.Y.; Feichtenhofer, C.; Darrell, T.; Xie, S. A ConvNet for the 2020s. *arXiv* **2022**, arXiv:2201.03545.
69. Huang, G.; Liu, Z.; Van Der Maaten, L.; Weinberger, K.Q. Densely connected convolutional networks. In Proceedings of the IEEE Conference on Computer Vision and Pattern Recognition, Honolulu, HI, USA, 21–26 July 2017; pp. 4700–4708.
70. Tan, M.; Le, Q. Efficientnet: Rethinking model scaling for convolutional neural networks. In Proceedings of the International Conference on Machine Learning (PMLR), Long Beach, CA, USA, 10–15 June 2019; pp. 6105–6114.
71. Tolstikhin, I.O.; Houlsby, N.; Kolesnikov, A.; Beyer, L.; Zhai, X.; Unterthiner, T.; Yung, J.; Steiner, A.; Keysers, D.; Uszkoreit, J.; et al. MLP-mixer: An all-mlp architecture for vision. *Adv. Neural Inf. Process. Syst.* **2021**, *34*, 24261–24272.
72. Wightman, R. PyTorch Image Models. 2019. Available online: https://github.com/rwightman/pytorch-image-models (accessed on 8 March 2023).
73. He, K.; Zhang, X.; Ren, S.; Sun, J. Deep residual learning for image recognition. In Proceedings of the IEEE Conference on Computer Vision and Pattern Recognition, Las Vegas, NV, USA, 27–30 June 2016; pp. 770–778.
74. Liu, Z.; Hu, H.; Lin, Y.; Yao, Z.; Xie, Z.; Wei, Y.; Ning, J.; Cao, Y.; Zhang, Z.; Dong, L.; et al. Swin Transformer V2: Scaling Up Capacity and Resolution. In Proceedings of the 2022 IEEE/CVF Conference on Computer Vision and Pattern Recognition (CVPR), New Orleans, LA, USA, 18–24 June 2022; pp. 11999–12009. [CrossRef]
75. Simonyan, K.; Zisserman, A. Very deep convolutional networks for large-scale image recognition. *arXiv* **2014**, arXiv:1409.1556.
76. Dosovitskiy, A.; Beyer, L.; Kolesnikov, A.; Weissenborn, D.; Zhai, X.; Unterthiner, T.; Dehghani, M.; Minderer, M.; Heigold, G.; Gelly, S.; et al. An image is worth 16x16 words: Transformers for image recognition at scale. *arXiv* **2020**, arXiv:2010.11929.
77. Chen, L.C.; Papandreou, G.; Schroff, F.; Adam, H. Rethinking Atrous Convolution for Semantic Image Segmentation. *arXiv* **2017**, arXiv:1706.05587.
78. Chen, L.C.; Zhu, Y.; Papandreou, G.; Schroff, F.; Adam, H. Encoder-Decoder with Atrous Separable Convolution for Semantic Image Segmentation. In *Computer Vision—ECCV 2018*; Ferrari, V., Hebert, M., Sminchisescu, C., Weiss, Y., Eds.; Springer International Publishing: Cham, Switzerland, 2018; pp. 833–851.
79. Iakubovskii, P. Segmentation Models Pytorch. 2019. Available online: https://github.com/qubvel/segmentation_models.pytorch (accessed on 8 March 2023).
80. Long, J.; Shelhamer, E.; Darrell, T. Fully convolutional networks for semantic segmentation. In Proceedings of the 2015 IEEE Conference on Computer Vision and Pattern Recognition (CVPR), Boston, MA, USA, 7–12 June 2015; pp. 3431–3440. [CrossRef]
81. Sun, K.; Zhao, Y.; Jiang, B.; Cheng, T.; Xiao, B.; Liu, D.; Mu, Y.; Wang, X.; Liu, W.; Wang, J. High-Resolution Representations for Labeling Pixels and Regions. *arXiv* **2019**, *arXiv:1904.04514*.
82. Ronneberger, O.; Fischer, P.; Brox, T. U-Net: Convolutional Networks for Biomedical Image Segmentation. In *Medical Image Computing and Computer-Assisted Intervention—MICCAI 2015*; Navab, N., Hornegger, J., Wells, W.M., Frangi, A.F., Eds.; Springer International Publishing: Cham, Switzerland, 2015; pp. 234–241.
83. Lin, T.; Goyal, P.; Girshick, R.; He, K.; Dollar, P. Focal Loss for Dense Object Detection. *IEEE Trans. Pattern Anal. Mach. Intell.* **2020**, *42*, 318–327. [CrossRef]
84. Li, Y.; Xie, S.; Chen, X.; Dollar, P.; He, K.; Girshick, R. Benchmarking Detection Transfer Learning with Vision Transformers. *arXiv* **2021**, arXiv:2111.11429. https://doi.org/10.48550/arXiv.2111.11429.
85. Ismail Fawaz, H.; Lucas, B.; Forestier, G.; Pelletier, C.; Schmidt, D.F.; Weber, J.; Webb, G.I.; Idoumghar, L.; Muller, P.A.; Petitjean, F. InceptionTime: Finding AlexNet for time series classification. *Data Min. Knowl. Discov.* **2020**, *34*, 1936–1962. [CrossRef]
86. Rußwurm, M.; Pelletier, C.; Zollner, M.; Lefèvre, S.; Körner, M. BreizhCrops: A Time Series Dataset for Crop Type Mapping. *arXiv* **2020**, arXiv:1905.11893.
87. Rußwurm, M.; Körner, M. Multi-Temporal Land Cover Classification with Sequential Recurrent Encoders. *ISPRS Int. J. Geo-Inf.* **2018**, *7*, 129. [CrossRef]
88. Wang, Z.; Yan, W.; Oates, T. Time series classification from scratch with deep neural networks: A strong baseline. In Proceedings of the 2017 International Joint Conference on Neural Networks (IJCNN), Anchorage, AK, USA, 14–19 May 2017; pp. 1578–1585.
89. Tang, W.; Long, G.; Liu, L.; Zhou, T.; Jiang, J.; Blumenstein, M. Rethinking 1D-CNN for Time Series Classification: A Stronger Baseline. *arXiv* **2021**, arXiv:2002.10061.
90. Turkoglu, M.O.; D'Aronco, S.; Wegner, J.; Schindler, K. Gating Revisited: Deep Multi-layer RNNs That Can Be Trained. *IEEE Trans. Pattern Anal. Mach. Intell.* **2021**, *44*, 4081–4092. [CrossRef]
91. Pelletier, C.; Webb, G.I.; Petitjean, F. Temporal Convolutional Neural Network for the Classification of Satellite Image Time Series. *Remote Sens.* **2019**, *11*, 523. [CrossRef]
92. Rußwurm, M.; Körner, M. Self-attention for raw optical Satellite Time Series Classification. *ISPRS J. Photogramm. Remote Sens.* **2020**, *169*, 421–435. [CrossRef]

93. Chen, H.; Chandrasekar, V.; Tan, H.; Cifelli, R. Rainfall Estimation From Ground Radar and TRMM Precipitation Radar Using Hybrid Deep Neural Networks. *Geophys. Rese. Lett.* **2019**, *46*, 10669–10678. [CrossRef]
94. Weng, Q.; Mao, Z.; Lin, J.; Guo, W. Land-Use Classification via Extreme Learning Classifier Based on Deep Convolutional Features. *IEEE Geosci. Remote Sens. Lett.* **2017**, *14*, 704–708. [CrossRef]
95. Castelluccio, M.; Poggi, G.; Sansone, C.; Verdoliva, L. Land Use Classification in Remote Sensing Images by Convolutional Neural Networks. *arXiv* **2015**, arXiv:1508.00092. https://doi.org/10.48550/arXiv.1508.00092.
96. Papoutsis, I.; Bountos, N.I.; Zavras, A.; Michail, D.; Tryfonopoulos, C. Efficient deep learning models for land cover image classification. *arXiv* **2022**, arXiv:2111.09451.
97. Devlin, J.; Chang, M.W.; Lee, K.; Toutanova, K. Bert: Pre-training of deep bidirectional transformers for language understanding. *arXiv* **2018**, arXiv:1810.04805.
98. Liu, Z.; Lin, Y.; Cao, Y.; Hu, H.; Wei, Y.; Zhang, Z.; Lin, S.; Guo, B. Swin Transformer: Hierarchical Vision Transformer using Shifted Windows. In Proceedings of the 2021 IEEE/CVF International Conference on Computer Vision (ICCV), Montreal, BC, Canada, 11–17 October 2021; pp. 9992–10002. [CrossRef]
99. Scheibenreif, L.; Hanna, J.; Mommert, M.; Borth, D. Self-supervised Vision Transformers for Land-cover Segmentation and Classification. In Proceedings of the 2022 IEEE/CVF Conference on Computer Vision and Pattern Recognition Workshops (CVPRW), New Orleans, LA, USA, 19–20 June 2022; pp. 1421–1430. [CrossRef]
100. Zhang, C.; Wang, L.; Cheng, S.; Li, Y. SwinSUNet: Pure Transformer Network for Remote Sensing Image Change Detection. *IEEE Trans. Geosci. Remote Sens.* **2022**, *60*, 5224713. [CrossRef]
101. Wang, D.; Zhang, J.; Du, B.; Xia, G.S.; Tao, D. An Empirical Study of Remote Sensing Pretraining. *IEEE Trans. Geosci. Remote Sens.* 2022, *Early Access*. [CrossRef]
102. Liu, S.; He, C.; Bai, H.; Zhang, Y.; Cheng, J. Light-weight attention semantic segmentation network for high-resolution remote sensing images. In Proceedings of the IGARSS 2020—2020 IEEE International Geoscience and Remote Sensing Symposium, Waikoloa, HI, USA, 26 September–2 October 2020; pp. 2595–2598.
103. Xu, Z.; Zhang, W.; Zhang, T.; Yang, Z.; Li, J. Efficient Transformer for Remote Sensing Image Segmentation. *Remote Sens.* **2021**, *13*, 3585. [CrossRef]
104. Alhichri, H.; Alswayed, A.S.; Bazi, Y.; Ammour, N.; Alajlan, N.A. Classification of Remote Sensing Images Using EfficientNet-B3 CNN Model With Attention. *IEEE Access* **2021**, *9*, 14078–14094. [CrossRef]
105. Meng, Z.; Zhao, F.; Liang, M. SS-MLP: A Novel Spectral-Spatial MLP Architecture for Hyperspectral Image Classification. *Remote Sens.* **2021**, *13*, 4060. [CrossRef]
106. Gong, N.; Zhang, C.; Zhou, H.; Zhang, K.; Wu, Z.; Zhang, X. Classification of hyperspectral images via improved cycle-MLP. *IET Comput. Vis.* **2022**, *16*, 468–478. [CrossRef]
107. Solórzano, J.V.; Mas, J.F.; Gao, Y.; Gallardo-Cruz, J.A. Land Use Land Cover Classification with U-Net: Advantages of Combining Sentinel-1 and Sentinel-2 Imagery. *Remote Sens.* **2021**, *13*, 3600. [CrossRef]
108. Cheng, Z.; Fu, D. Remote Sensing Image Segmentation Method based on HRNET. In Proceedings of the IGARSS 2020—2020 IEEE International Geoscience and Remote Sensing Symposium, Waikoloa, HI, USA, 26 September–2 October 2020; pp. 6750–6753. [CrossRef]
109. Chen, L.C.; Papandreou, G.; Kokkinos, I.; Murphy, K.; Yuille, A. DeepLab: Semantic Image Segmentation with Deep Convolutional Nets, Atrous Convolution, and Fully Connected CRFs. *IEEE Trans. Pattern Anal. Mach. Intell.* **2016**, *40*, 834–848. [CrossRef]
110. Ramirez, W.; Achanccaray, P.; Mendoza, L.F.; Pacheco, M.A.C. Deep Convolutional Neural Networks for Weed Detection in Agricultural Crops Using Optical Aerial Images. In Proceedings of the 2020 IEEE Latin American GRSS & ISPRS Remote Sensing Conference (LAGIRS), Santiago, Chile, 22–26 March 2020; pp. 133–137. [CrossRef]
111. Liu, M.; Fu, B.; Xie, S.; He, H.; Lan, F.; Li, Y.; Lou, P.; Fan, D. Comparison of multi-source satellite images for classifying marsh vegetation using DeepLabV3 Plus deep learning algorithm. *Ecol. Indic.* **2021**, *125*, 107562. [CrossRef]
112. Ren, S.; He, K.; Girshick, R.; Sun, J. Faster R-CNN: Towards Real-Time Object Detection with Region Proposal Networks. In *NIPS'15: Proceedings of the 28th International Conference on Neural Information Processing Systems—Volume 1*; MIT Press: Cambridge, MA, USA, 2015; pp. 91–99.
113. Breiman, L. Random Forests. *Mach. Learn.* **2001**, *45*, 5–32. [CrossRef]
114. Zhong, L.; Hu, L.; Zhou, H. Deep learning based multi-temporal crop classification. *Remote Sens. Environ.* **2019**, *221*, 430–443. [CrossRef]
115. Hochreiter, S.; Schmidhuber, J. Long Short-term Memory. *Neural Comput.* **1997**, *9*, 1735–1780. [CrossRef]
116. Sun, Z.; Di, L.; Fang, H. Using long short-term memory recurrent neural network in land cover classification on Landsat and Cropland data layer time series. *Int. J. Remote Sens.* **2018**, *40*, 593–614. [CrossRef]
117. Ndikumana, E.; Ho Tong Minh, D.; Baghdadi, N.; Courault, D.; Hossard, L. Deep Recurrent Neural Network for Agricultural Classification using multitemporal SAR Sentinel-1 for Camargue, France. *Remote Sens.* **2018**, *10*, 1217. [CrossRef]
118. Selvaraju, R.R.; Cogswell, M.; Das, A.; Vedantam, R.; Parikh, D.; Batra, D. Grad-CAM: Visual Explanations from Deep Networks via Gradient-Based Localization. In Proceedings of the 2017 IEEE International Conference on Computer Vision (ICCV), Venice, Italy, 22–29 October 2017; pp. 618–626.
119. Boguszewski, A.; Batorski, D.; Ziemba-Jankowska, N.; Zambrzycka, A.; Dziedzic, T. LandCover.ai: Dataset for Automatic Mapping of Buildings, Woodlands and Water from Aerial Imagery. *arXiv* **2020**, arXiv:2005.02264.

120. Lin, T.Y.; Maire, M.; Belongie, S.; Hays, J.; Perona, P.; Ramanan, D.; Dollar, P.; Zitnick, L. Microsoft COCO: Common Objects in Context. In Proceedings of the Computer Vision—ECCV 2014: 13th European Conference, Zurich, Switzerland, 6–12 September 2014; Springer: Cham, Switzerland, 2014.
121. Everingham, M.; Gool, L.V.; Williams, C.K.I.; Winn, J.; Zisserman, A. The Pascal Visual Object Classes (VOC) Challenge. *Int. J. Comput. Vis.* **2009**, *88*, 303–308. [CrossRef]
122. Lu, X.; Zhang, Y.; Yuan, Y.; Feng, Y. Gated and Axis-Concentrated Localization Network for Remote Sensing Object Detection. *IEEE Trans. Geosci. Remote Sens.* **2020**, *58*, 179–192. [CrossRef]
123. Tan, M.; Le, Q.V. EfficientNetV2: Smaller Models and Faster Training. In Proceedings of the 38th International Conference on Machine Learning (ICML 2021), Virtual Event, 18–24 July 2021; 2021; Volume 139, pp. 10096–10106.
124. Wilkinson, M.D.; Dumontier, M.; Aalbersberg, I.J.; Appleton, G.; Axton, M.; Baak, A.; Blomberg, N.; Boiten, J.W.; da Silva Santos, L.B.; Bourne, P.E.; et al. The FAIR Guiding Principles for scientific data management and stewardship. *Sci. Data* **2016**, *3*, 160018. [CrossRef]
125. Wang, Y.; Albrecht, C.; Ait Ali Braham, N.; Mou, L.; Zhu, X. Self-Supervised Learning in Remote Sensing: A Review. *IEEE Geosci. Remote Sens. Mag.* **2022**, *10*, 213–247. [CrossRef]
126. Akiva, P.; Purri, M.; Leotta, M. Self-Supervised Material and Texture Representation Learning for Remote Sensing Tasks. In Proceedings of the IEEE/CVF Conference on Computer Vision and Pattern Recognition (CVPR), New Orleans, LA, USA, 18–24 June 2022; pp. 8203–8215.
127. Manas, O.; Lacoste, A.; i Nieto, X.G.; Vazquez, D.; Rodriguez, P. Seasonal Contrast: Unsupervised Pre-Training from Uncurated Remote Sensing Data. In Proceedings of the 2021 IEEE/CVF International Conference on Computer Vision (ICCV), Montreal, BC, Canada, 11–17 October 2021; IEEE Computer Society: Los Alamitos, CA, USA, 2021; pp. 9394–9403. [CrossRef]
128. Wang, Y.; Braham, N.A.A.; Xiong, Z.; Liu, C.; Albrecht, C.M.; Zhu, X.X. SSL4EO-S12: A Large-Scale Multi-Modal, Multi-Temporal Dataset for Self-Supervised Learning in Earth Observation. *arXiv* **2022**, arXiv:2211.07044.
129. Buslaev, A.; Iglovikov, V.I.; Khvedchenya, E.; Parinov, A.; Druzhinin, M.; Kalinin, A.A. Albumentations: Fast and flexible image augmentations. *Information* **2020**, *11*, 125. [CrossRef]
130. Oliphant, T.E. *A Guide to NumPy*; Trelgol Publishing: Austin, TX, USA, 2006; Volume 1.
131. Kramer, O. Scikit-learn. In *Machine Learning for Evolution Strategies*; Springer: Cham, Switzerland, 2016; pp. 45–53.
132. Szymanski, P.; Kajdanowicz, T. Scikit-multilearn: A scikit-based Python environment for performing multi-label classification. *J. Mach. Learn. Res.* **2019**, *20*, 209–230.
133. Bisong, E. Matplotlib and seaborn. In *Building Machine Learning and Deep Learning Models on Google Cloud Platform*; Springer: Berkeley, CA, USA, 2019; pp. 151–165.
134. zipp: A pathlib-Compatible Zipfile Object Wrapper. 2023. Available online: https://doc.sagemath.org/html/en/reference/spkg/zipp.html (accessed on 8 March 2023).
135. dill: Serialize All of Python. 2023. Available online: https://pypi.org/project/dill/ (accessed on 8 March 2023).
136. Lmdb: A Universal Python Binding for the LMDB 'Lightning' Database. 2023. Available online: https://lmdb.readthedocs.io/en/release/ (accessed on 8 March 2023).
137. tifffile: Storing NumPY Arrays in TIFF and Read Image and Metadata from TIFF-Like Files. 2023. Available online: https://pypi.org/project/tifffile/ (accessed on 8 March 2023).
138. h5py: A Pythonic Interface to the HDF5 Binary Data Format. 2023. Available online: https://www.h5py.org/ (accessed on 8 March 2023).
139. Click: Command Line Interface Creation Kit. 2023. Available online: https://click.palletsprojects.com/en/8.1.x/ (accessed on 8 March 2023).
140. Munch: A Dictionary Supporting Attribute-Style Access. 2023. Available online: https://morioh.com/p/bbdd8605be66 (accessed on 8 March 2023).
141. Marshmallow: Simplified Object Serialization. 2023. Available online: https://marshmallow.readthedocs.io/en/stable/ (accessed on 8 March 2023).
142. Detlefsen, N.S.; Borovec, J.; Schock, J.; Harsh, A.; Koker, T.; Liello, L.D.; Stancl, D.; Quan, C.; Grechkin, M.; Falcon, W. TorchMetrics—Measuring Reproducibility in PyTorch. *J. Open Source Softw.* **2022**, *7*, 4101. [CrossRef]
143. Sechidis, K.; Tsoumakas, G.; Vlahavas, I. On the Stratification of Multi-Label Data. In Proceedings of the 2011 European Conference on Machine Learning and Knowledge Discovery in Databases—Volume Part III, Athens, Greece, 5–9 September 2011; Springer: Berlin/Heidelberg, Germany, 2011; pp. 145–158.
144. Zhai, X.; Puigcerver, J.; Kolesnikov, A.; Ruyssen, P.; Riquelme, C.; Lucic, M.; Djolonga, J.; Pinto, A.S.; Neumann, M.; Dosovitskiy, A.; et al. A Large-scale Study of Representation Learning with the Visual Task Adaptation Benchmark. *arXiv* **2019**, arXiv:1910.04867.
145. Risojevic, V.; Stojnic, V. Do we still need ImageNet pre-training in remote sensing scene classification? *arXiv* **2021**, arXiv:2111.03690. [CrossRef]
146. Kingma, D.P.; Ba, J. Adam: A method for stochastic optimization. *arXiv* **2014**, arXiv:1412.6980.

remote sensing

MDPI

Article

Dynamic High-Resolution Network for Semantic Segmentation in Remote-Sensing Images

Shichen Guo [1,2], Qi Yang [2,3], Shiming Xiang [2,3], Pengfei Wang [1] and Xuezhi Wang [1,*]

1 Computer Network Information Center, Chinese Academy of Sciences, Beijing 100083, China; guoshichen@cnic.cn (S.G.); pfwang@cnic.cn (P.W.)
2 University of Chinese Academy of Sciences, Beijing 100049, China; yangqi2021@ia.ac.cn (Q.Y.); smxiang@nlpr.ia.ac.cn (S.X.)
3 State Key Laboratory of Multimodal Artificial Intelligence Systems (MAIS), Institute of Automation, Chinese Academy of Sciences, Beijing 100190, China
* Correspondence: wxz@cnic.cn

Abstract: Semantic segmentation of remote-sensing (RS) images is one of the most fundamental tasks in the understanding of a remote-sensing scene. However, high-resolution RS images contain plentiful detailed information about ground objects, which scatter everywhere spatially and have variable sizes, styles, and visual appearances. Due to the high similarity between classes and diversity within classes, it is challenging to obtain satisfactory and accurate semantic segmentation results. This paper proposes a Dynamic High-Resolution Network (DyHRNet) to solve this problem. Our proposed network takes HRNet as a super-architecture, aiming to leverage the important connections and channels by further investigating the parallel streams at different resolution representations of the original HRNet. The learning task is conducted under the framework of a neural architecture search (NAS) and channel-wise attention module. Specifically, the Accelerated Proximal Gradient (APG) algorithm is introduced to iteratively solve the sparse regularization subproblem from the perspective of neural architecture search. In this way, valuable connections are selected for cross-resolution feature fusion. In addition, a channel-wise attention module is designed to weight the channel contributions for feature aggregation. Finally, DyHRNet fully realizes the dynamic advantages of data adaptability by combining the APG algorithm and channel-wise attention module simultaneously. Compared with nine classical or state-of-the-art models (FCN, UNet, PSPNet, DeepLabV3+, OCRNet, SETR, SegFormer, HRNet+FCN, and HRNet+OCR), DyHRNet has shown high performance on three public challenging RS image datasets (Vaihingen, Potsdam, and LoveDA). Furthermore, the visual segmentation results, the learned structures, the iteration process analysis, and the ablation study all demonstrate the effectiveness of our proposed model.

Keywords: semantic segmentation; remote-sensing image; neural architecture search; sparse regularization; HRNet

Citation: Guo, S.; Yang, Q.; Xiang, S.; Wang, P.; Wang, X. Dynamic High-Resolution Network for Semantic Segmentation in Remote-Sensing Images. *Remote Sens.* **2023**, *15*, 2293. https://doi.org/10.3390/rs15092293

Academic Editor: Michael K. Ng

Received: 7 March 2023
Revised: 12 April 2023
Accepted: 25 April 2023
Published: 26 April 2023

1. Introduction

With the rapid development of remote-sensing (RS) technologies, a large number of RS images are taken by different devices every day. In practice, it is an urgent need to understand well the contents recorded in these images. As a fundamental approach to analyzing RS images in many systems, the task of semantic segmentation is to divide the input image into regions with explicit category labels for downstream tasks. Compared with the task conducted on natural images, segmenting RS images could be more complex. RS images are all taken at a far distance, and there are a lot of confusing objects scattered spatially here and there, with significant variations in size, style, and visual appearance. This causes high intra-class scatter and low inter-class variance in pattern analysis, making it more difficult to achieve satisfactory performance.

Recent years have witnessed the great successes of deep learning in the field of RS image processing. Along this technical line, a fundamental job is to design a good architecture for data adaptability. There are many multi-scale objects for high-resolution RS images with rich ground details. Thus, when designing neural architecture, it is necessary to consider how to achieve multi-scale feature fusion well for fine semantic segmentation. To this end, some thoughtful designs have been demonstrated in the literature [1–7]. The practices have indicated that designing neural architectures that can extract and maintain the representations simultaneously with high, medium, and low resolutions is very important for segmenting the RS objects of different scales well.

To achieve good multi-scale feature fusion, architecture design can be incorporated into the AutoML framework [8–14]. One goal of AutoML is to construct neural architectures automatically by computing itself. Algorithmically, neural architecture search (NAS) methods have been invented to attend to this need. Intrinsically, NAS is an NP-hard problem. Thus, differentiable architecture search algorithms [10,13–16] have gained great attention in recent years due to their relatively low computing complexity. However, there still exist limitations to be overcome for semantic segmentation [17,18]. They tend to a give large chance to the skip connections [19,20] without guaranteeing that the learned architectures have explicit streams to extract the representations with different resolutions.

In the literature, the High-Resolution Network (HRNet) [21,22] is a famous neural architecture for high-resolution representation learning. Technically, it was initially designed for pose estimation [21], and its usage has been demonstrated later in many visual recognition tasks [7,22]. In addition, some variants have been developed, including the higher HRNet [23], the lightweight HRNet [24,25], the dynamic HRNet [26], the HRNet with transformer [27], and so on. Architecturally, the HRNet contains four parallel streams with different resolution representations. It also offers a new mechanism of cross-resolution interaction via dense connections between the streams at different stages. With feature mapping among different resolutions, the HRNet is enabled to capture rich multi-scale information. Therefore, such an architecture could attend well to the needs for the segmentation of RS objects.

Beyond directly applying the primary HRNet [21,22] to RS images, in this study, we start by analyzing the dense connections contained in it. When performing the cross-resolution feature fusion, the HRNet does not consider the contributions of the dense connections and channels, i.e., all of them are equally used. This motivates us to select the important ones by addressing the task in the NAS framework, hoping to enhance its representative capability further. Based on the above observations, in this paper, a Dynamic High-Resolution Network (DyHRNet) is proposed for semantic segmentation in RS images. The DyHRNet is initially constructed on and learned later from the primary HRNet to use its parallel streams with different resolution representations.

The key idea behind the DyHRNet is to evaluate the importance of dense connections and channels for cross-resolution feature fusion. This task is addressed in the NAS framework with channel-wise attention. Mathematically, to avoid solving an NP-hard problem, we choose to relax the 0/1 contributions to be soft ones. With a series of sparse regularizations posed on the learning model, unimportant or useless connections will be identified by assigning low or zero contributions. In this way, the architecture of the primary HRNet is dynamically changed for data adaptability. In addition, a channel-wise attention is designed to evaluate the channel contributions to cross-resolution feature fusion, which further enhances the representation capability of the proposed DyHRNet. Finally, the sparse regularization and the channel-wise attention are combined into a compact optimization model for end-to-end learning. The contributions and the main work are summarized as follows:

- A Dynamic High-Resolution Network (DyHRNet) is proposed for semantic segmentation in RS images. The neural architecture of the DyHRNet is constructed on and learned from the primary HRNet. This task is formulated as a problem of neural

architecture search (NAS) with channel-wise attention. Mathematically, a compact learning model with sparse regularization is developed to achieve this goal.
- Within the Stochastic Gradient Descent (SGD) approach framework for end-to-end training, the sparse regularization subproblem is iteratively solved by the Accelerated Proximal Gradient (APG) algorithm. As a result, the important connections between the parallel streams in the HRNet are selected for cross-resolution feature fusion.
- A mechanism of channel-wise attention is proposed to evaluate the channel contributions for cross-resolution feature aggregation. The attention module has a native structure in which it is easy to format the importance score for channel mapping. As a result, the channel contributions in HRNet are automatically modulated to enhance the representation capability of the DyHRNet.
- The performance of DyHRNet for segmenting RS images has been evaluated on three challenging public benchmarks, including the ISPRS 2D semantic segmentation challenge Vaihingen and Potsdam dataset and the LoveDA dataset. The extensive experiment results with numerical scores and visual segmentation, the learned structures, the iteration process analysis, and the ablation study all demonstrate the effectiveness of the proposed model.

The article is organized as follows: Section 1 describes the background information, the motivation, the objective, and the predictions of this study. Section 2 describes the related works. The details of the proposed method are introduced in Section 3. Experimental results are reported in Section 4. Discussions are given in Section 5, followed by the conclusions in Section 6.

2. Related Works

2.1. Semantic Segmentation for RS Images

With the great success of deep learning on semantic segmentation for natural images, tremendous efforts have been made by researchers to transfer deep models for RS images [28,29]. Architecturally, most of the models have been formulated with convolution, pooling, and up-sampling operations, such as the Fully Convolutional Network (FCN) [30], the UNet [31], the Pyramid Scene Parsing Network (PSPNet) [32], the DeepLab [33], the OCRNet [34], and so on. Later, some frameworks were constructed on the transformer. In this family, the SEgmentation TRansformer (SETR) [35] and the SegFormer [36] are two famous models. With the usage of encoder–decoder backbones, some variants have been constructed for this issue [1,4,37–40]. For example, the cascaded network with context information fusion was developed to extract confusing artificial objects [1]. The shuffling network is employed to enhance the feature learning ability [38]. These studies have primarily enhanced the semantic segmentation performance for RS images.

Later, more complex models were considered for segmenting RS images. Specifically, Diakogiannis et al. [41] developed an encoder–decoder with multi-tasking inference sequentially on object boundary, segmentation masks, and reconstruction of the input. Zhang et al. [6] employed a high-resolution network with different branches to extract features at both local and global levels. Xu et al. [7] constructed a high-resolution context extraction network to fuse multi-scale contextual information. Liu et al. [5] constructed a new multi-scale U-shaped CNN for extracting buildings in high-resolution RS images, rendering a novel proposal for this issue with multi-task learning to obtain precise masks and help avoid over-fitting. Tang et al. [40] developed a novel self-supervised contrastive learning framework for semantic segmentation in aerial imagery. Within their framework, the distinct characteristic lies in contrastive learning, which is performed both at the feature level and at the semantic level. Furthermore, with the use of the local mutual information that is embedded into the semantic level of contrastive learning, the representation power of the proposed model is largely enhanced for segmentation [40]. In addition, attention modules in different views [42–44] have been designed for fine segmentation. The transformer has recently been employed as the backbone of this task [45,46]. These models achieve good segmentation in different methods of local, global, and multi-scale feature fusion.

In the literature, there are a few works on the semantic segmentation of RS images under the NAS frameworks. Zhang et al. [47] employed a directed acyclic graph with tricks of Gumbel-max operations under a differentiable searching framework. Later, Wang et al. [48] proposed the decoupling NAS framework with a hierarchical search space for RS objects at the path level, connection level, and cell level. Broni-Bediako et al. [49] developed an evolutionary NAS method for this task. In their framework, gene expression programming, and cellular encoding were employed to represent the encoding scheme for block-building. In summary, although these approaches achieve good performance on accuracy, high computational complexity degrades their real-world applications.

2.2. Neural Architecture Search

Recently, constructing neural architectures automatically via NAS has received significant interest in both academia and industry [8–11,13,14]. There are in total three families of NAS methods, namely evolution-based NASs, RL-based NASs, and gradient-based NASs. For example, Ghiasi et al. [11] developed a NAS framework to search for better architectures of a feature pyramid network for object detection. In their work, a novel scalable search space is constructed to cover all cross-scale connections, and a combination of top-down and bottom-up connections is achieved via NAS tricks to fuse multi-scale features [11]. For another example, Weng et al. [12] designed three types of primitive operations on a search space to search U-like backbones for semantic segmentation. In this way, U-like backbone networks can be automatically constructed by stacking the same number of the searched down-sampling cells and up-sampling cells, rendering good performance for semantic segmentation [12]. In the literature, the proposals for NAS within scalable search spaces are rich, demonstrating bright performance enhancements for various types of visual computing tasks.

Documentation about NAS is rich. Here, we only give a brief review of gradient-based NAS methods, which are related to our work in this paper. Since the differentiable architecture search (DARTS) framework was released in 2017 [10], it has become a famous pipeline with gradient-based searching strategies due to their relatively low computational complexities and competitive performance. To reduce the gap between search and evaluation, tricks with progressive differentiable NAS [50], the combination of evaluation and search [51], Gumbel–Softmax [13], and path-level selection [52] have been proposed to achieve the goal of structure generation. However, the DARTS-based algorithms are prone to yield structures much more with skip connections, which limits their power for real-world applications.

As the current task of NAS is to identify a sub-structure or from a previously defined big structure, pruning tricks have been applied to network generation. Earlier, network pruning was performed for model acceleration or compression [53–56]. Recently, sparse representation has also been introduced to this issue. Yang et al. [57] addressed this task as a problem of sparse coding, where differentiable search is achieved within a lower-dimensional space. In addition, Zhang et al. [14] developed a direct sparse optimization to achieve the goal of model pruning. However, network pruning should be strictly performed on a specific architecture, without the ability to generate new topology and operations.

3. Method

Our task is to develop a Dynamic HRNet (DyHRNet) for the fine segmentation of RS objects. The task will be formulated as a NAS problem with channel-wise attention. Formally, a compact learning model with sparse regularization is developed to achieve this goal. The details are described in the following subsections.

3.1. Problem Formulation

Figure 1 demonstrates the super-architecture constructed according to the rule used in the original HRNet [21,22]. Totally it consists of four parallel streams with different resolution representations, where each row corresponds to a stream of representations with the same resolution. It offers a new mechanism for cross-resolution interaction via dense connections between the streams at different stages. With feature mapping among different resolutions, the HRNet is enabled to capture rich multi-scale information. Therefore, such an architecture could attend well to our needs for RS images.

For clarity, the architecture can be further divided into four stages. The first stage is at the highest resolution. It contains four convolutional layers, which are recorded together by block $\mathbf{O}_{1,1}$ for simplicity. The next three stages cover different streams with high, medium, and low resolutions. More specifically, the second stage contains one group of dense connections, shown as the dash lines between the representations $\mathbf{O}_{1,2}$, $\mathbf{R}_{1,2}$, $\mathbf{O}_{2,2}$, and $\mathbf{R}_{2,2}$ in Figure 1. In addition, there are four and three groups of dense connections, respectively, in the third and fourth stages (Architecturally, one can set any number of groups of dense connections if needed in practice. Without loss of generality, here we take them as those suggested by the original HRNet). Clearly, with these dense connections, multi-scale features are fused. Thus, such an architecture is suitable for segmenting RS images, where objects with different scales locate here and there in the image.

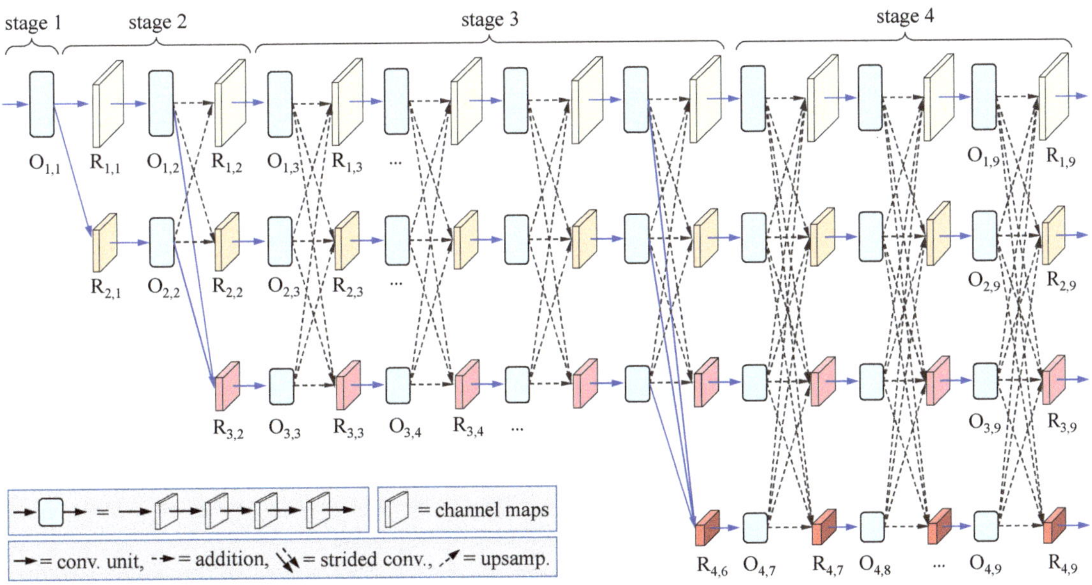

Figure 1. The primary HRNet used as a super-architecture to develop the Dynamic HRNet (DyHR-Net). Here the connections marked by the dash lines will be selected via sparse optimization.

We denote the representation (namely the output of the four convolutional layers) at the i-th row and j-th column in Figure 1 by $\mathbf{O}_{i,j}$, and the result of the feature fusion at the same position by $\mathbf{R}_{i,j}$. In the original HRNet, $\mathbf{R}_{i,j}$ is computed as follows:

$$\mathbf{R}_{i,j} = \text{ReLU}\left(\sum_{k=1}^{P} f_{k,j}^{(i)}\left(\mathbf{O}_{k,j}\right)\right), \tag{1}$$

where ReLU stands for the rectified linear unit, $\{f_{k,j}^{(i)}(\cdot)\}$ are the transformation functions and P is the number of streams with different resolutions, which changes within different stages. More specifically, as shown in Figure 1, in the second stage, $i = 1, 2$, $j = 2$ and $P = 2$; in the third stage, $1 \leq i \leq 3$, $3 \leq j \leq 6$, and $P = 3$; and in the fourth stage, $1 \leq i \leq 4$, $7 \leq j \leq 9$, and $P = 4$. In addition, for the function $f_{k,j}^{(i)}(\cdot)$, in the case of $k = j$, it is an identity mapping; in the case of $k < j$, it is a 3×3 convolution with stride 2; and in the case of $k > j$, it is a bilinear up-sampling operation with 1×1 convolution for feature alignment.

As can be seen from Equation (1), the representations in $\{\mathbf{O}_{k,j}\}$ are all equally treated without considering their importance. In other words, there is a lack of a mechanism to evaluate the contributions of the dense connections. Therefore, we cast this task in the NAS framework, which allows us to select those useful connections. Then, Equation (1) is reformulated as follows:

$$\mathbf{R}_{i,j} = \text{ReLU}\left(\sum_{k=1}^{P} s_{k,j}^{(i)}\left(f_{k,j}^{(i)}\left(\mathbf{O}_{k,j}\right)\right)\right), \quad s.t. \quad S_{k,j}^{(i)} \in \{0,1\}, \tag{2}$$

where $\{S_{k,j}^{(i)}\}$ are the selection parameters to be optimized. Technically, in the case of $S_{k,j}^{(i)} = 1$, the candidate link between $\mathbf{O}_{k,j}$ and $\mathbf{R}_{i,j}$ will be selected; and in the case of $S_{k,j}^{(i)} = 0$, the candidate link is useless, and will be discarded. This formulation attends to the task of link search in the NAS work setting. However, it is an NP-hard problem.

To solve this problem, we relax the search space to be a continuous one by allowing each $S_{k,j}^{(i)}$ as a non-negative scaling factor. Then, it turns out that

$$\mathbf{R}_{i,j} = \text{ReLU}\left(\sum_{k=1}^{P} s_{k,j}^{(i)}\left(f_{k,j}^{(i)}\left(\mathbf{O}_{k,j}\right)\right)\right), \quad s.t. \quad S_{k,j}^{(i)} \geq 0, \text{ and } \sum_{k=1}^{P} S_{k,j}^{(i)} < \lambda_j^{(i)}, \tag{3}$$

where $\{S_{k,j}^{(i)}\}$ are the continuous weighting parameters to be learned, the inequality constraint is introduced to force the sparsity of connections, and λ_j^i controls the amount of shrinkage for the sparse estimation. That is, a small λ_j will force sparser. Algorithmically, the formulation in Equation (3) is a convex relaxing to that in Equation (2).

Please note that the formulation in Equation (3) exhibits another flexible mechanism in that the channel-wise importance can be evaluated jointly. By considering the contributions of channels in each $\mathbf{O}_{k,j}$, it can be rewritten as follows:

$$\mathbf{R}_{i,j} = \text{ReLU}\left(\sum_{k=1}^{P} s_{k,j}^{(i)}\left(f_{k,j}^{(i)}\left(\mathbf{a}_{k,j}^{(i)} \otimes \mathbf{O}_{k,j}\right)\right)\right), \quad s.t. \quad S_{k,j}^{(i)} \geq 0, \text{ and } \sum_{k=1}^{P} S_{k,j}^{(i)} < \lambda_j^{(i)}, \tag{4}$$

where $\mathbf{a}_{k,j}^{(i)}$ is a weighting vector with a length equal to the number of the channels in $\mathbf{O}_{k,j}$, and $\otimes$ stands for the channel-wise product. Technically, channel-wise attention will be designed to fulfill this task (see Section 3.3).

In Equation (4), the weights in $\{s_{k,j}^{(i)}\}$ are learned via the NAS trick, and those in $\{\mathbf{a}_{k,j}^{(i)}\}$ are evaluated via the channel attention. All these weights are positive, which will be modulated by data-driven learning. They may be very small or even zero. In particular, after model training, the cross-resolution connections with zero $s_{k,j}^{(i)}$ and the channels with zero $\mathbf{a}_{k,j}^{(i)}$ will be deleted for prediction.

Now, we can explain the term "dynamic" in our work. In the literature, dynamic models could be developed at different levels of model adaptability in a way of data-driven learning, e.g., at the levels of input data [58], lightweight structures [25], adaptive weights of operations [26], and so on. By contrast, in our work setting, here we first explain its meaning given neural architecture design. The operation in Equation (1) indicates that the neural architecture will remain unchanged before and after training in the original HRNet. With the implementation guided by Equation (4), the connections could be maintained or cut dynamically during and after training. After the model is well trained under the NAS framework, the connections with $S_{k,j}^{(i)} = 0$ will be deleted from the original structure, yielding a new architecture for segmenting RS images. Then, we explain its meaning given channel-wise importance. The operator "$\otimes$" in Equation (4) will be performed channel by channel with different weights. In this way, the dynamic merit will be demonstrated in the use of channel-wise importance, which will be learned to modulate its contribution. As a whole, the NAS trick and the channel-wise attention will be combined via Equation (4) to develop a dynamic HRNet.

The structure in Figure 1 will be employed as a candidate backbone for architecture optimization. Thus, it is necessary to contain a head module to output semantic segmentation results. Accordingly, our DyHRNet has four groups of parameters to be optimized. The first group collects the parameters in all of the convolution kernels in Figure 1, which are recorded together by variable $\mathbf{W}$. The second group consists of those in the channel-wise attention module used to estimate $\{\mathbf{a}_{k,j}^{(i)}\}$, which are collected into variable $\mathbf{U}$. The third group includes those in the head part (namely the decoder module), which are collected into variable $\mathbf{H}$. The fourth group collects all $\{S_{k,j}^{(i)}\}$ in Equation (3), which are collected orderly into vector $\mathbf{s}$. Then, we have the following optimization problem for segmenting RS objects:

$$\begin{aligned} \min_{\mathbf{s}} \quad & L_{validation}(DyHRNet(\mathbf{W}^*, \mathbf{U}^*, \mathbf{H}^*, \mathbf{s})) \\ s.t. \quad & \sum_{(k,j,i)\in\mathcal{N}} S_{k,j}^{(i)} \le \lambda, \\ & S_{k,j}^{(i)} \ge 0, \\ & (\mathbf{W}^*, \mathbf{U}^*, \mathbf{H}^*) = \arg\min_{\mathbf{W},\mathbf{U},\mathbf{H}} L_{train}(DyHRNet(\mathbf{W}, \mathbf{U}, \mathbf{H}, \mathbf{s})), \end{aligned} \tag{5}$$

where $L_{train}(\cdot)$ is the loss calculated on the training samples, λ shrinks together all the controlling factors $\lambda_j^{(i)}$ in Equation (4), and set $\mathcal{N}$ collects the triples $\{(i,j,k)\}$ according to the dash-line links in Figure 1.

According to the relationship between the shrinkage constraints and the regularization representation for λ used in LASSO [59], Problem (5) can be reformulated as follows:

$$\begin{aligned} \min_{\mathbf{s}} \quad & L_{validation}(DyHRNet(\mathbf{W}^*, \mathbf{U}^*, \mathbf{H}^*, \mathbf{s})) + \lambda\|\mathbf{s}\|_1, \\ s.t. \quad & (\mathbf{W}^*, \mathbf{U}^*, \mathbf{H}^*) = \arg\min_{\mathbf{W},\mathbf{U},\mathbf{H}} L_{train}(DyHRNet(\mathbf{W}, \mathbf{U}, \mathbf{H}, \mathbf{s})), \end{aligned} \tag{6}$$

where $\|\mathbf{s}\|_1$ denotes the L_1 norm of $\mathbf{s}$.

In Problem (6), there are two subtasks that should be solved iteratively. One is to optimize $\mathbf{W}$, $\mathbf{U}$ and $\mathbf{H}$, given $\mathbf{s}$; and another is to optimize $\mathbf{s}$, given $\mathbf{W}$, $\mathbf{U}$ and $\mathbf{H}$. The former can be learned by the algorithm of back propagation of gradients. The latter is difficult to deal with because of the term of $\|\mathbf{s}\|_1$. In the following subsections, we will describe how to solve $\mathbf{s}$ and how to design the channel-wise attention module to calculate $\{\mathbf{a}_{k,j}^{(i)}\}$ in Equation (4).

3.2. Solving the Sparse Regularization Subproblem with Accelerated Proximal Gradient Algorithm

In this subsection, we use one of the dense connection units at stage 4 of HRNet to illustrate the whole sparse optimization process in the order from Figure 2a–c. To be more specific, Figure 2a depicts the dense cross-resolution connections of the original HRNet, which are also the candidates to be selected by the APG algorithm. At the beginning of the training stage, the weights of these connections are all set to 1.0 to guarantee that all of them have an equal probability to be selected. Figure 2b visualizes a group of learned weights using a solid line and a dashed line. The thicker the solid line is, the greater the weight is and the more important the connection is. In particular, the dashed lines indicate those connections are of zero importance, which could be directly cut off. Figure 2c shows the finally selected connections, which will be used at the inference stage of DyHRNet.

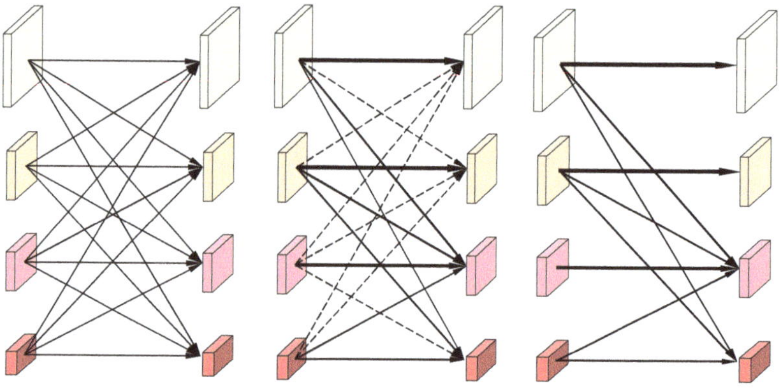

Figure 2. The process of sparse optimization is described by taking one group of the dense connections as an example in the order from (**a**) to (**b**) and (**c**). (**a**) The dense connections; (**b**) The weighted and pruned connections; (**c**) The final connections.

Unfortunately, solving the variable **s** in Problem (6) is a challenging task due to the sparse regularization term. One natural selection is to employ the traditional LASSO algorithm [59] to solve it. However, it is uneasy to unfold the mapping function $DyHRNet(\cdot)$ for deduction since it is a hierarchically composite function along the architecture of the DyHRNet. In addition, this could be more difficult since the loss function is defined on all the training samples, and the learning is data-driven in the stochastic work setting.

Alternatively, we employ the Accelerated Proximal Gradient (APG) algorithm [14,60] to optimize **s**. This algorithm has a theoretically sound foundation defined by the proximal algorithms. For convenience, a new function $f(\mathbf{s})$ is introduced to denote the objective function in Problem (6):

$$f(\mathbf{s}) = g(\mathbf{s}) + \lambda \|\mathbf{s}\|_1, \tag{7}$$

where

$$g(\mathbf{s}) = L_{validation}(DyHRNet(\mathbf{W}^*, \mathbf{A}^*, \mathbf{H}^*, \mathbf{s})). \tag{8}$$

Please note that, based on Equation (4), $g(\mathbf{s})$ is a differentiable function with respect to **s**. Now, we make a quadratic approximation to $g(\mathbf{s})$ around current **s**. Then, it follows that

$$g_v(\mathbf{z}) = g(\mathbf{s}) + \nabla g(\mathbf{s})^T(\mathbf{z} - \mathbf{s}) + \frac{1}{2v}\|\mathbf{z} - \mathbf{s}\|_2^2, \tag{9}$$

where $\nabla g(\mathbf{s})$ is the gradient vector of function $g(\cdot)$ at **s**, and v is a positive factor. Thus, based on Equation (9), the task of minimizing $f(\mathbf{s})$ is now updated as

$$\min_{\mathbf{z}} \; g_v(\mathbf{z}) + \lambda \|\mathbf{z}\|_1. \tag{10}$$

Equivalently, the optimum to Problem (10) can be obtained via the following problem:

$$\text{prox}_{v,\lambda}(\mathbf{y}) = \arg \min_{\mathbf{z}} \; \frac{1}{2v}\|\mathbf{z} - \mathbf{y}\|_2^2 + \lambda \|\mathbf{z}\|_1, \tag{11}$$

where $\mathbf{y} = \mathbf{s} - v\nabla g(\mathbf{s})$, which is known at current iteration. In addition, here "prox" is known as the proximal operator [60] and gives the optimum to Problem (10).

By further introducing the soft-thresholding operator [60], it turns out that

$$\left[\text{prox}_{v,\lambda}(\mathbf{y})\right]_m = \begin{cases} y_m - \lambda, & y_m \geq \lambda, \\ 0, & |y_m| < \lambda, \\ y_m + \lambda, & y_m \leq -\lambda, \end{cases} \tag{12}$$

where $[\cdot]_m$ stands for the m-th entity of the vector and y_m is the m-th entity of $\mathbf{y}$.

Now the original function $f(\mathbf{s})$ in Equation (7) can be minimized iteratively. With a momentum term to obtain a smooth solution path, we have

$$\mathbf{s}^{(t)} = \text{prox}_{v,\lambda}\left(\mathbf{y}^{(t)} - \eta_t \nabla g(\mathbf{y}^{(t)})\right), \tag{13}$$

in which

$$\mathbf{y}^{(t)} = \mathbf{s}^{(t-1)} + v_t(\mathbf{s}^{(t-1)} - \mathbf{s}^{(t-2)}), \tag{14}$$

where the superscript (t) indicates the t-th iteration, η_t is a learning ratio and v_t is a contribution factor for historical solution. According to the suggestion given in [60], v_t can be taken as $t/(t+3)$. As the number of iterations increases, it tends to be 1. Thus, v_t is fixed as 0.9 during iteration in our work.

In the first two iterations, both $\mathbf{s}^{(0)}$ and $\mathbf{s}^{(1)}$ are set to be a vector with all entities equal to 1. This means that all the connections in Figure 1 will be initially considered. When $\mathbf{s}$ is iteratively solved, all the connections will be assigned different weights to indicate their contributions to the final task.

3.3. Channel-Wise Attention for Feature Aggregation

As mentioned in Section 3.1, we introduce a channel-wise weighting operation in Equation (4) to develop the mechanism of the dynamic channel and enhance the flexibility of feature aggregation. Intrinsically, this can be addressed as an attention mechanism, which has been widely used in deep neural networks [61,62]. A similar idea has also been applied to the dynamic lightweight HRNet for pose estimation [25]. In this way, channel-wise attention can give larger weights to those important channels and lower weights to those unnecessary ones.

The main task here is to construct the modules to estimate the weighting vectors $\{\mathbf{a}_{k,j}^{(i)}\}$ in Equation (4). Motivated by the kernel aggregation used in [25], our module will be constructed on the representations $\mathbf{O}_{k,j}$ for dense links.

Without loss of generality, we take one group of dense connections in the fourth stage in Figure 1 as an example to explain how to design the attention module. Figure 3 illustrates the detailed layers. Given the representation $\mathbf{O}_{k,j}$, the channel features will be extracted by the Global Averaged Pooling (GAP). In this way, $\mathbf{O}_{k,j}$ will be transformed from a tensor to be a vector with a length equal to the number of the channels in $\mathbf{O}_{k,j}$. Then, it is pushed into the first Fully Connected (FC) layer, followed by the ReLU operation and the second FC layer. The final weighting vector with a length equal to the number of the channels in $\mathbf{O}_{k,j}$ will be output by the Sigmoid layer. Formally, we have

$$\mathbf{a}_{k,j}^{(i)} = \sigma^{(i)}\left(\text{FC}\left(\text{ReLU}\left(\text{FC}\left(\text{GAP}\left(\mathbf{O}_{k,j}\right)\right)\right)\right)\right), \quad i = 1,2;\ \text{or}\ 1 \le i \le 3;\ \text{or}\ 1 \le i \le 4, \tag{15}$$

where $\sigma(\cdot)^{(i)}$ stands for the final layer with Sigmoid as its activation function.

Finally, it is worth pointing out that the above module "GAP-FC-ReLU-FC-Sigmoid" will be re-used a few times. As illustrated in Figure 3, it will be copied four times to calculate four weighting vectors $\{\mathbf{a}_{k,j}^{(i)}\}$ with $1 \le i \le 4$, all taking $\mathbf{O}_{k,j}$ as their input. In this way, the channel importance of the feature maps is considered with adequate nonlinearity for semantic segmentation.

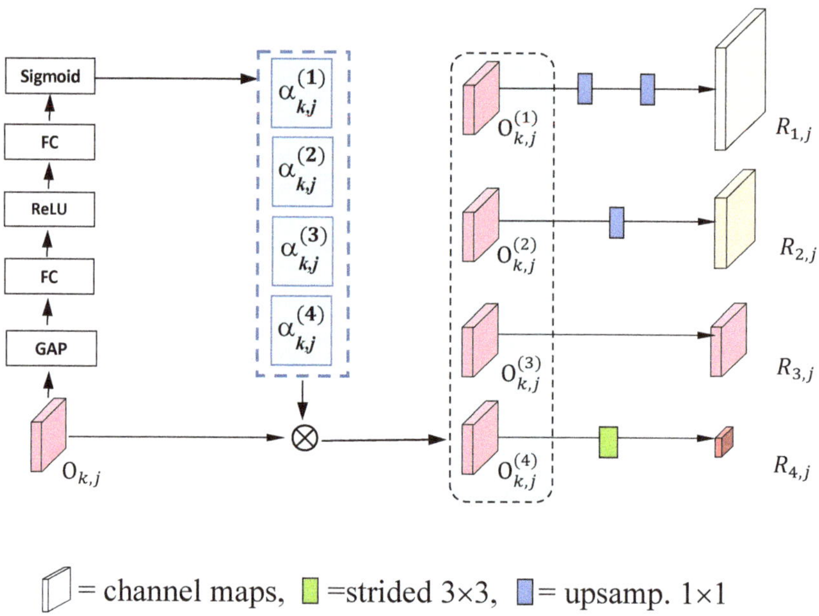

Figure 3. As example design of the channel-wise attention module at the fourth stage in Figure 1. Each module is comprised of five layers of "GAP-FC-ReLU-FC-Sigmoid". In this example, it will be copied four times with the same input to calculate $\mathbf{a}^{(i)}_{k,j}$, $1 \le i \le 4$.

3.4. The DyHRNet Neural Architecture

Based on the descriptions in Sections 3.2 and 3.3, we can now combine them to develop our DyHRNet for semantic segmentation of RS images. Please note that its backbone is optimized from the super-architecture in Figure 1 by applying the APG algorithm described in Section 3.2. It outputs the four representations $\mathbf{R}_{1,9}$, $\mathbf{R}_{2,9}$, $\mathbf{R}_{3,9}$ and $\mathbf{R}_{4,9}$ with different sizes. Accordingly, the the latter three representations will be bilinearly up-sampled, respectively, to be one with size equal to $\mathbf{R}_{1,9}$. Then, they are concatenated together and further transformed by a 1×1 convolution operation. Finally, we employ the object-contextual representation (OCR) scheme [34] (In the literature, there are many existing modules that can fulfill this task. Based on the empirical observations in [22], we follow the proposal to take the OCR scheme as the head part in our network.) as the decoder to format the output for semantic segmentation.

For clarity, Figure 4 demonstrates the overview of our DyHRNet. It consists of three parts. The first part is the encoder learned from the super-architecture, as demonstrated in Figure 1, which is responsible for feature extraction from the input images. The second part is just a concatenation unit followed by a 1×1 convolution operation. This treatment achieves multilevel feature fusion for decoding. The third part is the head sub-module of the OCR network [34] employed as the decoder to filter out the abstract features for semantic segmentation. As a whole, these three parts are combined as a whole dynamic network for end-to-end training.

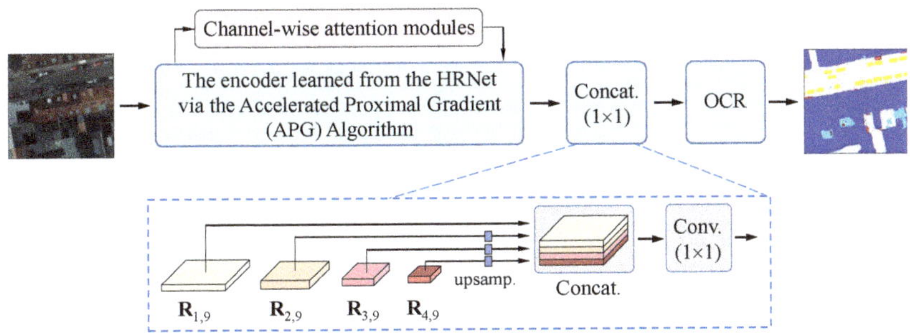

Figure 4. Overview of the neural architecture of DyHRNet.

3.5. Training and Inference

The original RS images and their ground-truth segmentations are taken to train the model in the learning stage. For each image $\mathcal{I}$ and its ground-truth $\mathcal{Y}$, we denote the predicted segmentation by $\hat{\mathcal{Y}}$. The loss function in Problem (6) is defined as follows:

$$L_{loss}(DyHRNet(\boldsymbol{\theta}, \mathbf{s})) = -\frac{1}{w \times h \times n} \sum_{\mathcal{I} \in trainset} \sum_{z_i \in I} \sum_{k=1}^{C} \delta(y_i = k) \log p_k(z_i), \tag{16}$$

where $\boldsymbol{\theta}$ collects all of the parameters in $\mathbf{W}$, $\mathbf{U}$ and $\mathbf{H}$, "trainset" indicates the training subset, C is the number of categories, $\delta(\cdot)$ is the truth function, z_i is the i-th pixel in image $\mathcal{I}$, y_i is the ground-truth label of z_i and $p_k(z_i)$ is the output probability at the k-th channel for pixel z_i, w and h are the width and height of the training images, and n is the total number of the images in the training set.

In Problem (6), there are two sub-problems to be solved. Technically, we solve them iteratively by fixing $\boldsymbol{\theta}$ or $\mathbf{s}$ once a time for another. Algorithm 1 lists the steps of how to train the DyHRNet. The learning rate η takes for gradient update when using the stochastic gradient descent (SGD) strategy to train the model. Except for the OCR module, there are in total more than 170 convolution operations in the DyHRNet. Batch normalization is performed after each convolution operation to guarantee convergence.

Algorithm 1 Training algorithm for the proposed DyHRNet.

Input: RS images with ground-truth segmentations, regularization parameters λ, learning rate η, and maximum number of iterations T.
Output: Parameter $\boldsymbol{\theta} = (\mathbf{W}, \mathbf{U}, \mathbf{H})$ and parameter $\mathbf{s}$.

Initialize $\boldsymbol{\theta}$ and $\mathbf{s}$.
Train $\boldsymbol{\theta}$ using SGD for several epochs with mini-batches and batch normalization by fixing $\mathbf{s}$, and obtain $\boldsymbol{\theta}^{(0)}$.
Let $t \leftarrow 0$.
while $t < T$ **do**
  Fix $\boldsymbol{\theta}^{(t)}$, update parameter $\mathbf{s}$ iteratively using the APG algorithm in Section 3.2, with Equations (12)–(14), and obtain $\mathbf{s}^{(t)}$.
  Fix $\mathbf{s}^{(t)}$, update parameter $\boldsymbol{\theta}$ iteratively using SGD for several epochs with mini-batches and batch normalization, and obtain $\boldsymbol{\theta}^{(t)}$.
  if $L_{loss}(DyHRNet(\boldsymbol{\theta}, \mathbf{s}))$ *converges* **then**
    Stop
  end
  $t \leftarrow t + 1$.
end

When performing the convergence check in Step 8 in Algorithm 1, the convergence condition is that the loss of the network maintains unchanged at two adjacent iterations. After the model is trained, it can be used for RS images with sizes larger than the training images. In this case, the image will be divided into several overlapped patches for segmentation, where the class probabilities of the pixels in the overlapped regions will be averaged to make the final inference.

4. Experiments

4.1. Data Description

The performance of the proposed DyNRNet has been evaluated on three public challenging benchmark datasets. The details of the datasets are described as follows:

Vaihingen: The Vaihingen dataset includes 33 images collected by an aerial camera. The size of the images is about 2494 × 2064 pixels on average, and the Ground Sampling Distance (GSD) or the spatial resolution on the ground is 9 cm. This dataset was constructed in a relatively small village with many buildings and roads. Each sample contains three images with true orthophoto (TOP), digital surface model (DSM), and ground truth. In this dataset, the TOP is composed of red and green bands. Ground truth contains six categories: impervious surface, building, low vegetation, tree, car, and cluster/background. In our experimental setup, only the TOP and ground truth of each sample were used without the DSM. A total of 344 samples are randomly obtained from 15 images for training, and 398 samples are randomly picked out from the remaining 18 images for testing. The samples are all cropped into patches of 512 × 512 pixels.

Potsdam: The Potsdam dataset has 38 samples with fine spatial resolution. On average, the size of the images is about 6000 × 6000 pixels, and the GSD is 5 cm. The samples in this dataset were taken from a scene in Potsdam City. Each sample contains three images with TOP, DSM, and ground truth, respectively. Each TOP includes the red, green, and blue bands. It has the same categories as those in the Vaihingen dataset. The DSM is not used for learning. A total of 3456 samples are randomly obtained from 24 images for training, and 2016 samples are randomly picked out from the remaining 14 images for testing. The samples are all cropped into patches of 512 × 512 pixels.

LoveDA: The LoveDA dataset contains 5987 samples [63]. Each image has 1024 × 1024 pixels, and the GSD is 0.3 m. The images in this dataset are collected in two senses: urban and rural. Each image includes red, green, and blue bands. The ground truth contains seven categories: building, road, water, barren, forest, agriculture and background. A total of 2522 images are taken for training, and 1669 images are used for testing.

In the phase of model training, data augmentation tricks are employed to enlarge the training samples, including random horizontal flipping, random vertical flipping, and random scaling from a range in {0.5, 0.75, 1.0, 1.25, 1.5} (all with equal probability). In addition, the brightness and contrast of the input image are also randomly changed for training. Then, patches with 512 × 512 pixels are cropped from these images. They are finally organized as a training and testing dataset for model training, evaluation, and comparison.

4.2. Compared Models and Experiment Settings

The proposed DyHRNet was compared with the nine classic or state-of-the-art (SOTA) deep learning models for semantic segmentation. These models achieve multi-scale feature fusion for fine segmentation in different ways. For convenience, we summarize them as follows:

- **FCN**: It is a seminal work for semantic segmentation [30]. Currently, it is usually taken as a baseline for comparison. In our experiments, we use the ResNet-101 [64] as its encoder.
- **UNet**: This model contains a contracting path and a symmetric expanding path for multi-scale feature fusion [31]. It is initially designed for biomedical image seg-

mentation, and later widely applied to other types of images. Here, the backbone UNet-S5-D16 is taken as its encoder.

- **PSPNet**: It consists of a pyramid parsing module for global prior representation [32]. The concatenation of multi-scale pyramid representations is transformed to obtain the final per-pixel prediction. The ResNet-101 is employed as its encoder.
- **DeepLabV3+**: This model has an encoder–decoder structure [33]. In DeepLabV3+, the Xception model is modified to extract dense feature maps, and the depthwise separable convolution is employed to design the atrous spatial pyramid pooling and decoder modules. In the experiments, the ResNet-101 is employed as its encoder.
- **OCRNet**: This model fulfills the task of semantic segmentation via three main steps [34]. First, the contextual pixels are divided into a set of soft object regions. Second, the representations of the pixels in each object region are aggregated to obtain object-level representation. Finally, the representation of each pixel is augmented by object-contextual representation (OCR). The ResNet-101 is taken as its encoder.
- **SETR**: The SEgmentation TRansformer (SETR) is a SOTA model [35]. It is a pure transformer. Each image is encoded as a sequence of patches. The VIT-L is employed as its encoder.
- **SegFormer**: It is also a SOTA model constructed on transformers [36]. It has a hierarchically structured transformer encoder to learn the multi-scale features. In addition, the decoder is directly constructed on a lightweight multilayer perceptron, which aggregates information from different layers. In our implementation, the MIT-B5 is taken as its encoder.
- **HRNet+FCN**: In this model, the encoder is the standard HRNet [22]. The decoder is the same as that used in FCN [30]. Please note that only the up-sampling framework of FCN is inherited for segmentation. The standard HRNet is used as its encoder.
- **HRNet+OCR**: It is a SOTA model. The encoder is the standard HRNet [22] and the decoder is the head subnetwork used in OCRNet [34] for segmentation.
- **DyHRNet (Our)**: The pipeline of our method consists of three stages. First, we perform step 3 in Algorithm 1 to train the completely connected network for several epochs to obtain a good initialization. Second, steps 4–11 in Algorithm 1 are implemented to search from the dense connections and obtain channel-wise attention. Third, the final architecture is re-trained with all training data for experimental comparison.

In the experiments, all the above nine models to be compared were performed in the experiment settings suggested in the corresponding paper within the Pytorch framework. In addition, the guidance given by the authors in their works is followed to initialize the hyper-parameters. All models are trained with the SGD strategy.

The base learning rate η was initially set as 0.01, the momentum is set to 0.9, and the weight decay is taken as 0.004 during iterations. In addition, the "poly" learning rate policy is employed to adjust the learning rate [65]. At the t-th iteration, it is taken as

$$\eta_t = \eta(1 - t/T)^{0.9}, \tag{17}$$

where T is the pre-defined total number of iterations in Algorithm 1. In this way, the sequence of iteration points could be smoother for convergence. It was set to be 40,000 for the Vaihingen dataset. For the larger Potsdam and the LoveDA datasets, it was taken as 80,000. The regularization parameters λ in Problem (6) are set to be 0.01 in our implementation.

In the experiments, the size of each mini-batch was taken as 8 for the Vaihingen dataset, and 16 for the Potsdam and LoveDA datasets, when training the model parameters via Algorithm 1. The ResNet-101 [64] was pre-trained on the ImageNet dataset [66]. In our implementation, the original HRNet was also pre-trained on this dataset to obtain a good initialization (https://github.com/HRNet/HRNet-Image-Classification/releases/download/PretrainedWeights/HRNet_W48_C_ssld_pretrained.pth (accessed on 6 March 2023)).

To comprehensively evaluate the different models, two overall benchmark metrics were employed, namely the Overall Accuracy (OA) score of the classification and the mean

Intersection over Union ($mIoU$) evaluated on the testing samples. In addition, the accuracy for each class will also be reported for comparison.

4.3. Experiment Results

To give a comprehensive comparison between the models, we list the quantitative scores of OA and $mIoU$ obtained by the ten models on the Vaihingen, Potsdam, and LoveDA datasets in the right panels in Tables 1–3, respectively. All these values are achieved on the corresponding test images and averaged as a whole on the categories. Specifically, two methods are implemented to output the final scores. One is to calculate them directly on the images in the testing dataset. Another is to evaluate them via the flip and multi-scale (MS) testing, which means that the testing images are randomly flipped and/or resized to augment the samples.

Table 1. Quantitative comparison results on the Vaihingen testing set. The digits are the percent scores (%). † means that the scores are obtained via the flip and MS testing. (Bold font represents the highest performance of the class).

Method	Imp. Surf.	Building	Low Veg.	Tree	Car	OA	$mIoU$	OA †	$mIoU$ †
FCN [30]	84.99	91.31	70.31	79.57	76.18	89.76	80.47	90.34	81.87
UNet [31]	82.78	87.41	67.69	78.17	65.90	88.15	76.39	89.44	79.15
PSPNet [32]	85.83	91.49	71.37	79.90	75.14	90.16	80.75	90.76	82.38
DeepLabV3+ [33]	86.20	91.61	71.43	79.74	75.18	90.23	80.83	90.81	82.19
OCRNet [34]	84.70	90.57	69.74	78.83	66.71	89.36	78.11	90.23	79.99
SETR [35]	83.12	88.21	67.07	77.86	55.31	88.13	74.31	89.04	75.72
SegFormer [36]	86.72	**92.42**	72.30	80.53	78.54	90.68	82.10	91.07	83.06
HRNet+FCN [30]	85.91	91.91	71.03	79.90	76.40	90.17	81.03	90.75	82.35
HRNet+OCR [34]	86.77	91.43	73.51	80.65	78.34	90.54	82.14	91.48	83.73
DyHRNet (Ours)	**87.06**	92.26	**73.68**	**80.83**	**82.76**	**90.96**	**83.32**	**91.70**	**84.34**

Table 2. Quantitative comparison results on the Potsdam testing set. The digits are the percent scores (%). † means that the scores are obtained via the flip and MS testing. (Bold font represents the highest performance of the class).

Method	Imp. Surf.	Building	Low Veg.	Tree	Car	OA	$mIoU$	OA †	$mIoU$ †
FCN [30]	87.00	93.58	75.77	78.90	92.44	90.44	85.54	90.82	86.23
UNet [31]	83.63	89.08	73.28	77.75	89.69	88.36	82.69	89.14	84.02
PSPNet [32]	87.44	94.03	76.64	79.33	93.02	90.80	86.09	91.29	86.81
DeepLabV3+ [33]	87.40	93.82	76.60	79.28	93.03	90.80	86.03	91.27	86.72
OCRNet [34]	85.17	90.22	75.31	76.96	89.83	89.33	83.50	90.21	84.92
SETR [35]	78.28	84.78	66.60	66.17	77.33	84.25	74.63	85.54	76.77
SegFormer [36]	87.54	94.04	**78.15**	80.06	92.22	91.18	86.40	91.61	87.19
HRNet+FCN [30]	81.22	93.75	76.86	79.54	92.65	90.80	84.80	91.36	86.89
HRNet+OCR [34]	86.31	93.03	77.04	78.86	90.58	90.57	85.16	91.25	86.62
DyHRNet (Ours)	**87.77**	**94.07**	77.93	**80.59**	**93.05**	**91.20**	**86.68**	**91.79**	**87.56**

As can be seen from the comparative results in these tables, our model DyHRNet largely outperforms the seminal FCN and the famous UNet, which were initially developed for semantic segmentation. It is also superior to the powerful PSPNet, DeepLabV3+, and OCRNet models, which were all designed with the tricks of multi-scale information fusion in large receptive fields. In addition, it outperforms the SOTA models, including SETR and SegFormer, in which the frontier technique of Transforms is employed to construct the models. In parallel, our model is developed from the original HRNet. The two combinations, HRNet+FCN and HRNet+OCR, were compared against our model. By contrast, the HRNet+FCN employs a simple head network for semantic segmentation, while the HRNet+OCR uses a complex head network for this task. As can be seen from these tables, our model achieves better results, demonstrating its effectiveness for RS image segmentation.

Table 3. Quantitative comparison results on the LoveDA testing set. The digits are the percent scores (%). † means that the scores are obtained via the flip and MS testing. Here, "Back." stands for "Background", "Build." stands for "Building", "Agri." stands for "Agriculture". (Bold font represents the highest performance of the class).

Method	Back.	Build.	Road	Water	Barren	Forest	Agri.	*OA*	*mIoU*	*OA* †	*mIoU* †
FCN [30]	52.75	62.63	53.62	66.06	22.38	38.97	49.54	68.11	49.42	67.47	42.66
UNet [31]	49.06	57.82	47.87	47.69	25.90	37.66	43.65	63.23	44.24	64.05	44.04
PSPNet [32]	55.14	64.24	55.54	68.03	27.01	41.56	51.53	70.27	51.86	69.98	51.34
DeepLabV3+ [33]	54.19	64.39	55.67	68.14	27.17	41.44	49.29	69.50	51.47	69.60	51.32
OCRNet [34]	53.10	51.79	54.56	59.71	23.70	35.69	46.99	66.56	46.51	65.54	45.21
SETR [35]	47.98	57.24	40.37	59.70	20.23	39.54	36.39	62.50	43.06	62.01	42.65
SegFormer [36]	52.82	**65.50**	**56.63**	70.64	29.29	**41.63**	51.93	69.96	52.63	70.07	52.25
HRNet+FCN [30]	54.37	60.97	56.08	68.54	26.77	41.11	52.09	69.87	51.42	70.06	51.64
HRNet+OCR [34]	**54.62**	63.47	52.76	70.54	34.68	35.55	51.02	69.89	51.81	69.24	50.27
DyHRNet (Ours)	54.03	62.83	55.82	**72.31**	**35.67**	38.04	**58.92**	**71.55**	**53.95**	**70.98**	**53.72**

Furthermore, most images render class imbalance at the pixel level in these datasets, i.e., some objects (for example, buildings) occupy large regions, while the small objects (for example, cars) have small regions here and there in images. Thus, we employ the metric $mIoU$ to measure the goodness of the segmentation, respectively, on the category level. As can be seen from the left panels in Tables 1–3, our model achieves better scores for most categories in these datasets. It renders a significant performance enhancement on small objects. Table 1 shows that DyHRNet achieves 4.42% higher accuracy than the second model (SegFormer) on the car category in the Vaihingen dataset. Such an enhancement can also be witnessed in Table 2 on the tiny objects, including the car and the tree in the scenes. In addition, as witnessed in Table 3, our model also obtains the SOTA performance on the LoveDA dataset.

Figure 5 illustrates the radar charts on the three datasets to further compare the performances of the ten models, category by category. The points in these charts stand for the corresponding $mIoU$ scores, which are obtained via data augmentation (flip and MS testing) tricks on the testing dataset. From these figures, it is seen that the curves obtained by our model always locate at the outer region, indicating that it achieves higher performance compared with the nine models.

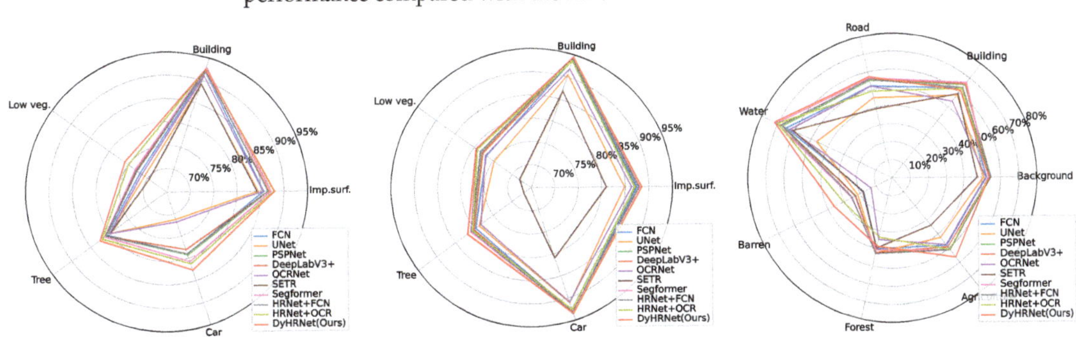

(a) (b) (c)

Figure 5. Comparisons category by category on the three datasets via Radar chart. The digits are the $mIoU$ scores, obtained via the flip and MS testing. (**a**) Vaihingen; (**b**) Potsdam; (**c**) LoveDA.

Finally, Figures 6–8 demonstrate the segmentation results of a few images obtained by the ten models, including FCN, UNet, PSPNet, DeepLabV3+, OCRNet, SETR, SegFormer, HRNet+FCN, HRNet+OCR and DyHRNet.

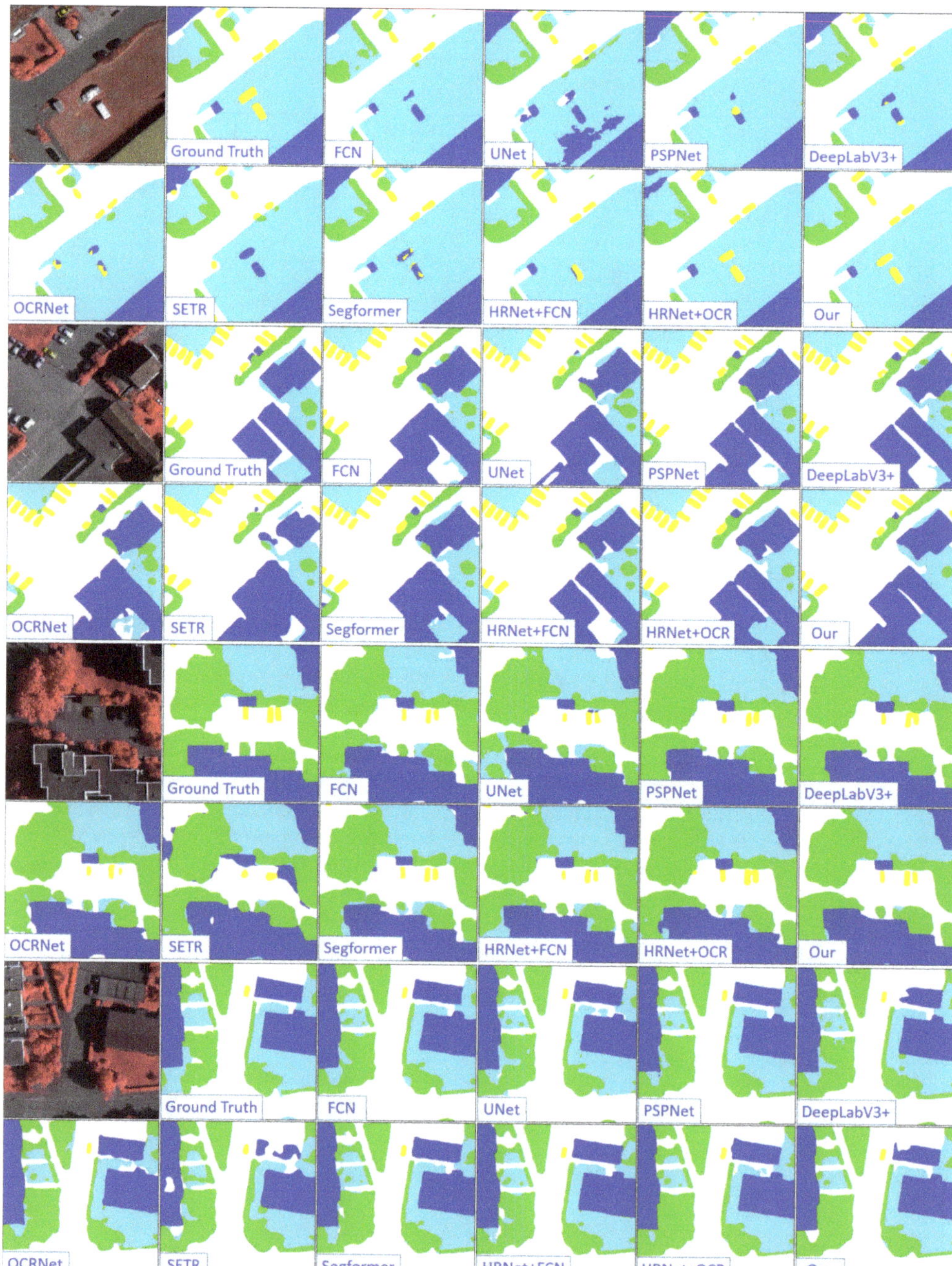

Figure 6. Visual comparisons between our method and other related methods on the Vaihingen dataset. The label includes six categories: impervious surface (white), building (blue), low vegetation (cyan), tree (green), and car (yellow).

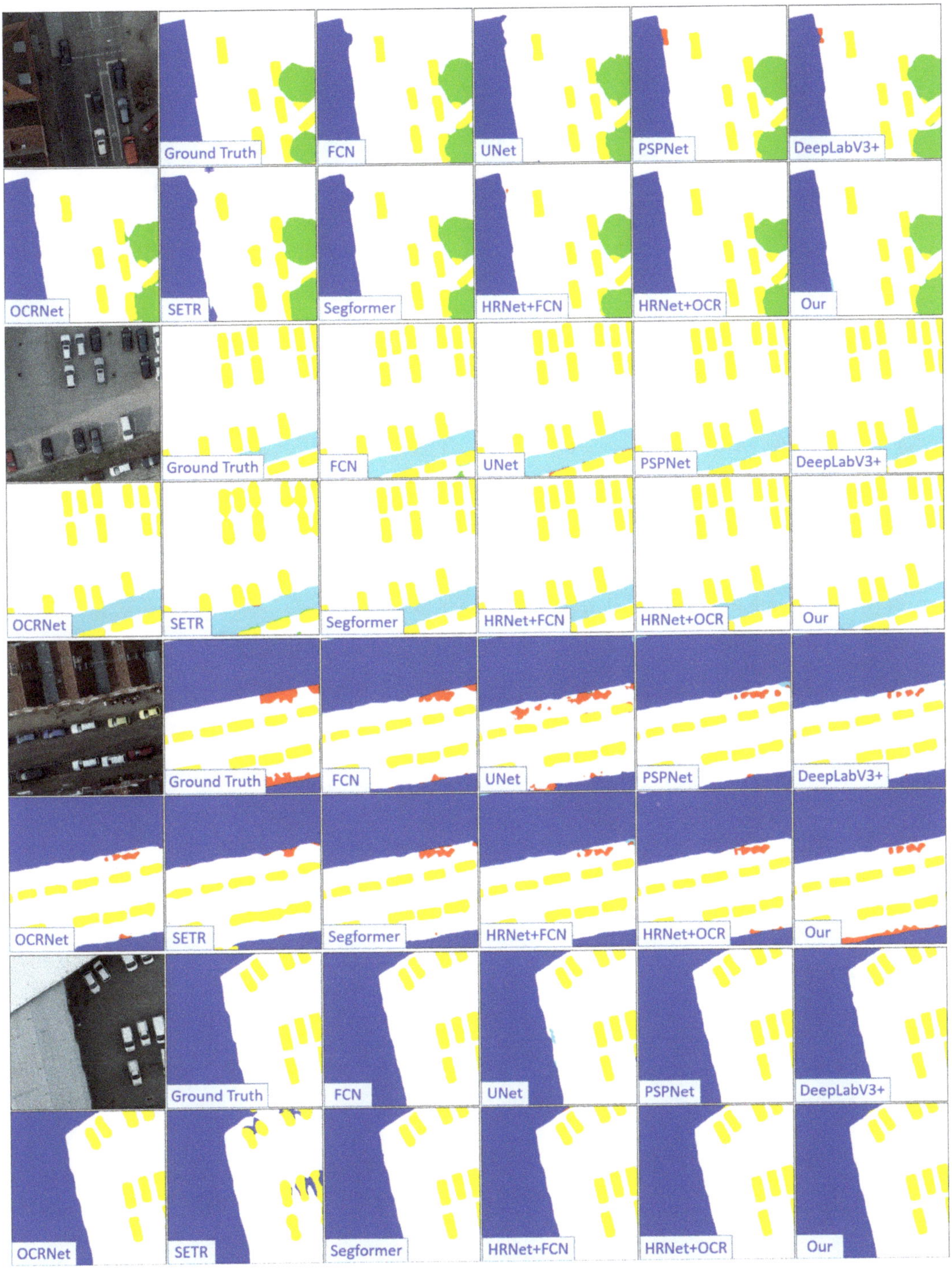

Figure 7. Visual comparisons between our method and other related methods on the Potsdam dataset. The label includes six categories: impervious surface (white), building (blue), low vegetation (cyan), tree (green), car (yellow), and clutter/background (red). Here, backgrounds are directly shown as they were considered to be masks when training the models.

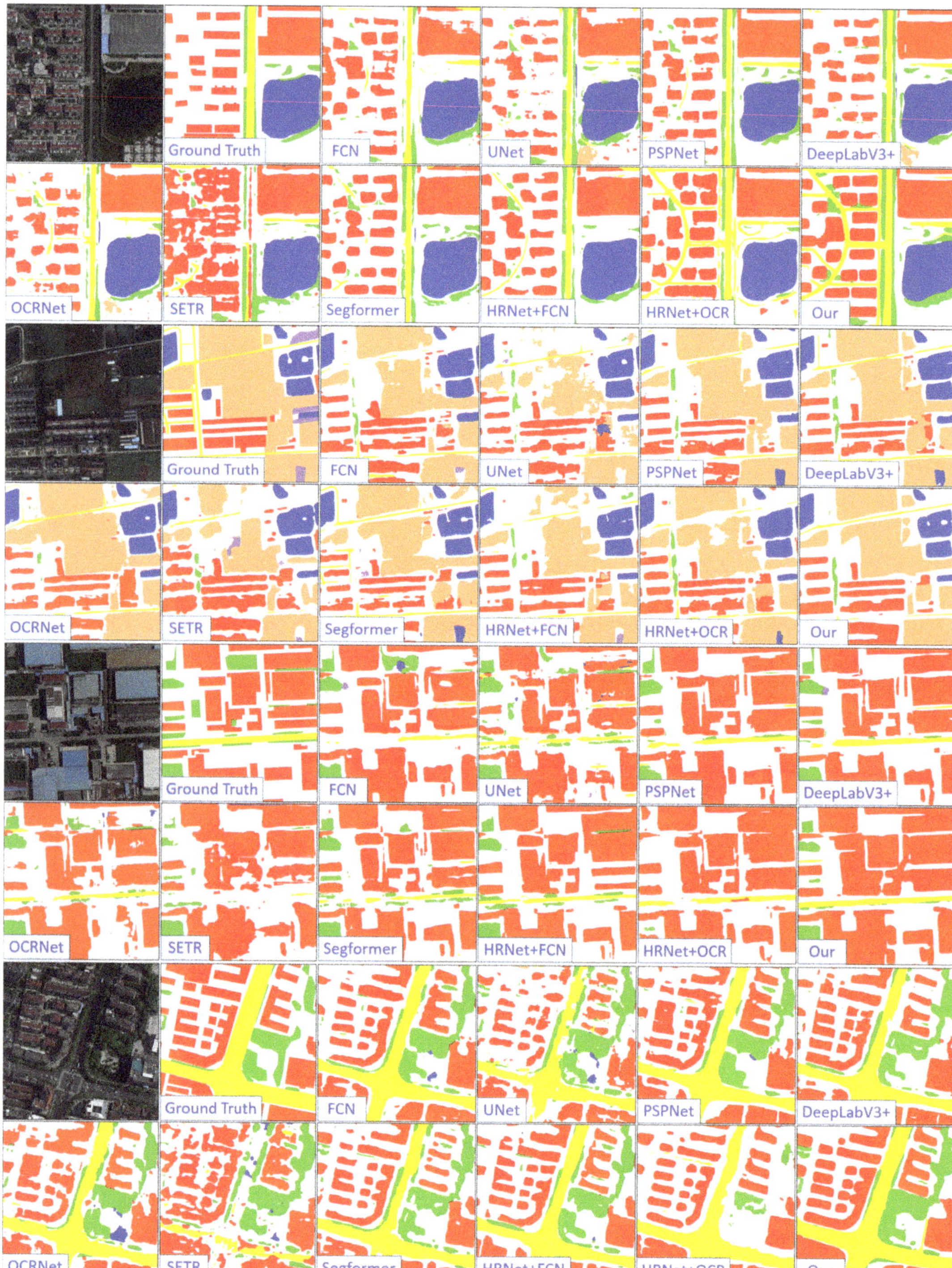

Figure 8. Visual comparisons between our method and other related methods on the LoveDA dataset. The label includes six categories: Background (white), Building (red), Road (yellow), Water (blue), Barren (plum), Forest (green), and Agriculture (orange).

As can be seen from these figures, the FCN and UNet render limited performances in segmenting the tiny objects. For the segmentation results obtained by the PSPNet, the boundaries of objects do not keep well, which can be seen from those regions of some buildings in the LoveDA dataset. DeepLabV3+, OCRNet, SETR, SegFormer, HRNet+FCN, and HRNet+OCR indeed improve the quality of segmentation, but false negatives can also be perceived from the segmentation results. This fact can be witnessed clearly in the Vaihingen dataset. In contrast, our model performs better and shows satisfactory edge preserving, typically on the Vaihingen and Potsdam datasets, where coherent segmentations are obtained on fine-structured buildings and tiny objects. In addition, on the challenging LoveDA dataset, for example, in the first image in Figure 8, the roads in the left region are all manually labeled as background, but our model segments well for these regions.

In summary, the above comparisons show that our DyHRNet is capable of segmenting confusing artificial objects in high-resolution RS images. In addition, it also shows good quality segmentations with satisfactory edge preserving for confusing size-variable objects and tiny objects such as cars and trees without performing post-processing. This indicates that our DyHRNet learned from the densely connected HRNet with channel-wise attention has powerful data adaptability for RS images.

4.4. Ablation Study

This subsection reports the ablation experiments to evaluate the importance of different components proposed in our method. Please note that in our work, there are two fundamental designs. One is the dynamic dense connections achieved with the APG algorithm in Section 3.2, and another is the dynamic channels in Section 3.3. Table 4 reports four combinations with or without using these two designs on the Vaihingen dataset. It is seen that performing both fundamental modelings helps enhance the performance. This indicates the validation of our proposed approach.

When learning our DyHRNet, there are two subtasks: training the neural network and searching from the dense connections. They are solved alternatively by fixing one for another in Algorithm 1. To further evaluate the importance of the APG algorithm for selecting out the important connections, we conducted another group of ablation experiments, in which the APG algorithm is performed once a time every two or four times during iterations. Table 5 reports the performances obtained in this experimental setting. It is seen that the performances decrease in most cases when the interval number increases, compared with the original step-by-step for these two subtasks. This indicates that the APG algorithm for sparse selection from the dense connections plays a significant role in data-driven learning for the semantic segmentation of RS images.

Table 4. The ablation study of our proposed DyHRNet on the Vaihingen dataset with different combinations. The digits are the percent scores (%). † means that the scores are obtained via the flip and MS testing.

Channel-Wise Attention	APG Sparse Optimization	*OA*	*mIoU*	*OA* †	*mIoU* †
✗	✗	90.54	82.14	91.48	83.73
✓	✗	90.92	83.11	91.55	84.12
✗	✓	90.94	83.23	91.54	84.15
✓	✓	90.96	83.32	91.70	84.34

Table 5. The ablation study of our proposed DyHRNet on the Vaihingen dataset. The digits are the percent scores (%). † means that the scores are obtained via the flip and MS testing.

Train/Search Step	*OA*	*mIoU*	*OA* †	*mIoU* †
1/1	90.96	83.32	91.70	84.34
2/1	90.46 (↓)	81.29 (↓)	91.20 (↓)	82.66 (↓)
4/1	90.40 (↓)	81.46 (↓)	90.77 (↓)	81.58 (↓)

Finally, the DyHRNet is used as the backbone of the models with different heads for semantic segmentation. Specifically, the heads in FCN and OCR are combined with the DyHRNet, respectively, for experimental evaluation. This treatment generates two models, known as "DyHRNet+FCN" and "DyHRNet+OCR". The latter is the model used in Section 4.2. For convenience, the original HRNet is also employed for comparison. The experimental settings are the same used in Section 4.2. Table 6 reports the experimental results on the Vaihingen dataset. It is seen that the performance of "DyHRNet+FCN" is better than that of "HRNet+FCN", while the performance of "DyHRNet+OCR" is better than that of "HRNet+OCR". As our model is learned from the original HRNet, this fact indicates the usage of our proposed method.

Table 6. The ablation study of the original HRNet and our DyHRNet with different segmentation heads on the Vaihingen dataset. The digits are the percent scores (%). † means that the scores are obtained via the flip and MS testing.

Backbone	FCN	OCR	*OA*	*mIoU*	*OA* †	*mIoU* †
HRNet	✓		90.17	81.03	90.75	82.35
		✓	90.54	82.14	91.48	83.73
DyHRNet	✓		90.80	82.72	91.45	83.82
		✓	90.96	83.32	91.70	84.34

4.5. Computational Efficiency

Here we analyze the computational efficiency of the models. To give a comprehensive analysis, the following models are compared, including the FCN, UNet, PSPNet, DeepLabV3+, OCRNet, SETR, SegFormer, HRNet+FCN, HRNet+OCR, DyHRNet +FCN, and DyHRNet+OCR. In Table 7, the performance of the architecture with a combination of the learned backbone DyHRNet and the head of the FCN, namely "DyNRnet+FCN", is additionally reported for comparison. The factors related to the computational efficiency are listed in Table 7, including the number of the parameters and the number of the FLoating-point OPerations (FLOPs) with Giga Multiplier ACcumulators (GMACs) in the model.

Table 7. Computational efficiency, including the total number of the FLoating-point OPerations (FLOPs) and the total number of the parameters in the model. The backbones and the mIoU scores calculated on the Vaihingen testing dataset are also listed here for comparison.

Method	Backbone	#Params	GFLOPs	*mIoU* (%)
FCN [30]	ResNet-101	68.48M	275.38	80.47
PSPNet [32]	ResNet-101	67.96M	256.14	80.75
DeepLabV3+ [33]	ResNet-101	62.57M	253.93	80.83
OCRNet [34]	ResNet-101	55.51M	230.57	78.11
UNet [31]	UNet-S5-D16	29.06M	202.63	76.39
SETR [35]	VIT-L	318.45M	260.85	74.31
SegFormer [36]	MIT-B5	82.01M	52.45	82.10
HRNet+FCN [30]	HRNetV2-W48	65.85M	93.43	81.03
HRNet+OCR [34]	HRNetV2-W48	70.36M	162.21	82.14
DyHRNet+FCN	DyHRNet	63.88M	91.57	82.72
DyHRNet+OCR	DyHRNet	66.44M	158.81	83.32

In addition, the backbones and the mIoU scores calculated on the Vaihingen testing dataset are also listed for comparison. In the experiments, the FCN, PSPNet, DeepLabV3+, and OCRNet all take the "ResNet-101" as their backbone. By contrast, UNet, SETR, SegFormer, and HRNet all have their own backbone. It is worth pointing out that the downsampling operations in UNet and ResNet are different from each other. Furthermore, they also have different output channels in consecutive stages. Thus, the "ResNet-101" is not selected as the backbone for UNet in our experiments.

In general, the number of parameters and the number of the FLOPs are two important factors to evaluate the computation scale in deep models. In particular, FLOPs can directly reflect the computational complexity of the model, which is related to the size of the input image and the neural architecture. As can be seen in Table 7, the computation scale in the DyHRNet was reduced to a medium level, but it achieves the best performance on the Vaihingen dataset. For example, by contrast to the FCN, our model has the parameters at the same level, but the computational scale is largely reduced to 57.6%. In addition, from Table 7, it is seen that the numbers of parameters and FLOPs in DyHRNet+FCN are both smaller than those in HRNet+FCN. This fact can also be witnessed when HRNet+OCR and DyHRNet+OCR are compared to each other. Thus, it can be concluded that the decrease of the computations in our model occurs in the backbone. This is due to the architecture learning from the HRNet, where those cross-resolution connections and channels with zero contributions will be ignored. This indicates the effectiveness of our method.

5. Discussions

5.1. The Learned Structure and Iteration Process Analysis

Algorithmically, the main task in this study is to solve the optimization problem in (6). In this task, the key job is to search for the important connections among the eight groups of dense connections in the original HRNet. This is achieved using the APG algorithm with sparse selection tricks. Figure 9a–c illustrate the learned weights of the eight groups of dense connections. For example, in the first panel in Figure 9a–c, there is a 2×2 matrix, including four weights. This panel corresponds to the group of dense connections in stage 2 in Figure 1. For clarity, we take Figure 9a as an example to explain the details. In the first panel, the value "0.021" is the learned weight of the connection between $R_{1,1}$ and $O_{1,2}$, "0.026" is that of the connection between $R_{1,1}$ and $O_{2,2}$, "0.000" is that of the connection between $R_{2,1}$ and $O_{1,2}$, and "0.202" is that of the connection between $R_{2,1}$ and $O_{2,2}$. Based on Figure 9a–c, it is seen that there are many connections with weights tending to zero. This fact indicates that our search approach plays an active role in the original architecture of HRNet, helping enhance the performance of the model for semantic segmentation with end-to-end training.

Based on the visualizations in Figure 9a–c, one interesting phenomenon is that all the connections at the same horizontal level retain relatively higher weights. This means that the feature maps at the same resolution should be maintained, which is more important for nonlinear feature learning. Small weights are always learned from cross-resolutions, and most of them correspond to the connections from high to low resolution. This may be because there are many tiny objects in the RS images, avoiding discarding the details of the tiny objects and performing those down-sampling operations.

As demonstrated in Figure 1, a total of 88 connections are employed as candidates to be selected by the APG algorithm described in Section 3.2. Figure 10 shows the $mIoU$ scores and the total weights of the 88 connections on the Vaihingen, Potsdam, and LoveDA datasets in the iterations. It is seen that the $mIoU$ scores of the learned models after 10,000 iterations will stay at the same level without rendering significant changes. However, the total sum of the learned weights is drastically decreased along with the increase of iterations. In the beginning, all the weights are taken as 1.0. Thus, there is a point marked as "88.0" in each figure. Then, it reduces and converges to keep at a stationary level. This means that the original dense connections have different contributions, far below the identical ones with the same importance. This fact also indicates that the connections with zero weights can be deleted from the original HRNet in the way of end-to-end learning.

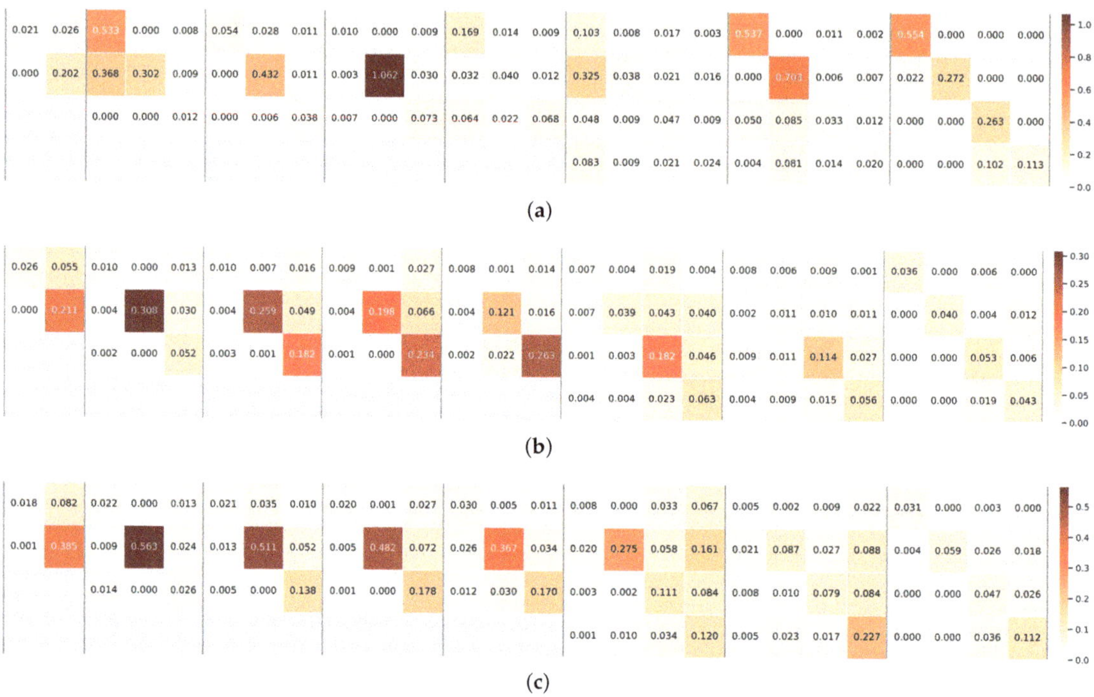

Figure 9. Visualization of the obtained weights of the dense connections. In each panel separated by vertical lines, there are a group of weights, corresponding to the connections in Figure 1. The larger the weight, the more important the connection; (**a**) Vaihingen; (**b**) Potsdam; (**c**) LoveDA.

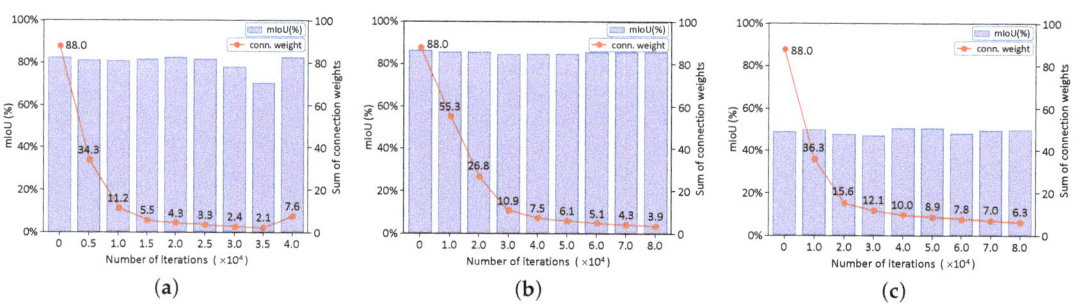

Figure 10. The mIoU score and the overall weight sum learned from the dense connections with iterations; (**a**) Vaihingen; (**b**) Potsdam; (**c**) LoveDA.

5.2. The Behavior of the Accelerated Proximal Gradient Algorithm

In Algorithm 1, one of the main tasks is to identify the important cross-resolution connections in the original HRNet for semantic segmentation. To this end, the APG algorithm is employed to solve the sparse regularization subproblem in Problem (6). Algorithmically, it starts with the original HRNet to modulate the weights of the cross-resolution connections, which are all initialized as 1.0 for iterations. This gives an equal chance for all the connections to be evaluated. In this way, a top-down training strategy is performed, in which all the weights are gradually nullified to small scores. After training, the connections with zero weights will be discarded for prediction.

Alternatively, a bottom-up training strategy could be implemented for the APG algorithm. To this end, we first assign a small weight to the connections, and then pre-train the original HRNet on the ImageNet dataset. Then, Algorithm 1 is implemented. Table 8 reports the experimental results on the Vaihingen dataset. Specifically, in Table 8, "Init-0.1", "Init-0.5", and "Init-1.0" correspond to the cases with all weights initialized to 0.1, 0.5, and 1.0, respectively. In this group of experiments, the maximum number of iterations is taken as 40,000, and all the training samples described in Section 4.1 are taken to learn the model. In the experiments, it is observed that small initial weights indeed help speed up the convergence, but the performance decreases drastically. As can be seen from Table 8, the models trained with small initial weights perform unsatisfactorily, compared to that with all weights set to be 1.0 for learning.

Table 8. Quantitative comparison results on the Vaihingen testing set with different initial weights to the cross-resolution connections for the AGP algorithm. The digits are the percent scores (%). † means that the scores are obtained via the flip and MS testing.

Weight Initialization	Imp. Surf.	Building	Low Veg.	Tree	Car	*OA*	*mIoU*	*OA* †	*mIoU* †
Init-0.1	75.85	84.48	65.93	74.19	55.74	85.63	71.24	82.46	66.60
Init-0.5	83.06	87.41	68.75	78.26	61.39	88.38	75.77	89.03	77.01
Init-1.0	87.06	92.26	73.68	80.83	82.76	90.96	83.32	91.70	84.34

To further investigate the behavior of the AGP algorithm, experiments with different ratios of training data and different numbers of iterations are conducted on the Vaihingen dataset. The goal is to demonstrate whether the weights of the cross-resolution connections learned by the AGP algorithm change drastically. Specifically, in the first group of experiments, the ratios are set as 10%, 50%, and 100% of all the total training samples, respectively. In this group, the maximum number of iterations, namely parameter T in Algorithm 1, is taken as 40,000. In another group of experiments, T is set as 10,000, 20,000, and 40,000, respectively. In this case, all the training samples are employed to learn the model. In this group, the weights of the cross-resolution connections are initialized to 1.0 for the APG algorithm. Figure 11 visualizes the weights of the cross-resolution connections in the DyHRNet, which are obtained, respectively, with these experimental settings. It is seen that all the weights are dropped below 1.0. Table 9 reports the performances of the learned models. By considering together the weights visualized in Figure 9a and the performance scores, which are obtained with 100% training samples and $T = 40,000$, there are no significant changes both in performance scores and in weight values.

Table 9. Quantitative comparisons on the Vaihingen testing set with different ratios of training samples and different numbers of iterations. The digits are the percent scores (%). † means that the scores are obtained via the flip and MS testing. Here, "100% + 40,000" means the model is learned with 100% of the training samples, and the maximum number of iterations (T) in Algorithm 1 is set to be 40,000.

	Imp. Surf.	Building	Low Veg.	Tree	Car	*OA*	*mIoU*	*OA* †	*mIoU* †
#training (10%)	86.49	91.81	71.51	79.72	79.11	90.41	81.73	91.10	82.07
#training (50%)	86.80	91.85	72.74	80.39	80.39	90.70	82.43	91.41	83.57
T = 10,000	86.31	92.00	72.46	80.68	81.75	90.68	82.64	91.37	83.84
T = 20,000	87.53	92.70	73.28	80.83	82.16	91.10	83.30	91.55	84.11
100% + 40,000	87.06	92.26	73.68	80.83	82.76	90.96	83.32	91.70	84.34

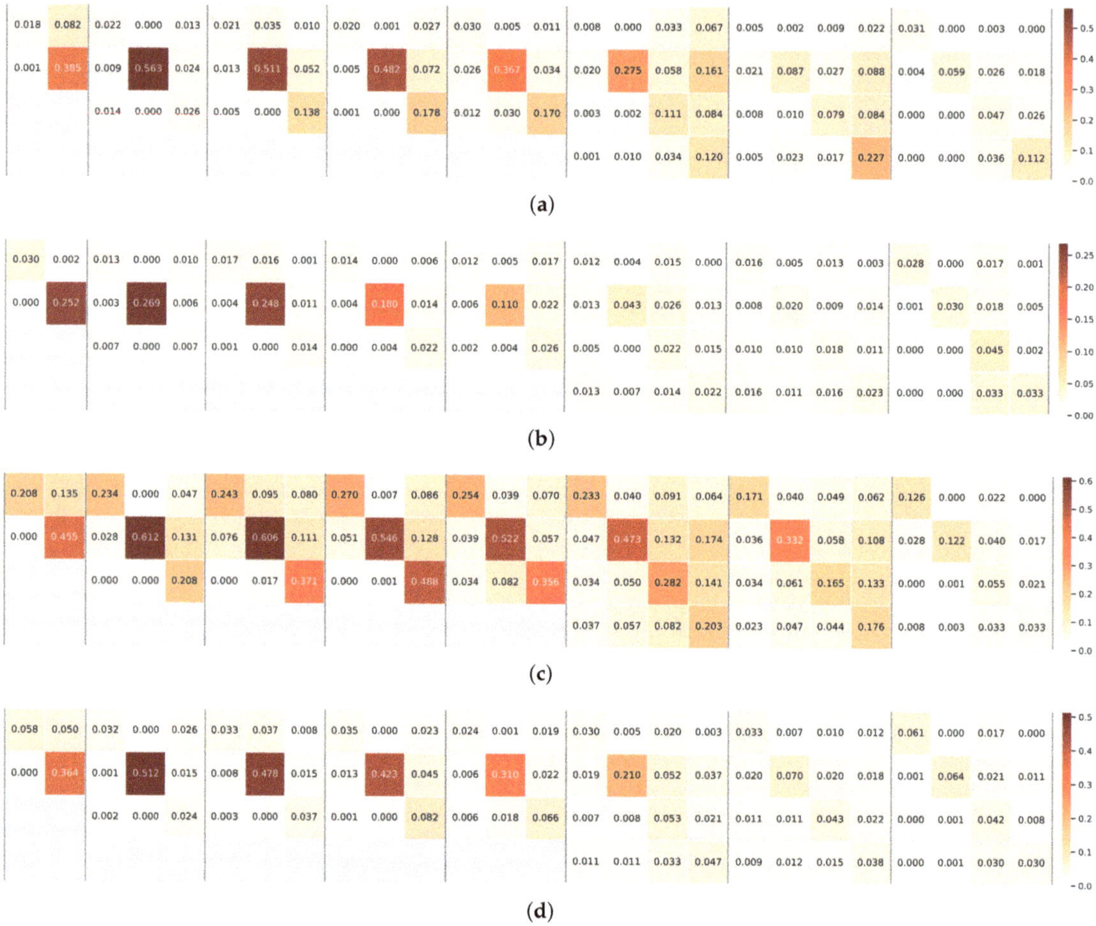

Figure 11. Visualization of the weights of the dense connections learned from the Vaihingen dataset with different settings. In each panel separated by vertical lines, there are a group of weights, corresponding to the connections in Figure 1. The weights in (**c**,**d**) are learned with all of the training samples used in Table 1. (**a**) The weights learned with 10% training samples; (**b**) The weights learned with 50% training samples; (**c**) The weights learned with 10,000 iterations; (**d**). The weights were learned with 20,000 iterations.

5.3. Implications and Limitations

In this study, we have conducted experimental evaluations on the three challenging public datasets. In these RS images, many artificial objects with different sizes and confusing appearances are located here and there. Achieving consistent and accurate semantic segmentation is a challenging task. The primary function of the DyHRNet is to enhance the quality of semantic segmentation via cross-resolution feature fusion. To this end, the structure of the original HRNet has been exploited by evaluating the contributions of the dense connections and the channels related to the cross-resolution feature fusion. With the problem formulation addressed under the NAS framework with channel-wise attention, the goal is well achieved. This means that the structure in the HRNet with parallel streams of high-, medium-, and low-resolution representation attends well to the needs of segmenting RS objects of multi-scales. In addition, compared with the original HRNet, the reduced parameters in our DyHRNet indicate that not all the connections and channels contribute equally to the task. In other words, redundant connections and channels exist for performing the cross-resolution feature fusion. Therefore, one can consider designing lightweight or dynamic HRNet or more general architectures by keeping the advantages of the multi-scale structure design rendered in the HRNet for this or similar tasks in the fields of RS image processing.

Methodologically, the proposed DyHRNet renders enhancements on both architecture design and segmentation performance. However, it has several limitations, which are described as follows:

- As outlined in Table 7, our model has more than 66 million parameters, and the computational scale is up to about 158 GFLOPs. Thus, high-performance computing resources with GPUs are needed to fulfill the computing task. This indicates that releasing it on edge computing devices is difficult with its current version. However, the sum of the total weights demonstrated in Figure 10 indicates many connections with small contributions. Thus, the scale of the models could be reduced by model pruning.
- The performance of the DyHRNet could be further improved for the objects with rich visual appearances and those with blurring edges, for example, the buildings and forests in the LoveDA dataset. On the one hand, more training samples are needed to guarantee the generalization of the model. On the other hand, some prior knowledge with constrained forms or regularization terms in the loss function could be introduced to guide the model training.
- The current version of the DyHRNet is not general given the dynamic architecture design. This is because the tricks with NAS are only applied to dense connections. However, many convolutional operations are contained in the blocks $\mathbf{O}_{i,j}$ in Figure 1. This indicates that one can perform the NAS on all operations to learn more general architecture for different needs in practice.

6. Conclusions

This work proposes a Dynamic High-Resolution Network (DyHRNet) for the semantic segmentation of RS images. The DyHRNet is an architecture-learnable model under a neural architecture search (NAS) framework with channel-wise attention. It takes the primary HRNet as its super-architecture, which has four parallel streams to retain the different resolutions for multi-scale feature fusion simultaneously. The learning task has been explicitly formulated with a series of sparse regularizations, where the Accelerated Proximal Gradient (APG) algorithm is introduced to solve the sparse optimization model. In contrast to the static HRNet, the dynamic merits of the DyHRNet with data adaptability lie in the following two aspects. On the one hand, the structure of the DyHRNet is dynamically adjusted for cross-resolution feature fusion by identifying those unimportant connections and ignoring those with zero contributions in the original HRNet. On the other hand, the contributions of channel-wise contribution for feature fusion are modulated automatically to enhance the representation capability of the proposed model.

Extensive experiments have been conducted on three public challenging RS image datasets. Nine classical or SOTA models have been employed to compare with our model. Comparative experiment results with numerical scores, visual segmentations, the learned structures, the iteration process analysis, and the ablation study demonstrate the advantages and effectiveness of the proposed DyHRNet. In the future, we aim to design a lightweight HRNet to perform the semantic segmentation of RS images in edge computing devices.

Author Contributions: Conceptualization, S.G. and S.X.; methodology, Q.Y. and S.X.; software, Q.Y. and S.G.; validation, S.G., Q.Y., S.X., P.W. and X.W.; investigation, S.G. and S.X.; writing—original draft preparation, S.G. and S.X.; writing—review and editing, S.G., S.X. and P.W.; formal analysis, S.G.; visualization, S.G. and Q.Y.; data curation, Q.Y.; supervision, S.X., P.W. and X.W.; resources, P.W. and X.W.; project administration, P.W. and X.W.; funding acquisition, X.W. All authors have read and agreed to the published version of the manuscript.

Funding: This work was supported by Key Research Program of Frontier Sciences, CAS (grant number: ZDBS-LY-DQC016), National Key Research and Development Program of China(grant number: 2022YFF1301803), and National Natural Science Foundation of China (NSFC) under grant 62076242.

Data Availability Statement: Three public datasets (i.e., the Vaihingen, Potsdam, LoveDA) were included in this study. Both the Vaihingen dataset and the Potsdam dataset were obtained via the official website: https://www.isprs.org/education/benchmarks/UrbanSemLab/default.aspx (accessed on 6 March 2023). The LoveDA dataset was downloaded from the webpage: https://github.com/Junjue-Wang/LoveDA (accessed on 6 March 2023).

Conflicts of Interest: The authors declare no conflict of interest.

References

1. Liu, Y.; Fan, B.; Wanga, L.; Bai, J.; Xiang, S.; Pan, C. Semantic Labeling in very High Resolution Images via A Self-cascaded Convolutional Neural Network. *ISPRS J. Photogramm. Remote Sens.* **2018**, *145*, 78–95. [CrossRef]
2. Li, L.; Yao, J.; Liu, Y.; Yuan, W.; Shi, S.; Yuan, S. Optimal Seamline Detection for Orthoimage Mosaicking by Combining Deep Convolutional Neural Network and Graph Cuts. *Remote Sens.* **2017**, *9*, 701. [CrossRef]
3. Panboonyuen, T.; Jitkajornwanich, K.; Lawawirojwong, S.; Srestasathiern, P.; Vateekul, P. Semantic Segmentation on Remotely Sensed Images Using an Enhanced Global Convolutional Network with Channel Attention and Domain Specific Transfer Learning. *Remote Sens.* **2019**, *11*, 83. [CrossRef]
4. Guo, S.; Jin, Q.; Wang, H.; Wang, X.; Wang, Y.; Xiang, S. Learnable Gated Convolutional Neural Network for Semantic Segmentation in Remote-Sensing Images. *Remote Sens.* **2019**, *11*, 1922. [CrossRef]
5. Liu, Y.; Chen, D.; Ma, A.; Zhong, Y.; Fang, F.; Xu, K. Multiscale U-Shaped CNN Building Instance Extraction Framework with Edge Constraint for High-Spatial-Resolution Remote Sensing Imagery. *IEEE Trans. Geosci. Remote Sens.* **2021**, *59*, 6106–6120. [CrossRef]
6. Zhang, J.; Lin, S.; Ding, L.; Bruzzone, L. Multi-Scale Context Aggregation for Semantic Segmentation of Remote Sensing Images. *Remote Sens.* **2020**, *12*, 701. [CrossRef]
7. Xu, Z.; Zhang, W.; Zhang, T.; Li, J. HRCNet: High-Resolution Context Extraction Network for Semantic Segmentation of Remote Sensing Images. *Remote Sens.* **2020**, *13*, 71. [CrossRef]
8. Elsken, T.; Metzen, J.H.; Hutter, F. Neural Architecture Search: A Survey. *J. Mach. Learn. Res.* **2019**, *20*, 55.
9. Bello, I.; Pham, H.; Le, Q.V.; Norouzi, M.; Bengio, S. Neural Combinatorial Optimization with Reinforcement Learning. In Proceedings of the ICLR Workshop Track, Toulon, France, 24–26 April 2017.
10. Liu, H.; Simonyan, K.; Yang, Y. DARTS: Differentiable Architecture Search. In Proceedings of the International Conference on Learning Representations, New Orleans, LA, USA, 6–9 May 2019.
11. Ghiasi, G.; Lin, T.Y.; Le, Q.V. NAS-FPN: Learning Scalable Feature Pyramid Architecture for Object Detection. In Proceedings of the IEEE/CVF International Conference on Computer Vision and Pattern Recognition, Long Beach, CA, USA, 15–20 June 2019; pp. 7036–7045.
12. Weng, Y.; Zhou, T.; Li, Y.; Qiu, X. NAS-Unet: Neural Architecture Search for Medical Image Segmentation. *IEEE Access* **2019**, *7*, 44247–44257. [CrossRef]
13. Zhang, X.; Chang, J.; Guo, Y.; Meng, G.; Xiang, S.; Lin, Z.; Pan, C. DATA: Differentiable ArchiTecture Approximation with Distribution Guided Sampling. *IEEE Trans. Pattern Anal. Mach. Intell.* **2021**, *43*, 2905–2920. [CrossRef]
14. Zhang, X.; Huang, Z.; Wang, N.; Xiang, S.; Pan, C. You Only Search Once: Single Shot Neural Architecture Search via Direct Sparse Optimization. *IEEE Trans. Pattern Anal. Mach. Intell.* **2021**, *43*, 2891–2904. [CrossRef]
15. Luo, R.; Tian, F.; Qin, T.; Chen, E.; Liu, T. Neural Architecture Optimization. In Proceedings of the Annual Conference on Neural Information Processing Systems, NeurIPS 2018, Montreal, QC, Canada, 3–8 December 2018; pp. 7827–7838.
16. Xie, S.; Zheng, H.; Liu, C.; Lin, L. SNAS: Stochastic Neural Architecture Search. In Proceedings of the International Conference on Learning Representations, New Orleans, LA, USA, 6–9 May 2019.
17. Liu, C.; Chen, L.; Schroff, F.; Adam, H.; Hua, W.; Yuille, A.L.; Fei-Fei, L. Auto-DeepLab: Hierarchical Neural Architecture Search for Semantic Image Segmentation. In Proceedings of the IEEE/CVF Conference on Computer Vision and Pattern Recognition, Long Beach, CA, USA, 15–20 June 2019; pp. 82–92.
18. Ulkua, I.; Akagunduz, E. A Survey on Deep Learning-based Architectures for Semantic Segmentation on 2D Images. *Appl. Artif. Intell.* **2022**, *36*, e2032924. [CrossRef]
19. Liang, H.; Zhang, S.; Sun, J.; He, X.; Huang, W.; Zhuang, K.; Li, Z. Darts+: Improved differentiable architecture search with early stopping. *arXiv* **2019**, arXiv:1909.06035.
20. Zela, A.; Elsken, T.; Saikia, T.; Marrakchi, Y.; Brox, T.; Hutter, F. Understanding and robustifying differentiable architecture search. In Proceedings of the International Conference on Learning Representations, Addis Ababa, Ethiopia, 26–30 April 2020.
21. Sun, K.; Xiao, B.; Liu, D.; Wang, J. Deep High-Resolution Representation Learning for Human Pose Estimation. In Proceedings of the IEEE/CVF International Conference on Computer Vision and Pattern Recognition, Long Beach, CA, USA, 15–20 June 2019; pp. 5693–5703.

22. Wang, J.; Sun, K.; Cheng, T.; Jiang, B.; Deng, C.; Zhao, Y.; Liu, D.; Mu, Y.; Tan, M.; Wang, X.; et al. Deep High-Resolution Representation Learning for Visual Recognition. *IEEE Trans. Pattern Anal. Mach. Intell.* **2021**, *43*, 3349–3364. [CrossRef] [PubMed]
23. Cheng, B.; Xiao, B.; Wang, J.; Shi, H.; Huang, T.S.; Zhang, L. HigherHRNet: Scale-Aware Representation Learning for Bottom-Up Human Pose Estimation. In Proceedings of the IEEE/CVF International Conference on Computer Vision and Pattern Recognition, Long Beach, CA, USA, 15–20 June 2020; pp. 5386–5395.
24. Yu, C.; Xiao, B.; Gao, C.; Yuan, L.; Zhang, L.; Sang, N.; Wang, J. Lite-HRNet: A Lightweight High-Resolution Network. In Proceedings of the IEEE/CVF International Conference on Computer Vision and Pattern Recognition, Virtual, 19–25 June 2021; pp. 10440–10450.
25. Li, Q.; Zhang, Z.; Xiao, F.; Zhang, F.; Bhanu, B. Dite-HRNet: Dynamic Lightweight High-Resolution Network for Human Pose Estimationn. In Proceedings of the International Joint Conference on Artificial Intelligence, Vienna, Austria, 23–29 July 2022; pp. 1095–1101.
26. Ding, M.; Zhang, S.; Yang, J. Learning a Dynamic High-Resolution Network for Multi-Scale Pedestrian Detection. In Proceedings of the International Conference on Pattern Recognition, Curico, Chile, 17–19 March 2021; pp. 9076–9082.
27. Yuan, Y.; Fu, R.; Huang, L.; Lin, W.; Zhang, C.; Chen, X.; Wang, J. HRFormer: High-Resolution Transformer for Dense Prediction. *arXiv* **2021**, arXiv:2110.09408v3.
28. Yuan, X.; Shi, J.; Gu, L. A review of deep learning methods for semantic segmentation of remote sensing imagery. *Expert Syst. Appl.* **2021**, *169*, 114417. [CrossRef]
29. Neupane, B.; Horanont, T.; Aryal, J. Deep Learning-Based Semantic Segmentation of Urban Features in Satellite Images: A Review and Meta-Analysis. *Remote Sens.* **2021**, *13*, 808. [CrossRef]
30. Long, J.; Shelhamer, E.; Darrell, T. Fully convolutional networks for semantic segmentation. *IEEE Trans. Pattern Anal. Mach. Intelligence.* **2015**, *79*, 1337–1342.
31. Ronneberger, O.; Fischer, P.; Brox, T. U-net: Convolutional networks for biomedical image segmentation. In Proceedings of the International Conference on Medical Image Computing and Computer-Assisted Intervention, Munich, Germany, 18–22 September 2015; pp. 234–241.
32. Zhao, H.; Shi, J.; Qi, X.; Wang, X.; Jia, J. Pyramid Scene Parsing Network. In Proceedings of the IEEE/CVF Conference on Computer Vision and Pattern Recognition, Honolulu, HI, USA, 21–26 July 2017; pp. 2881–2890.
33. Chen, L.C.; Zhu, Y.; Papandreou, G.; Schroff, F.; Adam, H. Encoder-Decoder with Atrous Separable Convolution for Semantic Image Segmentation. In Proceedings of the European Conference on Computer Vision, Munich, Germany, 8–14 September 2018; pp. 833–851.
34. Yuan, Y.; Chen, X.; Wang, J. Object-Contextual Representations for Semantic Segmentation. In Proceedings of the European Conference on Computer Vision, Honolulu, HI, USA, 21–26 July 2020; pp. 173–190.
35. Zheng, S.; Lu, J.; Zhao, H.; Zhu, X.; Luo, Z.; Wang, Y. Rethinking Semantic Segmentation From a Sequence-to-Sequence Perspective with Transformers. In Proceedings of the IEEE/CVF Conference on Computer Vision and Pattern Recognition, Nashville, TN, USA, 19–25 June 2021; pp. 6881–6890.
36. Xie, E.; Wang, W.; Yu, Z.; Anandkumar, A.; Alvarez, J.M.; Luo1, P. SegFormer: Simple and Efficient Design for Semantic Segmentation with Transformers. In Proceedings of the Annual Conference on Neural Information Processing Systems, NeurIPS 2021, Online, 6–14 December 2021; pp. 12077–12090.
37. Alshehhi, R.; Marpu, P.R.; Woon, W.L.; Mura, M.D. Simultaneous extraction of roads and buildings in remote sensing imagery with convolutional neural networks. *ISPRS J. Photogramm. Remote Sens.* **2017**, *130*, 139–149. [CrossRef]
38. Chen, K.; Fu, K.; Yan, M.; Gao, X.; Sun, X.; Wei, X. Semantic Segmentation of Aerial Images With Shuffling Convolutional Neural Networks. *IEEE Geosci. Remote Sens. Lett.* **2018**, *15*, 173–177. [CrossRef]
39. Chen, G.; Zhang, X.; Wang, Q.; Dai, F.; Gong, Y.; Zhu, K. Symmetrical Dense-Shortcut Deep Fully Convolutional Networks for Semantic Segmentation of Very-High-Resolution Remote Sensing Images. *IEEE J. Sel. Top. Appl. Earth Obs. Remote Sens.* **2018**, *11*, 1633–1644.
[CrossRef]
40. Tang, M.; Georgiou, K.; Qi, H.; Champion, C.; Bosch, M. Semantic Segmentation in Aerial Imagery Using Multi-level Contrastive Learning with Local Consistency. In Proceedings of the IEEE/CVF Winter Conference on Applications of Computer Vision, Waikoloa, HI, USA, 2–7 January 2023; pp. 3798–3807.
41. Diakogiannis, F.I.; Waldner, F.; Caccetta, P.; Wu, C. ResUNet-a: A deep learning framework for semantic segmentation of remotely sensed data. *ISPRS J. Photogramm. Remote Sens.* **2020**, *162*, 94–114. [CrossRef]
42. Ding, L.; Tang, H.; Bruzzone, L. LANet: Local Attention Embedding to Improve the Semantic Segmentation of Remote Sensing Images. *IEEE Trans. Geosci. Remote Sens.* **2021**, *59*, 426–435. [CrossRef]
43. Li, R.; Zheng, S.; Zhang, C.; Duan, C.; Su, J.; Wang, L.; Atkinson, P.M. Multiattention Network for Semantic Segmentation of Fine-Resolution Remote Sensing Images. *IEEE Trans. Geosci. Remote Sens.* **2022**, *60*, 5607713. [CrossRef]
44. Zhao, Q.; Liu, J.; Li, Y.; Zhang, H. Semantic Segmentation With Attention Mechanism for Remote Sensing Images. *IEEE Trans. Geosci. Remote Sens.* **2022**, *60*, 5403913. [CrossRef]
45. Panboonyuen, T.; Jitkajornwanich, K.; Lawawirojwong, S.; Srestasathiern, P.; Vateekul, P. Transformer-Based Decoder Designs for Semantic Segmentation on Remotely Sensed Images. *Remote Sens.* **2021**, *13*, 5100. [CrossRef]

46. Wang, L.; Li, R.; Duan, C.; Zhang, C.; Meng, X.; Fang, S. A Novel Transformer Based Semantic Segmentation Scheme for Fine-Resolution Remote Sensing Images. *IEEE Geosci. Remote Sens. Lett.* **2022**, *59*, 6506105. [CrossRef]
47. Zhang, M.; Jing, W.; Lin, J.; Fang, N.; Wei, W.; Wozniak, M.; Damasevicius, R. NAS-HRIS: Automatic Design and Architecture Search of Neural Network for Semantic Segmentation in Remote Sensing Images. *Sensors* **2020**, *20*, 5292. [CrossRef]
48. Wang, Y.; Li, Y.; Chen, W.; Li, Y.; Dang, B. DNAS: Decoupling Neural Architecture Search for High-Resolution Remote Sensing Image Semantic Segmentation. *Remote Sens.* **2022**, *14*, 3864. [CrossRef]
49. Broni-Bediako, C.; Murata, Y.; Mormille, L.H.; Atsumi, M. Evolutionary NAS for Aerial Image Ssegmentation with Gene Expression Programming of Cellular Encoding. *Neural Comput. Appl.* **2022**, *34*, 14185–14204. [CrossRef]
50. Chen, X.; Xie, L.; Wu, J.; Tian, Q. Progressive differentiable architecture search: Bridging the depth gap between search and evaluation. In Proceedings of the IEEE Conference on Computer Vision, Seoul, Republic of Korea, 27 October–2 November 2019; pp. 1294–1303.
51. Yang, Y.; You, S.; Li, H.; Wang, F.; Qian, C.; Lin, Z. Towards improving the consistency, efficiency, and flexibility of differentiable neural architecture search. In Proceedings of the IEEE Conference on Computer Vision and Pattern Recognition, Nashville, TN, USA, 20–25 June 2021; pp. 6667–6676.
52. Cai, H.; Zhu, L.; Han, S. ProxylessNAS: Direct Neural Architecture Search on Target Task and Hardware. In Proceedings of the International Conference on Learning Representations, New Orleans, LA, USA, 6–9 May 2019.
53. Han, S.; Pool, J.; Tran, J.; Dally, W. Learning both weights and connections for efficient neural network. In Proceedings of the Annual Conference on Neural Information Processing Systems, Montreal, QC, Canada, 7–12 December 2015; pp. 1135–1143.
54. Guo, Y.; Yao, A.; Chen, Y. Dynamic network surgery for efficient DNNs. In Proceedings of the Annual Conference on Neural Information Processing Systems, Barcelona, Spain, 5–10 December 2016; pp. 1379–1387.
55. Wang, J.; Xu, C.; Yang, X.; Zurada, J.M. A novel pruning algorithm for smoothing feedforward neural networks based on group lasso method. *IEEE Trans. Neural Netw. Learn. Syst.* **2017**, *29*, 2012–2024. [CrossRef]
56. Chen, S.; Zhao, Q. Shallowing deep networks: Layer-wise pruning based on feature representations. *IEEE Trans. Pattern Anal. Mach. Intell.* **2019**, *41*, 3048–3056. [CrossRef]
57. Yang, Y.; Li, H.; You, S.; Wang, F.; Qian, C.; Lin, Z. ISTA-NAS: Efficient and Consistent Neural Architecture Search by Sparse Coding. In Proceedings of the Annual Conference on Neural Information Processing Systems Advances in Neural Information Processing Systems, Virtual, 6–12 December 2020; pp. 1–9.
58. Cui, Y.; Yang, L.; Liu, D. Dynamic Proposals for Efficient Object Detection. *arXiv* **2022**, arXiv:2207.05252v1.
59. Tibshirani, R. Regression Shrinkage and Selection via the Lasso. *J. R. Stat. Soc. Ser. B* **1996**, *58*, 267–288. [CrossRef]
60. Parikh, N.; Boyd, S. Proximal Algorithms. *Found. Trends Optim.* **2014**, *1*, 127–239. [CrossRef]
61. Li, X.; Wang, W.; Hu, X.; Yang, J. Selective Kernel Networks. In Proceedings of the IEEE/CVF International Conference on Computer Vision and Pattern Recognition, Long Beach, CA, USA, 15–20 June 2019; pp. 510–519.
62. Zhao, Y.; Chen, J.; Zhang, Z.; Zhang, R. BA-Net: Bridge Attention for Deep Convolutional Neural Networks. In Proceedings of the European Conference on Computer Vision, Tel Aviv, Israel, 23–24 October 2022; pp. 297–312.
63. Wang, J.; Zheng, Z.; Ma, A.; Lu, X.; Zhong, Y. LoveDA: A Remote Sensing Land-Cover Dataset for Domain Adaptive Semantic Segmentation. *arXiv* **2021**, arXiv:2110.08733.
64. He, K.; Zhang, X.; Ren, S.; Sun, J. Deep residual learning for image recognition. In Proceedings of the IEEE Conference on Computer Vision and Pattern Recognition, Las Vegas, NV, USA, 27–30 June 2016; pp. 770–778.
65. Liu, W.; Rabinovich, A.; Berg, A.C. ParseNet: Looking Wider to See Better. *arXiv* **2015**, arXiv:1506.04579.
66. Russakovsky, O.; Deng, J.; Su, H.; Krause, J.; Satheesh, S.; Ma, S.; Huang, Z.; Karpathy, A.; Khosla, A.; Bernstein, M. ImageNet large scale visual recognition challenge. *Int. J. Comput. Vision.* **2015**, *115*, 211–252. [CrossRef]

remote sensing MDPI

Article

A Multi-Feature Fusion and Attention Network for Multi-Scale Object Detection in Remote Sensing Images

Yong Cheng [1], Wei Wang [2], Wenjie Zhang [3,*], Ling Yang [1], Jun Wang [1], Huan Ni [4], Tingzhao Guan [1], Jiaxin He [2], Yakang Gu [1] and Ngoc Nguyen Tran [5,6]

1 School of Software, Nanjing University of Information Science & Technology, Nanjing 210044, China
2 School of Automation, Nanjing University of Information Science & Technology, Nanjing 210044, China
3 School of Geographical Sciences, Nanjing University of Information Science & Technology, Nanjing 210044, China
4 School of Remote Sensing & Geomatics Engineering, Nanjing University of Information Science & Technology, Nanjing 210044, China
5 School of Information and Communication Technology, Hanoi University of Science and Technology, Hanoi 100803, Vietnam
6 School of Life Science, University of Technology Sydney, Ultimo 2007, Australia
* Correspondence: zhangwenjie@nuist.edu.cn

Abstract: Accurate multi-scale object detection in remote sensing images poses a challenge due to the complexity of transferring deep features to shallow features among multi-scale objects. Therefore, this study developed a multi-feature fusion and attention network (MFANet) based on YOLOX. By reparameterizing the backbone, fusing multi-branch convolution and attention mechanisms, and optimizing the loss function, the MFANet strengthened the feature extraction of objects at different sizes and increased the detection accuracy. The ablation experiment was carried out on the NWPU VHR-10 dataset. Our results showed that the overall performance of the improved network was around 2.94% higher than the average performance of every single module. Based on the comparison experiments, the improved MFANet demonstrated a high mean average precision of 98.78% for 9 classes of objects in the NWPU VHR-10 10-class detection dataset and 94.91% for 11 classes in the DIOR 20-class detection dataset. Overall, MFANet achieved an mAP of 96.63% and 87.88% acting on the NWPU VHR-10 and DIOR datasets, respectively. This method can promote the development of multi-scale object detection in remote sensing images and has the potential to serve and expand intelligent system research in related fields such as object tracking, semantic segmentation, and scene understanding.

Keywords: remote sensing images; multi-scale object detection; multi-feature fusion and attention network; multi-branch convolution; attention mechanism; loss function

Citation: Cheng, Y.; Wang, W.; Zhang, W.; Yang, L.; Wang, J.; Ni, H.; Guan, T.; He, J.; Gu, Y.; Tran, N.N. A Multi-Feature Fusion and Attention Network for Multi-Scale Object Detection in Remote Sensing Images. *Remote Sens.* **2023**, *15*, 2096. https://doi.org/10.3390/rs15082096

Academic Editors: Jukka Heikkonen, Fahimeh Farahnakian and Pouya Jafarzadeh

Received: 21 February 2023
Revised: 7 April 2023
Accepted: 12 April 2023
Published: 16 April 2023

1. Introduction

The multi-scale object feature recognition of remote sensing images plays a vital role in many fields, including military and civilian. In the military field, remote sensing images can be used to detect and identify ships at sea and then analyze the locations of ship objects to ensure naval defense security [1,2]. In the civilian sector, they can help predict changes in animal habitats and environmental quality [3,4]. Object detection technology in remote sensing images is significant for ocean monitoring, weather monitoring, military navigation, urban planning, and layout. Therefore, how to further improve the multi-scale object detection of remote sensing images has become the focus of research.

Multi-scale object detection in remote sensing images is generally performed in high-resolution images, which provides high-definition information on object features. Among multi-scale objects, large and medium ones are more feature-rich and easier to detect. However, small objects are generally difficult to be characterized and detected effectively. Early object detection methods in remote sensing images mainly used manual feature modeling combined with classifiers for classification to determine the object class. However, traditional detection methods were inefficient and costly in terms of time and labor, making it difficult to meet the needs of practical applications [5]. In recent years, the wide application of deep learning in the fields of video processing [6], image set classification [7,8], image encryption [9], and image recognition [10] has made rapid and accurate object feature recognition possible. Convolutional neural networks, such as RCNN [11] and Faster RCNN [12], have been able to extract high-level semantics from images with better robustness and expressiveness than artificial features, significantly improving detection accuracy. Zhu et al. [13] proposed a multi-layer feature fusion model based on Faster RCNN to improve small object characterization effectively. Shivappriya et al. [14] introduced the additive activation function into Faster RCNN to solve model overfitting problems and improve recognition performance. Although these algorithms improved the detection accuracy of remote sensing images to some extent, Faster RCNN had more parameters and a slower detection speed because it was a two-stage detection algorithm that needed to extract the candidate regions first and then classify and regress the candidate regions.

Single-stage detection algorithms are generally preferred, with representative algorithms such as You Only Look Once (YOLO) [15–19]. This algorithm obtains prediction results directly from the input image. It transforms the object detection problem into a regression problem, significantly improving the detection speed and meeting the demand for real-time detectability of remote sensing images. However, the semantic information in the YOLO about shallow features is weak. After high-fold features compress the input image in the deeper convolution layers, some object information is lost and the sensitivity of detecting objects will gradually decrease. Therefore, Laban et al. [20] proposed the method of an anchor expansion based on YOLOv3 to improve the ability to detect small category goals. Hong et al. [21] improved the anchor frame based on the K-means algorithm with linear scaling while introducing Gaussian parameters into YOLOv3 to enhance the accuracy of multi-scale object detection. Zhou et al. [22] introduced a frequency channel attention network in YOLOv5 to detect small targets in remote sensing images. For object multi-scale variation and dense object distribution characteristics, Wang et al. [23] designed the SPB module and PANet sampling strategy based on YOLOv5. The mean average precision (mAP) was improved by 5.3% compared with the baseline. To address the difficulty of small target object detection in remote sensing images, Han et al. [24] improved YOLO by increasing the residual connection and cross-layer attention to enhance the detection ability of the model for small targets in remote sensing images. Wu et al. [25] proposed the combination of a transformer encoder and a reparameterized backbone based on YOLOX, which effectively improved dense oil tank detection and classification. To enhance the feature learning ability of the network, Yang et al. [26] used efficient channel attention in YOLOX and combined adaptively spatial feature fusion with the neck network to finally achieve high-accuracy object detection in remote sensing images.

YOLOX, the latest version, has improved detection accuracy and speed, and its performance has reached new heights [19]. The anchorless-based network does not require manual anchor scale and aspect ratio setting. It is more suitable for remote sensing images and multi-target detection, for which YOLOX is chosen to conduct further research in this paper. YOLOX uses ordinary convolution, which is imperfect for verifying high-dimensional semantic information, for feature extraction. The problem of partial loss of object information in deep features can lead to lower detection accuracy. The improved feature map scaling and design feature fusion achieved good detection results in the multi-scale object detection task [27]. Therefore, this paper explored further research using YOLOX to improve multi-scale object detection accuracy. To achieve this, the paper proposed a multi-feature fusion and attention network (MFANet) based on YOLOX that effectively detected multi-scale objects while considering the detection accuracy of small objects. The study showed that multi-scale high-level feature extraction and multi-layer pyramidal feature fusion are effective for more accurate target detection. The proposed MFANet incorporated

RepVGG, detail channels, Res-RFBs, and CA modules to help the model extract remote sensing objects more accurately. The paper presented ablation and comparison experiments on two publicly available datasets, NWPU VHR-10 and DIOR. It showed that MFANet achieves 96.63% and 87.88%, respectively, and could handle multi-scale object detection tasks on remote sensing images better than existing methods.

This paper proposed a multi-feature fusion and attention network (MFANet) based on YOLOX to enhance multi-scale object detection accuracy in remote sensing images. The main contributions of this paper included:

(1) To enhance feature extraction of multi-scale objects in remote sensing images, MFANet introduced a structurally reparameterized VGG-like technique (RepVGG) to reparameterize a new backbone and improve multi-object detection accuracy without increasing computation time.

(2) Detailed enhancement channels were introduced in path aggregation feature pyramid networks (PAFPN) to express a great deal of object information. Combining with residual connections, this paper formed a new multi-branch convolutional module (Res-RFBs) to improve the recognition rate of multi-scale objects in remote sensing images. The coordinate attention (CA) mechanism was introduced to reduce the interference of background information and enhance the perception of remote sensing objects by the neural network.

(3) To address the shortcomings of the baseline in the object localization and identification problem, generalized intersection over union (GIoU) was used to optimize the loss, speed up the convergence of the model, and reduce the target miss rate.

2. Methods

2.1. The Structure of the Network

In 2021, Megvii Inc. (Beijing, China) proposed a new object detection network, YOLOX, which exceeds the performance of YOLOv3 and has certain advantages compared with YOLOv5 [19]. The algorithm does not use anchor points and performs dynamic sample matching for objects of different sizes, integrating the previous data enhancement and decoupled head. Its detection speed and effectiveness are improved. First, an image of size 640×640 is used as the input layer, and data enhancement is performed using Mosaic and Mixup. The pre-processed image is then fed into the CSPDarknet53 backbone for feature extraction, resulting in three feature layers with different resolutions derived from Dark3, Dark4, and Dark5. These layers are then fused using the path aggregation feature pyramid network (PAFPN) to enhance the information content. The fused feature layers, P3, P4, and P5, are obtained through upsampling, downsampling, and enhanced feature extraction of the three-resolution images and passed on to the three decoupled heads for accurate object prediction in images.

YOLOX excels at object detection. Among the YOLOX-derived models, YOLOX-s has the advantages of a low number of parameters and easy deployment. Therefore, this study chose to improve on YOLOX-s. The improved structure of the network is shown in Figure 1.

2.2. RepVGG Block

RepVGG is a simple and superior convolutional network. It decouples the model's training and inference time structures using structural reparameterization [28], fully balancing speed and accuracy, and is suitable for real-time detection in remote sensing images. The model's overall structure is a stack of more than 3×3 convolutional layers, divided into 5 parts. The first layer of each part is a downsample with stride = 2. Each convolutional layer uses Relu [29] as the activation function. During training, the RepVGG block is mainly obtained by adding the 3 deviation vectors to obtain the final deviation, extending the fused 1×1 conv and identity with complementary zeros to 3×3 conv, and then adding the 3 3×3 conv to obtain the final 3×3 convolutional layer, as shown in Figure 2. Based on this study, this paper used the RepVGG block to optimize the backbone of the baseline to improve the model for multi-feature extraction of the object.

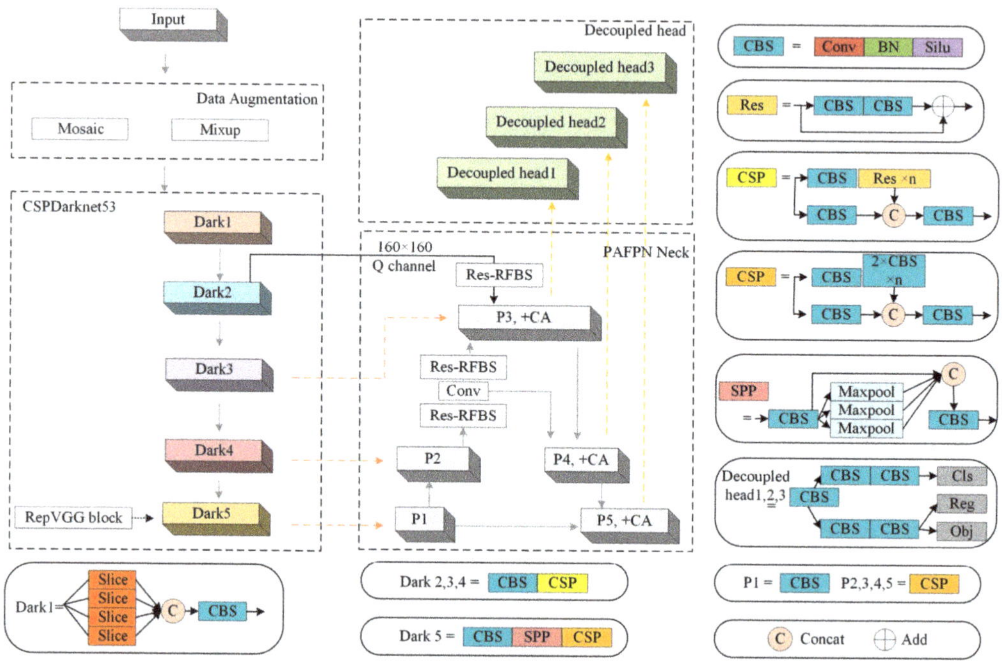

Figure 1. Improved network structure.

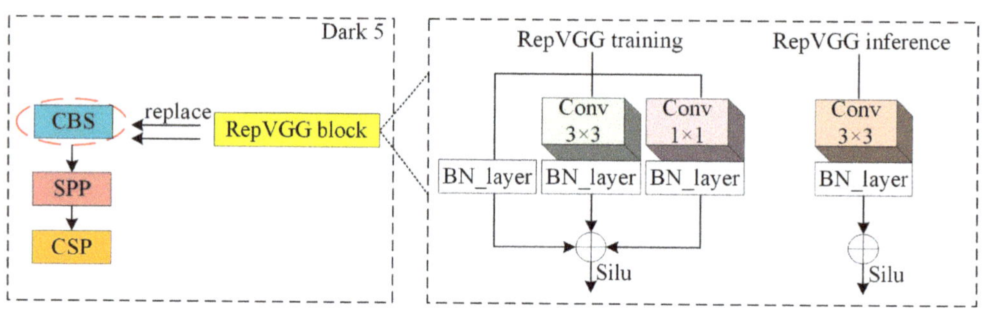

Figure 2. RepVGG block visualization.

In addition, Silu [30] is used instead of the Relu activation function. The Relu activation function is set to zero when the negative gradient is negative, causing some neurons to "necrotize" and affecting network convergence. Silu has the characteristics of no upper bound and lower bound, smooth, and non-monotonic, avoiding negative gradient zeroing to reduce neuronal "necrosis", and better gradient descent than Relu. A comparative plot of the Relu and Silu activation functions is shown in Figure 3. It can be seen that when the function is in a negative gradient, the Relu is set to zero, causing the neural network to fail to learn useful knowledge and neuron "necrosis". In contrast, the Silu function avoids zeroing the negative gradient, retains part of the buffer to reduce neuron "necrosis", and has a better overall gradient descent than Relu. Therefore, this section introduced an improved RepVGG block instead of the partial convolutional layer to optimize the backbone and improve the model's extraction of multi-scale target features.

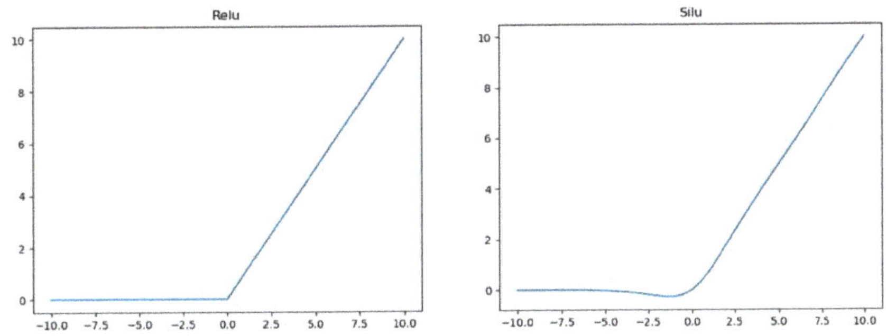

Figure 3. Relu and Silu activation function graphs.

2.3. Improved Feature Detection

The PAFPN pools the feature maps into 80 × 80, 40 × 40, and 20 × 20 resolutions, which are used to detect objects of different sizes. Low resolution detects large objects, medium resolution detects medium-sized objects, and high resolution detects small objects. The resolution of YOLOX for small object detection changes from 640 × 640 to 80 × 80 after convolution, and the lower resolution does not easily capture small object information, resulting in a weaker ability to detect small objects. To avoid the loss of important details in the transmission process of PAFPN, this paper introduced a smaller feature detection channel, 160 × 160, based on the traditional feature detection channel, which will directly input more small object feature information into the feature detection and fuse more network features, which was noted as the Q channel in this paper.

Meanwhile, as the convolution continues, the perceptual field gradually becomes smaller and the detection of multi-scale objects declines progressively. In PAFPN, a new multi-branch convolution module named Res-RFBs was proposed in combination with residual connections, which enhanced the screening of valuable features in this part of the network, as shown in Figure 4.

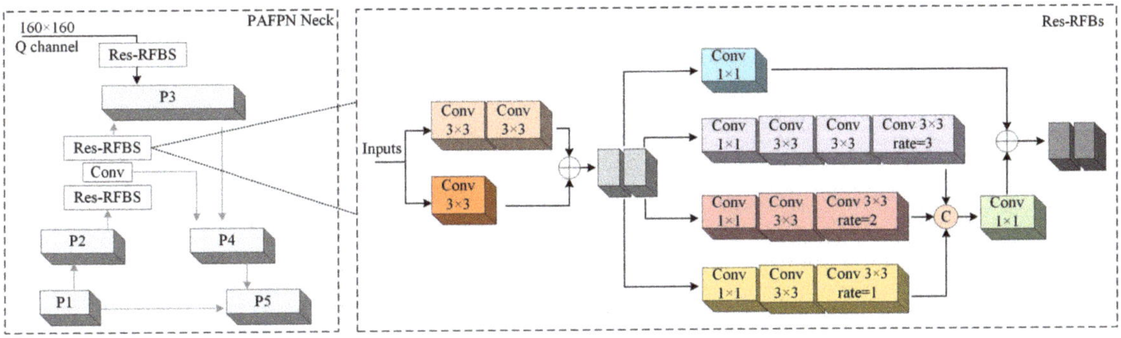

Figure 4. Improved feature detection structure.

ResNet is a deep neural network architecture that effectively solves the gradient disappearance problem in deep neural networks by adding cross-layer connections [31]. ResNet Block is designed to pass the input through two convolutional layers to obtain an output and then add this output to the input, which can further learn deeper features. For this reason, the residual connection was introduced in this paper to improve the network. The improved residual connection was divided into 2 paths: 1 goes through a 3 × 3 and 3 × 3 convolution, and the other is directly shorted with a 3 × 3 convolution. The two are added together and then output. Introducing residual connections into successive convolutions could enhance feature reuse, on the one hand, and avoid the problem of deep network degradation on the other. The receptive field block (RFB) [32] imitates the receptive field of human vision and enhances the feature expression ability of the network. Adding dilated convolution based on inception increases the receptive field and fuses more information from the image. The RFB structure is mainly composed of three branches, which are interconnected to achieve the fusion of different features. Based on the original RFB, each branch added a layer of 3 × 3 convolutions and replaced the 5 × 5 convolution of the original third branch with a 3 × 3 convolution. The expansion coefficients of rate = 1, rate = 2, and rate = 3 were used to increase the receptive field of multi-scale objects and further improve the detection accuracy of the model. The improved RFB module is shown in Figure 4. This multi-branch convolution module sampled the input features into four mutually independent channels. Within the shortcut channel, the feature map was not additionally processed. In the previous three channels, the convolutions of different numbers and expansion rates were superimposed according to the design to express the feature information of various receptive fields.

2.4. Coordinate Attention Mechanism

The coordinate attention (CA) mechanism is an attention mechanism that enhances the perceptual ability of neural networks by embedding spatial coordinate information into them to better capture the correlation between different locations. CA is divided into two steps: embedding coordinate information and generating coordinate attention, which encodes channel relationships and long-term dependencies using precise location information to fully capture the region of interest and the relationship between channels [33]. The mechanism aggregates the input feature maps along the X and Y directions through two global average pooling operations and then encodes the information through dimension transformation. Finally, the spatial information and channel features are weighted and fused, considering the channel and location information. Therefore, CA can better focus on the object of interest, as shown in Figure 5.

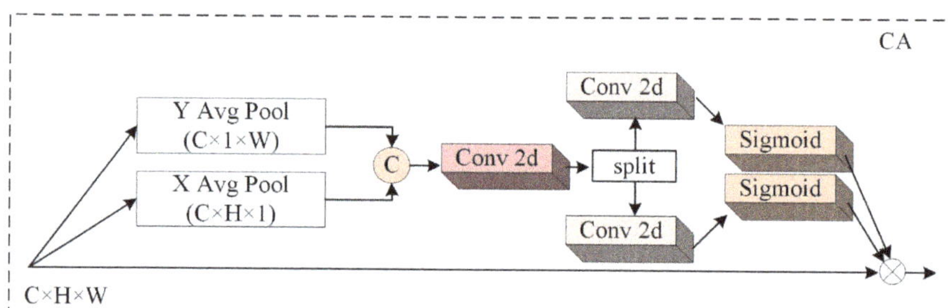

Figure 5. CA module.

In the actual recognition process of remote sensing images, due to the complexity of the image scene, the existing network often cannot eliminate redundant interference information, and the object to be detected is small and densely distributed. In the detection process, the convolutional network needs to process the cells divided by each image. Additionally, many calculations cannot perceive the object well, resulting in missed and false detection problems. Therefore, based on the improved feature extraction network in the previous section, this paper introduced the CA module before the decoupled head. The features could cover more parts of the object to be identified, reduce the interference of background information, make the network focus on essential details of interest, and enhance the expressiveness to improve detection accuracy.

2.5. Loss Function Improvement

The loss function of YOLOX consists of IoU loss (L_{IoU}), category loss (L_{Cls}), and confidence loss (L_{Obj}), which can be expressed as $L_{Loss} = L_{IoU} + L_{Cls} + L_{Obj}$.

Among them, *IoU* refers to the intersection and union ratio, a commonly used indicator in object detection, reflecting the detection effect of the predicted and real detection boxes. The calculation formula is shown in (1):

$$IoU = \frac{A \cap B}{A \cup B} \tag{1}$$

In Formula (1), A represents the prediction box and B represents the real box. IoU is the concept of ratio. In the calculation process using the IoU function, if the prediction box and the real box do not intersect, the degree of coincidence between the two cannot be reflected. In the process of prediction region regression, when the IoU value between the prediction box after the regression and the real box is zero, the problem of target miss rate is caused by the failure of the prediction region to return. In contrast, generalized intersection over union (GIoU) [34] satisfies the basic requirements of the loss function by being concerned not only with the overlapping regions but also with other non-overlapping regions, which can better reflect the coincidence degree between the two objects and accelerate the convergence rate of the model. GIoU first finds the minimum shape A_c to surround the prediction box and the real box. In order to compare two specific geometric shape types, A_c can come from the same type. Finally, the ratio between the area occupied by A_c is calculated and then divided by the total area occupied by A_c, as shown in Formula (2). Therefore, this paper replaced the IoU loss function with the *GIoU* loss function.

$$GIoU = IoU - \frac{|A_c - U|}{|A_c|} \tag{2}$$

In Equation (2), A_c represents the minimum closure area of the prediction box and the real box, and U represents $A \cup B$. For the GIoU loss function, L_{GIoU} can be expressed as:

$$L_{GIoU} = 1 - GIoU = 1 - IoU + \frac{|A_c - U|}{|A_c|} \tag{3}$$

The category loss contains the category information of the remote sensing images, and the confidence loss includes the background information of the image. The category loss and confidence loss are calculated using the bcewithlog_loss function to speed up the model convergence. The loss function is finally shown as (4):

$$L_{Loss} = L_{GIoU} + L_{Cls} + L_{Obj} \tag{4}$$

3. Experiment

3.1. Experimental Environment

The experimental operating system was Windows 10, the GPU was NVIDIA GeForce RTX 3060, and the memory was 12 G. The deep learning framework was Pytorch 1.7.1 and Cuda 11.6. The training had two stages: the freezing stage and the thawing stage. The SGD optimizer was used to adjust the learning rate using the cosine annealing strategy while using pre-training weights.

3.2. Data Set

This experiment uses the NWPU VHR-10 [35] and DIOR [36] datasets.

The NWPU VHR-10 is a high-resolution remote sensing image dataset with spatial resolution ranging from 0.5 m to 2 m. It contains 10 categories of objects and 800 images, with a total number of 3651 target instances. The short names, C1–C10, for our experiment categories were: Tennis court, Harbor, Ground track field, Basketball court, Airplane,

Storage tank, Baseball field, Ship, Vehicle, and Bridge. Figure 6 shows the remote sensing images and objects of the NWPU VHR-10 dataset, where the boxed positions are the objects.

Figure 6. Remote sensing images and objects of NWPU VHR-10 dataset.

DIOR is a large-scale benchmark data set for object detection in optical remote sensing images. It is divided into 20 object classes, including 23,463 remote sensing images and 190,288 instances. It has high similarity and diversity in different imaging conditions, weather, seasons, and image quality. The short names C1–C20 for categories in our experiment were defined as Airplane, Airport, Baseball field, Basketball court, Bridge, Chimney, Dam, Expressway service area, Expressway toll station, Golf field, Ground track field, Harbor, Overpass, Ship, Stadium, Storage tank, Tennis court, Train station, Vehicle, and Windmill. Figure 7 shows the remote sensing images and objects of the DIOR dataset, where the boxed positions are the objects.

3.3. Evaluation Metrics

To accurately evaluate the effect of the proposed method on remote sensing image detection, this study selected the mean average precision (mAP), precision rate (P), recall rate (R), and frame per second (FPS) as evaluation indicators. The calculation formula is shown in Formulas (5)–(7). When the accuracy and recall rates are compared separately, ambiguity will occur. Therefore, the experiment used the mAP to evaluate the model's effectiveness by comprehensively considering the precision and recall rates. FPS refers to the number of frames detected per second to measure the real-time performance of the model.

$$P = \frac{TP}{TP + FP} \tag{5}$$

$$R = \frac{TP}{TP + FN} \tag{6}$$

$$mAP = \frac{\sum_{i=1}^{k} AP_i}{k} \tag{7}$$

In Formulas (5) and (6), *TP* represents the number of correct predictions, *FP* represents the number of false predictions, and *FN* represents the number of missing predictions; in Formula (7), *k* is the category, and the calculation formula of average precision (*AP*) can be expressed as:

$$AP = \int_{0}^{1} \mathrm{p(r)dr} \quad (8)$$

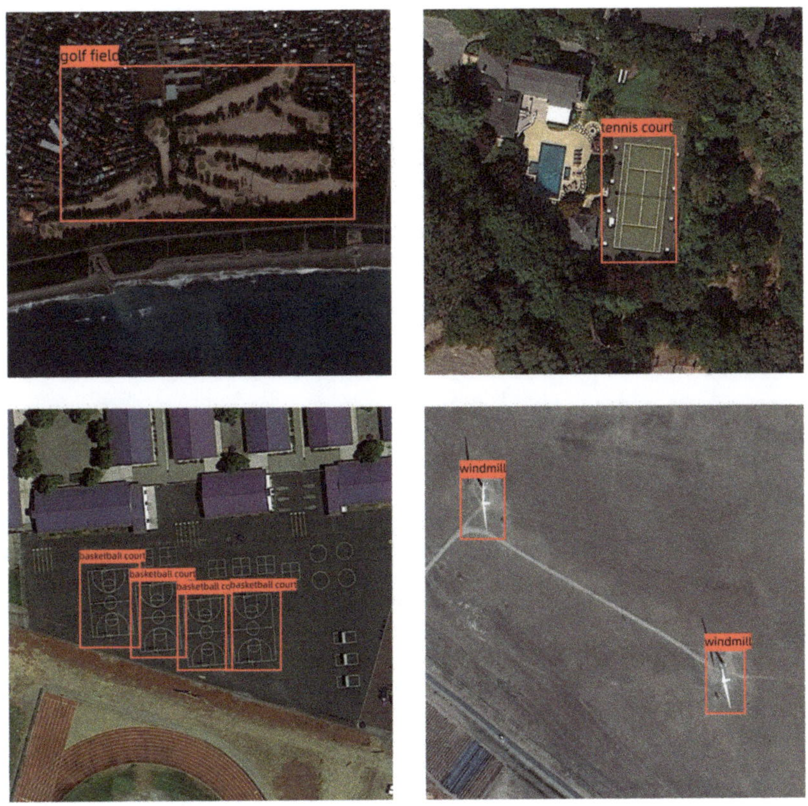

Figure 7. Remote sensing images and objects of DIOR dataset.

3.4. Ablation Experiment

In the improvement strategy for the backbone, if it is a direct addition of RepVGG modules, it may not necessarily have the desired effect. To explore the effectiveness of the RepVGG addition position, this paper added the RepVGG block to the backbone to determine the RepVGG block addition position. The different addition positions are shown in Figure 8.

Table 1 shows that the best mAP of remote sensing image detection was achieved using the RepVGG block instead of the fourth convolutional layer in the backbone. Compared with the experiments conducted by Relu, the Silu function was more stimulating to the feature extraction performance of the model and avoided some neuron necrosis. The different results with different addition positions are because the number of composite convolutional layers and the amount of information reorganization are different, thus bringing different gains to the model. After an experimental demonstration, the RepVGG block was used instead of the fourth convolutional layer in the backbone.

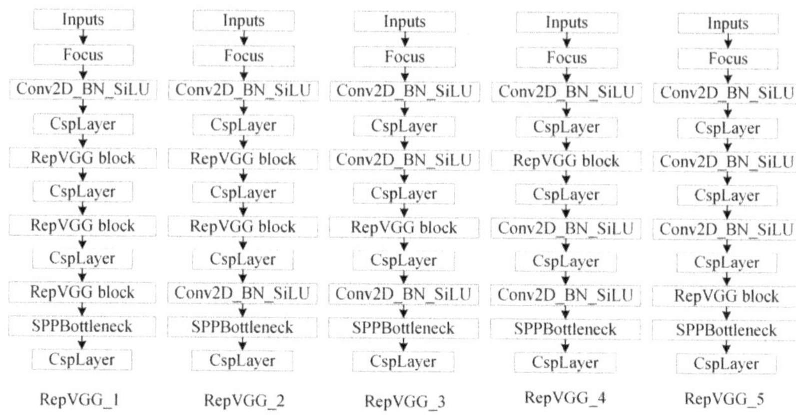

Figure 8. The RepVGG block of backbone is added at different locations.

Table 1. Experimental results of RepVGG at different locations. P stands for precision rate, R stands for recall rate, FPS stands for frame per second, and mAP stands for mean average precision.

Location	P/(%)	R/(%)	mAP/(%)	FPS/(f/s)
RepVGG_1(Relu)	91.55	86.40	90.14	45.64
RepVGG_1	91.94	87.20	92.19	46.89
RepVGG_2	92.91	88.07	93.07	46.37
RepVGG_3	91.03	90.58	93.34	48.39
RepVGG_4	91.17	91.93	92.99	47.45
RepVGG_5	90.67	89.78	93.53	48.50

To explore the effectiveness of the improved module, ablation experiments were conducted on the YOLOX-s-based NWPU VHR-10 dataset for the RepVGG block, Q, multi-branch convolution, CA attention, and GIoU loss function, respectively. The experimental projects were carried out by sequentially adding each proposed module; the results are shown in Table 2.

Table 2. Ablation experiment of the improved module. P stands for precision rate, R stands for recall rate, FPS stands for frame per second, and mAP stands for mean average precision.

RepVGG	Q+ Res-RFBs	CA	GIoU	P/(%)	R/(%)	mAP/(%)	FPS/(f/s)
-	-	-	-	89.68	90.37	92.23	48.02
√	-	-	-	90.67	89.78	93.53	48.50
-	√	-	-	93.45	91.66	94.98	33.61
-	-	√	-	90.51	93.28	94.14	41.26
-	-	-	√	93.12	90.65	93.56	35.61
√	√	√	√	94.09	94.94	96.63	30.09

As shown in Table 2, the mAP of using the RepVGG block was increased by 1.3%, and the FPS was increased by 1% compared with the base network, which indicated that the system performance was further improved. As seen in Table 1, the experiments obtained better results using the Silu activation function than Relu. The RepVGG block made extensive use of 3 × 3 convolution. It used a multi-branch network for training by structural reparameterization and fused the multi-branch into a single branch for prediction, facilitating network acceleration. Figure 9a showed the original input image, and a new layer of 160 × 160 input channels was introduced in the PAFPN, which could express more small object information compared with the original network, and more object information was obtained compared with that shown in Figure 9b,c. After adding Res-RFBs on the basis of introducing Q (160 × 160) input channels, the receptive field could be further expanded to enhance the detailed expression of the model effectively. As seen in Figure 9, the object feature information in the image was not effectively detected in the baseline feature heat map, resulting in a scattered region of interest for the network and object features that could not be extracted effectively. In the multi-scale feature heat map, it could be clearly observed that the features of different objects were enhanced after adding multi-scale convolution especially for ships and dense storage tanks. Information was enhanced in the image of the object, thus proving the effectiveness of multi-branch convolution for improving the features of the object. The results are shown in Figure 9d,e, which effectively enhanced the detection ability of multi-scale objects with dense distribution, and mAP was increased by 2.75%. The reason for the increase of 1.91% using the CA attention mechanism over the baseline was that the CA module considers both channel and direction-related location information to further strengthen the neural network's ability to perceive remote sensing objects and focus more on the object. The improved loss function increased the mAP by 1.33% and improved the model's performance. The reason was that the increased penalty measure of the GIoU function facilitates the network in making accurate judgments on remote sensing objects and compensates for the non-overlapping regions of the detection objects in the IoU loss function, which effectively reduces the target miss rate. After adding the RepVGG block, Q+ Res-RFBS, CA, and GIoU loss function, the detection accuracy reached 95.50%, which was 3.27% better than the baseline. The final detection accuracy of 96.63% was obtained after multiple training iterations. Overall, the improved modules enhanced detection accuracy, and the use of the above improvement strategies eventually brought a gain of 4.4 percentage points to the model, which proved the effectiveness of the improvement strategies.

3.5. Comparison with Other Algorithms

To further verify the effectiveness and rationality of the improved YOLOX for object detection in remote sensing images, this experiment used YOLOX, MFANet, and mainstream algorithms to train and test the detection accuracy of each algorithm in the NWPU VHR-10 and DIOR datasets. The experimental results are shown in Tables 3 and 4. In the NWPU VHR-10 dataset in Table 3, the methods of Faster RCNN, YOLOv4-tiny, YOLOv5, and YOLOX-s, Laban's [20], SCRDet [37], Fan's [38], Zhang's [39], and Xue's [40] networks were selected for comparison. The results showed that, compared with other models, the MFANet proposed in this paper had the best mAP, 96.63%, which was 17.15%, 7.49%, 4.88%, 3.23%, 1.04%, and 0.93% higher than Faster RCNN, YOLOv5, SCRDet [37], Fan's [38], Zhang's [39], and Xue's [40], respectively, and had better detection performance.

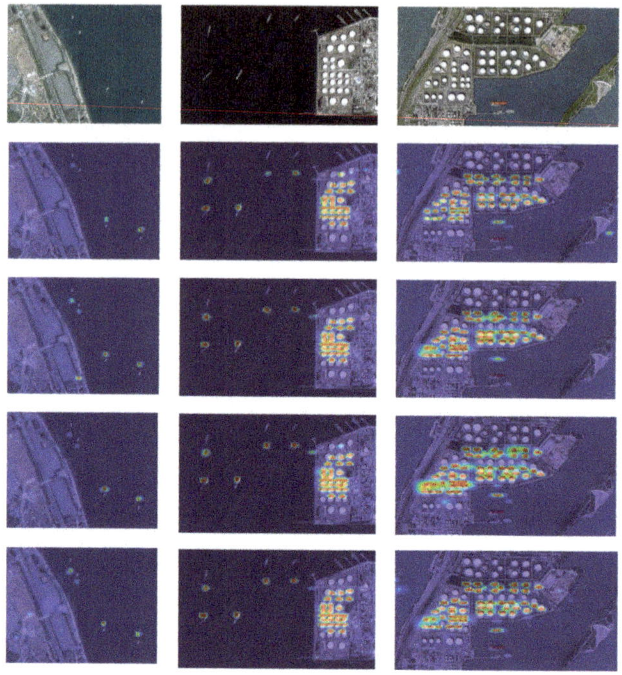

Figure 9. Visualization results of improved feature detection: (**a**) input image; (**b**) visualization results for YOLOX; (**c**) visualization results of adding 160 × 160 channels for YOLOX; (**d**) visualization results of adding Res-RFBs for YOLOX; (**e**) visualization results of adding 160 × 160 channels and Res-RFBs for YOLOX.

Table 3. Experimental results in the NWPU VHR-10 dataset. FPS stands for frame per second and mAP stands for mean average precision.

Method	mAP/(%)	FPS/(f/s)
Faster RCNN	79.48	10.59
Laban's [20]	78.00	-
YOLOv4-tiny	84.14	81.36
YOLOv5	89.14	54.91
SCRDet [37]	91.75	-
YOLOX-s	92.23	48.02
Fan's [38]	93.40	-
Zhang's [39]	95.59	30.07
Xue's [40]	95.70	-
MFANet	96.63	30.09

Table 4. Experimental results in the DIOR dataset. FPS stands for frame per second and mAP stands for mean average precision.

Method	mAP/(%)	FPS/(f/s)
Faster RCNN	57.35	10.42
AOPG [41]	64.41	-
LO-Det [42]	65.85	60.03
Li's [43]	66.71	-
YOLOv4-tiny	66.77	56.68
ASSD [44]	71.80	21.00
Yao's [45]	75.80	-
SCRDet++ [46]	77.80	-
YOLOv5	80.96	51.41
SPB-YOLO [23]	81.10	-
YOLOX-s	82.23	47.99
Zhou's [47]	84.30	-
YOLOX [24]	85.70	-
Ye's [48]	86.55	-
MFANet	87.88	29.45

It can be seen from Figure 10 that in the experiment of the NWPU VHR-10 dataset, when there was a complex scene to be detected, the object distribution was dense, or the object had a low resolution in the image, the effect of YOLOX-s detection was not good, and it was prone to problems such as missed detection and false detection. In Figure 10b, due to the interference of more background information in the image, the baseline network missed and made false detections of small objects, such as ships and vehicles. Compared with Figure 10b,c, the improved network improves the detection efficiency of objects, and the problems in Figure 10b are basically solved. The detection effect of this algorithm is significantly better than that of YOLOX. To compare the model detection effects in one step, this paper selected the AP and mAP results of YOLOv4-tiny, YOLOv5, YOLOX-s, Fan's, Zhang's, Xue's, and MFANet, and the results are shown in Table 5. For the object bridge, although the detection accuracy of MFANet was a little bit lower than the methods of Xue's and Zhang's results, it was higher than that of YOLOv4-tiny, YOLOv5, and YOLOX-s. It can also be seen that MFANet shows a significant improvement in detecting ships and vehicles compared to YOLOv4-tiny, YOLOv5, YOLOX-s, Fan's, Zhang's, and Xue's. It was 1% higher than Xue's when testing ships and 4% higher than Fan's when testing vehicles. In addition, the MFANet achieved better detection results on objects such as the Tennis court, Harbor, and Basketball court. When there is serious background interference in the target, such as the similarity in appearance between the refuse collection point and the parked vehicles alongside the road, the base detection network will miss the detection. In contrast, the improved network effectively improved the inspection accuracy of vehicles and avoided object misdetection.

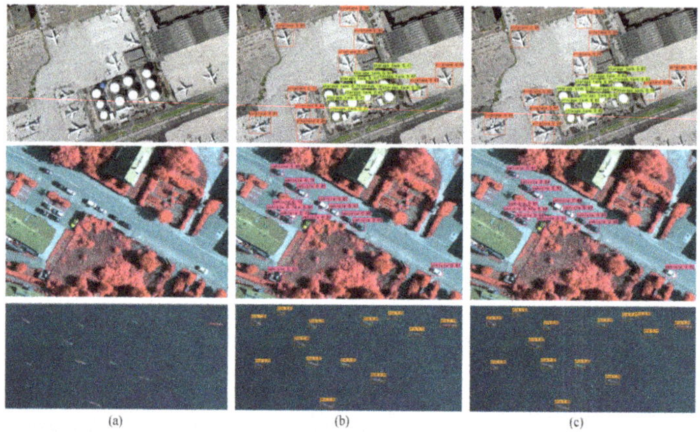

Figure 10. Detection results for the NWPU VHR-10 dataset: (a) input image; (b) for YOLOX-s; (c) for MFANet.

Table 5. AP and mAP of the different algorithms acting on multiple object categroies from NWPU VHR-10 data. AP stands for average precision and mAP stands for mean average precision.

Method	C1	C2	C3	C4	C5	C6	C7	C8	C9	C10	mAP
YOLOv4-tiny	0.91	0.98	0.99	0.66	1.00	0.71	0.99	0.76	0.82	0.59	0.84
YOLOv5	1.00	1.00	0.98	0.83	1.00	0.99	0.98	0.91	0.90	0.33	0.89
YOLOX-s	1.00	1.00	1.00	1.00	1.00	1.00	0.99	0.94	0.89	0.40	0.92
Fan's	1.00	1.00	0.91	0.91	1.00	0.90	0.94	0.91	0.90	0.91	0.93
Zhang's	1.00	1.00	1.00	1.00	1.00	0.90	1.00	0.90	0.87	0.90	0.96
Xue's	0.90	0.96	1.00	1.00	1.00	0.88	1.00	0.96	0.89	0.99	0.96
MFANet	1.00	1.00	1.00	1.00	1.00	1.00	0.98	0.97	0.94	0.77	0.96

The DIOR dataset has a significant difference in spatial resolution and cross-object scale, and its high inter-class similarity and class diversity increase the difficulty of detection. We can see that the YOLOv4-tiny part of the detection object was higher than YOLOX and MFANet, but MFANet still led the overall multi-object detection effect. YOLOX-s missed and mis-checked at the background of complex scenes that contained diverse categories of feature elements (Figure 11c). Compared with Figure 11d, it can be seen that the detection effect of the improved algorithm was significantly improved, and multi-scale objects were effectively detected. The improved model has strong robustness.

To compare the model detection effects in one step, this paper selected the AP and mAP results of ASSD, Yao's, SCRDet++, YOLOv5, YOLOX-s, Zhou's, and MFANet, which can be seen in Table 6. For the object Golf field, although the MFANet detection accuracy is not as good as Zhou's, it was higher than YOLOX-s, YOLOv5, SCRDet++, Yao's, and ASSD, by 3%, 15%, 1%, 6%, and 5%, respectively. It can be seen that in the detection of bridges and vehicles, compared with YOLOX-s, the MFANet had a significant improvement. In addition, we found that when the road and harbor samples were similar, the baseline network missed detection due to insufficient resolution and failed to identify the port object effectively. The optimized network with enhanced multi-scale feature extraction could effectively detect most objects and complete the object detection task.

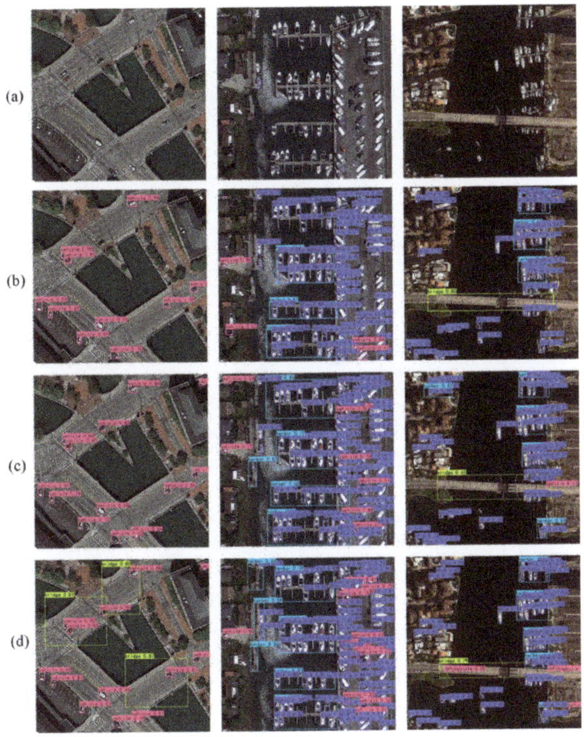

Figure 11. Detection results for the DIOR dataset: (**a**) input image; (**b**) for YOLOv4-tiny; (**c**) for YOLOX-s; (**d**) for MFANet.

Table 6. AP and mAP of the different algorithms acting on multiple object categroies from DIOR data. AP stands for average precision and mAP stands for mean average precision.

Method	C1	C2	C3	C4	C5	C6	C7	C8	C9	C10
ASSD	0.86	0.82	0.76	0.90	0.41	0.78	0.65	0.67	0.62	0.81
Yao's	0.91	0.75	0.93	0.83	0.47	0.92	0.63	0.68	0.61	0.80
SCRDet++	0.81	0.88	0.80	0.90	0.58	0.81	0.75	0.90	0.83	0.85
YOLOv5	0.96	0.86	0.97	0.86	0.48	0.86	0.75	0.86	0.77	0.71
YOLOX-s	0.96	0.88	0.96	0.83	0.48	0.78	0.78	0.94	0.79	0.83
Zhou's	0.98	0.90	0.95	0.93	0.62	0.91	0.68	0.96	0.86	0.87
MFANet	0.97	0.93	0.97	0.86	0.59	0.88	0.87	0.97	0.90	0.86
mAP	C11	C12	C13	C14	C15	C16	C17	C18	C19	C20
0.71	0.79	0.62	0.58	0.85	0.77	0.65	0.88	0.62	0.45	0.76
0.76	0.83	0.57	0.66	0.80	0.93	0.81	0.89	0.63	0.73	0.78
0.78	0.84	0.63	0.67	0.73	0.79	0.70	0.90	0.71	0.59	0.90
0.81	0.92	0.67	0.71	0.95	0.89	0.86	0.96	0.63	0.60	0.93
0.82	0.90	0.69	0.71	0.96	0.96	0.87	0.95	0.65	0.62	0.92
0.84	0.91	0.63	0.73	0.96	0.92	0.90	0.96	0.57	0.71	0.94
0.87	0.93	0.75	0.79	0.97	0.97	0.92	0.97	0.79	0.73	0.94

4. Discussion

The ablation experimental results show that the improved YOLOX in this paper can improve the model's multi-scale object recognition rate, and the mAP was improved by 4.4% compared to the YOLOX. Figure 9 shows that introducing a new layer of 160 × 160 input channels in the PAFPN can express more information about small objects than the original network. The addition of Res-RFBs was based on the introduction of detail-enhanced channels, which enhanced feature multiplexing and expanded the perceptual field, thus improving the detection accuracy of multi-scale objects by 2.75% compared to the mAP of the baseline. The results in Tables 1 and 2 show that the RepVGG block uses structural reparameterization to improve the extraction of multi-scale object features, and mAP was increased by 1.3%. The results in Table 2 and Figure 10 show that CA enhances the ability of the neural network to perceive remote sensing objects, with a 1.91% improvement in mAP. The results in Table 2 and Figures 10 and 11 show that the GIoU loss function reduces the target miss rate. The comparison experimental results show that the proposed method had a higher accuracy rate when compared with other mainstream object detection algorithms. On the DIOR dataset, mainstream algorithms such as Faster RCNN [12], YOLOv5, and YOLOX are used in this paper, while models such as AOPG [41], Li's [43], Yao's [45], SCRDet++ [46], Zhou's [47], and Ye's [48] are selected for comparison. The results in Table 4 show that the improved YOLOX model proposed in this paper had a better mAP than other models (30.53%, 6.92%, 6.78%, 3.58%, 2.18%, and 1.33% higher than Faster RCNN, YOLOv5, SPB-YOLO [23], Zhou's [47], YOLOX [24], and Ye's [48]), achieving advanced detection and classification performance. The NWPU VHR-10 dataset shows that the MFANet obtained a lower detection speed than YOLOv4-tiny, YOLOv5, etc., but a higher detection speed than Faster RCNN, Zhang's, etc. In addition, on the DIOR dataset, compared with LO-Det [42] and YOLOv5, although the detection speed was lower, the detection accuracy of the improved network in this paper was much higher than theirs: 22.03% and 6.92% higher than LO-Det and YOLOv5, respectively. Compared with ASSD [44], MFANet is leading in detection accuracy and speed. Comparing with Table 6, we find that for large objects, such as Airport, Expressway service areas, etc., with improved algorithm detection, the AP improved by 5% and 3%, respectively; for medium-sized objects, such as Harbor, Chimney, etc., with improved algorithm detection, the AP improved by 6% and 10%, respectively; for small objects, such as Bridge and Storage tank, the AP was enhanced by 11% and 5%, respectively, after improved algorithm detection. Overall, the experimental results verify the effectiveness of the improved network in detecting multi-scale objects.

At the same time, we find that in the area of small object distribution shown in Figure 11, some images of the small objects are blurred and carry too little feature information, resulting in the detector failing to effectively detect them, which affects the detection results. In Table 6, we can see that the AP for vehicles was lower than boats and airplanes, which is probably because the less contextually available feature information of small objects. In addition, the FPS of the improved algorithm proposed in this paper reached 30.09 and 29.45 on the NWPU VHR-10 and DIOR datasets, respectively, which were much higher than Faster RCNN. However, the FPS decreased compared to the original network. The reason for this is that the improved PAFPN makes the network structure complex, introducing many parameters and increasing the computational time consumption, thus slowing down the detector. A linear discriminant can cluster objects [49], and eliminating redundancy constraints can improve detection speed [50], providing a method for object detection in remote sensing images. Therefore, to improve some shortcomings of the algorithm in this paper, the following aspects can be considered. Firstly, using discriminant analysis and migration learning to improve the generalization of the network. Secondly, reducing the number of redundant parameters in the model while maintaining high efficiency and using deeper contextual feature information to achieve high-quality small object detection.

5. Conclusions

Aiming at the complex problem of multi-scale detection of remote sensing images, this research proposed the MFANet based on YOLOX. The MFANet used RepVGG to build a new backbone, and the detection accuracy of the backbone after reparameterization increased from 92.23% to 93.53%, which proved the effectiveness of its prediction; at the same time, the Silu activation function was selected and the detection accuracy increased by 2.05%. The choice of detail channel and multi-branch convolution should be combined with different datasets to determine the best object extraction performance. In this paper, the Q channel and three multi-branch convolutions were selected in PAFPN to achieve the best detection effect. In addition, this paper proved the role of the CA module in improving the image detection network by adding the CA module to the improved network, effectively reducing background interference and increasing the detection accuracy by 1.91%. Finally, the GIoU function was used to optimize the loss and the detection accuracy was increased by 1.33%, effectively avoiding missed object detection. The experiment was carried out on the NWPU VHR-10 and DIOR datasets. Compared with current object detection algorithms, the MFANet achieved higher detection accuracy. MFANet demonstrated a high mean average precision of 98.78% for 9 classes of objects in the NWPU VHR-10 10-class detection dataset and 94.91% for 11 classes of objects in the DIOR 20-class detection dataset. The overall performance of mAP was 96.63% and 87.88% for the NWPU VHR-10 and DIOR datasets, respectively. In summary, the combination of multi-branch feature fusion and an attention model is a superior approach to improving the accuracy of multi-scale object detection in remote sensing images. In the future, the feature extraction mechanism in MFANet can be further deepened and optimized, especially when many object categories are contained in remote sensing images.

Author Contributions: Y.C. and W.W. designed the study. Y.C. and W.W. conducted the analysis and wrote the manuscript. W.Z., L.Y., J.W., H.N., T.G., J.H., Y.G. and N.N.T. offered advice to improve the manuscript. Y.C., W.W. and W.Z. interpreted the results and revised the manuscript. All authors have read and agreed to the published version of the manuscript.

Funding: This research is supported by the National Natural Science Foundation of China under Grant no. 41975183 and Grant no. 41875184.

Data Availability Statement: The data (NWPU VHR-10 dataset and the DIOR dataset) used to support the results of this study are available from the respective authors upon request and can be found in the references [35,36].

Acknowledgments: We thank the anonymous reviewers for their comments and suggestions that improved this paper.

Conflicts of Interest: The authors declare no conflict of interest.

References

1. Li, W. Detection of Ship in Optical Remote Sensing Image of Median-Low Resolution. Master's Thesis, National University of Defense Technology, Changsha, China, November 2008.
2. Wang, Y.; Ma, L.; Tian, Y. State-of-the-art of Ship Detection and Recognition in Optical Remotely Sensed Imagery. *Acta Autom. Sin.* **2011**, *37*, 1029–1039.
3. Rajendran, G.B.; Kumarasamy, U.M.; Zarro, C.; Divakarachari, P.B.; Ullo, S.L. Land-Use and Land-Cover Classification Using a Human Group-Based Particle Swarm Optimization Algorithm with an LSTM Classifier on Hybrid Pre-Processing Remote-Sensing Images. *Remote Sens.* **2020**, *12*, 4135. [CrossRef]
4. Zhang, W.; Zhang, B.; Zhu, W.; Tang, X.; Li, F.; Liu, X.; Yu, Q. Comprehensive assessment of MODIS-derived near-surface air temperature using wide elevation-spanned measurements in China. *Sci. Total Environ.* **2021**, *800*, 149535. [CrossRef] [PubMed]
5. Nie, G.; Huang, H. A survey of object detection in optical remote sensing images. *Acta Autom. Sin.* **2021**, *47*, 1749–1768.
6. Parameshachari, B.; Gurumoorthy, S.; Frnda, J.; Nelson, S.C.; Balmuri, K.R. Cognitive linear discriminant regression computing technique for HTTP video services in SDN networks. *Soft Comput.* **2022**, *26*, 621–633. [CrossRef]
7. Wang, R.; Wu, X.; Kittler, J. SymNet: A simple symmetric positive definite manifold deep learning method for image set classification. *IEEE Trans. Neural Netw. Learn. Syst.* **2021**, *33*, 2208–2222. [CrossRef] [PubMed]
8. Gao, X.; Niu, S.; Wei, D.; Liu, X.; Wang, T.; Zhu, F.; Dong, J.; Sun, Q. Joint Metric Learning-Based Class-Specific Representation for Image Set Classification. *IEEE Trans. Neural Netw. Learn. Syst.* **2022**, 1–15. [CrossRef] [PubMed]
9. Parameshachari, B.; Panduranga, H. Medical image encryption using SCAN technique and chaotic tent map system. In *Recent Advances in Artificial Intelligence and Data Engineering*; Springer: Singapore, 2022; pp. 181–193.
10. Zhou, F.; Jin, L.; Dong, J. Review of Convolutional Neural Network. *Chin. J. Comput.* **2017**, *40*, 1229–1251.
11. Girshick, R.; Donahue, J.; Darrell, T.; Malik, J. Rich feature hierarchies for accurate object detection and semantic segmentation. In Proceedings of the IEEE Conference on Computer Vision and Pattern Recognition, Columbus, OH, USA, 23–28 June 2014; pp. 580–587.
12. Ren, S.; He, K.; Girshick, R.; Sun, J. Faster R-CNN: Towards Real-Time Object Detection with Region Proposal Networks. *IEEE Trans. Pattern Anal. Mach. Intell.* **2017**, *39*, 1137–1149. [CrossRef] [PubMed]

13. Zhu, M.; Xu, Y.; Ma, S.; Li, S.; Ma, H.; Han, Y. Effective airplane detection in remote sensing images based on multilayer feature fusion and improved nonmaximal suppression algorithm. *Remote Sens.* **2019**, *11*, 1062. [CrossRef]
14. Shivappriya, S.N.; Priyadarsini, M.J.P.; Stateczny, A.; Puttamadappa, C.; Parameshachari, B.D. Cascade Object Detection and Remote Sensing Object Detection Method Based on Trainable Activation Function. *Remote Sens.* **2021**, *13*, 200. [CrossRef]
15. Redmon, J.; Divvala, S.; Girshick, R.; Farhadi, A. You Only Look Once: Unified, Real-Time Object Detection. In Proceedings of the IEEE Conference on Computer Vision and Pattern Recognition, Las Vegas, NV, USA, 27–30 June 2016; pp. 779–788.
16. Redmon, J.; Farhadi, A. YOLO9000: Better, Faster, Stronger. In Proceedings of the IEEE Conference on Computer Vision and Pattern Recognition, Honolulu, HI, USA, 21–26 July 2017; pp. 6517–6525.
17. Redmon, J.; Farhadi, A. YOLOv3: An Incremental Improvement. *arXiv* **2018**, arXiv:1804.02767.
18. Bochkovskiy, A.; Wang, C.-Y.; Liao, H.-Y.M. YOLOv4: Optimal Speed and Accuracy of Object Detection. *arXiv* **2020**, arXiv:2004.10934.
19. Ge, Z.; Liu, S.; Wang, F.; Li, Z.; Sun, J. YOLOX: Exceeding YOLO Series in 2021. *arXiv* **2021**, arXiv:2107.08430.
20. Laban, N.; Abdellatif, B.; Ebeid, H.M.; Shedeed, H.A.; Tolba, M.F. Convolutional Neural Network with Dilated Anchors for Object Detection in Very High Resolution Satellite Images. In Proceedings of the International Conference on Computer Engineering and Systems (ICCES), Cairo, Egypt, 17 December 2019; pp. 34–39.
21. Hong, Z.; Yang, T.; Tong, X.; Zhang, Y.; Jiang, S.; Zhou, R.; Han, Y.; Wang, J.; Yang, S.; Liu, S. Multi-scale ship detection from SAR and optical imagery via a more accurate YOLOv3. *IEEE J. Sel. Top. Appl. Earth Obs. Remote Sens.* **2021**, *14*, 6083–6101. [CrossRef]
22. Zhou, H.; Guo, W. Improved YOLOv5 Network in Application of Remote Sensing Image Object Detection. *Remote Sens. Inf.* **2022**, *37*, 23–30.
23. Wang, X.; Li, W.; Guo, W.; Cao, K. SPB-YOLO: An Efficient Real-Time Detector For Unmanned Aerial Vehicle Images. In Proceedings of the International Conference on Artificial Intelligence in Information and Communication (ICAIIC), Jeju Island, Republic of Korea, 13–16 April 2021; pp. 099–104.
24. Han, X.; Li, F. Remote Sensing Small Object Detection Based on Cross-Layer Attention Enhancement. *Laser Optoelectron. Prog.* 2022, pp. 1–19. Available online: https://kns.cnki.net/kcms/detail/31.1690.TN.20220722.2132.050.html (accessed on 17 February 2023).
25. Wu, Q.; Zhang, B.; Xu, C.; Zhang, H.; Wang, C. Dense Oil Tank Detection and Classification via YOLOX-TR Network in Large-Scale SAR Images. *Remote Sens.* **2022**, *14*, 3246. [CrossRef]
26. Yang, L.; Yuan, G.; Zhou, H.; Liu, H.; Chen, J.; Wu, H. RS-YOLOX: A High-Precision Detector for Object Detection in Satellite Remote Sensing Images. *Appl. Sci.* **2022**, *12*, 8707. [CrossRef]
27. Guo, Q.; Yuan, C. Leveraging Spatial-Semantic Information in Object Detection and Segmentation. *Ruan Jian Xue Bao/J. Softw.* 2022, pp. 1–13. Available online: http://www.jos.org.cn/jos/article/abstract/6509 (accessed on 17 February 2023). (In Chinese).
28. Ding, X.; Zhang, X.; Ma, N.; Han, J.; Ding, G.; Sun, J. RepVGG: Making VGG-style ConvNets Great Again. In Proceedings of the IEEE Conference on Computer Vision and Pattern Recognition, Nashville, TN, USA, 20–25 June 2021; pp. 13728–13737.
29. Shang, W.; Sohn, K.; Almeida, D.; Lee, H. Understanding and improving convolutional neural networks via concatenated rectified linear units. In Proceedings of the 33rd International Conference on Machine Learning, New York, NY, USA, 19–24 June 2016; pp. 2217–2225.
30. Ramachandran, P.; Zoph, B.; Le, Q. Swish: A Self-Gated Activation Function. *arXiv* **2017**, arXiv:1710.05941.
31. He, K.; Zhang, X.; Ren, S.; Sun, J. Deep Residual Learning for Image Recognition. In Proceedings of the IEEE Conference on Computer Vision and Pattern Recognition, Las Vegas, NV, USA, 27–30 June 2016; pp. 770–778.
32. Liu, S.; Huang, D.; Wang, Y. Receptive field block net for accurate and fast object detection. In Proceedings of the European Conference on Computer Vision (ECCV), Munich, Germany, 8–14 September 2018; pp. 385–400. Available online: https://doi.org/10.48550/arXiv.1711.07767 (accessed on 28 August 2022).
33. Hou, Q.; Zhou, D.; Feng, J. Coordinate Attention for Efficient Mobile Network Design. In Proceedings of the IEEE Conference on Computer Vision and Pattern Recognition, Nashville, TN, USA, 20–25 June 2021; pp. 13708–13717.
34. Rezatofighi, H.; Tsoi, N.; Gwak, J.; Sadeghian, A.; Reid, I.; Savarese, S. Generalized Intersection Over Union: A Metric and a Loss for Bounding Box Regression. In Proceedings of the IEEE Conference on Computer Vision and Pattern Recognition, Long Beach, CA, USA, 15–20 June 2019; pp. 658–666.
35. Cheng, G.; Han, J. A survey on object detection in optical remote sensing images. *ISPRS J. Photogramm. Remote Sens.* **2016**, *117*, 11–28. [CrossRef]
36. Li, K.; Wan, G.; Cheng, G.; Meng, L.; Han, J. Object detection in optical remote sensing images: A survey and a new benchmark. *ISPRS J. Photogramm. Remote Sens.* **2020**, *159*, 296–307. [CrossRef]
37. Yang, X.; Yang, J.; Yan, J.; Zhang, Y.; Zhang, T.; Guo, Z.; Sun, X.; Fu, K. SCRDet: Towards More Robust Detection for Small, Cluttered and Rotated Objects. In Proceedings of the International Conference on Computer Vision (ICCV), Seoul, Republic of Korea, 27 October–2 November 2019; pp. 8231–8240.
38. Fan, X.; Yan, W.; Shi, P.; Zhang, X. Remote sensing image target detection based on a multi-scale deep feature fusion network. *Natl. Remote Sens. Bull.* **2022**, *26*, 2292–2303.
39. Zhang, J.; Wu, X.; Zhao, X.; Zhuo, L.; Zhang, J. Scene Constrained Object Detection Method in High-Resolution Remote Sensing Images by Relation-Aware Global Attention. *J. Electron. Inf. Technol.* **2022**, *44*, 2924–2931.
40. Xue, J.; Zhu, J.; Zhang, J.; Li, X.; Dou, S.; Mi, L.; Li, Z.; Yuan, X.; Li, C. Object Detection in Optical Remote Sensing Images Based on FFC-SSD Model. *Acta Opt. Sin.* **2022**, *42*, 138–148.

41. Cheng, G.; Wang, J.; Li, K.; Xie, X.; Lang, C.; Yao, Y.; Han, J. Anchor-Free Oriented Proposal Generator for Object Detection. *IEEE Trans. Geosci. Remote Sens.* **2022**, *60*, 5625411. [CrossRef]
42. Huang, Z.; Li, W.; Xia, X.; Wang, H.; Jie, F.; Tao, R. LO-Det: Lightweight Oriented Object Detection in Remote Sensing Images. *IEEE Trans. Geosci. Remote Sens.* **2022**, *60*, 1–15. [CrossRef]
43. Li, W.; Chen, Y.; Hu, K.; Zhu, J. Oriented reppoints for aerial object detection. In Proceedings of the IEEE Conference on Computer Vision and Pattern Recognition, New Orleans, LA, USA, 19–24 June 2022; pp. 1829–1838.
44. Xu, T.; Sun, X.; Diao, W.; Zhao, L.; Fu, K.; Wang, K. ASSD: Feature Aligned Single-Shot Detection for Multiscale Objects in Aerial Imagery. *IEEE Trans. Geosci. Remote Sens.* **2022**, *60*, 5607117. [CrossRef]
45. Yao, Y.; Cheng, G.; Xie, X.; Han, J. Optical remote sensing image object detection based on multi-resolution feature fusion. *Natl. Remote Sens. Bull.* **2021**, *25*, 1124–1137.
46. Yang, X.; Yan, J.; Liao, W.; Yang, X.; Tang, J.; He, T. SCRDet++: Detecting Small, Cluttered and Rotated Objects via Instance-Level Feature Denoising and Rotation Loss Smoothing. *IEEE Trans. Pattern Anal. Mach. Intell.* **2022**, *45*, 2384–2399. [CrossRef] [PubMed]
47. Zhou, L.; Zheng, C.; Yan, H.; Zuo, X.; Liu, Y.; Qiao, B.; Yang, Y. RepDarkNet: A Multi-Branched Detector for Small-Target Detection in Remote Sensing Images. *ISPRS Int. J. Geo-Inf.* **2022**, *11*, 158. [CrossRef]
48. Ye, Y.; Ren, Y.; Gao, X.; Wang, J. Remote sensing image target detection based on improved YOLOv4. *J. Optoelectron. Laser* **2022**, *33*, 607–613.
49. Zhu, F.; Gao, J.; Yang, J.; Ye, N. Neighborhood linear discriminant analysis. *Pattern Recognit.* **2022**, *123*, 108422. [CrossRef]
50. Zhu, F.; Ning, Y.; Chen, X.; Zhao, Y.; Gang, Y. On removing potential redundant constraints for SVOR learning. *Appl. Soft Comput.* **2021**, *102*, 106941. [CrossRef]

remote sensing

MDPI

Article

Learning Sparse Geometric Features for Building Segmentation from Low-Resolution Remote-Sensing Images

Zeping Liu [1] and Hong Tang [1,2,*]

1 Key Laboratory of Environmental Change and Natural Disaster of Ministry of Education, Beijing Normal University, Beijing 100875, China; liuzeping@mail.bnu.edu.cn
2 State Key Laboratory of Remote Sensing Science, Faculty of Geographical Science, Beijing Normal University, Beijing 100875, China
* Correspondence: hongtang@bnu.edu.cn

Abstract: High-resolution remote-sensing imagery has proven useful for building extraction. Unfortunately, due to the high acquisition costs and infrequent availability of high-resolution imagery, low-resolution images are more practical for large-scale mapping or change tracking of buildings. However, extracting buildings from low-resolution images is a challenging task. Compared with high-resolution images, low-resolution images pose two critical challenges in terms of building segmentation: the effects of fuzzy boundary details on buildings and the lack of local textures. In this study, we propose a sparse geometric feature attention network (SGFANet) based on multi-level feature fusion to address the aforementioned issues. From the perspective of the fuzzy effect, SGFANet enhances the representative boundary features by calculating the point-wise affinity of the selected feature points in a top-down manner. From the perspective of lacking local textures, we convert the top-down propagation from local to non-local by introducing the grounding transformer harvesting the global attention of the input image. SGFANet outperforms competing baselines on remote-sensing images collected worldwide and multiple sensors at 4 and 10 m resolution, thereby, improving the IoU by at least 0.66%. Notably, our method is robust and generalizable, which makes it useful for extending the accessibility and scalability of building dynamic tracking across developing areas (e.g., the Xiong'an New Area in China) by using low-resolution images.

Keywords: artificial intelligence; deep learning; remote sensing; semantic segmentation; building extraction

Citation: Liu, Z.; Tang, H. Learning Sparse Geometric Features for Building Segmentation from Low-Resolution Remote-Sensing Images. *Remote Sens.* **2023**, *15*, 1741. https://doi.org/10.3390/rs15071741

Academic Editors: Jukka Heikkonen, Fahimeh Farahnakian and Pouya Jafarzadeh

Received: 18 February 2023
Revised: 19 March 2023
Accepted: 21 March 2023
Published: 23 March 2023

1. Introduction

As building footprints are commonly applied in urban environments [1] for urban planning [2] and in rapid responses to natural disasters [3], the methods for effectively extracting buildings have become a popular research topic. Satellite remote sensing can observe a large area over a long time series; thus, current research efforts, particularly large-scale building mapping, primarily focus on using remote-sensing images as the data source [4]. The interest in the development of new methodologies for building segmentation is primarily motivated by high-resolution earth-observation technologies and several public high-resolution (HR) benchmark datasets, such as AIRS (0.075 m/pixel) [5], INRIA (0.3 m/pixel) [6] and the WHU Building Dataset (0.3 m/pixel) [7].

Thus, many new approaches in this field of research now focus on obtaining fine segmentation results from HR images (e.g., [4,8]). Unfortunately, HR images are captured infrequently at the same position on the Earth's surface (once a year or less), particularly in developing regions where such images are arguably more needed. In addition, HR images were less commonly captured historically, making it difficult to produce distribution maps of buildings in a specific event or disaster scenario.

Even if available, however, it is prohibitively expensive to purchase a large amount of HR data (e.g., \$23/km^2 in Digital Globe). Data with low-resolution (LR) ("High" and "low" resolutions are relative definitions and vary from task to task. In this study, to avoid narrative ambiguity with other studies, and considering previous studies and discussions, we define <4 m as HR and >4 m as LR in this paper. Further elaboration can be found in Section 4.1) are freely available and have shorter revisit periods (subweek), such as the images provided by Gaofen-2 satellite (4 m/pixel) and Sentinel-2 satellite (10 m/pixel). However, the methods designed for HR images may show severe degradation on LR images, which limits the scalability of these methods. These observations motivated us to develop a new method that is focused on building segmentation from remote-sensing images with a lower resolution.

Compared with HR remote-sensing images, building segmentation from LR images is more challenging. There are at least three reasons, one of which is a common issue in building segmentation:

1. There exists a large-scale variation of buildings in LR images (Figure 1A). This issue poses a multi-scale problem and makes it more difficult to locate and segment. This is a common issue in building segmentation.
2. The boundary details of buildings (i.e., edges and corners on buildings) are fuzzier in LR images. As shown in Figure 1B, the boundaries of buildings are fuzzier and even blend into the background, which causes difficulties for models to delineate boundaries accurately.
3. LR remote-sensing images always lack local textures due to low contrast in low resolution (Figure 1C). As a result, it is difficult to capture sufficient context information from a small patch of the image (e.g., the sliding window with a fixed size in a convolutional layer).

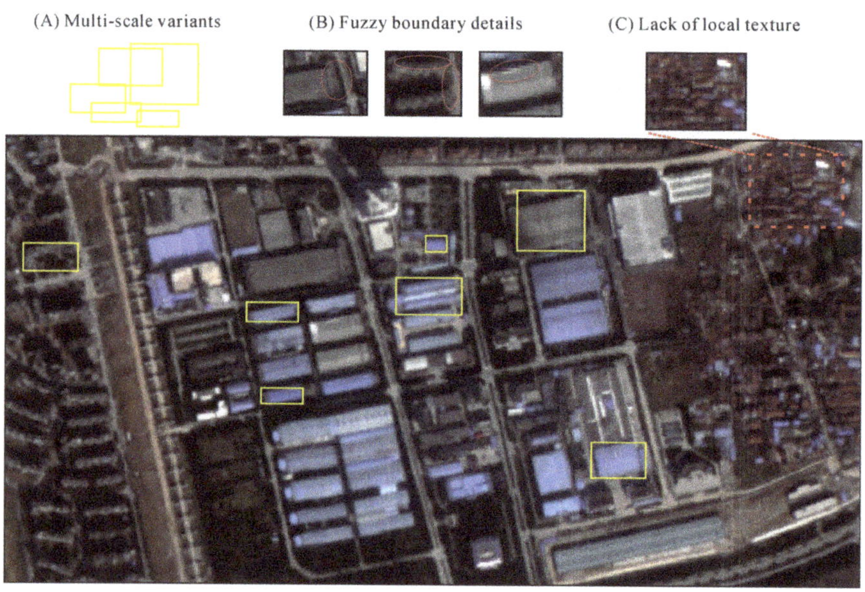

Figure 1. The main challenges of building segmentation in the LR remote-sensing imagery. (**A**) multi-scale variants. (**B**) fuzzy boundary details. (**C**) lack of local textures. The image was captured by the Gaofen-2 satellite with a resolution of 4 m/pixel. Red circles in subfigure (**B**) indicate the notable parts of the fuzzy boundary.

In the field of remote sensing, the task of building segmentation is widely recognized as the semantic segmentation task. Unlike other objects of interest, such as bodies of water or agricultural land, buildings exhibit two unique characteristics, namely diverse scales (Figure 1A) and regularized morphology (polygonal shapes). In addressing these unique issues, numerous studies have focused on designing specific modules to enhance the prediction of building boundaries at multi-scale levels.

For instance, Zhu et al. [9] and Lee et al. [10] embeded boundary information into multi-scale feature fusion. Additionally, some works detect building fragments and then utilize specific rules to compose building structure fragments [11–14]. However, due to the inherent fuzziness in both low-level features and boundary pixels, the resulting segmentation may not preserve the building with high fidelity, which can significantly limit its applicability in low-resolution remote-sensing images.

To obtain satisfactory results with low-resolution (LR) images, extensive research has been conducted in the adoption of a super resolution-then-semantic segmentation (SR-then-SS) pipeline. This pipeline involves first using super resolution (SR) techniques to restore high-resolution (HR) details from LR images, followed by using current semantic segmentation (SS) methods to extract buildings from the restored images.

To our knowledge, previous studies primarily focused on the development of effective SR modules and training strategies. For instance, recent studies by Zhang et al. [15], He et al. [16] and Kang et al. [17] have proposed innovative and effective SR methods. Additionally, investigations by Xu et al. [18] and Zhang et al. [19] have focused on the impact of the SR output when using HR images or labels as reference. However, few studies have been dedicated to the design of the SS component within the contemporary SR-then-SS framework.

In response to the limitations of current semantic segmentation (SS) methods in low-resolution (LR) imagery, this study is centered on the development of a model that can accurately extract buildings from LR images with a focus on two key aspects:

1. The proposed model aims to achieve higher accuracy than the existing methods for building extraction from LR images.
2. The proposed model is intended to outperform other SS methods when utilized as the SS module within the super resolution then semantic segmentation (SR-then-SS) framework.

To achieve these goals, we designed a novel deep-learning method for automatically extracting buildings from LR remote-sensing imagery. The overall architecture is built upon multi-level feature fusion to bridge the gap between multi-scale features. For fuzzy boundaries in LR buildings, densely propagating the boundary geometry contexts (e.g., edges and corners) in the network likely mixes the fuzzy context into predictions on LR images (see the demonstrations in Section 2.3).

Therefore, the proposed method uses a sparse propagation method, in which geometric contexts are propagated by a dedicated sampler (i.e., the sparse boundary fragment sampler module) and gated module (i.e., the gated fusion module). To compensate for the local texture, we enhanced the global attention of the feature map by introducing the grounding transformer (GT) [20]. Due to the sparse ways to manage geometry feature propagation, the proposed method is referred to as the "sparse geometry feature attention network" (SGFANet).

SGFANet addresses three issues to improve the accuracy of building segmentation from low-resolution images, which are tackled through the following contributions.

1. A sparse geometry feature attention network (SGFANet) is proposed for extracting buildings from LR remote-sensing imagery accurately, where feature pyramid networks are adopted to solve multi-scale problems.
2. To circumvent the effect of fuzzy boundary details on buildings in LR images, we propose the sparse boundary fragment sampler module (SBSM) and the gated fusion module (GFM) for point-wise affinity learning. The former makes the model more focused on the salient boundary fragment, and the latter is used to suppress the inferior multi-scale contexts.

3. To mitigate the lack of local texture in LR images, we convert the top-down propagation from local to non-local by introducing the grounding transformer (GT). The GT leverages the global attention of images to compensate for the local texture.

The remainder of this paper is organized as follows. We review the issue of interest in the literature in Section 2. Section 3 describes the algorithms of SGFANet, and Section 4 describes the experiments and analysis performed in this study. Section 5 describes a pilot application of the proposed method, and our conclusions are provided in Section 6.

2. Related Work

2.1. Deep Learning for Building Segmentation

With the development of the convolutional neural network (CNN) [21,22], there has been a great deal of progress in building segmentation. Different from other objects in remote-sensing images, buildings have regular boundaries and sharp corners; therefore, the extraction of buildings strongly depends on the accuracy of their boundary extraction. However, boundary areas only occupy a small proportion of the input image, which results in a small gradient in backpropagation. To address this issue, most studies have attempted to enhance the perception of building boundaries in deep-learning architecture. EANet [23] embeded a boundary learning branch in building segmentation to maintain both an accurate rooftop and its boundary.

Huang et al. [24] and Liu et al. [25] concatenated a boundary map and extracted features to facilitate the propagation of boundary information in an end-to-end manner. Some studies have also applied adversarial loss to refine predictions. For example, Zorzi et al. [26] and Ding et al. [27] applied an additional discriminator network to ameliorate the boundary. Additionally, some studies treat a building as a set of lines and apply specific rules [11–14,25] to compose building structure fragments (e.g., Nauata and Furukawa [11] and Liu et al. [14]), which first detected edge and corner primitives, and then composited them using the extracted topology.

However, these approaches may still have issues in practical applications. In remote-sensing images, buildings always have large-scale variation, particularly in urban areas, and the model should have the ability to generate variable-sized outputs to capture different scales of buildings. This topic (e.g., capturing variable-sized objects) has been extensively surveyed in object detection and medical image segmentation tasks, with FPN [28] and UNet [29] as representatives. Their concepts are similar (i.e., fusing features from different scales of the encoder in the decoding part). For buildings that have distinct geometric properties, the current research enhances the fusion of multi-scale building geometry. Wei et al. [30] and Chatterjee and Poullis [31] used dense connections to retain multi-level features.

Liu et al. [32] designed a spatial residual inception module along with a revised decoder to maintain both global and local information. ME-Net [33] took edge feature fusion a step further using the erosion module to crisp edges at different scales. Recently, CBR-Net [34] combined different scales of edge and rooftop features of buildings and used a coarse-to-fine prediction strategy to suppress irrelevant noise and achieved good results on several high-resolution benchmarks. Thus, the state-of-the-art (SOTA) methods use a multi-level feature fusion structure to precisely predict where a building is. Such a strategy and its limitations in LR applications are described in more detail in the following subsections.

2.2. Multi-Level Feature Fusion

Multi-level (i.e., low–high level) feature fusion is widely used in existing methods, which typically include a backbone network, which captures multi-level features and a feature fusion path. Generally, given an image $I \in R^{C \times H \times W}$, where C, H, and W are the channel dimension, height, and width, respectively, the backbone outputs a series of multi-level features $\{E_l | l = 2,3,4,5\}$, and E_l means the $1/2^l$ resolution with respect to the input image. In semantic segmentation, the feature map of the higher level is primarily used for richer semantics to further capture the contextual semantics. However, the higher semantic indicates a lower resolution without detailed spatial information, which affects the performance on smaller targets.

In contrast, lower-level feature maps from shallow layers have higher resolution but fewer semantics, which improves the performance in small targets. Thus, the fusion path is used to model the feature-wise relationship to harvest both the high semantic and high-resolution context. The conventional fusion design (i.e., the feature pyramid network (FPN) [35]) is formulated as follows:

$$D_{l-1} = \zeta(E_{l-1}) + \phi(D_l) \quad (1)$$

where ζ is the lateral connection implemented by a convolutional layer with a 1×1 kernel; ϕ denotes upsampling with a scale factor of 2; and D_{l-1} and $D_l \in R^{C\times H\times W}$ are the two adjacent features in the FPN. Through the lateral connection and top-down procedure in FPN, the feature map is augmented by the high semantics from the high-level feature map and the high spatial details from the shallow feature map in the backbone.

Evolved from this design, existing feature fusion structures [20,36,37] in recent years have been built upon the dense affinity function as shown below:

$$D_{l-1} = A(E_{l-1}, D_l)D_l \quad (2)$$

where A is the affinity function with a specific designation. For example, ICTNet [31] and UNet++ [38] used dense connections as the affinity function A. Li et al. [39] used A as a gate to filter useless contexts. EPUNet [40] denoted A as a series of CNNs, which were supervised by the edge ground truth. CBR-Net [34] enhances the high-level feature by adding prior knowledge of buildings, where A serves as a multitask classifier.

2.3. Issues in Current Research

Existing segmentation techniques based on multi-level fusion have been proven effective with building extraction, particularly in HR images. However, when these algorithms are applied to LR images, the extracted building is not always sufficiently accurate to serve building-oriented applications. As shown in the first row of Figure 2, all models exhibit considerable accuracy in high-resolution scenes with little variation.

However, as the image resolution decreases to 4 m/pixel in the second row, the IoU score drops by approximately 30%, indicating more delineation errors. Specifically, errors primarily occurred on the boundaries. In the third row, where the image resolution is further reduced to 8 m, the model has difficulty locating buildings, particularly small buildings. The results show a gap in accuracy in predictions with HR and LR images. The paragraphs below detail potential reasons for these misclassifications.

First, low–high level semantics may not be accurately preserved in LR images. Due to the limited resolution, LR images contain more ambiguous information (e.g., mixed pixels), indicating that the original image (i.e., the lowest-level feature) contains irrelevant noise. The fuzzy semantics of non-building objects or background have a detrimental effect on prediction; therefore, it is essential to suppress the irrelevant noise in a self-adaptive manner to improve the segmentation accuracy.

A second issue is the confusion of fuzzy boundary details. To involve the boundary information, one always needs to use the boundary ground truth to provide additional supervision. In LR images, however, the buildings always have fuzzy boundaries (as seen in Figure 1B), and some boundaries even blend into the building background, appearing as a non-edge pixel. This causes ambiguity between the boundary ground truth and the boundary pixel. A more effective approach is to learn the features sparsely by concentrating on only the salient and representative portions of boundaries because the dense information propagation guided by the edge ground truth may be confused with other irrelevant details.

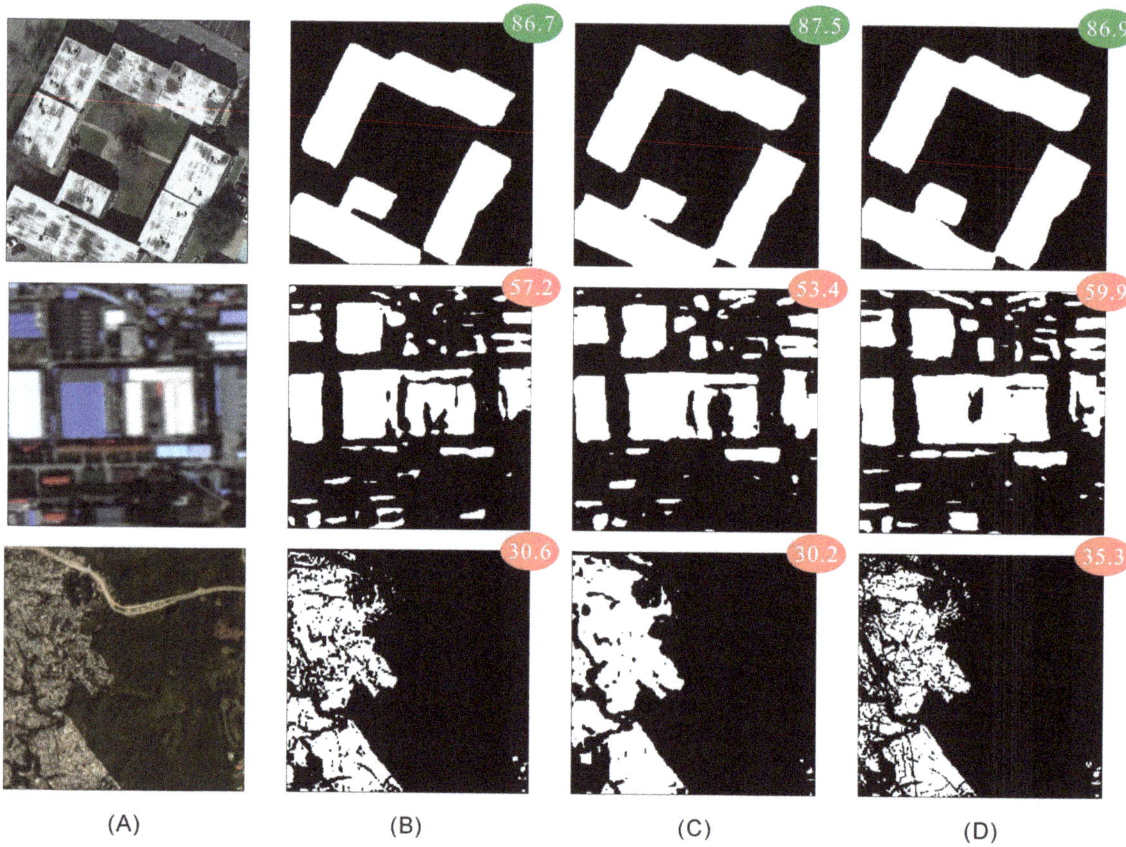

Figure 2. Real-world examples of the SOTA method. From top to bottom, the resolution of the images are 0.3, 4, and 8 m/pixel, respectively. The circle in the upper right of the images indicates the Intersect Over Union (IoU) score (%) for buildings. (**A**) Image. (**B**) Unet++. (**C**) EPUNet. (**D**) CBR-Net.

The third issue is the lack of local texture. Due to the low contrast, it is more challenging to capture texture details in LR images than it is in HR images. Akiva et al. [41] applied a pixel adaptive convolution layer [42] to introduce the texture from the input image, thus, refining deep features and encouraging the representation of pixels with similar signatures. Unfortunately, for building segmentation, involving the image texture would cause non-building targets to become prominent, leading to large intra-class variance. Utilizing non-local operation among different features to enlarge the model perception would be a more practical choice.

In this study, to address the fuzziness issue, we use the point-wise sparse propagating strategy instead of the dense propagating strategy (feature-wise) as in Equation (2). Specifically, we only calculate the affinity of representative feature points on the edges and corners of buildings (i.e., $D_{l-1}(p)$ and $D_l(p)$, p is the sampled pixels) in the two adjacent pyramid feature maps D_{l-1} and D_l. The positions of the selected feature points are learned by the model. Then, we apply the proposed gated fusion module as an affinity function A to harvest the specific context only on selected edges and corners of buildings to alleviate the side effect of the fuzzy contexts in LR images:

$$D_{l-1}(p) = A(E_{l-1}(p), D_l(p))D_l(p) \tag{3}$$

The differences among the three types of feature fusion design are shown in Figure 3. The innovation compared to other work is that we propagate the building feature contexts in a sparse way instead of in a dense way. Finally, for the lack of local textures, we harvest the global attention of the input using the recently introduced grounding transformer, in which the interactions between adjacent features are in a non-local style.

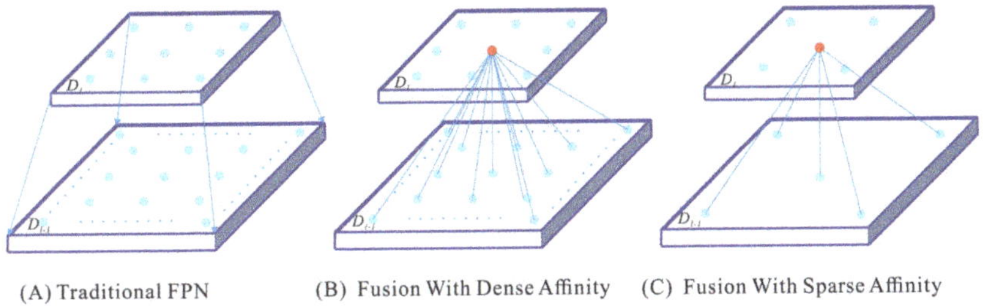

Figure 3. Schematic diagram of three types of feature fusion structures. (**A**) The FPN, where the top feature is directly propagated by the addition function. (**B**) The recent feature fusion structure with a dense affinity function, where the top feature is propagated by calculating the affinity on two adjacent layers. (**C**) The sparse affinity function, which only calculates the affinity of representative features.

3. Sparse Geometry Feature Attention Network

In this section, we describe, in detail, the proposed SGFANet, including the overview (Section 3.1), the modules (Section 3.2 and the loss function (Section 3.4)).

3.1. Overview

The proposed method is based on a bottom-up feature extractor and a top-down feature fusion path as shown in Figure 4. The choice of the feature extractor is not the focus of this study; thus, ResNet-50 [43] was implemented. The feature extractor produces a series of multi-level features that are denoted as $\{E_l | l = 2, 3, 4, 5\}$. To enhance contextual semantics, we applied the pyramid pooling module (PPM) to the highest feature map E_5 and obtained the top pyramid feature D_5. This is a widely used setup in various feature fusion methods [20,28,37,39].

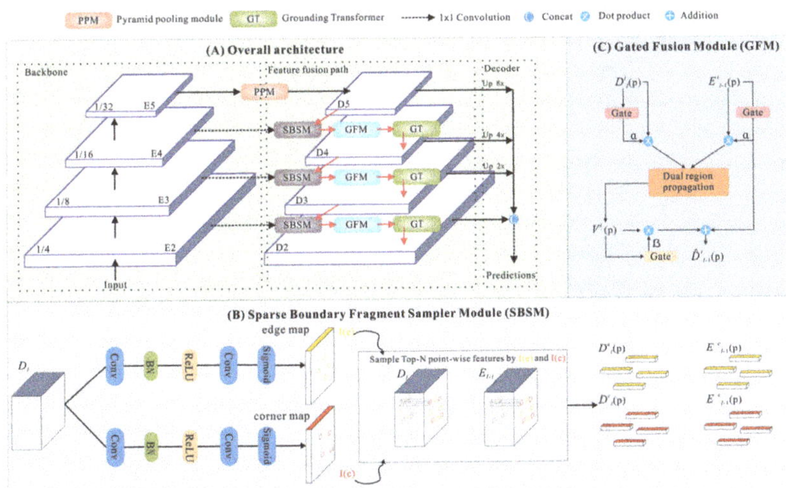

Figure 4. Frameworks of the proposed SGFANet. (**A**) The overall pipeline of the proposed SGFANet includes a bottom-up basic hierarchical feature extractor, a top-down feature fusion path composed of SBSM, GFM, and GT, and a decoder. (**B**) The sparse boundary fragment sampler module (SBSM) serves to sample the top-N representative feature points about the building boundaries (i.e., the edges and corners). N is a hyperparameter and can be different for edges and corners. (**C**) The gated fusion module (GFM), which is used to calculate the affinity of the selected point-wise features.

Then, we fused the set of E_l along with the building boundary feature in a top-down and sparse manner (i.e., focusing on the representative features to alleviate the irrelevant noises brought by fuzzy details). This process generates the pyramid feature set $\{D_l | l = 2, 3, 4\}$. Finally, we concatenated all pyramid features in D_l to generate the building prediction result.

3.2. Learning Sparse Geometry Features

This subsection posits that the multi-level feature set, denoted as E_l, and the highest-level pyramid feature, which is denoted as D_5, have been successfully acquired by the ResNet-50 backbone and PPM module. Then, the proposed top-down fusion method is described in detail.

3.2.1. Sparse Boundary Fragment Sampler Module

Different from other methods [23,40], we propagate the edge and corner contexts separately and sparsely in a top-down manner. In this study, we argue that the most representative feature points can be represented as top-N points from the possibility maps of both edges and corners. We define "Top-N" as "the first N points with the highest possibility in the possibility map", where N is a hyperparameter and can be different in edge and corner possibility maps (see more detail in Section 4.8).

As shown in Figure 4B, we propose the sparse boundary fragment sampler module (SBSM) to select the top-N feature points. The SBSM uses two independent branches over the pyramid feature D_l to obtain the edge and corner possibility map of the input image. These two branches are both implemented by a 3 × 3 convolutional layer, a batch normalization layer, a ReLU function, and another 3 × 3 convolutional layer followed by a sigmoid function and are supervised by the ground truth of the edge and corner distribution map. Then, the SBSM samples the top-N points according to their possibility value to obtain the representative edge indices I_e and corner indices I_c.

Before obtaining the representative feature points, we first refine E_{l-1} using a 1 × 1 convolutional layer. Next, the SBSM obtains the representative feature points from D_l and $E_l - 1$ according to I_e and I_c. We use the normalized grids and bilinear interpolation during the implementation. The representative edge feature points are denoted as $D_l^e(p)$ and $E_{l-1}^e(p)$, while the corners are denoted as $D_l^c(p)$ and $E_{l-1}^c(p)$.

3.2.2. Gated Fusion Module

After obtaining $D_l(p)$ and $E_{l-1}(p)$, the critical point is how to propagate them in the proposed fusion path. However, different levels of D_l have different capacities to capture the edge and corner spatial and contextual information. Simply designing a uniform propagation mechanism for each level of the pyramid feature would produce a semantic gap; thus, we design an attention mechanism that serves as an affinity function to reweight the feature points from different levels and propagate them.

We propose a gated fusion module, which contains two parts: (1) a gated operator and (2) dual region propagation. The gated operator is a channel attention mechanism focusing on reweighting features along the channel of each pyramid level. It first explicitly models the dependency of features along the channel and learns a descriptor to express the importance of each channel. Then, the descriptor enhances the useful channel and suppresses the inferior channel by dot production. The gated operator is as follows:

$$\begin{aligned} \tilde{D}_l^i(p) &= \alpha_l^i(p) \cdot D_l^i(p) \\ \tilde{E}_{l-1}^i(p) &= \alpha_{l-1}^i(p) \cdot E_{l-1}^i(p), \end{aligned} \tag{4}$$

where i is the index from I_e or I_c; $\alpha_l^i(p) \in [0, 1]$ and $\alpha_{l-1}^i(p) \in [0, 1]$ are the associated gate maps of each level l, which are obtained through linear projection from $D_l^i(p)$ and $E_{l-1}^i(p)$, respectively; and $\cdot$ denotes the dot multiplication broadcasting in the channel dimension. More detail is available in Figure 5.

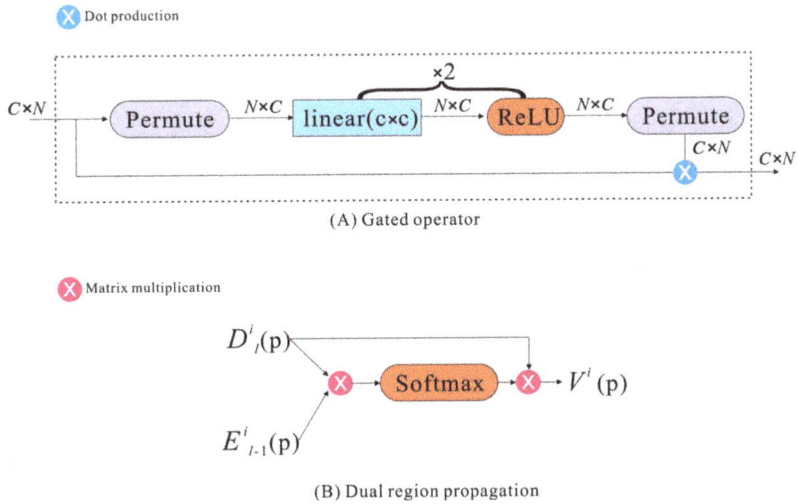

Figure 5. The detailed structure of (**A**) the gated operator and (**B**) dual region propagation. The input of our gated operator is with dimension of $C \times N$. SGFANet utilizes it in three features, e.g., $D_l(p)$, $D_{l-1}(p)$ and $V^i(p)$. The dual region propagation is a kind of self-attention mechanism, for projecting the feature $D^i_l(p)$ and $E^i_{l-1}(p)$ to one-feature embedding.

Then, we propagate those gated sampled features independently. For each group of sampled features (i.e., the edge or the corner), top-down propagation is realized by dual region propagation followed by another gated mechanism as shown in Equation (5):

$$V^i(p) = \delta(\tilde{D}^i_l(p) \times \tilde{E}^i_{l-1}(p)) \times \tilde{D}^i_l(p) \tag{5}$$

and (6),

$$\hat{D}^i_{l-1}(p) = \beta^i(p) \cdot V^i(p) + \tilde{E}^i_{l-1}(p) \tag{6}$$

where δ denotes the softmax function for value normalization; $V^i(p)$ is the affinity of $\tilde{D}^i_l(p)$ and $\tilde{E}^i_{l-1}(p)$; $\times$ denotes matrix multiplication; $\cdot$ denotes the dot product; and $\beta^i(p) \in [0,1]$ is another gated operator obtained from the linear projection of $V^i(p)$.

Equation (5) is the dual region propagation, which is a type of self-attention mechanism, and we only apply selected representative feature points to the calculation. Equation (6) is another gated operator that is used to filter out useless contexts. In this study, we use the residual design for easier training in Equation (6) (i.e., the add operation to accelerate gradient broadcast). The overall gated pipeline is shown in Figure 4C.

3.3. From Local to Non-Local Features

The CNN has been shown to be powerful in local features for its shared position-based kernels over a local and fixed-size window, which keeps the translation invariant and makes promising results in capturing local features, such as shapes [44,45]. Unfortunately, the LR image lacks local texture details due to the low contrast. If we focus on the local pattern, as conventional CNN does, the learning procedure in a low-resolution plane would be insufficient. To combat this issue, we transfer the local calculation to non-local operation using the recently introduced grounding transformer (GT) [20]. In this study, we use two convolution layers on D_l to obtain q and v and one convolution layer on E_{l-1} to obtain k. q, k, and v denote the key, query, and value in the transformer, respectively. Then, we implement the GT as:

$$\begin{aligned} s &= q \cdot k \\ w &= \delta(s) \\ \hat{D}_{l-1} &= w \times v \end{aligned} \tag{7}$$

where δ is the softmax function, and the output of the GT is denoted as $\hat{D}_{l-1}$, which harvests the nonlocal contexts (i.e., global) of two adjacent pyramid features (i.e., E_{l-1} and D_l). Finally, the refined output feature D_{l-1} is obtained by scattering the feature point $\hat{D}^i_{l-1}(p)$ into $\hat{D}_{l-1}$ according to the indices I_e and I_c. The top-down procedure is summarized in Algorithm 1.

Algorithm 1 The top-down procedure in SGFANet.

Require: $\{E_l | l = 2,3,4,5\}$: the extracted feature set from the backbone; N_c: the sampling number of corner points; N_e: the sampling number of edge points; and D_5: the highest-level pyramid feature.
Ensure: The pyramid feature set $\{D_l | l = 2,3,4\}$

```
for l in [5, 4, 3, 2] do
    edge_map, corner_map ← Sigmoid(Convolution(D_l))
    Obtain I_e and I_c from edge_map and corner_map
    D^e_l(p), D^c_l(p) ← Grid_sample(D_l, I_e), Grid_sample(D_l, I_c)
    E^e_{l-1}(p), E^c_{l-1}(p) ← Grid_sample(E_{l-1}, I_e), Grid_sample(E_{l-1}, I_c)
    D̃^i_l(p) ← Gated_operator(D^i_l(p)) with Equation (4)
    Ẽ^i_{l-1}(p) ← Gated_operator(E^i_{l-1}(p)) with Equation (4)
    D̂^i_{l-1}(p) ← Dual_region_propagation(Ẽ^i_{l-1}(p), Ẽ^i_{l-1}(p)) with Equations (5) and (6)
    D̂_{l-1} ← Grounding_transformer(E_{l-1}, D_l) with Equation (7)
    D_{l-1} ← Scatter the D̂^i_{l-1}(p) into D̂_{l-1} according to the I_e and I_c.
end for
```

As previously described, CNN is limited by its locality when resolving LR images. As we want to emphasize nonlocal attention to compensate for the lack of local texture, we use GT to fully interact with the pyramid feature in both spaces and scales. When using GT, the model encourages representation consistency for the building region and inhibits the false alarms of the building background as shown in Figure 6.

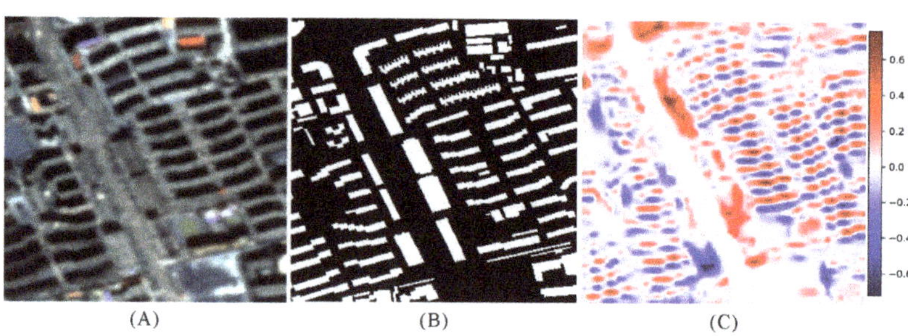

(A) (B) (C)

Figure 6. The impact of the GT can be better recognized by a heat map of the predicted score. The building predicted scores are the output of the sigmoid layer with the range [0, 1]. (**A**,**B**) The image and the corresponding label, respectively. (**C**) Heatmap obtained by visualizing the values of the difference between the prediction score of SGFANet and that of SGFANet without GT.

3.4. Decoder and Loss Function

The decoder is used to recover the spatial resolution of the output feature $\{D_l | l = 2,3,4,5\} \in R^{C \times H \times W}$. We adopted a simple and lightweight decoder in SGFANet to preserve the model's efficiency. SGFANet concatenates all the refined D_l by upsampling them to the same resolution (1/4 resolution of the input image) and performs the final prediction by a 1 × 1 convolutional layer and sigmoid function.

The overall output of SGFANet is threefold: the edge confidence map, the corner confidence map, and the final segmentation prediction map. We, thus, use binary cross entropy as a loss function, and these losses are weighted to 1 by default.

4. Experiments

4.1. Definition of LR and HR Images in This Paper

To segment individual buildings, the definitions of LR and HR are different from those of other tasks [46]. As shown in Figure 7, most buildings are under 144 m^2. Assuming a resolution of 4 m/pixel, up to nine pixels are used to render one building. Convolutional layers with 3 × 3 kernels are the foundation for contemporary deep-learning segmentation models. Due to the small size of the building, a sliding window could not be filled in one convolutional layer, introducing a great deal of background noise into building segmentation. In addition, the geometric components of buildings (e.g., the edges and corners) are small and could blend into the background. Accurate building segmentation could, thus, be strongly hampered by these confusing pixels. Xu et al. [18] and Zhang et al. [15] defined 1.2, 2, and 4 m/pixel as low resolution, while Shi et al. [47] defined 3 m/pixel as moderate resolution. In this study, the LR images are those with a resolution of greater than 4 m, while the HR images are the reverse.

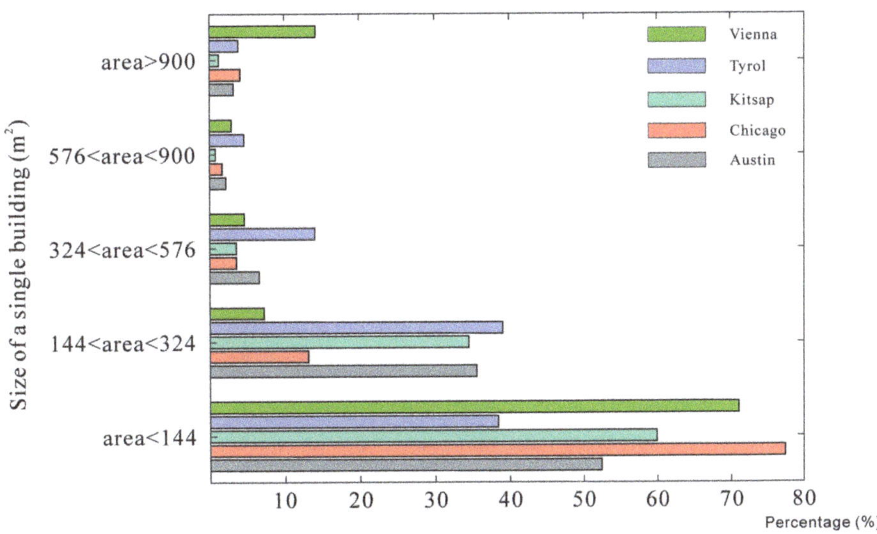

Figure 7. Statistical results of the single building size in the INRIA building dataset, including five cities, e.g., Vienna, Tyrol, Kitsap, Chicago, and Austin.

4.2. Datasets

To demonstrate the effectiveness of the proposed methods, we have three goals: (1) to assess the performance of SGFANet on LR images and other existing methods; (2) to assess the performance of SGFANet as an SS module under the existing SR-then-SS framework; and (3) to assess the effectiveness of the proposed module and evaluating how different combinations affect the model performance.

We used two datasets: the Multi-Temporal Urban Development Spacenet (i.e., Spacenet 7) and the DREAM-A+ dataset [18,19]. Spacenet 7 provides 4 m/pixel RGB imagery and the corresponding building footprint in GeoJSON format. The released images and the ground truth were captured in 60 locations worldwide. Each location provided 24 images (one per month) with a size of 1024 × 1024 between 2017 and 2020, covering an area of approximately 18 km^2. The DREAM-A+ dataset contains 4 m/pixel RGB imagery captured by the Gaofen-2 satellite sensor as well as the corresponding building footprints in Shapefile format. The images are presented as 256 × 256 image patches that were uniformly collected across 14 cities in China. The distribution of the data samples is shown in Figure 8.

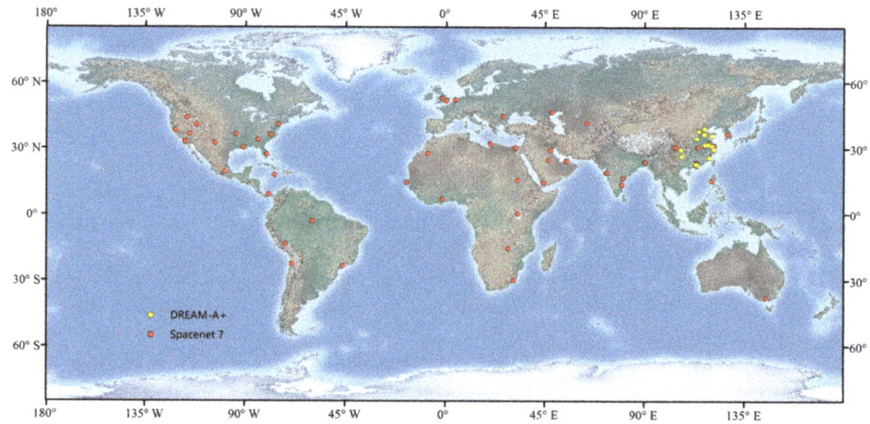

Figure 8. Distributions of images and the corresponding ground truths in the proposed experiment.

In addition, we collected the corresponding low-resolution imagery with RGB and NIR bands from Sentinel-2 (10 m/pixel). The images were obtained in the same period and location as the corresponding image in Spacenet 7 and DREAM-A+. Images with an excessive amount of clouds (e.g., 20%) were omitted from the Sentinel-2 dataset, reducing the amount of available imagery. Then, by rasterizing the related GeoJSON and Shapefile, we matched each Sentinel-2 image with a 2.5 m building ground truth.

With these datasets, we conducted three types of experiments. First, we evaluated SGFANet compared to existing methods on Spacenet 7 and DREAM-A+, which we merged into one dataset. This scenario is closer to practical applications (i.e., satellite images derived from multiple sensors and geographic locations), which requires higher generalization performance of the model. Second, we evaluated the SGFANet performance in the SR-then-SS framework on the collected Sentinel-2 imagery. Sentinel-2 imagery is publicly accessible from most places around the globe; therefore, it is practicable to use only this sensor. Third, we verified the proposed module utilizing the DREAM-A+ dataset. In this experiment, we were concerned with the effect of various modules on the model performance; thus, we used a single source image to make the experiment reliable.

4.3. Implementation Details

We cropped the 4 m/pixel images into a fixed size of 256 × 256 and the 10 m/pixel images into the size of 64 × 64. In the process of training, the AdamW [48] optimizer was used with a 0.001 learning rate, and the method was trained for 100 epochs on image tiles. The batch size was 16. In each mini-batch, we used standard data augmentation with a probability of 20%, including rotation of $[-20°, 20°]$, horizontal and vertical flip during training. For hyperparameter N, we adopted 128 and 32 for sampling points from the edges and corners, respectively.

The threshold of the sigmoid output was 0.5. The layers in Resnet-50 were initialized with the pre-trained weight on ImageNet. The training, validation, and test set were randomly divided in a ratio of 5:2:3. The number of image tiles in each set is shown in Table 1. During training, we assumed that the test set was invisible. After each epoch, we verified the accuracy of the model on the validation set and chose the model with the highest accuracy in the validation set for testing. Unless otherwise stated, the precision we report in the experiment is the precision of the test set. All experiments were implemented in PyTorch.

Table 1. Number of image tiles in the train set, validation set, and test set for each dataset.

	Train	Validation	Test
DREAM-A+ dataset	1950	780	1169
Spacenet7 dataset	6571	2628	3941
Sentinel-2 dataset	15,274	6109	9164

4.4. Accuracy Assessment

Four evaluation metrics were used to assess the performance of the models, including the intersection over union (IoU), overall accuracy (OA), F1 score (F1), and boundary F1 score (b-F1). After obtaining the prediction maps, true-positive (TP), false-positive (FP), true-negative (TN), and false-negative (FN) pixels were calculated. True positives mean the truly predicted pixels with positive labels, and false positives denote negative labels with false predictions. For IoU, OA, and F1, the positive pixels are the building regions. For b-F1, the positive pixels are building boundary pixels. The evaluation metrics are calculated as follows:

$$\begin{aligned} IoU &= \frac{TP}{TP + FP + FN} \\ OA &= \frac{TP + TN}{TP + FP + FN + TN} \\ F1 &= \frac{2 \times TP}{2 \times TP + FP + FN} \end{aligned} \tag{8}$$

4.5. Results of SGFANet

Using the collected 4 m/pixel images, we compare the proposed SGFANet with several literature works to assess its effectiveness on the semantic segmentation task, including state-of-the-art (SOTA) methods, the dense boundary propagation method, and the shape-learning method:

1. DeepLabV3+ [49] achieved SOTA results in the PASCAL VOC dataset. It is also a common baseline method in the semantic segmentation field.
2. Unet++ [38] is the SOTA architecture among variants of the Unet. Its multi-scale architecture makes it effective in capturing various sized targets and is, therefore, often applied in building extraction.
3. ICTNet [31] was the winner in the 2019 INRIA competition. It utilizes dense-connection block [50] to extract building features. Instead, our method adopts a sparse propagation strategy supervised by the building boundary label.
4. CBR-Net [34] achieved SOTA results in the WHU building dataset. It is also the most recent SOTA algorithm.
5. PFNet [37] achieved SOTA results in the iSAID dataset. It is not dedicated to extracting buildings; however, it uses a sampling strategy similar to ours. The greatest difference is that it samples mainly to tackle the imbalance between the foreground and background pixels, while we are more refined, targeting only the building boundary and corner pixels.
6. EPUNet [40] is a dense boundary propagation method. Compared with ICTNet, it introduces building boundaries as supervision. Compared with our method, it propagates the boundary contexts densely (i.e., without any context filtering).
7. ASLNet [27] is a shape-learning method that applies the adversarial loss as a boundary regularizer. It is designed to extract a more regularized building shape.

For the comparison methods, we used the hyperparameter settings from their respective studies (e.g., the dimensions of the convolution filters). To create fair comparisons, other training parameters, such as the backbone network, learning rate, and batch size, were kept consistent with the proposed approach.

4.5.1. Comparison with State-of-the-Art Methods

The quantitative evaluations of the proposed SGFANet and other existing methods are listed in Table 2. The proposed SGFANet outperformed the other methods by a wide margin. DeepLabV3+, Unet++, and PFNet were initially proposed for segmentation on general segmentation tasks but do not consider special issues in segmentation on building extraction and achieved lower accuracies than SGFANet. ICTNet was designed to use the compact internal representation of the backbone network and dynamic attention mechanisms in the decoder network to extract buildings in HR images well.

Table 2. Comparison with the related results on the benchmark. The bold values in each column means the best entries.

	IoU(%)	OA(%)	F1(%)	b-F1(3PX)
DeepLabV3+	45.09	89.29	62.15	63.70
Unet++	46.12	89.32	63.13	65.87
ICTNet	46.69	89.32	63.66	66.89
CBR-Net	47.80	89.85	64.68	66.73
PFNet	47.26	89.72	64.19	64.92
EPUNet	45.84	89.14	62.87	65.45
ASLNet	40.47	88.79	58.29	60.18
SGFANet	**48.46**	**90.06**	**65.28**	**66.94**

However, ICTNet's lack of explicitly modeling the LR issues makes it inefficient compared to the proposed method. The CBR-Net is built upon a coarse-to-fine framework by simultaneously taking several sources of prior information about buildings (e.g., the boundary directions). Unfortunately, in LR images, this prior information is less marked than that in HR images due to less detailed spatial information, which poses a weak supervision problem for CBR-Net. Compared with CBR-Net, our method achieved 0.66%, 0.21%, 0.60%, and 0.21% improvement on IoU, OA, F1, and b-F1, respectively.

Qualitative results are shown in Figure 9. Due to the image's low resolution, the building in it is smaller and more difficult to distinguish. However, the proposed SGFANet still had better segmentation results in handling false-positives of dense building regions and a better capability to extract high-quality building boundaries compared with the other methods. Specifically, the proposed SGFANet had several advantages in different scenes. The method could accurately extract individual buildings from the first row to the third row of Figure 9.

When the building boundary becomes much fuzzier in the fourth row, the proposed method still produced accurate predictions, while the other compared methods failed to manage this change. The segmentation results of the proposed method have fine-grained boundaries in the fifth and sixth rows, while other methods may generate blob-like results. Notwithstanding the advancements achieved by our proposed technique, the extent of progress in the small-sized building extraction remains constrained as illustrated in (Figure 10). While SGFANet successfully identified small buildings, it faced difficulties in accurately outlining their boundaries. This inadequacy can be mainly attributed to the comparatively smaller dimensions of these buildings, resulting in an imbalanced training set with fewer pixels available for training.

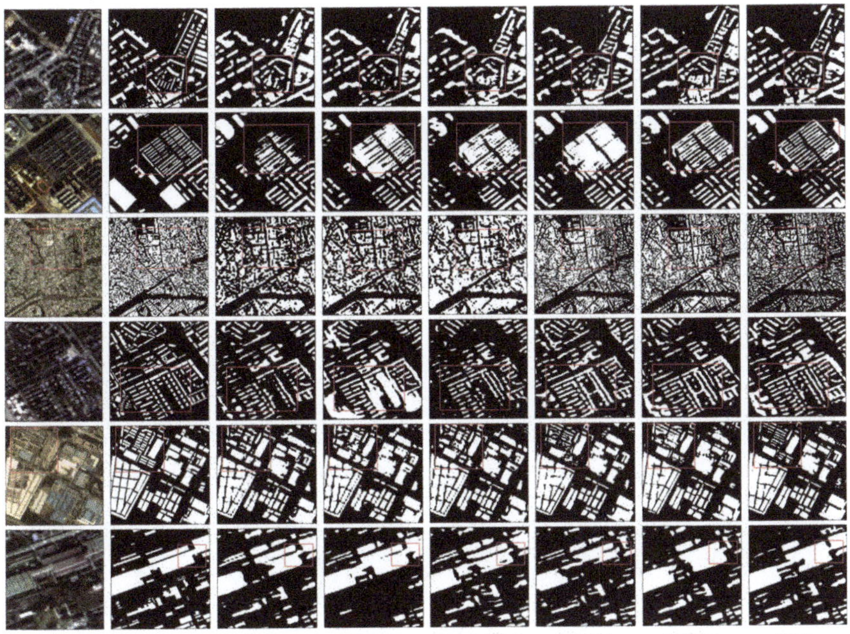

Figure 9. Examples of the building extraction results of the proposed SGFANet and other SOTA methods. The red squares indicate our notable improvement.

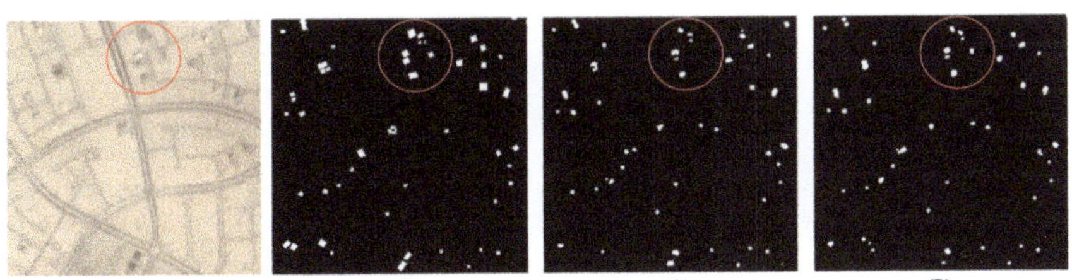

Figure 10. Example of the poor extraction result. (**A**) Image. (**B**) The ground truth. (**C**) CBR-Net. (**D**) SGFANet. A distinct area of inadequate extraction is highlighted by a red circle.

4.5.2. Comparison with Dense Boundary Propagation Methods

We further compare our method with a dense boundary context propagation method, the EPUNet [40], which adopts the opposite strategy to ours in handling the boundary supervisions. The EPUNet achieved promising results on HR remote sensing benchmarks, such as the WHU Building Dataset (0.3 m/pixel) and SYSU Building Dataset (0.8 m/pixel) with 7.47% and 2.86% accuracy gaps of IoU compared with DeepLabV3+ as reported in Guo et al. [40]. However, in our experiment, when the resolution was reduced to 4 m/pixel, the accuracy of EPUNet dropped by a large amount and approached that of Deeplabv3+ (see in Table 2).

We propose that the dense propagation of edge information brings too many fuzzy contexts to the prediction, and this phenomenon was more significant when the resolution decreases. The proposed SGFANet propagates the building boundary information (i.e., the edges and the corners) sparsely by the proposed SBSM and the GFM. As seen in Table 2, our result outperformed EPUNet by 2.62%, 0.92%, 2.41%, and 1.49% at IoU, OA, F1, and b-F1. Additionally, when buildings are densely distributed, the EPUNet tends to obtain ambiguous predictions of building boundaries, while we could more reliably detect boundary pixels from this dense area (see in Figure 11).

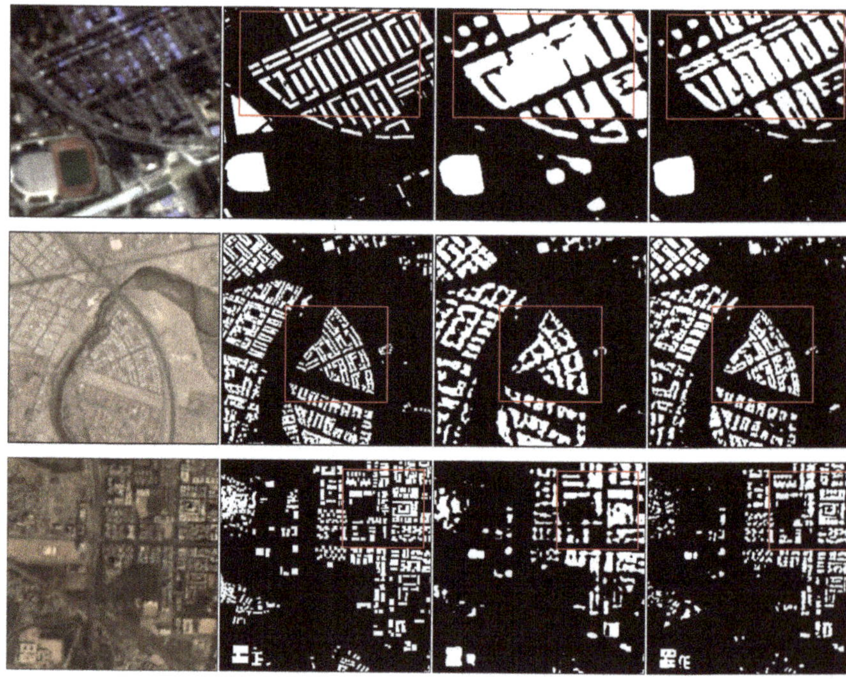

Figure 11. Sample images and results where the buildings are densely distributed. (**A**) Image tile. (**B**) Ground truth. (**C**) EPUNet. (**D**) SGFANet. EPUNet propagates all features of the building boundary without filtering, which limits the performance in LR imagery due to the fuzzy boundary details—particularly for the buildings in dense areas. The red squares indicate our notable improvement.

4.5.3. Comparison with Shape Learning Methods

Recent studies in building extraction have used shape constraints as an additional loss to optimize the training procedure, particularly when employing HR images [26,27,51]. ASLNet [27] uses adversarial loss [52] to determine whether the morphology of the model output is consistent with the ground truth and, thereby, achieves good performance that is comparable to those of other generative adversarial network (GAN)-based methods. However, ASLNet struggles to appropriately capture building shape when it meets with LR imagery. As shown in Table 2, our method improved 7.99%, 1.27%, 6.99%, and 6.76% in terms of the IoU, OA, F1, and b-F1, respectively. In Figure 12, the ASLNet did not converge as well as expected and therefore output blob-like results. This result likely occurred because the LR images typically have unclear building shape patterns due to confusing pixels, providing the model with less effective shape information for optimization.

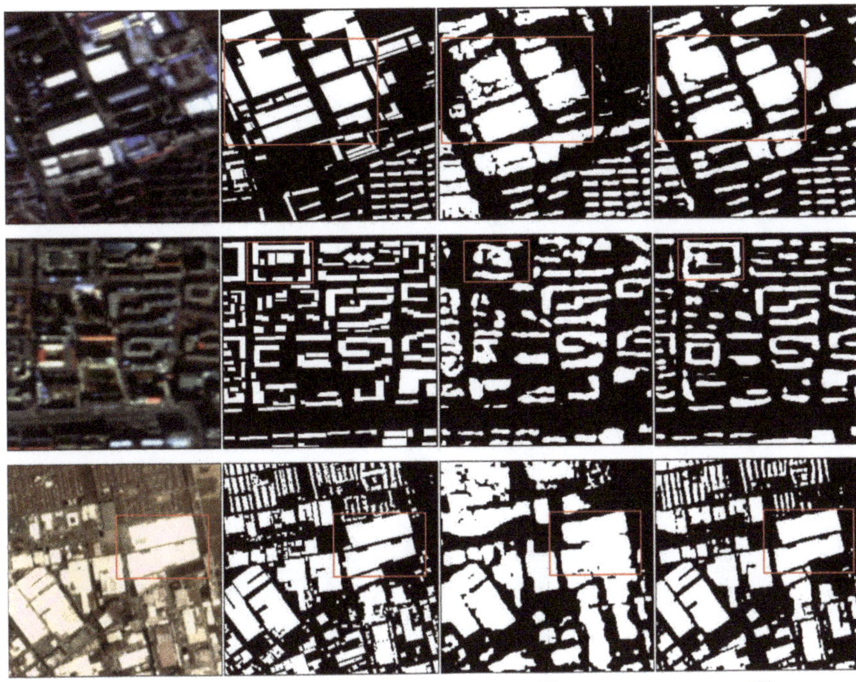

(A) (B) (C) (D)

Figure 12. Examples of the ability to capture building shapes. From left to right are (**A**) Image tile. (**B**) Ground truth. (**C**) ASLNet. (**D**) GFANet. ASLNet was proposed to find the regularized shape of buildings. However, it failed to predict accurate building boundaries compared with our method. The red squares indicate our notable improvement.

4.6. Super Resolution and then Semantic Segmentation

4.6.1. Framework Architecture

Without a loss of generality, we used the commonly used baseline method from the SR field of research (i.e., the efficient subpixel convolutional neural network (ESPCN) [53]) as our SR module. Following the previous well-established experimental setting [18], we used the SR as the front component and the SS as the rear component. Specifically, the input LR imagery was first processed by the SR to output the upsampling feature. Then, the SS was fed the generated upsampling feature to obtain an upsampling prediction result. In this experiment, we kept the SR module invariant and only replaced the SS module with the model mentioned in Section 4.5 and conducted experiments on the collected Sentinel-2 imagery.

Additionally, to comprehensively evaluate the performance, we introduce two additional comparison methods:

1. ESPC_NASUnet [18] realizes the SR by ESPCN and the SS by NASUnet [54].
2. FSRSS-Net [19] introduces successive deconvolution layers in Unet, thus, achieving super resolution results.

4.6.2. Results

The quantitative results and the qualitative results are shown in Table 3 and Figure 13, respectively. From the quantitative perspective, with the SR module plugged in at the front, SGFANet still outperformed existing semantic segmentation methods and performed markedly better than the other two existing methods, achieving 2.74%, 1.49%, and 3.16% improvements in IoU, OA, and F1. From the qualitative perspective, the proposed method maintained accurate building boundaries while improving the segmentation of single buildings in dense residential areas.

Table 3. Comparison with the related results based on the SR-then-SS framework. The bold values in each column means the best entries.

	SR Module	IoU (%)	OA (%)	F1 (%)
DeepLabV3+	ESPCN	31.07	82.90	47.41
Unet++	ESPCN	32.04	83.57	48.53
ICTNet	ESPCN	32.19	82.75	48.71
CBR-Net	ESPCN	32.72	83.09	49.53
PFNet	ESPCN	29.44	**84.11**	45.49
EPUNet	ESPCN	31.83	83.07	48.29
ASLNet	ESPCN	28.12	82.56	43.90
SGFANet (ours)	ESPCN	**33.11**	84.00	**49.75**
ESPC_NASUnet	ESPCN	30.37	82.51	46.59
FSRSS-Net	-	27.28	83.66	42.87

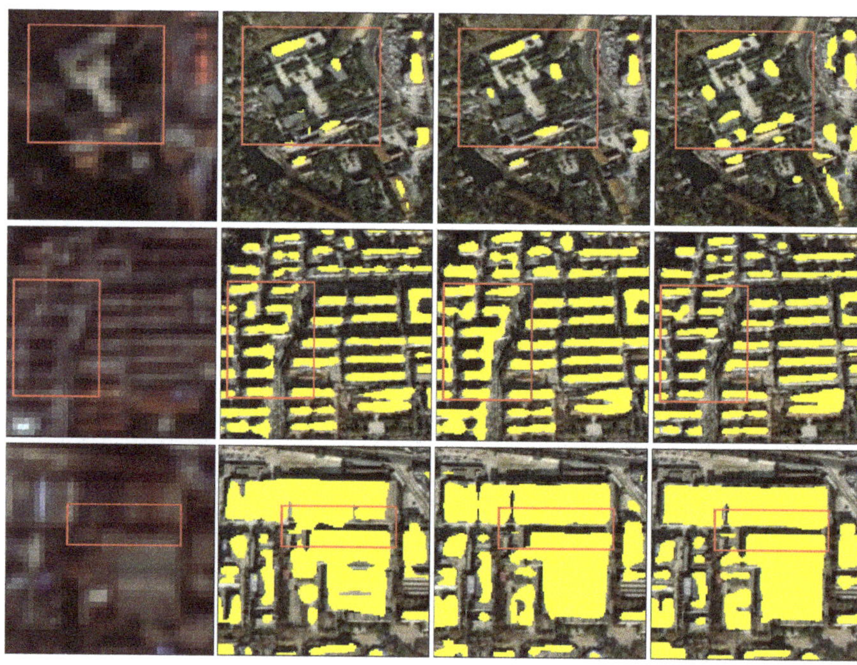

Figure 13. Examples of the super resolution and then semantic segmentation results obtained by the different methods in Sentinel-2 images. (**A**) The image tile (10 m/pixel). (**B–D**) The building extraction results (2.5 m/pixel) from ESPC_NASUnet, FSRSS-Net, and our SGFANet under the SR-then-SS framework, respectively. For better visual effects, we used the 2.5 m image from Esri community as the base map. The red squares indicate our notable improvement.

4.7. Model Efficiency

Figure 14 shows the trade-off between speed and accuracy. SGFANet outperformed methods with comparable parameter sizes, such as PFNet and CBR-Net, and achieved a higher efficiency improvement than did ICTNet (1.77 % in IoU). Thus, the proposed method achieved the optimal balance between speed and accuracy on the proposed benchmark.

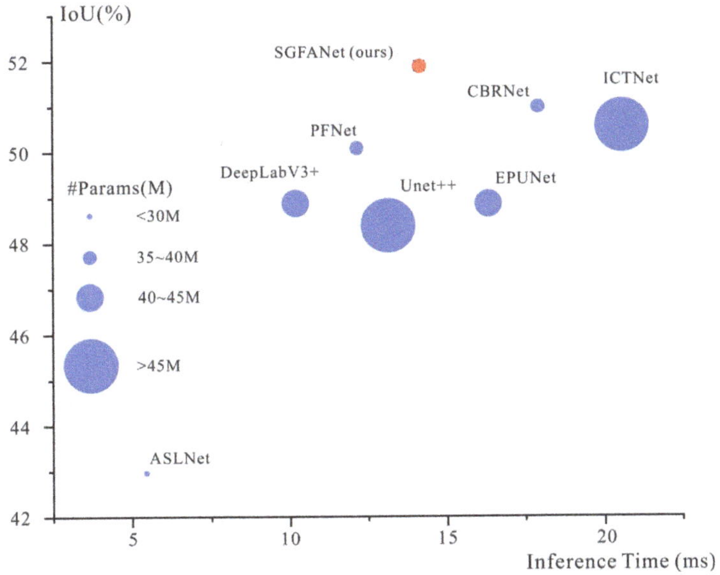

Figure 14. Speed versus accuracy on the DREAM-A+ dataset. The radii of circles represent the number of parameters (million). All the methods were tested with one V-100 GPU for a fair comparison.

4.8. Sampling of Edge and Corner Points

The number of sampled feature points about the edge (N_e) and the corner (N_c) impact the model performance. Due to the presence of these two parameters, it would be laborious to find their optimal combination during training. Although several parameter optimization approaches [55], such as grid search, random search [56], and genetic algorithms [57], have been shown to be useful in deep learning, directly using them would markedly increase the computational load.

In this study, we propose three straightforward and intuitive strategies for selecting the optimal number of sampled edges and corners: (1) sample only the edge points first and look for the best N_e, and then adjust N_c based on the best N_e; (2) sample only the corner points first and look for the best N_c, and then N_e is adjusted based on the best N_c; and (3) keep the ratio of N_e and N_c constant and adjust N_e and N_c concurrently.

We conducted experiments with these strategies separately on the DREAM-A+ dataset. As reported in Figure 15, different strategies obtained different trained results with different times. Table 4 reports the results on DREAM-A+ *test* of these three sets of parameters, where "S1", "S2", and "S3" denote the adjusting strategy used, and the column "Time" is the rough cost time to adjust N_e and N_C on a single V-100 GPU. Considering the trade-off between accuracy and time cost, we recommend the third strategy, as mentioned above, for selecting the optimal N_e and N_c combination.

In Figure 15, if less appropriate parameter values are chosen, the accuracy of SGFANet decreases considerably. Thus, three configurations degrade the accuracy: (1) only sampling the corner or edge points; (2) undersampling the edge and corner points; and (3) oversampling the edge and corner points. These results verify that both edge and corner supervised information are equally important and indicates that sparse sampling (i.e., only sampling corners and edges simultaneously) is the bottleneck of LR image building segmentation because too little sampling leads to insufficient supervision, and too much sampling introduces too many fuzzy boundary contexts from LR images.

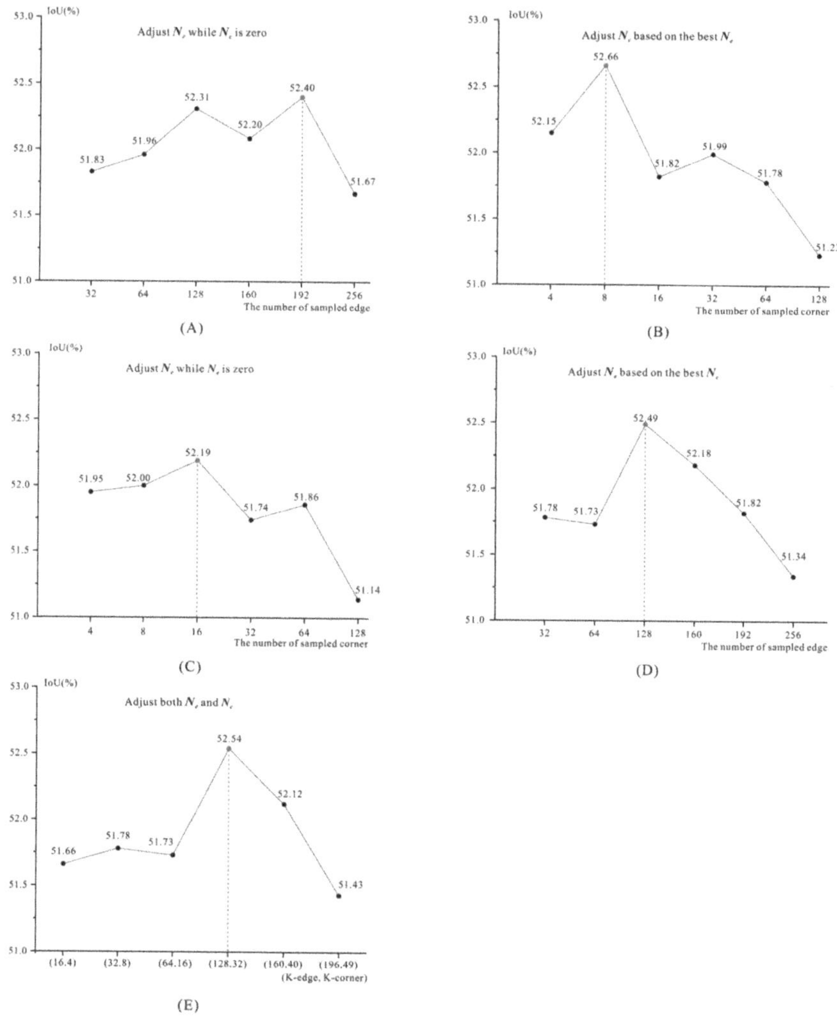

Figure 15. The IoU on DREAM-A+ *val* of different combinations of sampled edges (N_e) and corners (N_c) under different strategies. First, we sampled the edge points while N_c remained zero (shown in (**A**)). Then, based on the best N_e, we sampled the corner (shown in (**B**)). The best combination was 192 and 8 for N_e and N_c, respectively. Second, we sampled the corners (shown in (**C**)) and then sampled the edges (shown in (**D**)), and the best combination was 128 and 16. Third, we kept the ratio of N_e and N_c to 4:1, and the best combination was 128 and 32 as shown in (**E**).

Table 4. Comparisons on DREAM-A+ of different N_e and N_c adjusting strategies. The bold values in each column means the best entries.

	IoU(%)	OA(%)	F1(%)	Time (h)
SGFANet-S1 (192, 8)	**51.94**	**78.80**	**68.54**	36
SGFANet-S2 (128, 16)	51.72	78.52	68.04	36
SGFANet-S3 (128, 32)	51.88	78.70	68.32	**19**

4.9. Ablation Study

In this subsection, we present comprehensive experiments to analyze the proposed modules in SGFANet. All experiments were conducted on the DREAM-A+ dataset.

The baseline of the ablation studies was the vanilla FPN architecture [35]. We ablated four modules in SGFANet: the pyramid pooling module (PPM), sparse boundary sampling module (SBSM), gated fusion module (GFM), and grounding transformer (GT). Table 5 reports the ablation results, where the baseline from the proposed method is the initial row. From top to bottom, the proposed modules are added in different combinations for the module analysis.

The PPM is essential for the feature fusion architecture, which achieved 0.26% improvement in IoU. The SBSM and GFM improved the IoU by 2.51% due to better boundary prediction as verified in Table 6. The primary improvement of GT lies in the undersegmentation of the building area. As a result, baseline+PPM+SBSM+GFM+GT achieved an IoU result with 0.67% improvement and an F1 result with 0.58% improvement compared with the baseline+PPM+SBSM+GFM. The gradual improvement in accuracy is indicative of the complementary property of the proposed method.

Table 5. Ablation experimental results. The bold values in each column means the best entries.

+PPM	+SBSM	+GFM	+GT	IoU(%)	OA(%)	F1(%)	Description
-	-	-	-	48.44	78.07	62.91	The baseline of the ablation studies
✓	-	-	-	48.70	78.35	65.50	Add PPM to the top layer in the top-down procedure, the propagation is feature-wise
✓	-	-	✓	49.68	78.68	66.38	Append GT and PPM into the top-down procedure, the propagation is feature-wise
✓	✓	-	✓	50.31	78.69	66.94	Append GT and PPM into the top-down procedure, the propagation is point-wise realized by the SBSM
✓	✓	✓	-	51.21	79.13	67.74	Append GFM and PPM into the top-down procedure, the further contexts filtering is included in the point-wise propagation
✓	✓	✓	✓	**51.88**	**78.70**	**68.32**	Our method

Table 6. Boundary F1 score. The bold values in each column means the best entries.

	IoU	b-F1(3px)	b-F1(9px)	b-F1(12px)
baseline + PPM	48.70	55.37	77.19	79.65
+GT	49.68	55.63	76.83	79.29
+GT + SBSM	50.31	56.33	77.73	80.18
+GT + SBSM + GFM	**51.88**	**60.44**	**81.13**	**83.38**

To demonstrate the effectiveness of the proposed modules, we verified the boundary improvements using the boundary F1-score metric (b-F1) with three different pixel thresholds in Table 6. Adding GT did not greatly improve the boundary predictions, which suggests that GT should primarily be used to capture the global attention of the image. By further applying SBSM, the boundary accuracy improved almost 1%. Moreover, as noted in the last row, adopting GFM resulted in much better results compared with the other designations. Empirically, using the gated operator is essential to preserve high-quality boundary information due to its ability to filter the boundary contexts in a top-down manner.

In Figure 16, we show two examples of the locations of the sampled edges and corner points on the original images by visualizing the points from the feature fusion path in the last stage. The positions of the sampled points indicate the representative and salient parts of the building boundary learned by the model. The GT was introduced to capture the non-local (i.e., global) attention of the image. We perform subtraction between the prediction score of SGFANet and that of SGFANet without the grounding transformer.

As shown in Figure 6, GT can significantly improve the predicted score of the building region and inhibit the false alarms of the building background. Since the improvement is homogeneous on building predictions, in terms of evaluation metrics, GT brings the improvement of IoU rather than the b-F1 score as Table 6 indicates.

(A)

(B)

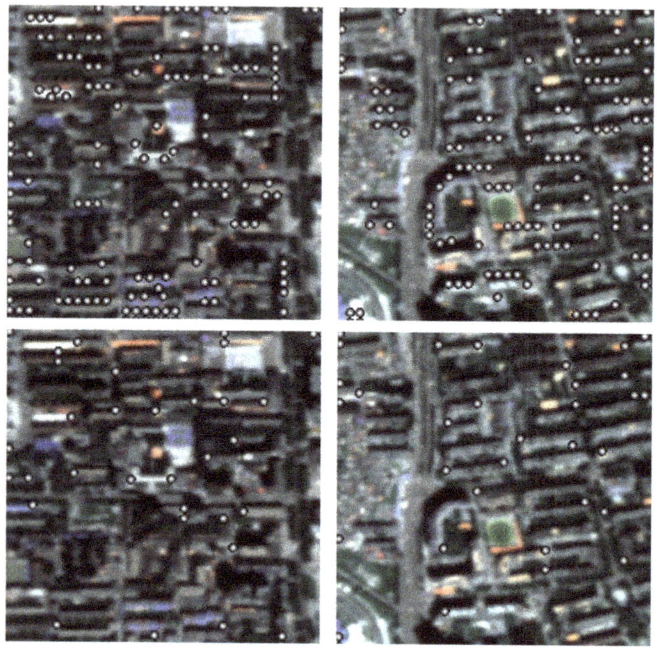

Figure 16. Visualization of the locations of sampled points. (**A**) Edge points. (**B**) Corner points.

As a conclusion of the ablation study, the proposed modules in SGFANet have three benefits: (1) the sampling process in the SBSM and the gated operator in the GFM preserved high-quality boundaries in building segmentation by alleviating the fuzzy boundary detail from LR remote-sensing imagery; (2) the GT improved the building region predictions and inhibited background noise by introducing global attention to the input; and (3) SBSM, GFM, and GT were complementary to each other, and using these modules concurrently achieved much better performance than using the proposed baseline method.

5. Pilot Application: Dynamic Building Change of the Xiong'an New Area in China

In our benchmark, SGFANet achieved promising results in building segmentation from low-resolution remote-sensing images and improved the extraction accuracy in the SR-then-SS framework, thus, assisting in extending the accessibility and scalability of building dynamic tracking. We demonstrate this capability in this study with one example by presenting the dynamic building change in the Xiong'an New Area in China.

The Xiong'an New Area (XNA) plan [58] was initiated by the Chinese government on 1 April 2017. The XNA plan is a national strategy aimed at relieving the pressure on the Chinese capital city Beijing by migrating "noncore" functions. Large-scale construction activity has occurred in this 2000 square kilometer area since 2017.

Traditionally, there are two alternatives for building tracking. The first is to use HR images (e.g., the metric) to identify single-building-level changes [59], and the second is to use LR imagery (e.g., the decametric) to identify human settlement changes [60] or land-cover changes [61]. Existing research indicates that the high acquisition cost and low time availability of HR images make them problematic for large-scale applications. In addition, for LR imagery, it is challenging to detect changes at the level of individual buildings. Therefore, we would like to address the aforementioned issue using free LR data to enable the tracking of HR building changes.

We used publicly available Sentinel-2 images collected from the GEE platform to implement building-change detection at the annual scale from 2016 to 2021. We extracted the buildings for each year utilizing SGFANet with ESPCN (Section 4.6). Figure 17 shows the final building-change-detection results obtained after a time series analysis of the building distribution maps. The building construction and demolition activities were concentrated in three primary regions, as the red rectangle indicates.

Figure 17A shows the process of the demolition of some small villages and the construction of modern apartments. Between 2019 and 2021, some small settlements were progressively abandoned, and high buildings were created during 2020–2021. In Figure 17B, the process is nearly identical, with demolition occupying the vast majority of the area beginning in 2020, indicating that more construction is planned for the area. Figure 17C shows the building process of Xiong'an Railway Station, which was under construction in 2018 and operational in 2020.

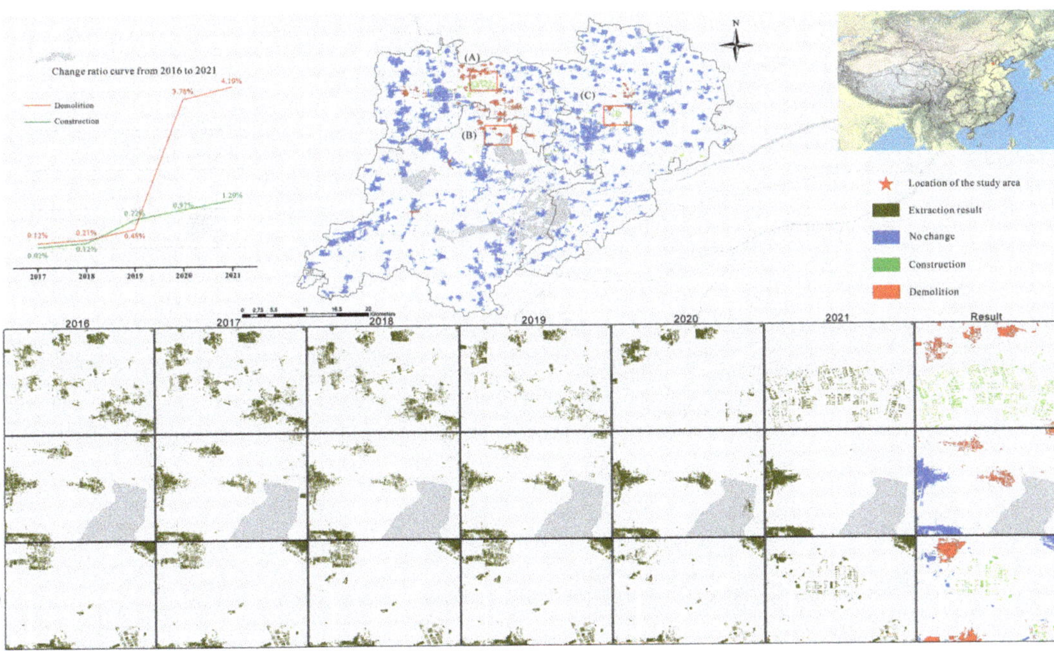

Figure 17. Dynamic building-change-tracking results (2016–2021) of the Xiong'an New area in China. The subplots (**A–C**) are three detail examples to show the building dynamic.

As reported by the statistical curve in Figure 17, construction began as early as 2018–2019, while demolition activity peaked in 2020–2021. Compared to 2016, 1.2% of buildings were constructed, and 4.1% were demolished. The trend of the curve indicates that the construction and demolition process will continue in the future.

To verify these results, we calculated the temporal correlation coefficients (2016–2021) on the number of foreground pixels of the results with three well-acknowledged products: (1) the Dynamic World product (https://dynamicworld.app/, accessed on 23 March 2023), providing global land coverage at a 10 m resolution from 2015 to the present, including the "built-up area" category; (2) global artificial impervious area (GAIA) [62], providing an impervious surface at 30 m from 1985 to 2018; and (3) MCD12Q1 (https://lpdaac.usgs.gov/products, accessed on 23 March 2023), providing global land-cover types at yearly intervals (2001–2020), including the "urban and built-up lands" category.

As reported in Table 7, the results show good temporal consistency with Dynamic World ($R^2 = 0.7702$), indicating the reliability of the proposed results. However, the result with GAIA is lower ($R^2 = 0.6864$), which might be caused by GAIA being constructed based on a one-way conversion assumption [63] (i.e., that the impervious surface will increase from year to year). The result with MCD12Q1 is the lowest ($R^2 = 0.4907$), which is likely due to its limited expression in local areas due to its low resolution (500 m).

Table 7. Correlation coefficients (R^2) between our results and other thematic related products.

	Resolution	Correlation Coefficients
Dynamic World	10 m	0.7702
GAIA	30 m	0.6864
MCD12Q1	500 m	0.4907

Thus, we demonstrated the dynamic tracking of building changes at an annual scale of 2.5 m in a developing region using LR images. Compared to using HR images, the proposed approach can save individual users approximately $200,000 in data expenditures. In the future, we plan to use multiple data sources to achieve even higher accuracies (both spatial and temporal) for the dynamic monitoring of buildings, thus, offering data support for human activity investigations.

6. Conclusions

In this study, we argued that the fuzzy boundary detail (i.e., the edges and corners) and the lack of local textures are the bottlenecks of building segmentation in LR remote-sensing imagery. To mitigate these bottlenecks, we proposed the sparse geometric feature attention network (SGFANet), which learns representative edges and corners on buildings to resolve the fuzzy effect and converts the top-down propagation from local to nonlocal to harvest the global attention of the input image, thus, alleviating the lack of local textures.

Comprehensive experiments showed that SGFANet obtained good accuracy compared with other existing methods. A pilot application using SGFANet in Xiong'an New Area, China demonstrated the potential of the proposed approach for future large-scale tracking of buildings. We anticipate that the proposed method could help expand the accessibility of large-scale building segmentation and building-change tracking on low-resolution remote-sensing images.

Author Contributions: Conceptualization, H.T. and Z.L.; methodology, Z.L.; software, Z.L.; validation, H.T. and Z.L.; writing—original draft preparation, Z.L.; writing—review and editing, H.T. All authors have read and agreed to the published version of the manuscript.

Funding: This study was supported by the National Natural Science Foundation of China under Grant Nos. 42192584 and 41971280 and by the Key Laboratory of Environmental Change and Natural Disaster of Ministry of Education, Beijing Normal University (Project No. 2022-KF-07).

Data Availability Statement: The code is available at https://github.com/zpl99/SGFANet, accessed on 23 March 2023.

Conflicts of Interest: The authors declare no conflict of interest. The funders had no role in the design of the study; in the collection, analyses, or interpretation of data; in the writing of the manuscript; or in the decision to publish the results.

References

1. Wong, C.H.H.; Cai, M.; Ren, C.; Huang, Y.; Liao, C.; Yin, S. Modelling building energy use at urban scale: A review on their account for the urban environment. *Build. Environ.* **2021**, *205*, 108235. [CrossRef]
2. Ma, R.; Li, X.; Chen, J. An elastic urban morpho-blocks (EUM) modeling method for urban building morphological analysis and feature clustering. *Build. Environ.* **2021**, *192*, 107646. [CrossRef]
3. Chen, J.; Tang, H.; Ge, J.; Pan, Y. Rapid Assessment of Building Damage Using Multi-Source Data: A Case Study of April 2015 Nepal Earthquake. *Remote Sens.* **2022**, *14*, 1358. [CrossRef]

4. Li, J.; Huang, X.; Tu, L.; Zhang, T.; Wang, L. A review of building detection from very high resolution optical remote sensing images. *GISci. Remote Sens.* **2022**, *59*, 1199–1225. [CrossRef]
5. Chen, Q.; Wang, L.; Wu, Y.; Wu, G.; Guo, Z.; Waslander, S.L. Aerial imagery for roof segmentation: A large-scale dataset towards automatic mapping of buildings. *ISPRS J. Photogramm. Remote Sens.* **2019**, *147*, 42–55. [CrossRef]
6. Maggiori, E.; Tarabalka, Y.; Charpiat, G.; Alliez, P. Can semantic labeling methods generalize to any city? The inria aerial image labeling benchmark. In Proceedings of the 2017 IEEE International Geoscience and Remote Sensing Symposium (IGARSS), IEEE, Fort Worth, TX, USA, 23–28 July 2017; pp. 3226–3229.
7. Ji, S.; Wei, S.; Lu, M. Fully convolutional networks for multisource building extraction from an open aerial and satellite imagery data set. *IEEE Trans. Geosci. Remote Sens.* **2018**, *57*, 574–586. [CrossRef]
8. Jing, W.; Lin, J.; Lu, H.; Chen, G.; Song, H. Learning holistic and discriminative features via an efficient external memory module for building extraction in remote sensing images. *Build. Environ.* **2022**, *222*, 109332. [CrossRef]
9. Zhu, Y.; Liang, Z.; Yan, J.; Chen, G.; Wang, X. ED-Net: Automatic building extraction from high-resolution aerial images with boundary information. *IEEE J. Sel. Top. Appl. Earth Obs. Remote Sens.* **2021**, *14*, 4595–4606. [CrossRef]
10. Lee, K.; Kim, J.H.; Lee, H.; Park, J.; Choi, J.P.; Hwang, J.Y. Boundary-Oriented Binary Building Segmentation Model With Two Scheme Learning for Aerial Images. *IEEE Trans. Geosci. Remote Sens.* **2021**, *60*, 5604517. [CrossRef]
11. Nauata, N.; Furukawa, Y. Vectorizing world buildings: Planar graph reconstruction by primitive detection and relationship inference. In Proceedings of the European Conference on Computer Vision, Glasgow, UK, 23–28 August 2020; Springer: Berlin/Heidelberg, Germany, 2020; pp. 711–726.
12. Girard, N.; Smirnov, D.; Solomon, J.; Tarabalka, Y. Polygonal building extraction by frame field learning. In Proceedings of the IEEE/CVF Conference on Computer Vision and Pattern Recognition, Nashville, TN, USA, 22–25 June 2021; pp. 5891–5900.
13. Zhu, Y.; Huang, B.; Gao, J.; Huang, E.; Chen, H. Adaptive Polygon Generation Algorithm for Automatic Building Extraction. *IEEE Trans. Geosci. Remote Sens.* **2021**, *60*, 1–14. [CrossRef]
14. Li, W.; Zhao, W.; Zhong, H.; He, C.; Lin, D. Joint semantic–geometric learning for polygonal building segmentation. In Proceedings of the AAAI Conference on Artificial Intelligence, Online, 2–9 February 2021; Volume 35; pp. 1958–1965.
15. Zhang, L.; Dong, R.; Yuan, S.; Li, W.; Zheng, J.; Fu, H. Making low-resolution satellite images reborn: A deep learning approach for super-resolution building extraction. *Remote Sens.* **2021**, *13*, 2872. [CrossRef]
16. He, Y.; Wang, D.; Lai, N.; Zhang, W.; Meng, C.; Burke, M.; Lobell, D.; Ermon, S. Spatial-Temporal Super-Resolution of Satellite Imagery via Conditional Pixel Synthesis. *Adv. Neural Inf. Process. Syst.* **2021**, *34*, 27903–27915.
17. Kang, X.; Li, J.; Duan, P.; Ma, F.; Li, S. Multilayer Degradation Representation-Guided Blind Super-Resolution for Remote Sensing Images. *IEEE Trans. Geosci. Remote Sens.* **2022**, *60*, 5534612. [CrossRef]
18. Xu, P.; Tang, H.; Ge, J.; Feng, L. ESPC_NASUnet: An End-to-End Super-Resolution Semantic Segmentation Network for Mapping Buildings From Remote Sensing Images. *IEEE J. Sel. Top. Appl. Earth Obs. Remote Sens.* **2021**, *14*, 5421–5435. [CrossRef]
19. Zhang, T.; Tang, H.; Ding, Y.; Li, P.; Ji, C.; Xu, P. FSRSS-Net: High-resolution mapping of buildings from middle-resolution satellite images using a super-resolution semantic segmentation network. *Remote Sens.* **2021**, *13*, 2290. [CrossRef]
20. Zhang, D.; Zhang, H.; Tang, J.; Wang, M.; Hua, X.; Sun, Q. Feature pyramid transformer. In Proceedings of the European Conference on Computer Vision, Glasgow, UK, 23–28 August 2020; Springer: Berlin/Heidelberg, Germany, 2020; pp. 323–339.
21. Krizhevsky, A.; Sutskever, I.; Hinton, G.E. Imagenet classification with deep convolutional neural networks. In *Advances in Neural Information Processing Systems*; MIT Press: Cambridge, MA, USA, 2012; Volume 25.
22. Long, J.; Shelhamer, E.; Darrell, T. Fully convolutional networks for semantic segmentation. In Proceedings of the IEEE Conference on Computer Vision and Pattern Recognition, Boston, MA, USA, 7–12 June 2015; pp. 3431–3440.
23. Yang, G.; Zhang, Q.; Zhang, G. EANet: Edge-aware network for the extraction of buildings from aerial images. *Remote Sens.* **2020**, *12*, 2161. [CrossRef]
24. Huang, W.; Liu, Z.; Tang, H.; Ge, J. Sequentially Delineation of Rooftops with Holes from VHR Aerial Images Using a Convolutional Recurrent Neural Network. *Remote Sens.* **2021**, *13*, 4271. [CrossRef]
25. Liu, Z.; Tang, H.; Huang, W. Building Outline Delineation From VHR Remote Sensing Images Using the Convolutional Recurrent Neural Network Embedded With Line Segment Information. *IEEE Trans. Geosci. Remote Sens.* **2022**, *60*, 4705713. [CrossRef]
26. Zorzi, S.; Bittner, K.; Fraundorfer, F. Machine-learned regularization and polygonization of building segmentation masks. In Proceedings of the 2020 25th International Conference on Pattern Recognition (ICPR), IEEE, Milan, Italy, 10–15 January 2021; pp. 3098–3105.
27. Ding, L.; Tang, H.; Liu, Y.; Shi, Y.; Zhu, X.X.; Bruzzone, L. Adversarial shape learning for building extraction in VHR remote sensing images. *IEEE Trans. Image Process.* **2021**, *31*, 678–690. [CrossRef] [PubMed]
28. Zhao, H.; Shi, J.; Qi, X.; Wang, X.; Jia, J. Pyramid scene parsing network. In Proceedings of the IEEE Conference on Computer Vision and Pattern Recognition, Honolulu, HI, USA, 21–26 July 2017; pp. 2881–2890.
29. Ronneberger, O.; Fischer, P.; Brox, T. U-net: Convolutional networks for biomedical image segmentation. In Proceedings of the International Conference on Medical Image Computing and Computer-Assisted Intervention, Munich, Germany, 5–9 October 2015; Springer: Berlin/Heidelberg, Germany, 2015; pp. 234–241.
30. Wei, S.; Ji, S.; Lu, M. Toward automatic building footprint delineation from aerial images using CNN and regularization. *IEEE Trans. Geosci. Remote Sens.* **2019**, *58*, 2178–2189. [CrossRef]

31. Chatterjee, B.; Poullis, C. On building classification from remote sensor imagery using deep neural networks and the relation between classification and reconstruction accuracy using border localization as proxy. In Proceedings of the 2019 16th Conference on Computer and Robot Vision (CRV), IEEE, Kingston, ON, Canada, 29–31 May 2019; pp. 41–48.
32. Liu, P.; Liu, X.; Liu, M.; Shi, Q.; Yang, J.; Xu, X.; Zhang, Y. Building footprint extraction from high-resolution images via spatial residual inception convolutional neural network. *Remote Sens.* **2019**, *11*, 830. [CrossRef]
33. Wen, X.; Li, X.; Zhang, C.; Han, W.; Li, E.; Liu, W.; Zhang, L. ME-Net: A multi-scale erosion network for crisp building edge detection from very high resolution remote sensing imagery. *Remote Sens.* **2021**, *13*, 3826. [CrossRef]
34. Guo, H.; Du, B.; Zhang, L.; Su, X. A coarse-to-fine boundary refinement network for building footprint extraction from remote sensing imagery. *ISPRS J. Photogramm. Remote Sens.* **2022**, *183*, 240–252. [CrossRef]
35. Lin, T.Y.; Dollár, P.; Girshick, R.; He, K.; Hariharan, B.; Belongie, S. Feature pyramid networks for object detection. In Proceedings of the IEEE Conference on Computer Vision and Pattern Recognition, Honolulu, HI, USA, 21–26 July 2017; pp. 2117–2125.
36. Li, X.; You, A.; Zhu, Z.; Zhao, H.; Yang, M.; Yang, K.; Tan, S.; Tong, Y. Semantic flow for fast and accurate scene parsing. In Proceedings of the European Conference on Computer Vision, Glasgow, UK, 23–28 August 2020; Springer: Berlin/Heidelberg, Germany, 2020; pp. 775–793.
37. Li, X.; He, H.; Li, X.; Li, D.; Cheng, G.; Shi, J.; Weng, L.; Tong, Y.; Lin, Z. Pointflow: Flowing semantics through points for aerial image segmentation. In Proceedings of the IEEE/CVF Conference on Computer Vision and Pattern Recognition, Nashville, TN, USA, 20–25 June 2021; pp. 4217–4226.
38. Zhou, Z.; Siddiquee, M.M.R.; Tajbakhsh, N.; Liang, J. Unet++: Redesigning skip connections to exploit multiscale features in image segmentation. *IEEE Trans. Med. Imaging* **2019**, *39*, 1856–1867. [CrossRef] [PubMed]
39. Li, X.; Zhao, H.; Han, L.; Tong, Y.; Tan, S.; Yang, K. Gated fully fusion for semantic segmentation. In Proceedings of the AAAI Conference on Artificial Intelligence, New York, NY, USA, 7–12 February 2020; Volume 34, pp. 11418–11425.
40. Guo, H.; Shi, Q.; Marinoni, A.; Du, B.; Zhang, L. Deep building footprint update network: A semi-supervised method for updating existing building footprint from bi-temporal remote sensing images. *Remote Sens. Environ.* **2021**, *264*, 112589. [CrossRef]
41. Akiva, P.; Purri, M.; Leotta, M. Self-supervised material and texture representation learning for remote sensing tasks. In Proceedings of the IEEE/CVF Conference on Computer Vision and Pattern Recognition, New Orleans, LA, USA, 18–24 June 2022; pp. 8203–8215.
42. Su, H.; Jampani, V.; Sun, D.; Gallo, O.; Learned-Miller, E.; Kautz, J. Pixel-adaptive convolutional neural networks. In Proceedings of the IEEE/CVF Conference on Computer Vision and Pattern Recognition, Long Beach, CA, USA, 16–17 June 2019; pp. 11166–11175.
43. He, K.; Zhang, X.; Ren, S.; Sun, J. Deep residual learning for image recognition. In Proceedings of the IEEE Conference on Computer Vision and Pattern Recognition, Las Vegas, NV, USA, 26 June–1 July 2016; pp. 770–778.
44. Azulay, A.; Weiss, Y. Why do deep convolutional networks generalize so poorly to small image transformations? *J. Mach. Learn. Res.* **2019**, *20*, 1–25.
45. Gulati, A.; Qin, J.; Chiu, C.C.; Parmar, N.; Zhang, Y.; Yu, J.; Han, W.; Wang, S.; Zhang, Z.; Wu, Y.; et al. Conformer: Convolution-augmented Transformer for Speech Recognition. In Proceedings of the Interspeech 2020, Shanghai, China, 25–29 October 2020; pp. 5036–5040. [CrossRef]
46. Ma, L.; Liu, Y.; Zhang, X.; Ye, Y.; Yin, G.; Johnson, B.A. Deep learning in remote sensing applications: A meta-analysis and review. *ISPRS J. Photogramm. Remote Sens.* **2019**, *152*, 166–177. [CrossRef]
47. Shi, Y.; Li, Q.; Zhu, X.X. Building segmentation through a gated graph convolutional neural network with deep structured feature embedding. *ISPRS J. Photogramm. Remote Sens.* **2020**, *159*, 184–197. [CrossRef]
48. Loshchilov, I.; Hutter, F. Decoupled weight decay regularization. *arXiv* **2017**, arXiv:1711.05101.
49. Chen, L.C.; Zhu, Y.; Papandreou, G.; Schroff, F.; Adam, H. Encoder-decoder with atrous separable convolution for semantic image segmentation. In Proceedings of the European Conference on Computer Vision (ECCV), Munich, Germany, 8–14 September 2018; pp. 801–818.
50. Huang, G.; Liu, Z.; Van Der Maaten, L.; Weinberger, K.Q. Densely connected convolutional networks. In Proceedings of the IEEE Conference on Computer Vision and Pattern Recognition, Honolulu, HI, USA, 21–26 July 2017; pp. 4700–4708.
51. Li, Q.; Zorzi, S.; Shi, Y.; Fraundorfer, F.; Zhu, X.X. RegGAN: An End-to-End Network for Building Footprint Generation with Boundary Regularization. *Remote Sens.* **2022**, *14*, 1835. [CrossRef]
52. Goodfellow, I.; Pouget-Abadie, J.; Mirza, M.; Xu, B.; Warde-Farley, D.; Ozair, S.; Courville, A.; Bengio, Y. Generative adversarial nets. In *Advances in Neural Information Processing Systems*; MIT Press: Cambridge, MA, USA, 2014; Volume 27.
53. Shi, W.; Caballero, J.; Huszár, F.; Totz, J.; Aitken, A.P.; Bishop, R.; Rueckert, D.; Wang, Z. Real-time single image and video super-resolution using an efficient sub-pixel convolutional neural network. In Proceedings of the IEEE Conference on Computer Vision and Pattern Recognition, Las Vegas, NV, USA, 26 June–1 July 2016; pp. 1874–1883.
54. Weng, Y.; Zhou, T.; Li, Y.; Qiu, X. Nas-unet: Neural architecture search for medical image segmentation. *IEEE Access* **2019**, *7*, 44247–44257. [CrossRef]
55. Bergstra, J.; Bardenet, R.; Bengio, Y.; Kégl, B. Algorithms for hyper-parameter optimization. In *Advances in Neural Information Processing Systems*; MIT Press: Cambridge, MA, USA, 2011; Volume 24.
56. Bergstra, J.; Bengio, Y. Random search for hyper-parameter optimization. *J. Mach. Learn. Res.* **2012**, *13*, 281–305.

57. Young, S.R.; Rose, D.C.; Karnowski, T.P.; Lim, S.H.; Patton, R.M. Optimizing deep learning hyper-parameters through an evolutionary algorithm. In Proceedings of the Workshop on Machine Learning in High-Performance Computing Environments, Austin, TX, USA, 15 November 2015; pp. 1–5.
58. Zou, Y.; Zhao, W. Making a new area in Xiong'an: Incentives and challenges of China's "Millennium Plan". *Geoforum* **2018**, *88*, 45–48. [CrossRef]
59. Zheng, H.; Gong, M.; Liu, T.; Jiang, F.; Zhan, T.; Lu, D.; Zhang, M. HFA-Net: High frequency attention siamese network for building change detection in VHR remote sensing images. *Pattern Recognit.* **2022**, *129*, 108717. [CrossRef]
60. Marconcini, M.; Metz-Marconcini, A.; Üreyen, S.; Palacios-Lopez, D.; Hanke, W.; Bachofer, F.; Zeidler, J.; Esch, T.; Gorelick, N.; Kakarla, A.; et al. Outlining where humans live, the World Settlement Footprint 2015. *Sci. Data* **2020**, *7*, 242. [CrossRef]
61. Xu, L.; Herold, M.; Tsendbazar, N.E.; Masiliūnas, D.; Li, L.; Lesiv, M.; Fritz, S.; Verbesselt, J. Time series analysis for global land cover change monitoring: A comparison across sensors. *Remote Sens. Environ.* **2022**, *271*, 112905. [CrossRef]
62. Gong, P.; Li, X.; Wang, J.; Bai, Y.; Chen, B.; Hu, T.; Liu, X.; Xu, B.; Yang, J.; Zhang, W.; et al. Annual maps of global artificial impervious area (GAIA) between 1985 and 2018. *Remote Sens. Environ.* **2020**, *236*, 111510. [CrossRef]
63. Li, X.; Gong, P.; Liang, L. A 30-year (1984–2013) record of annual urban dynamics of Beijing City derived from Landsat data. *Remote Sens. Environ.* **2015**, *166*, 78–90. [CrossRef]

remote sensing

Article

A Lightweight and High-Accuracy Deep Learning Method for Grassland Grazing Livestock Detection Using UAV Imagery

Yuhang Wang [1,2], Lingling Ma [1,*], Qi Wang [1], Ning Wang [1], Dongliang Wang [3], Xinhong Wang [1], Qingchuan Zheng [4], Xiaoxin Hou [4] and Guangzhou Ouyang [1]

1 Key Laboratory of Quantitative Remote Sensing Information Technology, Aerospace Information Research Institute, Chinese Academy of Sciences, Beijing 100094, China
2 School of Electronic, Electrical and Communication Engineering, University of Chinese Academy of Sciences, Beijing 100049, China
3 Key Laboratory of Land Surface Pattern and Simulation, Institute of Sciences and Natural Resources Research, Chinese Academy of Sciences, Beijing 100101, China
4 Inner Mongolia North Heavy Industries Group Co., Ltd., Baotou 014033, China
* Correspondence: llma@aoe.ac.cn

Abstract: Unregulated livestock breeding and grazing can degrade grasslands and damage the ecological environment. The combination of remote sensing and artificial intelligence techniques is a more convenient and powerful means to acquire livestock information in a large area than traditional manual ground investigation. As a mainstream remote sensing platform, unmanned aerial vehicles (UAVs) can obtain high-resolution optical images to detect grazing livestock in grassland. However, grazing livestock objects in UAV images usually occupy very few pixels and tend to gather together, which makes them difficult to detect and count automatically. This paper proposes the GLDM (grazing livestock detection model), a lightweight and high-accuracy deep-learning model, for detecting grazing livestock in UAV images. The enhanced CSPDarknet (ECSP) and weighted aggregate feature re-extraction pyramid modules (WAFR) are constructed to improve the performance based on the YOLOX-nano network scheme. The dataset of different grazing livestock (12,901 instances) for deep learning was made from UAV images in the Hadatu Pasture of Hulunbuir, Inner Mongolia, China. The results show that the proposed method achieves a higher comprehensive detection precision than mainstream object detection models and has an advantage in model size. The mAP of the proposed method is 86.47%, with the model parameter 5.7 M. The average recall and average precision can be above 85% at the same time. The counting accuracy of grazing livestock in the testing dataset, when converted to a unified sheep unit, reached 99%. The scale applicability of the model is also discussed, and the GLDM could perform well with the image resolution varying from 2.5 to 10 cm. The proposed method, the GLDM, was better for detecting grassland grazing livestock in UAV images, combining remote sensing, AI, and grassland ecological applications with broad application prospects.

Keywords: unmanned aerial vehicle (UAV); deep learning; object detection; grassland grazing livestock; remote sensing image

Citation: Wang, Y.; Ma, L.; Wang, Q.; Wang, N.; Wang, D.; Wang, X.; Zheng, Q.; Hou, X.; Ouyang, G. A Lightweight and High-Accuracy Deep Learning Method for Grassland Grazing Livestock Detection Using UAV Imagery. *Remote Sens.* **2023**, *15*, 1593. https://doi.org/10.3390/rs15061593

Academic Editors: Jukka Heikkonen, Fahimeh Farahnakian and Pouya Jafarzadeh

Received: 7 February 2023
Revised: 12 March 2023
Accepted: 13 March 2023
Published: 15 March 2023

1. Introduction

Overgrazing destroys grassland ecological functions. The survey of grazing animals is of great significance in maintaining the balance of grass and livestock. Investigating the geographical and temporal distribution of different grazing livestock (sheep, cattle, horses, etc.) provides the basic and indispensable information for grassland ecological management [1].

Satellites and manned aircraft are usually used in early animal surveys. Spaceborne remote sensing data with low and medium spatial resolution (1–60 m) have been used for indirect animal surveys since the early 1980s [2], mainly by detecting signs indicating the presence of animals in the area, such as fecal counts [3–5], food removal, and burrow counts [6,7]. Submeter very-high-resolution (VHR) spaceborne imagery has potential in modeling the population dynamics of large (>0.6 m) wild animals at large spatial and temporal scales, but has difficulty discerning small (<0.6 m) animals at the species level, although high-resolution commercial satellites, such as WorldView-3 and WorldView-4, have reached ground resolution of up to 0.31 m in panchromatic mode [2]. Although satellites have the advantages of wide coverage and not disturbing animals, they are limited by the weather, and the resolution is still not high enough to finely distinguish animal objects. Manned aircraft have also been widely used for wild animal surveys, such as kangaroo censuses in New South Wales, Australia [8] and polar bear censuses in the seasonally ice-free Foxe Basin, Canada [9]. Although manned aircraft are flexible in terms of survey time and area, they are relatively expensive [10], require qualified pilots, and possibly have individual biases when used in real-time censuses [8].

In recent years, unmanned aerial vehicles (UAVs), a convenient and low-cost remote sensing platform, have been widely used in various fields, including wild animal surveys. Compared with manned helicopters, UAVs are more flexible and quieter, keeping the distance between the observer and the animal, ensuring the safety of field investigators in dangerous environments, and avoiding human interference with animal habitats. Previous surveys relied on manually observing and counting from large numbers of images. Researchers developed a series of automatic and semiautomatic object detection methods to improve efficiency. Moreover, some scholars [11] compared the factors affecting the detection probability of ground observation, manual inspection, and automatic detection from UAV images. They concluded that the combination of drone-captured imagery and machine learning does not suffer from the same biases that affect conventional ground surveys and could better provide information for managing the ecological population [11].

Studies have shown that some simple threshold-based methods are still sufficient for detecting and counting animals with similar grayscale values and significant differences from the background. For example, using threshold segmentation and template matching techniques, Gonzalez et al. [12] developed an algorithm to count and track koalas and deer in UAV RGB and thermal imaging videos. However, against complex backgrounds, these methods' accuracy will usually be greatly affected. As higher-resolution images become available, researchers developed various algorithms based on machine learning to extract more complex features. Xue et al. [13] developed a semi-supervised object-based method that combined a wavelet algorithm and an adaptive network-based fuzzy neural network (ANFIS) to detect and count wildebeests and zebras in a single VHR GeoEye-1 panchromatic image of open savanna. The accuracy of this method is significantly higher than that of the traditional threshold-based method (0.79 vs. 0.58). Torney et al. [14] developed a method via rotation-invariant object descriptors combined with machine learning algorithms to detect and count wildebeests in aerial images collected in the Serengeti National Park, Tanzania. The algorithm was more accurate for the total count than both manual counts, while the per-image error rates were greater than manual counts, and the recognition accuracy was 74.15%. Rey et al. [10] proposed a semiautomated data-driven active learning system jointly based on an object proposal strategy with an ensemble of exemplar support vector machine (EESVM) models to detect large mammals, including common elands, greater kudus, and gemsboks, in the semiarid African savanna from 6500 RGB UAS images, achieving a recall of 75% for a precision of 10%. The author believes that recall is much more important than precision in this application. Although machine learning methods based on non-deep neural networks can still produce good detection results in simple cases, these methods usually cannot fully mine complex animal features.

Deep learning technology, such as convolutional neural networks (CNNs), has developed rapidly in recent years and achieved great success in computer vision. Compared with traditional methods, which only extract shallow image features, convolutional neural networks can automatically learn much richer semantic information and high-level image features with higher learning efficiency. It more comprehensively describes the differences between various types of objects. The CNN-based object detection algorithm includes anchor-free and anchor-based models. Anchor-based models include Faster R-CNN [15], RetinaNet [16], YOLOv3 [17], YOLOv7 [18], etc. These models need to adjust the hyperparameter settings of the anchor during the training procedure to better match the size of the objects in the dataset. The anchor-free model is more convenient without such a process, and the representative models include FCOS [19], CenterNet [20], YOLOX [21], etc.

Some researchers have also introduced the deep learning method to detect animal objects in UAV remote sensing images. For example, in 2017, Kellenberger et al. [22] used a two-branch CNN network structure to detect wild animals in the Kuzikus Wildlife Conservation Park in Namibia, and the precision and speed of the model were greatly improved compared with Fast RCNN. In 2018, Kellenberger et al. [23] studied how to extend CNN to large-scale wildlife census tasks. When the recall was set to 90%, false positives of the CNN were reduced by an order of magnitude, but the precision of the model was still lower. In short, the above two methods lack consideration for the comprehensive performance of the model and cannot guarantee good performance in both recall and precision. In 2020, Roosjen et al. [24] used the neural network resnet18 to automatically detect and count spotted wing drosophila, including sex prediction and discrimination. The results showed that UAV images have the potential to be researched and applied to integrated pest management (IPM) strategies. In 2020, Peng et al. [25] developed an automatic detection model for kiangs in Tibet, based on the improved Faster R-CNN and aiming at small object detection of the UAV images, and increased the *F*1 *score* from 0.85 to 0.94. However, the dataset in this study was relatively small, and there was only one type of animal object. The classifying ability of similar types of objects has not been verified.

In this study, the grassland grazing livestock detection in UAV images was different from that in natural images taken on the ground and other objects' detection of remote sensing images, which brings the following challenges to the algorithm design.

First, considering the surveying efficiency, the field of view angle tends to be large, and the image resolution is low. Therefore, animal objects only occupy very few pixels, making it difficult to extract useful and distinguishable features to perform detection. Moreover, grassland grazing livestock such as cattle, horses, and sheep share similar characteristics and are much more difficult to distinguish than those in the classic applications, vehicles, aircraft, and ships.

Second, the UAV images contain a large area of invalid complex background, with changeable illumination conditions, and many false objects exist, such as rocks, haystacks, and woods. Moreover, due to the posture changes of the animals, the animals of the same category may have a different appearance in imagery.

In addition, with the development of the deep learning network model, deeper networks with high precision consume a large amount of computer resources, making them hard to use in portable minimized platforms and for real-time processing. Therefore, the model size is an important index to be considered in the model construction, as well as the precision.

To solve the difficulties in the above aspects, in this paper, we propose an effective grazing livestock detection model—GLDM, based on YOLOX nano [21]. The rest of the paper is organized as follows.

1. A grazing livestock dataset based on UAV imagery data of the Hulunbuir grassland was established. We describe the dataset in detail and show some examples in Section 2.
2. The proposed model is elaborated in Section 3.
3. Comparison, ablation, and multi-scale adaptation experiments of the model had been conducted. The details and results of the experiments are shown in Section 4.
4. Finally, we summarize and conclude the paper in Section 5.

2. Materials and Methods

2.1. Materials

2.1.1. Study Area and the UAV Imagery

Experimental data in this section were captured in Hadatu Pasture (49°32′–49°59′N, 119°3′–120°15′E, about 1000 km^2), Hulunbuir City, Inner Mongolia, China, as shown in Figure 1. Hadatu Pasture has a large distribution of common livestock such as cattle, horses, and sheep. We used a fixed-wing unmanned aerial vehicle (UAV) to collect ground images in Hadatu Pasture from 24 to 29 July 2021. The UAV images covered 20% of the total area—about 200 square kilometers. The photography was taken from the overhead orthographic attitude. The forward overlap rate was 80%, and the flight altitude was about 300 m. The relative altitude of the flight changed with the rugged terrain field. The images' spatial resolution was about 3~7 cm. A total of 30 flight strips were flown, and 33,304 shots of RGB images were taken. The image set covered objects of various landforms, buildings, and large numbers of cattle, horses, and sheep in the area.

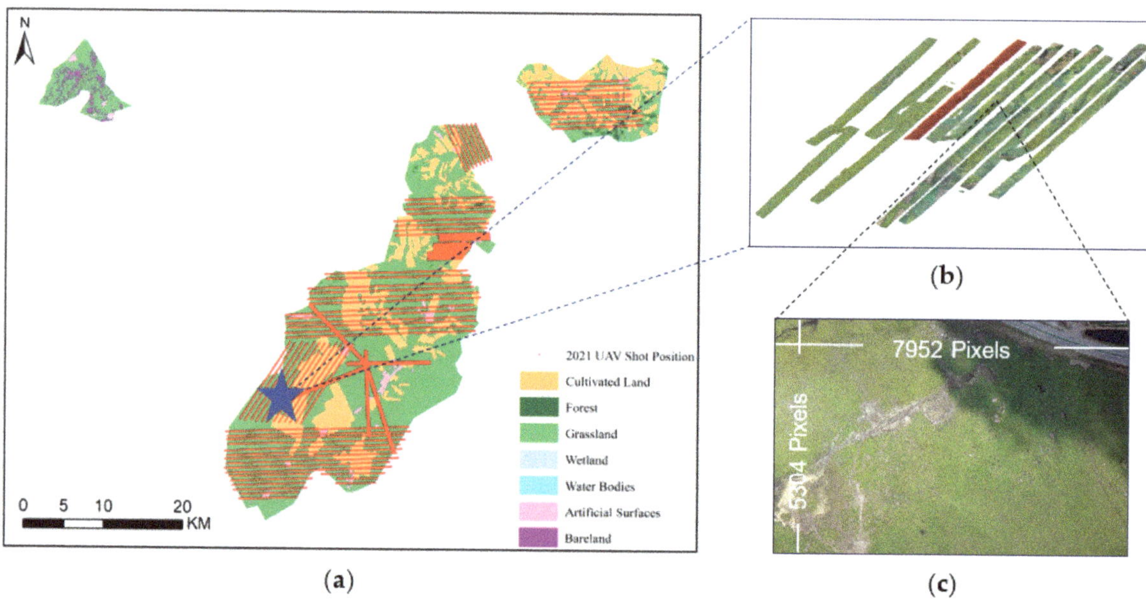

Figure 1. Data collection. (**a**) Map of Hadatu Pasture and UAV flight routes; (**b**) flight strip image captured and mosaiced by the UAV; (**c**) original image sample of the UAV.

2.1.2. Data Preparation

For the training and testing of the deep learning model, we used Labelme 4.6.0 [26] software to label the data on the original UAV images. This study used a rectangular box to label the ground truth, as shown in Figure 2c. The box was composed of an upper left point and a lower right point, using which the width and height of the object on the image could be obtained.

The size of the UAV images we obtained was 7952 × 5304. The original images were split into subpatches of 1024 × 1024 to suit the limited computer memory (as shown in Figure 2), and the width of the overlapping area was 100 pixels, which was wider than the largest object, to ensure that an object on the split line would not be lost in the dataset.

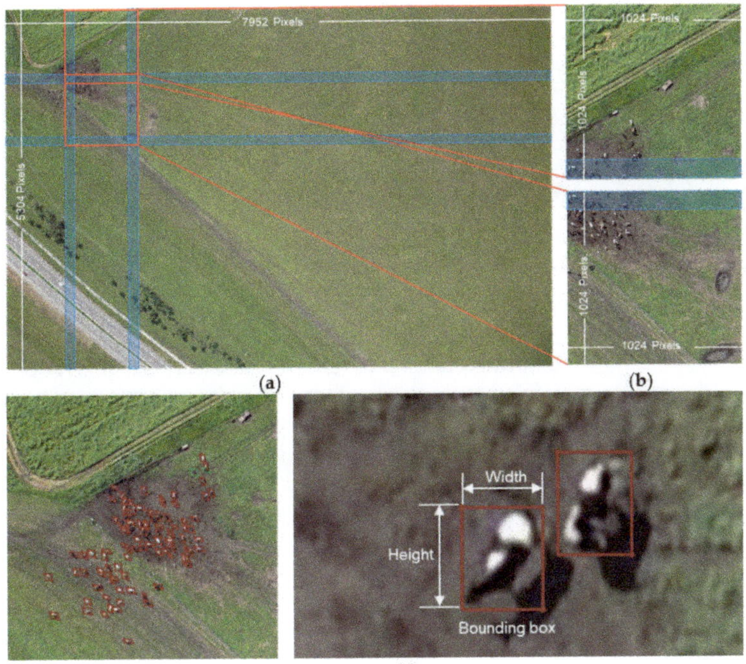

(c)

Figure 2. The split UAV image. (**a**) The original UAV image. The overlapped blue areas are 100 pixels wide to prevent the object from being split. (**b**) The subpatches after split are images in our dataset. The size of them is 1024 × 1024. (**c**) The annotation of samples using Labelme.

Seventy-five original UAV images with livestock were selected, and after the split, there were a total of 4050 image subpatches, including 469 animal patches (images containing animals), to build the dataset. The set was randomly divided into training, validation, and testing datasets with a ratio of about 7:1:2. More details about dataset allocation are shown in Table 1. Figure 3 shows the image samples in the dataset containing three types of animals: cattle, horses, and sheep. Figure 4 shows some object instances of the three types of animals in the dataset.

Table 1. The allocation of images into the training, validation, and testing datasets.

Datasets	Animal Patches	Cattle Instances	Horse Instances	Sheep Instances
Training	344	1511	1149	5354
Validation	39	169	87	1495
Testing	86	471	323	2342
Total	469	2151	1559	9191

(a) (b) (c)

Figure 3. Dataset image. (**a**) A patch of cattle; (**b**) a patch of horses; (**c**) a patch of sheep.

(a) cattle (b) horses (c) sheep

Figure 4. The object instances of the dataset. (**a**) cattle instances; (**b**) horse instances; (**c**) sheep instances.

Table 2 and Figure 5 are the statistical information of the animal samples in the dataset, as shown below. From Figure 5d, it can be seen that the side length of most object boxes is 10–20 pixels. The size of most sheep objects is in this range. The mean absolute size (MAS) is defined as the square root of the object box average area, while the mean relative size (MRS) is defined as the percentage of the mean area to the total patch area. The formula is as follows:

$$MAS = \sqrt{Average\ area} \tag{1}$$

$$MRS = \frac{Average\ area}{Patch\ area} \tag{2}$$

Table 2. Details of objects' sizes in the dataset.

Category	Min Width	Max Width	Average Width	Min Height	Max Height	Average Height	MAS	MRS	Number
cattle	11	73	34.87	9	87	33.37	33.99	0.11%	2151
horse	9	81	37.17	11	82	36.90	36.45	0.13%	1559
sheep	5	42	17.40	6	39	19.60	18.42	0.03%	9191

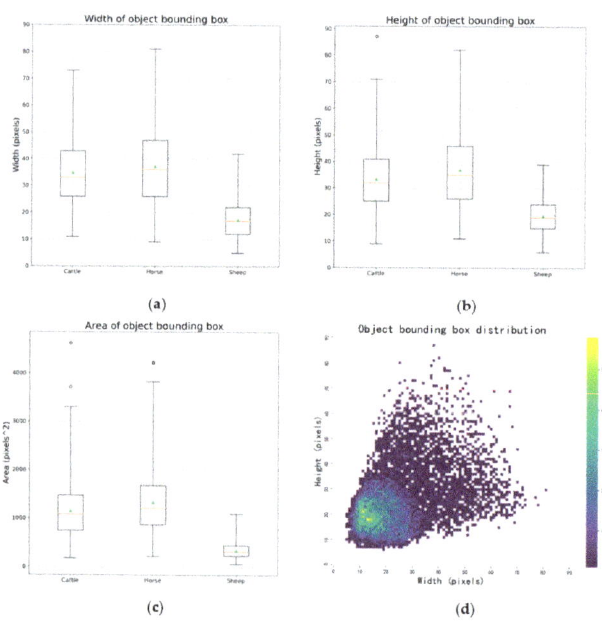

Figure 5. (**a**) The boxplot of object bounding box width. (**b**) The boxplot of object bounding box height. (**c**) The boxplot of object bounding box area. (**d**) The distribution of the object bounding boxes in the dataset. The horizontal axis is the width of the object bounding boxes and the vertical axis is the height. The color represents the number of boxes with the width and height in the coordinate system.

In order to test the multi-scale adaptability of the model, we made a multi-scale dataset with nine scales of 0.2, 0.25, 0.33, 0.5, 1, 2, 3, 4, and 5. Each scale set of the dataset consisted of 10 random animal patches, so there were a total of 90 patches, and each patch was still 1024 × 1024. Image scaling uses the bilinear interpolation method, in which an image with a scale smaller than 1 is a scaled-down image patch, and its surroundings are filled with zero values. In comparison, an image with a scale greater than 1 is the window for capturing the original image patch after scaling up, as shown in Figure 6. The details of the objects in the multi-scale dataset are shown in Table 3. Among them, the mean absolute size of the objects in the 0.2-scale set was already lower than 10 pixels, and the mean absolute size of the sheep was only 3.03 pixels. However, in the 5-scale set, the mean absolute size of sheep reached 72.04 pixels, and that of cattle reached 179.24 pixels. Generally, the multi-scale dataset scaled across an approximately 22–24 times change. Table 3 shows that the number of objects (scale > 1) was reduced due to the image being intercepted by the fixed window. The average size of the objects was not strictly proportional to the previous scale.

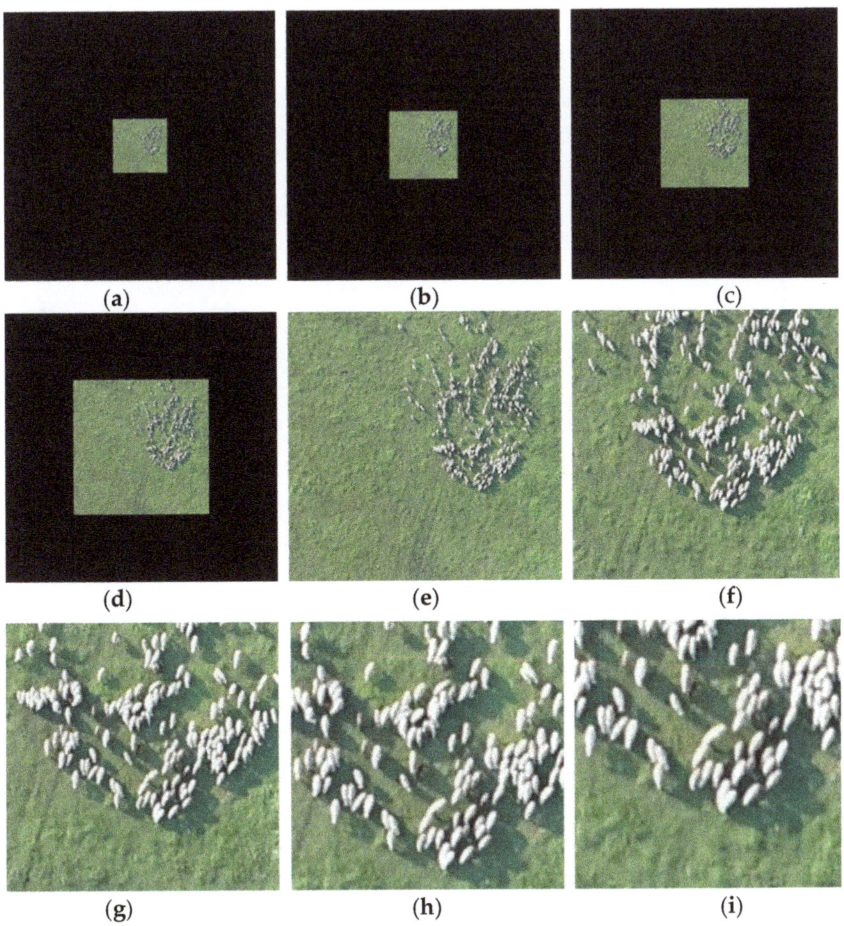

Figure 6. (**a–i**) The display of image patches of scale 0.2-5 in the multi-scale dataset. Among them (**a–d**) are image patches (scale < 1) surrounded by zero values. (**e**) The original image patch (scale = 1); (**f–i**) image patches (scale > 1) which are windows of 1024 × 1024 to intercept the enlarged images.

Table 3. Details of objects' size in multi-scale dataset.

Scale	Category	MAS	MRS	Number
0.2	cattle	8.12	0.79%	108
	horse	6.41	0.63%	107
	sheep	3.03	0.30%	605
0.25	cattle	10.15	0.99%	108
	horse	8.03	0.78%	107
	sheep	3.82	0.37%	605
0.33	cattle	13.38	1.31%	108
	horse	10.55	1.03%	107
	sheep	5.01	0.49%	605
0.5	cattle	20.31	1.98%	108
	horse	16.04	1.56%	107
	sheep	7.6	0.74%	605
1	cattle	40.59	3.96%	108
	horse	32.07	3.13%	107
	sheep	15.21	1.49%	605
2	cattle	79.55	7.77%	61
	horse	63.87	6.24%	92
	sheep	29.68	2.90%	347
3	cattle	114.73	11.20%	36
	horse	94.45	9.22%	78
	sheep	43.79	4.28%	195
4	cattle	150.94	14.74	28
	horse	122.43	11.96%	64
	sheep	59.23	5.78%	153
5	cattle	179.24	17.50%	23
	horse	149.89	14.64%	50
	sheep	72.04	7.04%	90

2.2. Proposed Method

Before selecting a model, we conducted a large number of model experiments. For details, see the experimental part later. According to the experiment results, the YOLOX series models were more suitable for our application because they have a higher recall rate than other models, which could achieve better performance when we calculated sheep units in the area. Moreover, this model was the state-of-the-art model in the past two years, integrating many effective strategies. The model's performance, such as accuracy, size, and speed, reached a good balance. Therefore, the GLDM (grazing livestock detection model) proposed in this paper was improved based on YOLOX.

YOLOX is an anchor-free model in the YOLO series proposed by Ge et al. [21] in July 2021. Another iconic model in the YOLO series after YOLOv5, the YOLOX model is based on different network depths and widths. The YOLOX model contains six different sizes: nano, tiny, s, m, l, and x, arranged from small to large. The nano version was designed for lightweight models with a weight size of only 3.9 MB. Regarding network structure, the YOLOX model used CSPDarknet in the backbone network, PANet [27] in the neck part, and proposed a decoupled head in the head part, improving the performance and convergence speed of the model. Mosaic and Mixup were used as data augmentation methods in training. The model also proposed to use a new sample matching method, SimOTA, to better improve the performance of the anchor-free model.

However, YOLOX had certain limitations in this study. First, the capability of detection for small objects with low resolution was inadequate, leading to missed detection. Second, the model was susceptible to confusion when distinguishing between different animal objects with similar body shapes. To facilitate the deployment of the model in the future, the nano in the YOLOX family was chosen as the baseline. The model was improved as follows.

1. An enhanced CSPDarknet (ECSP) was proposed as the backbone network of our model with three improvement tricks: a cascaded hybrid dilated convolutional module, stage compute ratio optimization, and input size optimization. The new backbone introduced context-related features and improved the feature extraction ability. This maximized the performance, especially the recall, with as few parameters as possible.
2. A weighted aggregation feature re-extraction pyramid module (WAFR) was also proposed as the neck part of our model, which made better use of the shallow features in the network and achieved effective multi-scale feature fusion.

The network structure of the GLDM is shown in Figure 7.

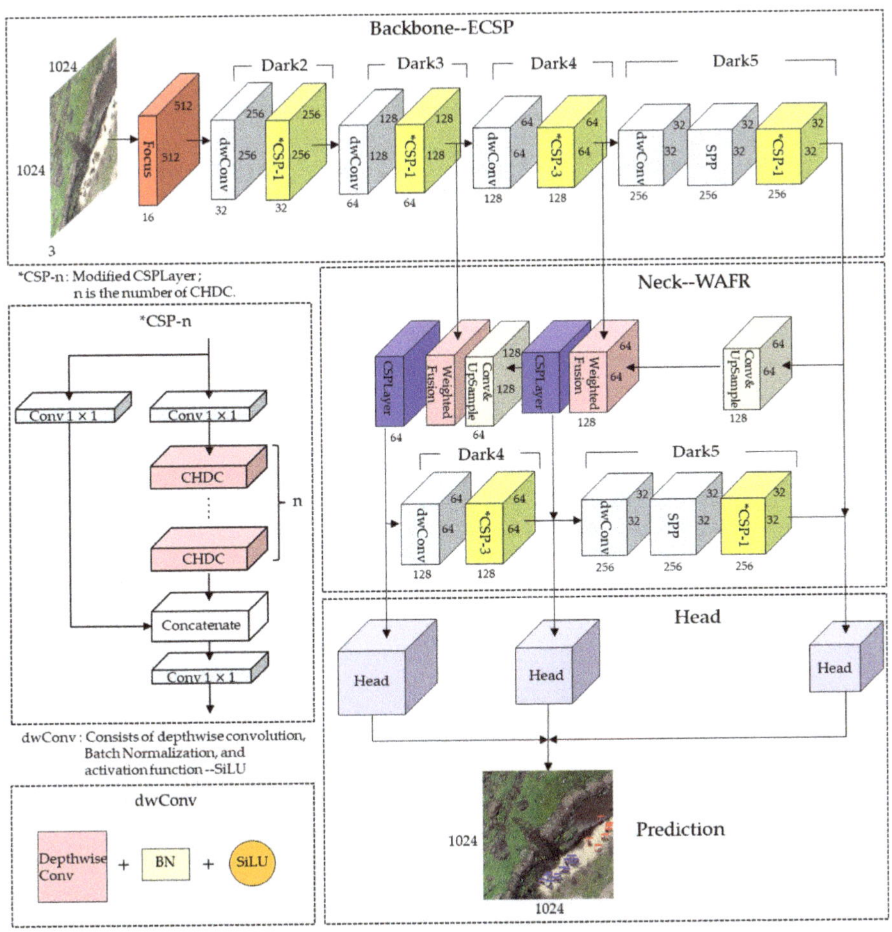

Figure 7. Network structure of the GLDM, roughly divided into parts of backbone, neck, and head. The backbone part is the proposed ECSP, and the neck part is the proposed WAFR.

2.2.1. Enhanced CSPDarknet

The backbone network is an essential part of an object detection model, which is responsible for the feature extraction of images. Modifying the backbone network to improve feature extraction ability is the key to improving the performance of the network. CSPDarknet, as an excellent backbone network after the proposal of CSPNet [28], was first used in YOLOv4 [29]. The subsequent YOLOv5 and YOLOX have continued to use this network. Therefore, based on the CSPDarknet, an enhanced CSPDarknet (ECSP) was proposed to further enhance its ability to extract small object features.

1. Cascade Hybrid Dilated Convolution Module

Introducing a large receptive field will bring more context-related features, improving the detection ability for small objects. Ordinary convolutional networks mainly obtain further feature maps through pooling operations, such as maximum pooling or average pooling, to increase the receptive field. However, this approach will make the size of the feature map smaller and inevitably lose information in the network downsampling process, which will seriously impact the detection accuracy, especially for small objects with fewer features in the original image. To avoid information loss, expand the receptive field, and increase the object context feature information, this study introduced the idea of dilated convolution. Dilated convolution was proposed by Yu et al. [30], and it was originally used for intensive prediction tasks such as semantic segmentation. This convolution can systematically aggregate multi-scale context information and exponentially expand the receptive field without reducing the resolution and without additional parameters. Inspired by [31], this study used dilated convolutions in combination with different dilation rates to expand the receptive field. However, the selection of the expansion rate here is not arbitrary. When the expansion rate of continuous hole convolution is the same, the features on the original image will be missed. Therefore, according to a design criteria of the hybrid dilated convolution module proposed by [31], combined with our large number of experiments, an effective cascaded hybrid dilated convolution module (CHDC) was proposed, as shown in Figure 8. The dilated rate of the convolution combination we designed is [1,2,5]. Convolution kernels with different dilated rates will sample features of different scales. After the combination and superposition, the receptive field will significantly exceed three consecutive standard convolutions, and there will be no missing samples in this combination. In this module, we uniquely use two sets of such hybrid dilated convolution for cascading. What is more, inspired by the residual idea of Resnet [32], we performed a shortcut connection to prevent gradient problems caused by too many convolutional layers. The shortcut adds the original information and the convolved information by matrix addition instead of concatenating. This module effectively improves the mAP, especially the recall in experiments.

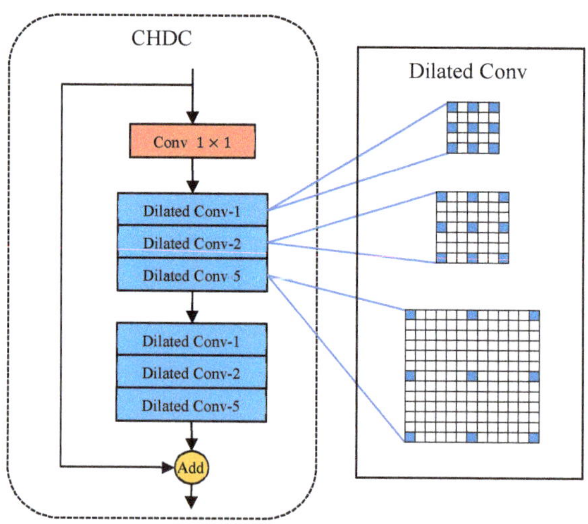

Figure 8. The designed cascaded hybrid dilated convolution module (CHDC), in which dilated conv-1, 2, and 5 are dilated convolutions with dilated rates of 1, 2, and 5, respectively, as shown on the right.

Figure 9 shows the comparison of the receptive fields of the two convolutions. The color from light to dark indicates the number of pixel calculations from less to more. Figure 10 is a schematic diagram of the sampling effect of the module in the image. For example, select a partial area of 224 × 224 in the original image, which is 61 × 61 after dark2 (4 times downsampling). Each grid in the figure represents a pixel, and the blue grid is a sampling point. It can be seen that the sampling coverage obtained after CHDC is wider than the original bottleneck in nano, and the associated information around the object can be obtained.

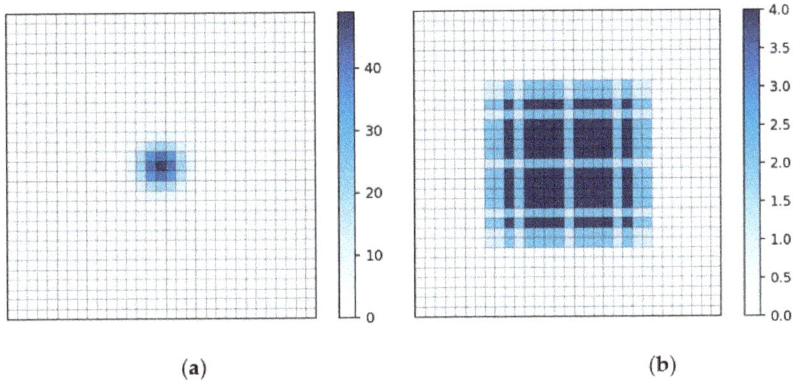

Figure 9. Comparison diagram of continuous convolution sampling. The color from light to dark indicates the number of pixel calculations from less to more. (**a**) The receptive field after three standard convolutions [1,1,1]; (**b**) the receptive field after the combination of dilated convolutions with expansion rates of [1,2,5].

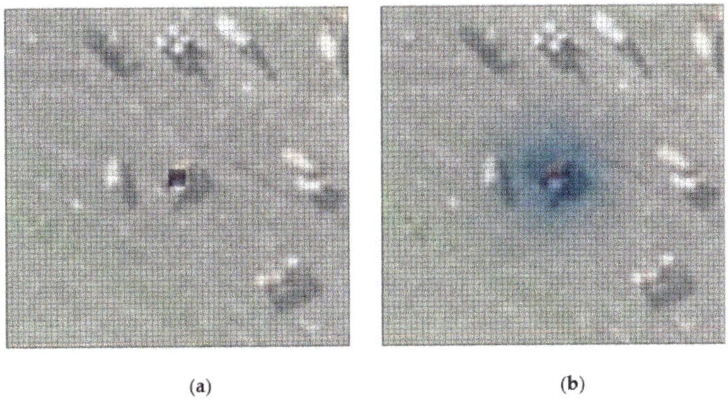

Figure 10. Schematic diagram of sampling on an image. (**a**) A schematic diagram of sampling after only one standard convolution; (**b**) a schematic diagram of sampling after CHDC with the sampling area larger, and the associated information around the object can be obtained.

2. Stage Compute Ratio Optimization

The stage settings in the backbone network can be traced back to Resnet, and the feature maps of each stage have different resolutions. The authors of [33] believe that the original design of the computation distribution across stages in Resnet is largely empirical. Different dark modules in CSPDarknet represent different stages, and the initial stage compute ratio is 1:3:3:1. In the past two years, Transformer has shown extraordinary performance in computer vision. Swin-T [34] also used similar ideas in the network but with a slightly different stage compute ratio of 1:1:3:1. The author of ConvNeXt [33] discovered and applied this ratio to Resnet, changing the number of blocks in each stage from (3, 4, 6, 3) in ResNet-50 to (3, 3, 9, 3). Here, we have carried out much experimental exploration based on this idea and finally adjusted the stage compute ratio in CSPDarknet to 1:1:3:1. Experiments have proved that this ratio not only reduces the model parameters but also improves the accuracy of the model. Overall, it is the best ratio at present. The specific experimental results are shown in Section 3.2.1.

3. Input Size Optimization

In order to ensure the consistency of the size and dimension of the output of the features by the network, we usually perform data size preprocessing before the image enters the backbone network; that is, bilinear interpolation operation, to force the image to be converted into the same size. For natural image detection tasks, to reduce GPU memory usage and reduce calculations, the YOLOX baseline usually compresses the input image to 640 × 640, which is larger than that. Although such an operation may cause image deformation and information loss, it is also beneficial, reducing the amount of calculation and the risk of model overfitting. However, for the small object detection in this study, if the algorithm compresses the image, the feature information of the small object will be lost before entering the model. This lost information cannot be recovered, which ultimately affects the detection performance of the model. Therefore, according to this study's image size and data characteristics, we bilinearly interpolated the input image into a size of 1024 × 1024 in this model because bilinear interpolation has a balance between effects and computation. That ensured the image lost no data before entering the network training and obtained more effective feature extraction in the subsequent network, effectively improving the model accuracy.

2.2.2. Weighted Aggregation Feature Re-Extraction Pyramid

Some researchers believe that low-level features from shallow layers of the network contain more fine-grained feature information and background noise, while features extracted from deeper layers contain more semantic information [35]. A modern detector consists of at least two parts: a backbone and a detection head. With the development of the model, there are many layer structures between the backbone and the head. This part of the structure is usually used to obtain feature maps at different stages for fusion to learn better features, which we call the model neck. FPN is a classic feature fusion structure with a top-down pathway proposed by Lin et al. [36] in 2017. It has been widely used in object detection models. Subsequently, many effective multi-scale feature fusion structures have emerged. Starting from YOLOv4, the YOLO model uses PANet [27] as a neck. PANet is a feature fusion structure divided into two processes: top-down and bottom-up. The structure fuses the features of different layers equally. Hence, the extraction effect for small object features is limited.

Therefore, to facilitate the detection of small objects in our dataset, we propose a new feature fusion structure: a weighted aggregation feature re-extraction pyramid (WAFR) with top-down and bottom-up two pathways as well. In this structure, the weighted fusion is started from the highest layer upwards, and the way of concatenating in the original PANet is abandoned, instead of the form of matrix addition. The fusion formula is as follows: where $r_{C4} : r_{C5}$ is 2:1, and $r_{C3} : r_{s4}$ is 2:1, the N3 layer after the top-down weighted fusion is re-extracted by the Dark4 and Dark5 in the backbone. Moreover, the S4 layer and the C5 layer are cross-fused during the extraction process. Finally, three feature map outputs with different scales are obtained. The structure is shown in Figure 11.

$$F = \frac{1}{r_i + r_j}(r_i F_i + r_j F_j) \quad (3)$$

F is the fused feature, F_i and F_j are the i and j feature layers, respectively, and r_i and r_j are the weights of the feature layer.

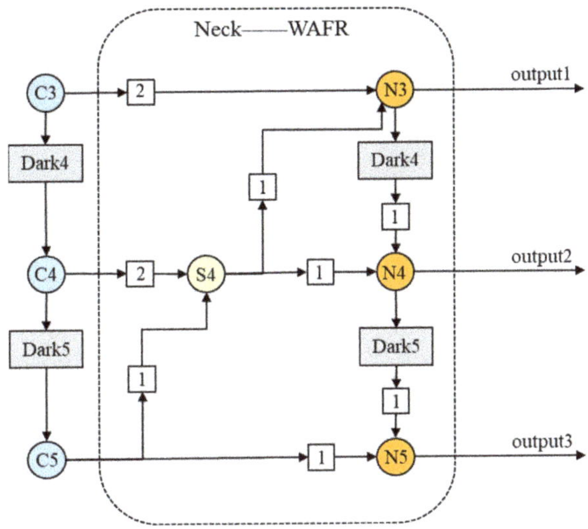

Figure 11. Weighted aggregation feature e-extraction pyramid (WAFR). In the figure, layer C5 is the deepest output of the backbone network, which has the most semantic information. Layer N3 is a top-down fusion feature map. Layers N4 and N5 are feature maps after re-extraction and cross-fusion of backbone network features.

2.2.3. Standard of Performance Evaluation

A series of indices were adopted to evaluate object detection performance from different aspects: the precision, the recall, the $F1$ *score*, and mean average precision (mAP).

The precision evaluates the accuracy in the total number of predictions, whereas the recall provides insight into how well the prediction covers the objects of interest [25]. The mathematical formula of precision and recall are defined as follows:

$$precision = \frac{TP}{TP + FP} \tag{4}$$

$$recall = \frac{TP}{TP + FN} \tag{5}$$

where TP, FP, and FN represent the true positive, false positive, and false negative.

$F1$ *score* and average precision (AP) were adopted to make a comprehensive evaluation of the results, since recall and precision reflect only one aspect of the model's performance [25]. $F1$ *score* is the harmonic mean of precision and recall, and the formula is defined as:

$$F1\ score = \frac{2}{\frac{1}{precision} + \frac{1}{recall}} \tag{6}$$

AP is defined as the area surrounded by the recall-precision curve, which is formulated as:

$$AP = \int_0^1 precision(recall)d(recall) \tag{7}$$

To evaluate the overall performance of the model on the dataset, mAP was adopted, which is formulated as:

$$mAP = \frac{1}{n}\sum_{i=1}^{n} AP_i \tag{8}$$

where n is the number of categories in the dataset.

In order to apply the model to the survey of grassland grazing livestock, we proposed a new evaluation index—livestock accuracy (LAC), which is the accuracy of the evaluation model when calculating the number of grazing livestock in an area. This evaluation index reflects the accounting accuracy of regional livestock volume (without distinguishing categories) under the premise of no prior knowledge and full trust model. The formula is defined as:

$$LAC = \frac{SU_{truth} - \left|SU_{predicted} - SU_{truth}\right|}{SU_{truth}} \tag{9}$$

$$SU_{predicted} = \sigma_{class}(TP_{class} + FP_{class}) \tag{10}$$

$$SU_{truth} = \sigma_{class} GT_{class} \tag{11}$$

where SU is the sheep unit, which is a unified conversion unit for calculating the amount of livestock. One sheep unit is a sheep with a live weight of 40 kg and its suckling lambs, and the daily eclipse of forage grass is 5.0–7.5 kg [37]. GT is the number of ground truth and σ_{class} is the conversion factor of the object class and sheep unit. For example, in this research area, 1 cattle unit can be converted into 5 sheep units. σ_{class} is defined as follows:

$$\sigma_{sheep} = 1,\ 1\ sheep = 1\ SU \tag{12}$$

$$\sigma_{cattle} = 5,\ 1\ \text{cattle} = 5\ SU \tag{13}$$

$$\sigma_{horse} = 5,\ 1\ \text{horse} = 5\ SU \tag{14}$$

3. Results

The model was run on an InterXeon(R) Gold 5118 CPU@2.30GHZ, NVIDIA RTX A6000 GPU, and Ubuntu 18.04.5 LTS system, using the Pytorch 1.10 deep learning framework. The model used the stochastic gradient descent (SGD) optimizer and was trained for 500 epochs. In the training strategy, we froze the backbone network at 0–50 epochs and trained with a batch size of 16. Then, we unfroze the backbone network, all model parameters participated in the training together, and the batch size was 8. The learning rate was initially set to 0.01, the minimum learning rate was set to 0.0001, the momentum was set to 0.937, and the weight decay was set to 0.0005. The learning rate drop method adopted was the *cos* drop method.

3.1. Algorithm Performance Comparison

To demonstrate the advantage of the proposed method, different deep learning object detection models are executed in this section, such as Faster R-CNN, RetinaNet, FCOS, and YOLO series, including the-state-of-art models. The above experimental models include the classic one-stage, two-stage, and anchor-based and anchor-free classic models as the object detection model. We used these models to make a horizontal comparison with our proposed GLDM and observed the performance of these models on our dataset from a large number of experiments, as shown as Table 4.

Faster RCNN is a milestone model in the RCNN series proposed by Ren et al. [15] in 2016. As a two-stage classic model, Faster RCNN once became the most accurate object detection model. However, the model is anchor-based, and the accuracy of the model depends on the setting of anchor hyperparameters, which requires prior knowledge. After some simple anchor hyperparameter adjustments, the experimental results' mAP only reached 21.75%.

RetinaNet is a one-stage detection model as well as an anchor-based model proposed by He et al. [16] in 2017. In order to solve the problem of class imbalance, the author proposed using the Focal Loss. Through experiments, the mAP of this model reached 35.78%, which was slightly better than that of the Faster RCNN model, but the detection effect for the category sheep was not satisfactory.

Table 4. Comparison of state-of-the-art object-detection models.

Model	Time	Category	AP	F1	Recall	Precision	mAP	Parameters [1]
Faster R-CNN	2016	cattle horse sheep	39.34% 25.37% 0.53%	0.48 0.36 0.53	54.99% 42.72% 0.60%	42.88% 31.15% 60.87%	21.75%	28.3M
RetinaNet	2017	cattle horse sheep	60.09% 47.26% 0.00%	0.52 0.42 0	36.73% 27.86% 0.00%	91.05% 85.71% 0.00%	35.78%	36.4M
YOLOv3	2018	cattle horse sheep	82.43% 69.06% 72.59%	0.78 0.67 0.75	71.76% 60.06% 74.72%	85.79% 74.62% 75.82%	74.69%	61.5M
FCOS	2019	cattle horse sheep	84.95% 82.06% 67.17%	0.85 0.8 0.55	83.23% 78.95% 40.73%	87.31% 81.99% 86.73%	78.06%	32.1M
YOLOX-nano	2021	cattle horse sheep	84.69% 75.72% 72.78%	0.83 0.76 0.77	82.80% 74.30% 71.69%	82.80% 77.67% 82.87%	77.73%	0.9M
YOLOX-x	2021	cattle horse sheep	89.61% 86.63% 85.32%	0.87 0.85 0.88	90.23% 87.93% 86.68%	83.66% 82.08% 88.53%	87.19%	99.0M
GLDM (ours)	2022	cattle horse sheep	88.52% 85.87% 85.03%	0.86 0.84 0.86	87.47% 83.90% 83.99%	84.25% 84.16% 87.42%	86.47%	5.7M

[1] The parameter quantity here is an approximate value, and the conversion relationship is taken as 1 M = 1000 k.

YOLOv3 was proposed by Redmon et al. [17] in 2018, using DarkNet-53 as the backbone network and absorbing the idea of FPN with high computing efficiency. As a one-stage anchor-based model, YOLOv3 introduced the most effective tricks in the industry at that time, and it is also a milestone object detection model. Through experiments, although the model size was larger, the accuracy was greatly improved: the mAP reached 74.69%.

FCOS is an anchor-free detection model. It was proposed by Tian et al. [19] in 2019. This model no longer depends on the anchor and avoids all hyperparameters related to the anchor that affect the final detection result, and it is a fully convolutional, one-stage model for pixel-by-pixel prediction. Experiments showed that compared with YOLOv3, the model achieved a higher mAP with a smaller number of parameters, reaching 78.06%.

The YOLOX-nano in the YOLOX series had a mAP of 77.73% with 0.9 M parameters, and the YOLOX-x was the model with the largest number of parameters and the highest accuracy in the series, with a mAP of 87.19%.

The mAP of the GLDM reached 86.47%, a bit lower than the YOLOX-x model. While improving the detection performance, we also considered the model size. The number of parameters of the proposed model was only 5.7 M, or only about 5.76% of that of YOLOX-x. The average precision (AP) for sheep detection exhibited a notable increase of 12.25%, indicating the effectiveness of our enhancements in the detection of small objects. In summary, we achieved a lighter and higher-precision unification on our UAV dataset, which is currently a more efficient detection model for grassland grazing livestock detection.

Figure 12 shows a comparison of the detection results from different models. The chosen models were Faster R-CNN, FCOS, and YOLOv3, which are the typical two-stage, anchor-free, and classic YOLO model, respectively. Faster R-CNN had a very poor detection effect on sheep, which had been greatly improved in YOLOv3. Although FCOS was not as good as YOLOv3 in detecting sheep, it was better in detecting horses than YOLOv3. It can also be seen from the figure that the above objects all had excellent detection results in our model.

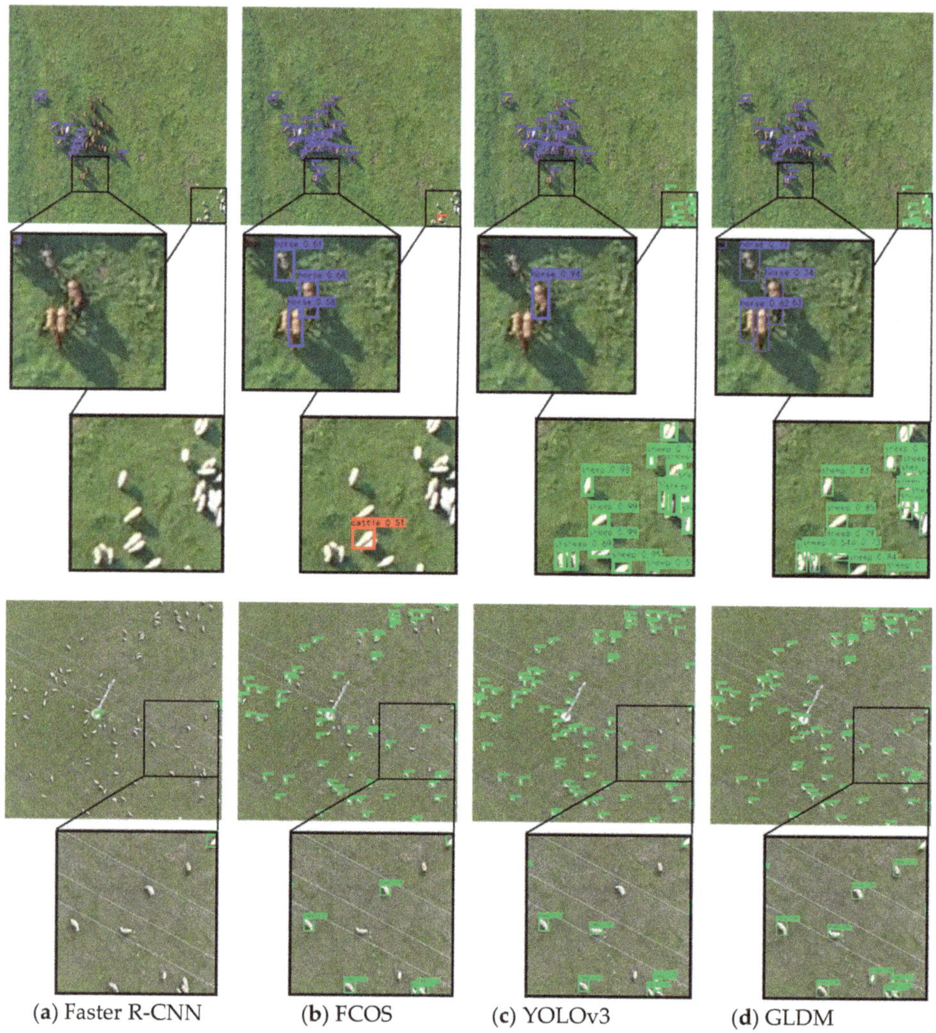

Figure 12. Visualization of the detection results for different methods. (**a**) Faster R-CNN; (**b**) FCOS; (**c**) YOLOv3; (**d**) our model (GLDM).

3.2. Ablation Experiments

To verify the effectiveness of the method proposed, ablation experiments were performed in this study. First, we carried out the ablation verification of the overall model and conducted experiments on the two improved modules, ECSP and WAFR in sequence. The experimental results are shown in Table 5. After using ECSP, the mAP of the model was improved by 8.25%. Due to the increase of CHDC in ECSP to the receptive field, and bilinear interpolation input, the recall rate was significantly improved. The average recall improved by 8.48%, which aligned with our expectations. Results with high recall are more beneficial for our application. After the model used WAFR, the mAP of the model increased by 8.74%, reaching 86.47%. The average recall was improved by 8.86%. Although the number of parameters changed and increased, the model accuracy improvement in exchange was worthwhile.

Table 5. Ablation experiments for the proposed method.

Baseline	Backbone	Neck	Average Recall	Average Precision	mAP	Parameters
nano			76.26%	81.11%	77.73%	0.9M
nano	ECSP		84.74%	84.93%	85.98%	3.8M
nano	ECSP	WAFR	85.12%	85.28%	86.47%	5.7M

3.2.1. Use of Enhanced CSPDarknet

The experimental results above proved that ECSP improved the mAP and recall of the model. To verify the effectiveness of the three tricks proposed in ECSP, as mentioned above, we decomposed ECSP and conducted more in-depth ablation experiments. As shown in Table 6, the experimental results proved that mAP had been improved by each trick.

Table 6. Ablation experiments for ECSP as the backbone.

Baseline	Bottleneck	Image Input	Block Rate	mAP
nano				77.73%
nano	CHDC			78.17%
nano	CHDC	1024 × 1024		85.04%
nano	CHDC	1024 × 1024	1:1:3:1	**85.98%**

As for improving our stage compute rate on the backbone, it is an optimal ratio that we have explored through many experiments. In Table 7 below, several representative experiments are given. The original ratio was 1:3:3:1, referring to the ratio of CHDC in Dark2-5. In the nano model, the depth factor was 1, so 1:3:3:1 was also the real number ratio of CHDC in each stage. First, the ratio was adjusted based on keeping the total number consistent. Although higher accuracy was obtained, the size of the model was sacrificed.

Table 7. Ablation experiments for stage compute rate in ECSP.

Baseline	Bottleneck	Image Input	Block Rate	mAP	Parameter
nano	CHDC	1024 × 1024	1:3:3:1	85.04%	3.949M
nano	CHDC	1024 × 1024	1:1:3:3	84.96%	5.642M
nano	CHDC	1024 × 1024	1:1:4:2	85.19%	4.965M
nano	CHDC	1024 × 1024	1:1:5:1	85.88%	4.288M
nano	CHDC	1024 × 1024	1:1:3:1	**85.98%**	**3.836M**

In the final experiment, the mAP of the model with a ratio of 1:1:3:1 was 0.94% higher than that of the original model. The parameters were reduced by 0.11 M, which achieved higher accuracy and smaller volume than the original model, and is currently the best choice.

3.2.2. Use of Weighted Aggregation Feature Re-Extraction Pyramid

In this part, we give the ablation tests for WAFR, as shown in Table 8 below. Experiments in the table verify the effect of WAFR on the model with or without ECSP. When not using our proposed ECSP, WAFR brought a 0.9% improvement to the model, and after using ECSP, WAFR brought a 0.49% improvement to the model. Experiments proved that WAFR was effective for model improvement.

3.2.3. Visualization of Results

In order to see the difference before and after the model improvement more clearly, we show in Figure 13 the original test image, the test result image before and after the model improvement, and the ground truth. The figure shows the detection results of typical sheep, horses, and cattle images. We can see an improvement in missed detection and false detection. In a dense scene like a flock of sheep, the improved model significantly improved the recall of sheep and detected some previously undetected objects efficiently. For horses, false detections before improvement were corrected. For the cattle in the figure, there may be two cattle with one detection box before the improvement, and the individuals were better distinguished after the improvement.

Table 8. Ablation experiments for WAFR.

Baseline	Backbone	Neck	mAP
nano	CSPDarknet	PAN	77.73%
nano	CSPDarknet	WAFR	78.63%
nano	ECSP	PAN	85.98%
nano	ECSP	WAFR	**86.47%**

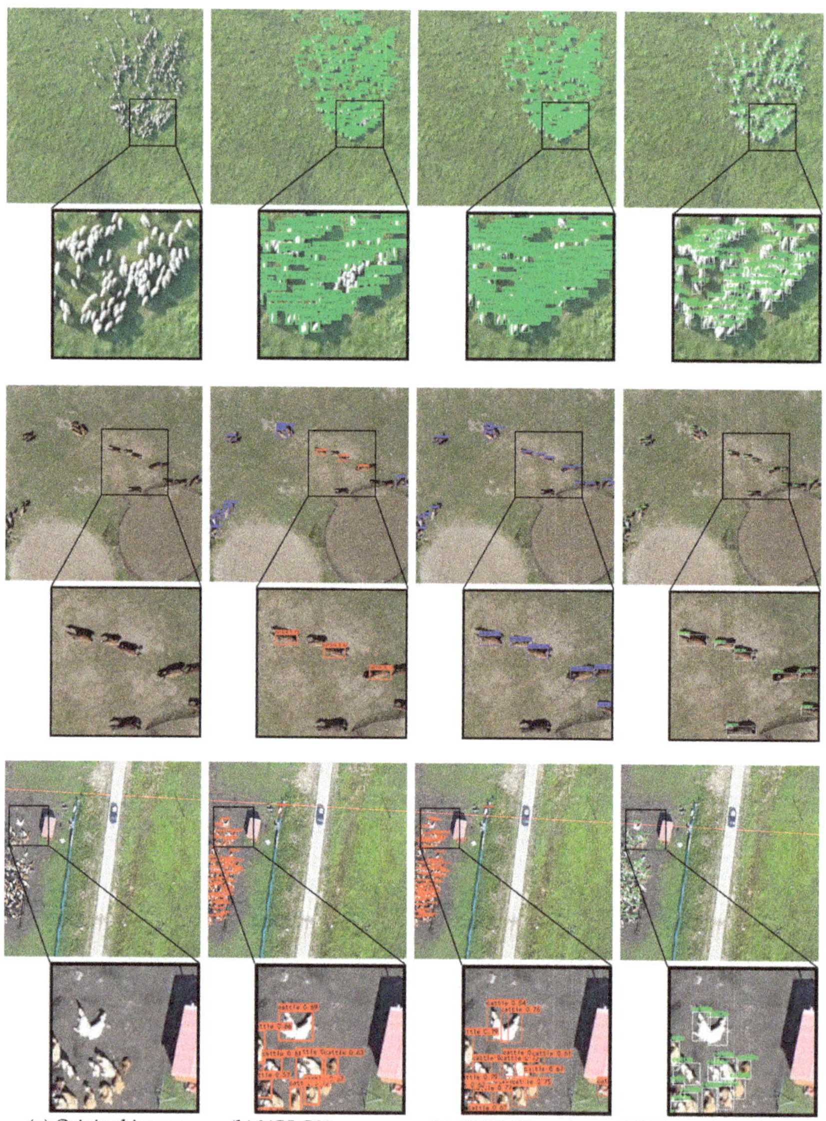

(**a**) Original image (**b**) YOLOX-nano (**c**) GLDM (ours) (**d**) Ground truth

Figure 13. Comparison of detection results before and after model improvement.

3.3. Model Application

In this part, we study the application of the model, including the scale adaptability of the model; that is, in what resolution range the model can still maintain a good accuracy, the model inference method of large-size remote sensing image, and the grassland livestock counting in the test dataset.

3.3.1. Model Scale Adaptability

The data used in this study were collected by UAVs flying at 300 m, although due to the ups and downs of the terrain, the actual shooting distance to the ground varied. However, generally speaking, the resolution was roughly within a range; that is, the size of similar objects did not change much. Since the model proposed was trained from such data, we needed to test the model's adaptability to large-scale changes. Therefore, we made a test dataset with multi-scale variation, as described in Section 2. The test results are shown in Table 9 and Figure 14. The general trend was that as the scale shrank or expanded, the detection performance decreased, but the performance of different categories was slightly different.

Table 9. AP of three animals' detection at different scales.

Scale	0.2	0.25	0.33	0.5	1	2	3	4	5
cattle AP	0.5047	0.6361	0.7955	0.9035	0.9142	0.8467	0.245	0.1038	0.029
horse AP	0.1479	0.4022	0.5168	0.7929	0.8919	0.9159	0.7309	0.3634	0.089
sheep AP	0.0255	0.1187	0.2602	0.6222	0.9074	0.9457	0.9127	0.7319	0.4122
mAP	0.226	0.3857	0.5242	0.7728	0.9045	0.9028	0.6296	0.3997	0.1767

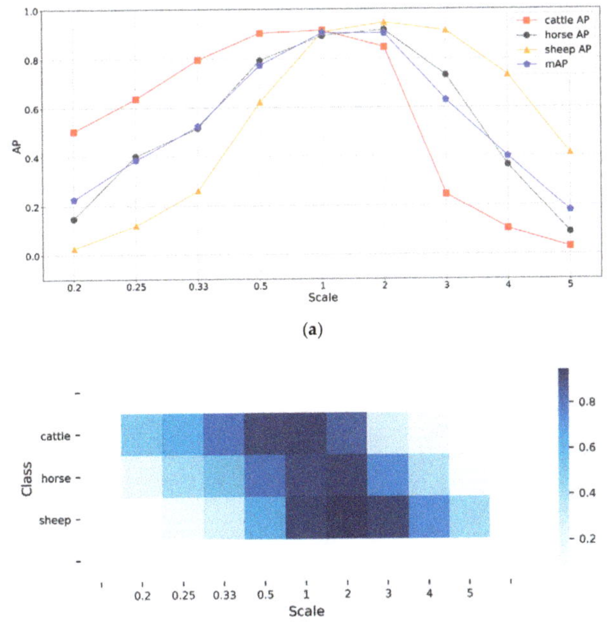

(a)

(b)

Figure 14. (**a**) The line graph of the AP values of cattle, horse, sheep, and mAP on the multi-scale test dataset; (**b**) the heat map of the AP values of the three types of objects on the multi-scale test dataset. This more intuitively reflects the difference in the detection of scale changes in the model for cattle, horses, and sheep.

From Figure 14b, it can be seen more intuitively that cattle and sheep have misplacement for scale changes. Cattle have stronger adaptability to scale reduction, and sheep have stronger adaptability to scale expansion. Assuming that the AP value of any category is required to be greater than or equal to 0.5, a scale change of 0.5–2 can meet the requirements. Performance drops off significantly outside of a scale change of 0.33-3.

Considering that the original image resolution of our dataset was about 5 cm, and corresponding to the above conclusion, our model still performed well between 2.5 and 10 cm through resolution conversion. The range could be expanded appropriately, but it was best to stay within 15 and 1.7 cm. Figure 15 below is the test dataset detection results at scales of 0.5, 1, and 2.

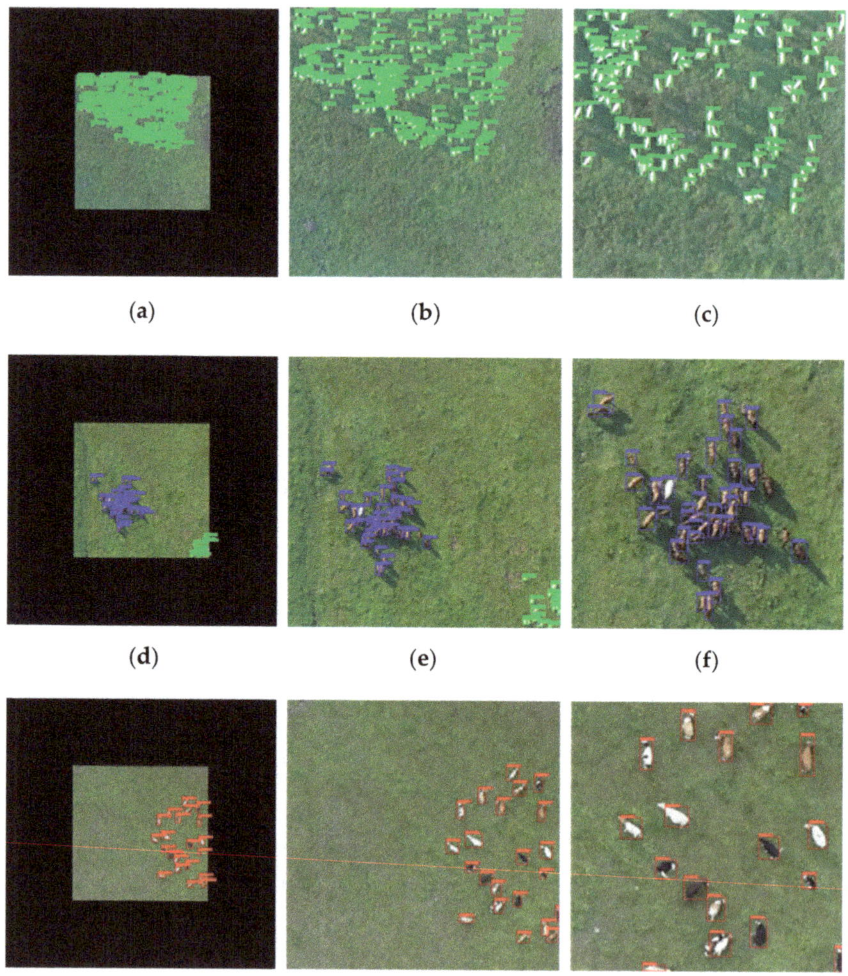

(**a**) (**b**) (**c**)

(**d**) (**e**) (**f**)

(**g**) (**h**) (**i**)

Figure 15. The detection results of the model at scales of 0.5, 1, and 2; (**a–c**) cattle at scales of 0.5, 1, and 2; (**d–f**) horses at scales of 0.5, 1, and 2; (**g–i**) sheep at scales of 0.5, 1, and 2.

3.3.2. Large-Size Remote Sensing Image Inference and Grassland Livestock Accounting

In practical applications, we often obtain large-size remote sensing images. For example, the original image size of the UAV in our research reached 7952 × 5304. If an image of this size is directly detected without a split, the entire image will be compressed to a fixed size before entering the network. The algorithm directly filters a large amount of image information in the preprocessing stage. Because the object to be detected is extremely small, the image entering the network may have no object information.

We used the sliding window detection method to achieve the inference process of large-size remote sensing images. The detection window slides from the upper left of the image until it slides to the lower right, traversing the entire image. The final large-size image detection result is obtained, as shown in Figure 16. The object in the figure, which is the original UAV image, has been effectively detected without split.

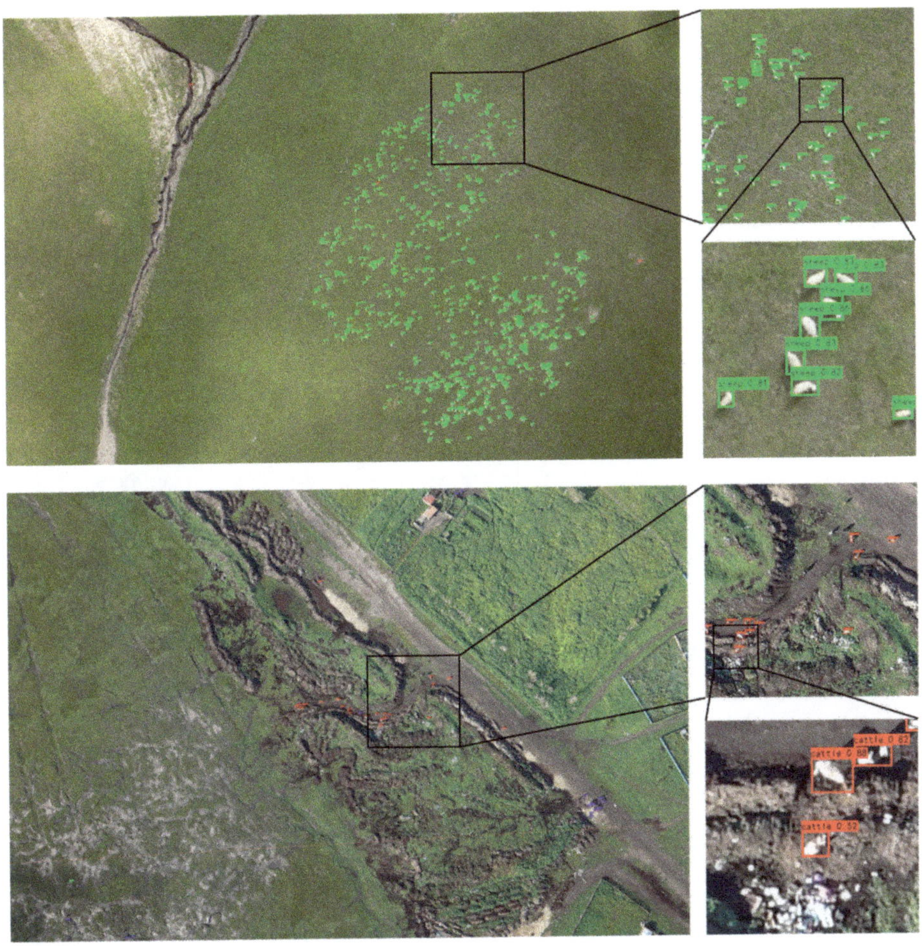

Figure 16. The result of the direct detection of the original image of the UAV.

We used the model in this paper to calculate the number of livestock in the testing dataset, and the detailed data are shown in Table 10. The test set had 3136 livestock samples, which could be converted into 6312 sheep units. The model detected 489 cattle objects, 322 horse objects, and 2261 sheep objects, which were converted into 6316 sheep units. Therefore, LAC (the accuracy of accounting for livestock in Section 2.2.3) using UAV images in this testing dataset reached 99% without prior knowledge.

Table 10. Livestock accounting in testing dataset.

Category	Truth	TP	FP	FN	LAC
cattle	471	412	77	59	
horse	323	271	51	52	0.99
sheep	2342	1967	294	375	

4. Discussion

The use of UAV images combined with deep learning for the automatic detection of grassland grazing livestock presents several technical difficulties, such as small objects, easy false detection, and easily missed detection. In this study, an improved deep learning detection model GLDM was designed based on the YOLOX nano model to deal with the above difficulties. The enhanced backbone network ECSP we proposed incorporates the hybrid dilated convolution idea to expand the receptive field to extract the context features of small objects. We also optimized the stage compute ratio and the input size. These techniques enhanced the feature extraction ability of small objects in the image, significantly improving the model's recall. Additionally, a feature fusion structure WAFR was designed to strengthen the network's utilization of objects' shallow features, further enhancing the detection and positioning capabilities of small objects. Experimental results have also shown that our proposed model achieves better performance than before, effectively addressing the challenges posed by small object detection, false detection, and missed detection. The proposed model can also be used for large-scale grassland livestock surveys.

Nonetheless, there remain several areas for improvement in this study. First, the computational efficiency of the model has not been optimized. Secondly, the model has not improved the detection of ultra-dense sheep in sheep pens, which is also a direction for future research. For future work, in addition to exploring the above issues, there are several directions that are worthy of further work.

1. Increase the number of labeled samples and add other object categories to explore possible long-tail object detection. This is a complex problem in the field of object detection that warrants further investigation.
2. Collect multi-angle and multi-scale data to expand the model's application scenarios. This will make the model more flexible and allow it to be applied to tasks such as target tracking in the future, as well as enabling the description of objects at ultra-high resolution.
3. Study the domain adaptation problem in transfer learning and explore the knowledge transfer of the model. This is critical for the inheritance and evolution of the model, as well as the reduction of data labeling costs.

5. Conclusions

In this study, a deep learning dataset based on UAV images was constructed, and a deep learning method GLDM for grassland grazing livestock detection was proposed to better integrate deep learning technology into remote sensing and serve grassland animal husbandry. The model's detection ability for small and confusing objects in UAV images was effectively strengthened by the enhanced backbone network ECSP and the feature fusion module WAFR. Experimental results show that this model has better detection performance and fewer model parameters than existing object detection algorithms. It performed well in the recall rate, and the mAP of the model reached 86.47%, which is suitable for detecting grassland grazing livestock in UAV images. Moreover, the performance of the model was investigated, and it was observed to maintain good performance in images with a spatial resolution ranging from 2.5 to 10 cm. The inference process of large-size remote sensing images was implemented without splitting. Overall, the proposed model can achieve remarkable results in livestock detection, effectively overcome the main practical problems, and can practicably perform livestock surveys using UAV remote sensing over extensive grassland.

Author Contributions: Conceptualization, Y.W. and Q.W.; methodology, Y.W. and Q.W.; software, Y.W. and Q.W.; validation, Y.W., Q.W. and L.M.; formal analysis, Y.W. and Q.W.; investigation, Y.W. and Q.W.; resources, Y.W., Q.W., N.W., D.W. and L.M.; data curation, Y.W. and Q.W.; writing—original draft preparation, Y.W.; writing—review and editing, Y.W., Q.W., X.W. and L.M.; visualization, Y.W.; supervision, Q.W., N.W. and L.M.; project administration, Q.W., N.W., Q.Z., X.H., L.M. and G.O.; funding acquisition, Q.W., N.W., L.M. and G.O. All authors have read and agreed to the published version of the manuscript.

Funding: This research was funded by the Strategic Priority Research Program of Chinese Academy of Sciences (Grant No. XDA26010201) and the Science and Technology Major Project of Inner Mongolia Autonomous Region of China (Grant No. 2021ZD0044).

Data Availability Statement: The data presented in this study are available on request from the corresponding author.

Conflicts of Interest: The authors declare no conflict of interest.

References

1. Wang, D.; Liao, X.; Zhang, Y.; Cong, N.; Ye, H.; Shao, Q.; Xin, X. Grassland Livestock Real-Time Detection and Weight Estimation Based on Unmanned Aircraft System Video Streams. *Chin. J. Ecol.* **2021**, *40*, 4099–4108.
2. Wang, D.; Shao, Q.; Yue, H. Surveying Wild Animals from Satellites, Manned Aircraft and Unmanned Aerial Systems (UASs): A Review. *Remote Sens.* **2019**, *11*, 1308. [CrossRef]
3. Fretwell, P.T.; Trathan, P.N. Penguins from Space: Faecal Stains Reveal the Location of Emperor Penguin Colonies. *Glob. Ecol. Biogeogr.* **2009**, *18*, 543–552. [CrossRef]
4. Schwaller, M.R.; Southwell, C.J.; Emmerson, L.M. Continental-Scale Mapping of Adélie Penguin Colonies from Landsat Imagery. *Remote Sens. Environ.* **2013**, *139*, 353–364. [CrossRef]
5. Schwaller, M.R.; Olson, C.E.; Ma, Z.; Zhu, Z.; Dahmer, P. A Remote Sensing Analysis of Adélie Penguin Rookeries. *Remote Sens. Environ.* **1989**, *28*, 199–206. [CrossRef]
6. Löffler, E.; Margules, C. Wombats Detected from Space. *Remote Sens. Environ.* **1980**, *9*, 47–56. [CrossRef]
7. Wilschut, L.I.; Heesterbeek, J.A.P.; Begon, M.; de Jong, S.M.; Ageyev, V.; Laudisoit, A.; Addink, E.A. Detecting Plague-Host Abundance from Space: Using a Spectral Vegetation Index to Identify Occupancy of Great Gerbil Burrows. *Int. J. Appl. Earth Obs. Geoinf.* **2018**, *64*, 249–255. [CrossRef]
8. Caughley, G.; Sinclair, R.; Scott-Kemmis, D. Experiments in Aerial Survey. *J. Wildl. Manag.* **1976**, *40*, 290–300. [CrossRef]
9. Stapleton, S.; Peacock, E.; Garshelis, D. Aerial Surveys Suggest Long-Term Stability in the Seasonally Ice-Free Foxe Basin (Nunavut) Polar Bear Population. *Mar. Mammal Sci.* **2016**, *32*, 181–201. [CrossRef]
10. Rey, N.; Volpi, M.; Joost, S.; Tuia, D. Detecting Animals in African Savanna with UAVs and the Crowds. *Remote Sens. Environ.* **2017**, *200*, 341–351. [CrossRef]
11. Corcoran, E.; Denman, S.; Hamilton, G. Evaluating New Technology for Biodiversity Monitoring: Are Drone Surveys Biased? *Ecol. Evol.* **2021**, *11*, 6649–6656. [CrossRef]
12. Gonzalez, L.F.; Montes, G.A.; Puig, E.; Johnson, S.; Mengersen, K.; Gaston, K.J. Unmanned Aerial Vehicles (UAVs) and Artificial Intelligence Revolutionizing Wildlife Monitoring and Conservation. *Sensors* **2016**, *16*, 97. [CrossRef]
13. Xue, Y.; Wang, T.; Skidmore, A.K. Automatic Counting of Large Mammals from Very High Resolution Panchromatic Satellite Imagery. *Remote Sens.* **2017**, *9*, 878. [CrossRef]
14. Torney, C.J.; Dobson, A.P.; Borner, F.; Lloyd-Jones, D.J.; Moyer, D.; Maliti, H.T.; Mwita, M.; Fredrick, H.; Borner, M.; Hopcraft, J.G.C. Assessing Rotation-Invariant Feature Classification for Automated Wildebeest Population Counts. *PLoS ONE* **2016**, *11*, e0156342. [CrossRef] [PubMed]
15. Ren, S.; He, K.; Girshick, R.; Sun, J. Faster R-CNN: Towards Real-Time Object Detection with Region Proposal Networks. In Proceedings of the Advances in Neural Information Processing Systems, Montreal, QC, Canada, 7–12 December 2015; Curran Associates, Inc.: Red Hook, NY, USA, 2015; Volume 28.
16. Lin, T.-Y.; Goyal, P.; Girshick, R.; He, K.; Dollar, P. Focal Loss for Dense Object Detection. In Proceedings of the IEEE International Conference on Computer Vision (ICCV), Venice, Italy, 22–29 October 2017; pp. 2980–2988.
17. Redmon, J.; Farhadi, A. YOLOv3: An Incremental Improvement. *arXiv* **2018**, arXiv:1804.02767.
18. Wang, C.-Y.; Bochkovskiy, A.; Liao, H.-Y.M. YOLOv7: Trainable Bag-of-Freebies Sets New State-of-the-Art for Real-Time Object Detectors. *arXiv* **2022**, arXiv:2207.02696.
19. Tian, Z.; Shen, C.; Chen, H.; He, T. FCOS: Fully Convolutional One-Stage Object Detection. In Proceedings of the IEEE/CVF International Conference on Computer Vision (ICCV), Seoul, Republic of Korea, 27 October–2 November 2019; pp. 9627–9636.
20. Zhou, X.; Wang, D.; Krähenbühl, P. Objects as Points. *arXiv* **2019**, arXiv:1904.07850.
21. Ge, Z.; Liu, S.; Wang, F.; Li, Z.; Sun, J. YOLOX: Exceeding YOLO Series in 2021. *arXiv* **2021**, arXiv:2107.08430.

22. Kellenberger, B.; Volpi, M.; Tuia, D. Fast Animal Detection in UAV Images Using Convolutional Neural Networks. In Proceedings of the 2017 IEEE International Geoscience and Remote Sensing Symposium (IGARSS), Fort Worth, TX, USA, 23–28 July 2017; pp. 866–869.
23. Kellenberger, B.; Marcos, D.; Tuia, D. Detecting Mammals in UAV Images: Best Practices to Address a Substantially Imbalanced Dataset with Deep Learning. *Remote Sens. Environ.* **2018**, *216*, 139–153. [CrossRef]
24. Roosjen, P.P.; Kellenberger, B.; Kooistra, L.; Green, D.R.; Fahrentrapp, J. Deep Learning for Automated Detection of Drosophila Suzukii: Potential for UAV-Based Monitoring. *Pest Manag. Sci.* **2020**, *76*, 2994–3002. [CrossRef]
25. Peng, J.; Wang, D.; Liao, X.; Shao, Q.; Sun, Z.; Yue, H.; Ye, H. Wild Animal Survey Using UAS Imagery and Deep Learning: Modified Faster R-CNN for Kiang Detection in Tibetan Plateau. *ISPRS J. Photogramm. Remote Sens.* **2020**, *169*, 364–376. [CrossRef]
26. Wada, K. Labelme: Image Polygonal Annotation with Python. Available online: https://github.com/wkentaro/labelme (accessed on 19 January 2023).
27. Liu, S.; Qi, L.; Qin, H.; Shi, J.; Jia, J. Path Aggregation Network for Instance Segmentation. In Proceedings of the IEEE Conference on Computer Vision and Pattern Recognition (CVPR), Salt Lake City, UT, USA, 18–23 June 2018; pp. 8759–8768.
28. Wang, C.-Y.; Liao, H.-Y.M.; Wu, Y.-H.; Chen, P.-Y.; Hsieh, J.-W.; Yeh, I.-H. CSPNet: A New Backbone That Can Enhance Learning Capability of CNN. In Proceedings of the IEEE/CVF Conference on Computer Vision and Pattern Recognition (CVPR) Workshops, Seattle, WA, USA, 14–19 June 2020; pp. 390–391.
29. Bochkovskiy, A.; Wang, C.-Y.; Liao, H.-Y.M. YOLOv4: Optimal Speed and Accuracy of Object Detection. *arXiv* **2020**, arXiv:2004.10934.
30. Yu, F.; Koltun, V. Multi-Scale Context Aggregation by Dilated Convolutions. *arXiv* **2016**, arXiv:1511.07122.
31. Wang, P.; Chen, P.; Yuan, Y.; Liu, D.; Huang, Z.; Hou, X.; Cottrell, G. Understanding Convolution for Semantic Segmentation. In Proceedings of the 2018 IEEE Winter Conference on Applications of Computer Vision (WACV), Lake Tahoe, NY, USA, 12–15 March 2018; pp. 1451–1460.
32. He, K.; Zhang, X.; Ren, S.; Sun, J. Deep Residual Learning for Image Recognition. In Proceedings of the IEEE Conference on Computer Vision and Pattern Recognition (CVPR), Las Vegas, NV, USA, 27–30 June 2016; pp. 770–778.
33. Liu, Z.; Mao, H.; Wu, C.-Y.; Feichtenhofer, C.; Darrell, T.; Xie, S. A ConvNet for the 2020s. *arXiv* **2022**, arXiv:2201.03545.
34. Liu, Z.; Lin, Y.; Cao, Y.; Hu, H.; Wei, Y.; Zhang, Z.; Lin, S.; Guo, B. Swin Transformer: Hierarchical Vision Transformer Using Shifted Windows. *arXiv* **2021**, arXiv:2103.14030.
35. Liu, J.-J.; Hou, Q.; Cheng, M.-M.; Feng, J.; Jiang, J. A Simple Pooling-Based Design for Real-Time Salient Object Detection. *arXiv* **2019**, arXiv:1904.09569.
36. Lin, T.-Y.; Dollar, P.; Girshick, R.; He, K.; Hariharan, B.; Belongie, S. Feature Pyramid Networks for Object Detection. *arXiv* **2017**, arXiv:1612.03144.
37. Xu, M. A Review of Grassland Carrying Capacity: Perspective and Dilemma for Research in China on "Forage—Livestock Balance". *Acta Prataculturae Sin.* **2014**, *23*, 321–329. [CrossRef]

remote sensing

MDPI

Article

Complex-Valued U-Net with Capsule Embedded for Semantic Segmentation of PolSAR Image

Lingjuan Yu [1,*], Qiqi Shao [1], Yuting Guo [1], Xiaochun Xie [2], Miaomiao Liang [1] and Wen Hong [3]

1 School of Information Engineering, Jiangxi University of Science and Technology, Ganzhou 341000, China
2 School of Physics and Electronic Information, Gannan Normal University, Ganzhou 341000, China
3 Aerospace Information Research Institute, Chinese Academy of Sciences, Beijing 100194, China
* Correspondence: yulingjuan@jxust.edu.cn

Abstract: In recent years, semantic segmentation with pixel-level classification has become one of the types of research focus in the field of polarimetric synthetic aperture radar (PolSAR) image interpretation. Fully convolutional network (FCN) can achieve end-to-end semantic segmentation, which provides a basic framework for subsequent improved networks. As a classic FCN-based network, U-Net has been applied to semantic segmentation of remote sensing images. Although good segmentation results have been obtained, scalar neurons have made it difficult for the network to obtain multiple properties of entities in the image. The vector neurons used in the capsule network can effectively solve this problem. In this paper, we propose a complex-valued (CV) U-Net with a CV capsule network embedded for semantic segmentation of a PolSAR image. The structure of CV U-Net is lightweight to match the small PolSAR data, and the embedded CV capsule network is designed to extract more abundant features of the PolSAR image than the CV U-Net. Furthermore, CV dynamic routing is proposed to realize the connection between capsules in two adjacent layers. Experiments on two airborne datasets and one Gaofen-3 dataset show that the proposed network is capable of distinguishing different types of land covers with a similar scattering mechanism and extracting complex boundaries between two adjacent land covers. The network achieves better segmentation performance than other state-of-art networks, especially when the training set size is small.

Keywords: semantic segmentation; complex-valued U-Net; complex-valued capsule network; polarimetric synthetic aperture radar

Citation: Yu, L.; Shao, Q.; Guo, Y.; Xie, X.; Liang, M.; Hong, W. Complex-Valued U-Net with Capsule Embedded for Semantic Segmentation of PolSAR Image. *Remote Sens.* **2023**, *15*, 1371. https://doi.org/10.3390/rs15051371

Academic Editors: Fahimeh Farahnakian, Jukka Heikkonen and Pouya Jafarzadeh

Received: 5 January 2023
Revised: 24 February 2023
Accepted: 26 February 2023
Published: 28 February 2023

1. Introduction

Polarimetric synthetic aperture radar (PolSAR) can obtain rich information of the observed targets [1–3]. It has been widely used in agricultural monitoring, natural disaster assessment, biomass statistics, etc. The technologies of PolSAR image interpretation can help us understand and identify targets from an image; thus, it has always been a concern for researchers. However, since the formation mechanism of a PolSAR image is very complex, PolSAR image interpretation is still a challenging task. In recent years, driven by deep learning, two kinds of image interpretation technologies, namely image classification [4–8] and semantic segmentation [9–11], have made remarkable achievements.

PolSAR image classification based on deep learning has been deeply and extensively studied. According to the supervision mode, it mainly has three kinds of methods, namely, supervised, unsupervised, and semi-supervised [12]. For the supervised methods, convolutional neural networks (CNNs) [5,13] were the most widely used. Furthermore, a recursive neural network [14] was used. For the unsupervised methods, a deep belief network [15], an auto-encoder combined with Wishart distribution [16], and a task-oriented generative adversarial network [17] were studied. For the semi-supervised methods, a self-training model [18] and graph-based models [19,20] were proposed. In addition, some deep active learning and other new techniques were explored [21,22].

The research about CNN-based PolSAR image classification mainly focused on the real-valued (RV) structures and the inputs of networks at first. In terms of RV CNN structure, a four-layer CNN [23], a dual-branch deep CNN [24], a multi-channel fusion CNN [25], and a lightweight 3D CNN [26] were proposed. In terms of RV CNN inputs, some polarimetric features were selected to accelerate the convergence speed of CNN [13]. Then, complete features obtained by the polarimetric scattering coding were also used as the inputs [27]. Moreover, the amplitude and phase of the polarimetric coherence matrix were extracted as the inputs of a multi-task CNN [28]. Although both the methods of improving the network structure and extracting the complete input information effectively improved the classification performance, the problem of information loss caused by the RV structures and inputs still existed. Subsequently, some complex-valued (CV) CNNs were proposed to directly use the CV polarimetric coherent matrix as the input, aiming to avoid information loss. For example, a three-layer CV CNN [5], a CV 3D CNN [29], a CV 3D CNN combined with conditional random field [30], and a CV PolSAR-tailored differentiable architecture network [7] were widely studied. Furthermore, recurrent CV CNN combined with semi-supervised learning was also proposed for the classification with a small number of samples [31]. In the above classification processes, the training samples were obtained by using a sliding window, and each of them was an image patch with a small size. After the classification of each image patch, the obtained category was determined as the category of the center pixel in the patch. This pixel-by-pixel process had the problem of computational redundancy.

Semantic segmentation based on deep learning can achieve pixel-level classification in an end-to-end manner. Among various deep learning technologies, semantic segmentation networks based on s fully convolutional network (FCN) [32] have been rapidly developed. There are U-Net [33], SegNet [34], PSPNet [35], DeepLabv3+ [36], RefineNet [37], and so on. These networks have been used for semantic segmentation of optical [34–37], medical [33], and remote sensing images [38–41]. In the field of PolSAR image interpretation, semantic segmentation can effectively avoid the repeated computation when compared with image classification. Figure 1 shows the main differences between PolSAR image classification and semantic segmentation. For PolSAR image classification, the input is a small image patch that can be represented by a polarimetric coherent matrix, polarimetric decomposition results, and so on. The output is the category of the central pixel in the image patch. For semantic segmentation of a PolSAR image, the input is a large image block including multiple classes of targets, and its representation is the same as that of image classification. The output is the categories of all pixels in the image. Obviously, semantic segmentation can significantly reduce the computing time.

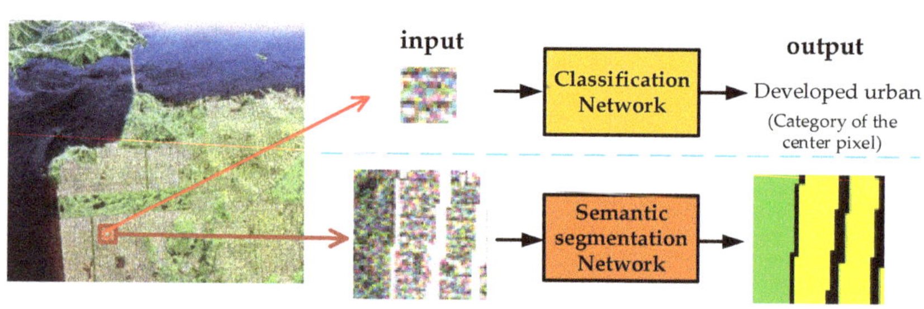

Figure 1. Differences between PolSAR image classification and semantic segmentation.

The research about semantic segmentation of PolSAR images started from FCN. Initially, FCN was only used to extract image features. Wang et al. used FCN to extract the spatial features and combined them with sparse and low-rank subspace features [42]. He et al. integrated FCN with a manifold graph embedding model to extract spatially polarized features [43]. After that, FCN, U-Net, SegNet, and DeepLabv3+ were applied in the semantic segmentation of PolSAR images in the real sense. Li et al. combined sliding window FCN with sparse coding to avoid the repeated calculations [44]. Mohammadimanesh et al. presented a new FCN with an encoder–decoder architecture for semantic segmentation of complex land cover ecosystems [45]. Pham et al. verified that SegNet could obtain promising segmentation results on very high-resolution PolSAR images [46]. Wu et al. used two structural modes (i.e., FCN and U-Net) and combined transfer learning strategies to perform semantic segmentation when the training set size was small [47]. Zhao et al. proposed a parallel dual-channel dilated FCN and used a semi-supervised fuzzy c-means clustering method to increase the number of the training samples [48]. Jing et al. proposed a polarimetric space reconstruction network, which included a scattering and polarimetric coherency coding module, a statistics enhancement module, and a dual self-attention convolutional network [49]. In order to fully utilize both amplitude and phase information of PolSAR data, Cao et al. presented a novel CV FCN [50], and Yu et al. proposed a lightweight CV Deeplabv3+ (L-CV-Deeplabv3+) [51].

The above CNN-based PolSAR image classification and FCN-based semantic segmentation greatly promoted the development of PolSAR image interpretation. However, there is still room for improvement due to the inherent defects of CNN. Taking face recognition as an example, even if the relative position of human eyes and mouth in an image is incorrect, CNN still recognizes it as a face because it is difficult for CNN to learn the relative positions between different entities in an image. A capsule network, which uses a vector neuron as the basic unit, can learn more information than CNN [52]. The amplitude of one activity vector neuron represents the probability of the existence of entities, and the orientation expresses instantiation parameters. Regarding the face recognition example, the capsule network can obtain the relative position between human eyes and the mouth, thus correctly recognizing the face. Furthermore, since vector neurons contain multiple properties of entities in the image, the capsule network requires fewer training samples than CNN during the training process. Up to now, the capsule network has also been widely used in optical [53,54], medical [55,56], remote sensing [57], and other image interpretation fields [58]. In the field of PolSAR image classification, Cheng et al. proposed a hierarchical capsule network to extract deep and shallow features. Experimental results on datasets from different platforms showed that the network had good generalization performance [59]. In the field of medical image semantic segmentation, Lalonde et al. proposed a segmented capsule network with U-shaped structure. Experimental results on pathological lungs showed that this network not only had better segmentation performance but also involved fewer parameters than U-Net [60].

In this paper, based on our previous work [61], we propose a CV U-Net with a capsule network embedded for semantic segmentation of PolSAR images. The reason for using U-Net as the backbone network is that its simple structure has good performance on small datasets when compared with other FCN-based segmentation networks. Considering that the PolSAR datasets used in this paper are small, the structure of U-Net is further lightweight to match these datasets. Moreover, U-Net is extended to the CV domain to directly use CV PolSAR data as the input. This CV network can mine the characteristics of the target from CV data, avoiding loss of information. In order to improve the feature extraction ability of the network, a CV capsule network is added behind the encoder of the CV U-Net. The CV capsule network mainly includes the CV primary capsules and the segmented capsules, which are different from those given in [52]. Inspired by [60], we also propose a locally constrained CV dynamic routing mechanism to realize the connection between capsules in two adjacent layers. The main contributions of this paper can be concluded as follows.

1. A lightweight CV U-Net is designed for semantic segmentation of PolSAR image. It uses a polarimetric coherence matrix as the input of the network, aiming to utilize both the amplitude and phase information of PolSAR data. The lightweight structure of the network can match the PolSAR datasets with a small number of training samples.
2. A CV capsule network is embedded between the encoder and decoder of CV U-Net to extract abundant features of PolSAR image. To make the CV capsule network suitable

for semantic segmentation, the segmented capsule is adopted to replace the digital capsule used in the image classification.

3. The locally constrained CV dynamic routing is proposed for the connection between capsules in two adjacent layers. The locally constrained characteristic helps to extend the dynamic routing to the connection of capsules with large sizes, and the routing consistency of the real part and imaginary part of the CV capsules improves the correctness of the extracted entity properties.
4. Experiments on two airborne datasets and one Gaofen-3 PolSAR dataset verify that the proposed network can achieve better segmentation performance than other RV and CV networks, especially when the training set size is small.

The rest of this paper is organized as follows. The theoretical background about PolSAR data, U-Net and the capsule network are introduced in Section 2. Section 3 proposes a CV U-Net with capsule embedded and illustrates the principle of locally constrained CV dynamic routing. Experimental results and analysis are shown in Section 4. Section 5 discusses the improvement of segmentation performance brought by the CV capsule network. Finally, the conclusion is given in Section 6.

2. Related Work

2.1. PolSAR Data

Because PolSAR systems can transmit and receive different polarimetric electromagnetic waves, they can obtain rich scattering information about the observed targets. H and V are denoted as the horizontal and vertical polarization modes, respectively. Then, a 2×2 complex polarimetric scattering matrix can be written by,

$$[S] = \begin{bmatrix} S_{HH} & S_{HV} \\ S_{VH} & S_{VV} \end{bmatrix} \tag{1}$$

where the subscripts x and y of S_{xy} ($x = H, V; y = H, V$) represent the polarization modes of the received and transmitted electromagnetic wave, respectively.

From the reciprocity theorem, S_{HV} is approximately equal to S_{VH} for monostatic SAR. Then, the scattering vector under Pauli basis can be expressed by,

$$\mathbf{k} = \frac{1}{\sqrt{2}}[S_{HH} + S_{VV}, S_{HH} - S_{VV}, 2S_{HV}]^T \tag{2}$$

Furthermore, the polarimetric coherence matrix $\mathbf{T}$ is calculated by,

$$\mathbf{T} = \left\langle \mathbf{k} \cdot \mathbf{k}^H \right\rangle = \begin{bmatrix} T_{11} & T_{12} & T_{13} \\ T_{21} & T_{22} & T_{23} \\ T_{31} & T_{32} & T_{33} \end{bmatrix} \tag{3}$$

2.2. U-Net

U-Net was first applied to semantic segmentation of medical images. The structure of U-Net proposed in [33] is shown in Figure 2. The upper part of the figure is the encoder, while the lower part is the decoder. They are almost symmetrical. The purpose of the encoder is to extract deep features, while the purpose of the decoder is to obtain target areas in the image and then determine their categories.

The encoder is also named the contracting path, because its feature sizes are gradually decreasing. It mainly includes four blocks, and each block contains two convolutional layers and one max pooling layer. From the second block, the first convolutional layer doubles the feature channels, and the maximum pooling layer halves the feature size. The decoder is also named the expanding path, because its feature sizes are gradually increasing. Corresponding to the encoder, it also includes four blocks. Each block contains one upsampling and convolutional layer and two convolutional layers. The upsampling and convolutional layer doubles the feature sizes and halves the feature channels. After the upsampling and convolution operation, the feature maps are concatenated with those copied and clipped from the corresponding encoder layer, aiming to restore good details of targets in the image. Then, the following convolutional layer halves the feature channels.

For all the convolutional operations in the network, the sizes of convolutional kernels are 3×3, and the convolutional stride is 1. After each convolutional operation, the ReLU activation function is used. The size of the pooling operation is 2×2, and the pooling stride is 2. For all the upsampling and convolution operations, the upsampling ratio is 2×2, and the size of the convolutional kernels is 2×2. After the last convolutional layer in the expansion path, there is one 1×1 convolution to obtain the categories of all targets in the image.

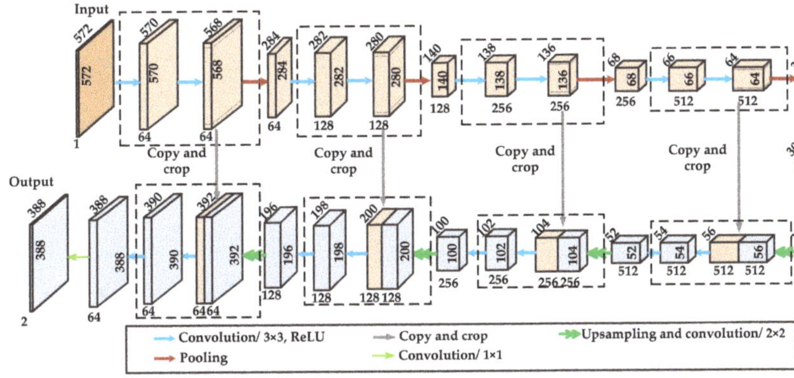

Figure 2. Architecture of U-Net.

2.3. Capsule Network

A capsule network can overcome the shortcomings of CNN and obtain multiple properties of entities in the image. The architecture of the capsule network proposed in [52] is shown in Figure 3. It mainly includes two convolutional layers and one fully connected layer. The first convolutional layer is used in extracting features. The convolutional kernel has a size of 9×9, and the convolutional stride is 1. After the convolution operation, a ReLU activation function is used. The second convolutional layer is used in obtaining the primary capsules. The convolutional kernel also has a size of 9×9, and the convolutional stride is 2. The output of this layer is $32 \times 6 \times 6$ primary capsules, and each primary capsule is a vector with size of 1×8. The final layer is used in obtaining the digital capsules, which connects the primary capsules through dynamic routing. The total number of digital capsules is 10, and each digital capsule is a vector with a size of 1×16. Finally, the category of sample is the number of the digital capsule whose value has the maximum under L2-norm.

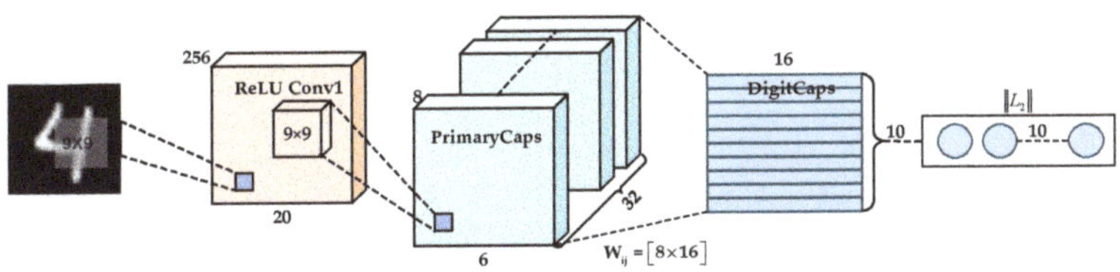

Figure 3. Architecture of a capsule network.

3. Methodology

A CV U-Net with a capsule embedded is proposed for semantic segmentation of the PolSAR image in this section. The architecture of this network is shown in Figure 4. It mainly includes the CV encoder, CV decoder and CV capsule network, where the CV capsule network is embedded between the encoder and the decoder. Furthermore, the locally constrained CV dynamic routing is proposed for the connection between capsules in two adjacent layers of the CV capsule network. The detailed structure of the three parts and the principle of the locally constrained CV dynamic routing are introduced as follows.

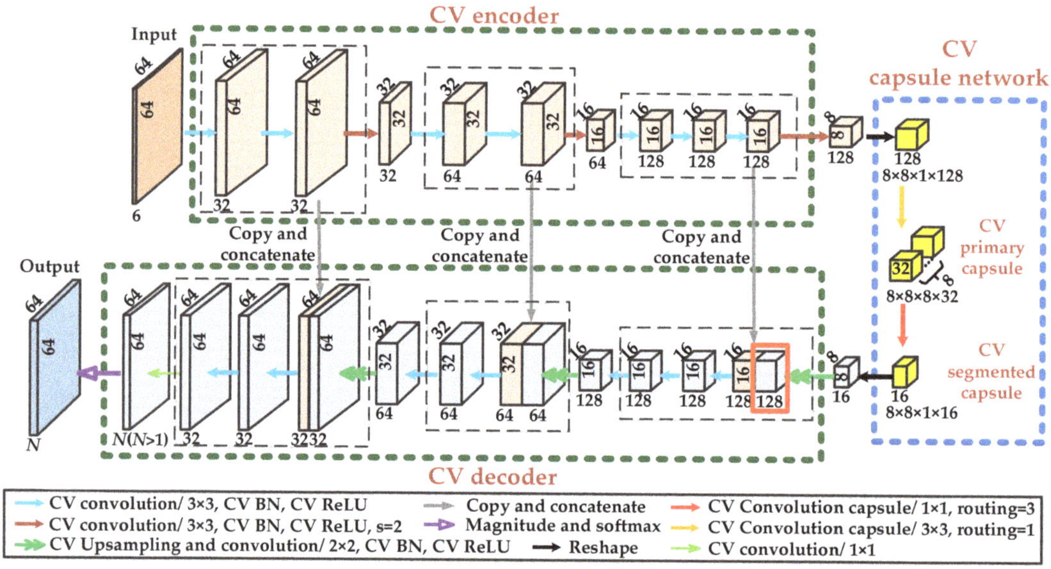

Figure 4. Architecture of CV U-Net with the capsule network embedded.

3.1. CV Encoder

The CV encoder shown in Figure 4 (the green dashed box at the top) includes three CV convolutional blocks. For the first block, there are three CV convolutional layers in total. Each of the first two CV convolutional layers has 32, 3 × 3 convolution kernels, and the convolutional stride is 1. The output sizes of these two convolutional layers are the same as the input. The third CV convolutional layer has 32, 3 × 3 convolution kernels, and the convolutional stride is 2. The output size of this layer is half of the input. For the second block, the structure and convolution parameters are the same as those of the first block except that the number of convolution channels of each layer is twice that of the corresponding layer in the first block. For the third block, there are a total of four CV convolutional layers. Compared with the second block, it adds a convolution layer and doubles the convolution channels for each corresponding layer. Both a CV ReLU activation function and a CV batch normalization operation are located after each convolution operation. Formulas about the CV convolution operation and the CV ReLU activation function as well as the CV batch normalization method are provided in [62].

Regarding the structure of the CV encoder, it was designed based on the RV encoder shown in Figure 2. However, there are several differences between the CV and RV encoders. First, the number of convolutional blocks and convolution channels of the CV encoder are less than those of the RV encoder, so as to make the network fit for the small PolSAR datasets. Second, all the network parameters and convolution operations involved in the CV encoder are extended to the CV domain, which aims to extract more abundant information from the CV input. Third, the CV convolution with a stride of 2 in the CV encoder replaces the pooling operation in the RV encoder, which can reduce the loss of information. Finally, the mode of convolution with a stride of 1 in the CV encoder is different from that in the RV encoder. The output size obtained in the CV encoder is consistent with the input.

3.2. CV Decoder

The CV decoder shown in Figure 4 (the green dashed box at the bottom) includes three CV deconvolutional blocks, which are symmetrical to the CV encoder. For the first deconvolutional block, there are one CV upsampling and convolutional layer and three CV convolutional layers. The CV upsampling and convolutional layer has an upsampling ratio of 2×2, and the output size is twice the input. Then, a copy and concatenation operation is performed to combine low-level feature maps from the encoder with deep-level feature maps from the decoder. Next, three convolutional layers all have 128, 3×3 convolution kernels with a stride of 1. The output sizes of these convolutional layers are the same as the input. For the second deconvolutional block, it reduces one CV convolutional layer and halves the convolution channels of each layer when compared with the previous block. For the third CV deconvolutional block, the structure and convolution parameters are the same as those of the second block except for the convolution channels. There are also a CV ReLU activation and a CV batch normalization operation after each of the aforementioned convolution operation. After these three deconvolutional blocks, there is one CV convolution operation with a size of 1×1, which aims to make the convolution channels consistent with the total number of categories in the image. Subsequently, magnitude operation is implemented to covert CV features into RV. Finally, the softmax operation is used to complete the classification. Formulas about the upsampling operation, the concatenation, and the magnitude operation are provided in [51].

Comparing the above CV decoder with the RV decoder shown in Figure 2, there are also four differences. The first two differences are the same as those analyzed in the CV encoder. The third difference is that the magnitude operation is added in the CV decoder to directly use the RV softmax function for the classification. The last one is that the segmentation task of two categories in the RV decoder is extended to N ($N > 1$) categories in the CV decoder.

3.3. CV Capsule Network

The CV capsule network shown in Figure 4 (the blue dashed box) is embedded between the encoder and the decoder. The design of this structure is inspired by [60]. It mainly includes two CV convolution capsule layers and two reshape operations. At first, 128 feature maps with size of 8×8 are reshaped into $8 \times 8 \times 1$, which can also be seen as 128×8 capsules with size of 1×8. Then, the CV primary capsules are obtained by the first convolution capsule layer. There are $32 \times 8 \times 8$ primary capsules, and each primary capsule is a vector with size of 1×8. The connection between the previous feature maps in a vector form and the CV primary capsules adopts the locally constrained CV dynamic routing way introduced in Section 3.4. Next, the CV segmented capsules are obtained by the second convolution capsule layer. There are 16×8 segmented capsules with size of 1×8. The locally constrained CV dynamic routing is also used to connect the CV primary capsules and the segmented capsules. Finally, the segmented capsule is reshaped into 16 feature maps with size of 8×8, and it is used as the input of the CV decoder.

From the above structure and operations of the CV capsule network, it is obviously different from the original capsule network shown in Figure 3. In order to make the CV capsule network suitable for the semantic segmentation, the segmented capsule replaces the digital capsule used in the image classification. Furthermore, the convolution capsule operation instead of the fully connected operation is used in obtaining the segmented capsule. Because the convolution operation can share weight parameters, it is beneficial to reduce the amount of calculation. Furthermore, the locally constrained CV dynamic routing is used for the connection between capsules in two adjacent layers. It is helpful to extend the dynamic routing to capsules with large sizes. More importantly, the routing consistency of the real part and imaginary part of the CV capsules can improve the correctness of the extracted entity properties from the PolSAR image.

3.4. Locally Constrained CV Dynamic Routing

For any layer l ($l \geq 1$) and its adjacent layer $l+1$ in the CV capsule network, suppose there are N_c types of CV child capsules at layer l and N_p types of CV parent capsules at layer $l+1$. Denote the CV child capsule types as $T^l = \left\{t_1^l, t_2^l, \ldots, t_i^l \ldots, t_{N_c}^l\right\}$ and the CV parent capsule types as $T^{l+1} = \left\{t_1^{l+1}, t_2^{l+1}, \ldots, t_j^{l+1} \ldots, t_{N_p}^{l+1}\right\}$. For each $t_i^l \in T^l$, there are $h^l \times w^l$ child capsules with dimension of z^l. For each $t_j^{l+1} \in T^{l+1}$, there are $h^{l+1} \times w^{l+1}$ parent capsules with dimension of z^{l+1}. All the parent capsules can also be expressed by $P = \left\{\boldsymbol{p}_{t_j^{l+1}11}, \ldots, \boldsymbol{p}_{t_j^{l+1}1w^{l+1}}, \ldots, \boldsymbol{p}_{t_j^{l+1}h^{l+1}1}, \ldots, \boldsymbol{p}_{t_j^{l+1}h^{l+1}w^{l+1}}\right\}$. Take a CV parent capsule $\boldsymbol{p}_{t_j^{l+1}xy} \in P$ $(1 \leq x \leq h^{l+1}; 1 \leq y \leq w^{l+1})$, for example, the principle of locally constrained CV dynamic routing is shown in Figure 5, and the detailed process of child capsules routing to parent capsules is analyzed as follows.

At first, in a user-defined kernel (the red solid box), the CV convolution operation implemented by matrix multiplication is used in obtaining the prediction vectors. For a type of t_i^l child capsules, a CV kernel with size of $k_h \times k_w \times z^l$ is defined as $\boldsymbol{u}_{t_i^l x_0 y_0}$, where (x_0, y_0) is the center of the kernel. The CV matrix with size of $k_h \times k_w \times z^l \times N_p \times z^{l+1}$ is denoted as $\boldsymbol{M}_{t_i^l}$, which is shared over all the kernels in the same type of CV child capsules. Therefore, the predicted CV vector $\hat{\boldsymbol{u}}_{t_j^{l+1}xy|t_i^l}$ can be calculated by,

$$\hat{\boldsymbol{u}}_{t_j^{l+1}xy|t_i^l} = \boldsymbol{M}_{t_i^l}\boldsymbol{u}_{t_i^l x_0 y_0} \tag{4}$$

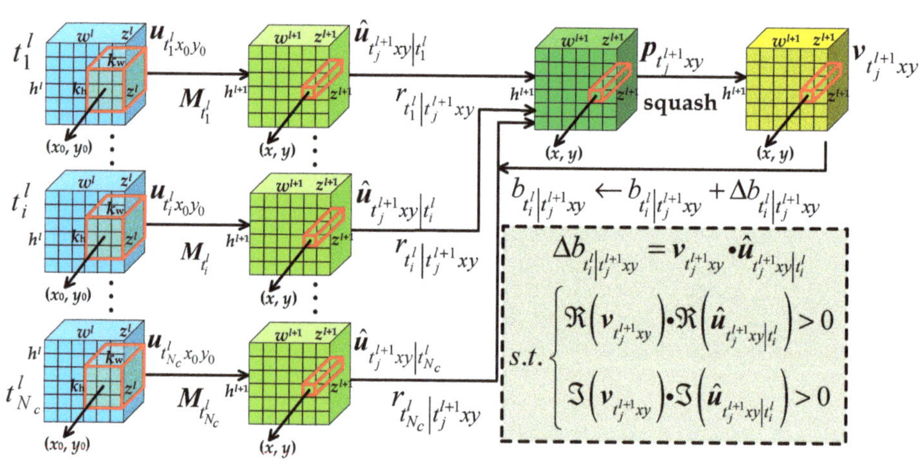

Figure 5. Locally constrained CV dynamic routing.

Then, the weighted sum over the predicted vectors from all the types of CV child capsules is implemented to obtain the CV parent capsule $\boldsymbol{p}_{t_j^{l+1}xy}$, which can be expressed by,

$$\boldsymbol{p}_{t_j^{l+1}xy} = \sum_i r_{t_i^l|t_j^{l+1}xy}\hat{\boldsymbol{u}}_{t_j^{l+1}xy|t_i^l} \tag{5}$$

where $r_{t_i^l|t_j^{l+1}xy}$ is the routing coefficient. It can be calculated by,

$$r_{t_i^l|t_j^{l+1}xy} = \frac{\exp\left(b_{t_i^l|t_j^{l+1}xy}\right)}{\sum\limits_k \exp\left(b_{t_i^l|t_j^{l+1}k}\right)} \tag{6}$$

where $b_{t_i^l|t_j^{l+1}xy}$ is the log prior probability.

Next, the squashed CV parent capsule $\boldsymbol{v}_{t_j^{l+1}xy}$ is obtained by,

$$\boldsymbol{v}_{t_j^{l+1}xy} = \frac{\left\|\boldsymbol{p}_{t_j^{l+1}xy}\right\|^2}{1+\left\|\boldsymbol{p}_{t_j^{l+1}xy}\right\|^2}\frac{\boldsymbol{p}_{t_j^{l+1}xy}}{\left\|\boldsymbol{p}_{t_j^{l+1}xy}\right\|} \tag{7}$$

where

$$\left\|\boldsymbol{p}_{t_j^{l+1}xy}\right\| = \sqrt{\left\|\Re\left(\boldsymbol{p}_{t_j^{l+1}xy}\right)\right\|^2+\left\|\Im\left(\boldsymbol{p}_{t_j^{l+1}xy}\right)\right\|^2}. \tag{8}$$

Finally, the coefficient $b_{t_i^l|t_j^{l+1}xy}$ is updated by,

$$b_{t_i^l|t_j^{l+1}xy} \leftarrow b_{t_i^l|t_j^{l+1}xy} + \Delta b_{t_i^l|t_j^{l+1}xy}. \tag{9}$$

where $\Delta b_{t_i^l|t_j^{l+1}xy}$ is obtained by,

$$\Delta b_{t_i^l|t_j^{l+1}xy} = \begin{cases} \boldsymbol{v}_{t_j^{l+1}xy}\cdot\hat{\boldsymbol{u}}_{t_j^{l+1}xy|t_i^l} & \begin{array}{c}\Re\left(\boldsymbol{v}_{t_j^{l+1}xy}\right)\cdot\Re\left(\hat{\boldsymbol{u}}_{t_j^{l+1}xy|t_i^l}\right) > 0\\ and\ \Im\left(\boldsymbol{v}_{t_j^{l+1}xy}\right)\cdot\Im\left(\hat{\boldsymbol{u}}_{t_j^{l+1}xy|t_i^l}\right) > 0\end{array} \\ 0 & otherwise \end{cases} \tag{10}$$

Equation (10) means that only when the product of the real parts of $\boldsymbol{v}_{t_j^{l+1}xy}$ and $\hat{\boldsymbol{u}}_{t_j^{l+1}xy|t_i^l}$ is greater than 0 and the product of the imaginary parts of these two CV vectors is also greater than 0, $\Delta b_{t_i^l|t_j^{l+1}xy}$ is equal to the dot product of $\boldsymbol{v}_{t_j^{l+1}xy}$ and $\hat{\boldsymbol{u}}_{t_j^{l+1}xy|t_i^l}$. Otherwise, $\Delta b_{t_i^l|t_j^{l+1}xy}$ is equal to 0. This update condition is stricter than that of RV dynamic routing, because the coefficient is updated only when the dynamic routing of the real part is consistent with that of the imaginary part.

The above process of locally constrained CV dynamic routing can also be summarized in Algorithm 1.

Algorithm 1 Locally Constrained CV Dynamic Routing

Procedure Routing ($\hat{\boldsymbol{u}}_{t_j^{l+1}xy|t_i^l}, d, l, k_h, k_w$)
  for all CV child capsule types t_i^l within a $k_h \times k_w$ kernel in layer l and a CV parent capsule $t_j^{l+1}xy$ in layer l + 1: $b_{t_i^l|t_j^{l+1}xy} \leftarrow 0$.
  while iteration < d **do**
    for all CV child capsule types t_i^l in layer l:
      $r_{t_i^l|t_j^{l+1}xy} \leftarrow softmax\left(b_{t_i^l|t_j^{l+1}xy}\right)$ ▷ softmax computes Equation (6)
    for the CV capsule $t_j^{l+1}xy$ in layer l + 1:
      $\boldsymbol{p}_{t_j^{l+1}xy} \leftarrow \sum_j r_{t_i^l|t_j^{l+1}xy}\hat{\boldsymbol{u}}_{t_j^{l+1}xy|t_i^l}$
    for the CV capsule $t_j^{l+1}xy$ in layer l + 1:
      $\boldsymbol{v}_{t_j^{l+1}xy} \leftarrow squash\left(\boldsymbol{p}_{t_j^{l+1}xy}\right)$ ▷ squash computes Equation (7)
    for all CV capsule types t_i^l and the CV capsule $t_j^{l+1}xy$:
      $b_{t_i^l|t_j^{l+1}xy} \leftarrow b_{t_i^l|t_j^{l+1}xy} + \Delta b_{t_i^l|t_j^{l+1}xy}$ ▷$\Delta b_{t_i^l|t_j^{l+1}xy}$ computes Equation (9)
  end while
return $\boldsymbol{v}_{t_j^{l+1}xy}$

4. Experiments and Analysis

Experimental datasets are briefly described in Section 4.1. Then, both the data preprocessing and the experimental setup are introduced in Section 4.2. Finally, the detailed experimental results and analysis are given in Section 4.3.

4.1. Experimental Datasets

Experiments were implemented on three fully polarimetric datasets. Two are collected by AIRSAR airborne platform, and one is collected by Gaofen-3. The detailed descriptions of datasets are as follows.

(1) Flevoland dataset: It was collected in the Flevoland area of the Netherlands in 1989. The Pauli RGB image of this L-band dataset is shown in Figure 6a, and its size is 1024×750. In the following experiments, 15 types of land covers are considered and others are regarded as backgrounds.
(2) San Francisco dataset: It was collected in the area of San Francisco Bay in 1988. The Pauli RGB image of this L-band dataset is shown in Figure 6b, and its size is 1024×900. In the following experiments, five types of land covers are considered and others are regarded as backgrounds.
(3) Hulunbuir dataset: It was collected in the Hulunbuir area of China. The Pauli RGB image of this C-band dataset is shown in Figure 6c, and its size is 1265×1147. In the following experiments, eight types of land covers are considered and others are regarded as backgrounds.

(a) (b) (c)

Figure 6. Pauli RGB image. (**a**) Flevoland dataset; (**b**) San Francisco dataset; (**c**) Hulunbuir dataset.

4.2. Data Preprocessing and Experimental Setup

In the following experiments, four networks are compared with the proposed network. They are U-Net, DeepLabv3+, CV U-Net, and L-CV-DeepLabv3+ [51]. The first two are RV networks, while the last two are CV networks. With the goal of achieving good segmentation results, all four networks are lightweight to match the above three PolSAR datasets. In Figure 4, we give the structure and parameters of the proposed network. For the sake of fairness, we design the structure of CV U-Net by removing the CV capsule network from the proposed network. In addition, we regard U-Net as the RV version of CV U-Net. As for L-CV-DeepLabv3+, its structure and parameters are given in [51]. Moreover, DeepLabv3+ is the RV version of L-CV-DeepLabv3+.

The inputs of the above networks are also RV and CV, correspondingly. For three CV networks, because the polarimetric coherence matrix $\mathbf{T}$ in Equation (3) is a Hermitian symmetric matrix, the upper triangular 6-channel CV data are directly used as the inputs. They can be expressed by $\{T_{11}, T_{22}, T_{33}, T_{12}, T_{13}, T_{23}\}$. For two RV networks, 9-channel RV data are used as the inputs to make them close to the CV input. They are expressed by $\{T_{11}, T_{22}, T_{33}, real(T_{12}), imag(T_{12}), real(T_{13}), imag(T_{13}), real(T_{23}), imag(T_{23})\}$.

The data preprocessing about the polarimetric coherence matrix is as follows. At first, each dataset is expanded in a mirror mode. The sizes of three expanded datasets are 1024×832, 1024×960 and 1280×1152, respectively. Then, we cut each expanded dataset into blocks without overlapping by using a sliding window with size of 64×64. Next, we build the training set by selecting 40% of the blocks and build the test set by using the remaining 60% of the blocks for each dataset. Finally, each train set is expanded by the scaling and rotating operations. The sizes of training sets before and after expansion are shown in Table 1, and the sizes of test sets are also given there.

Table 1. Sizes of training and test sets.

Dataset	Training		Test
	Before Expansion	**After Expansion**	
Flevoland dataset	51	909	75
San Francisco dataset	96	1710	144
Hulunbuir dataset	56	1003	83

All the experiments are implemented on the Ubuntu operating system. We use Anaconda 3 as the software environment and Python 3.6 as the programming language. For the hardware environment, the CPU mode is Intel core i7-10700K, the memory size is 16G, the GPU model is Nvidia GeForce RTX 2080, and the video memory is 8G. In the training process, we use Adam as the optimized algorithm. The learning rate is 1×10^{-4}, and the batch size is 16. In terms of performance indicators, we use intersection over union (IOU) to evaluate the segmentation effect of a single category, and we use mean intersection over union (MIOU), overall accuracy (OA), and mean pixel accuracy (MPA) to evaluate the overall segmentation effect of all categories. The calculation formulas of these indicators are provided in [51].

4.3. Experimental Results and Analysis

4.3.1. Experiments on Flevoland Dataset

The semantic segmentation results of the Flevoland dataset obtained by five networks are shown in Figure 7a–e. Figure 7f gives the ground truth. Comparing the segmentation results of CV networks with those of RV networks, we can find that the results shown in Figure 7c–e are better than those shown in Figure 7a,b, especially in the black box area marked with the number 1. In this area, rapeseed is seriously misclassified into pea, wheat1, wheat2 and wheat3 by two RV networks, because the scattering mechanisms of these land covers are close. For bare soil in this area, since its scattering mechanism is close to that of water, it is misclassified into water by two RV networks. However, this situation is improved by three CV networks. Therefore, the three CV networks have a stronger ability to distinguish land covers with similar scattering mechanisms than two RV networks, because CV networks can extract abundant information from CV data.

For the black box area marked with the number 2 shown in Figure 7a–e, we enlarge them to obtain Figure 8a–e, respectively. The enlarged ground truth and Pauli RGB image of this area are given in Figure 8f,g, respectively. From Figure 8a to Figure 8e, wheat1 has different degrees of incorrect segmentation, because its scattering mechanism is very close to that of wheat2 and wheat3. However, the error area for wheat1 in Figure 8e is much smaller than that in the other figures, which means that the proposed network has stronger discrimination ability on land covers with similar scattering mechanisms than other networks. Similarly, the rapeseed in this area is wrongly classified by all the networks except by the proposed network.

To quantitatively analyze the segmentation performance, we calculated four indicators, and they are shown in Table 2. It is easy to find that three CV networks achieve higher MIOUs, OAs and MPAs than two RV networks. In addition, the proposed network achieves the highest MIOU, OA and MPA among all the networks. Comparing the proposed network with CV U-Net, we can find that the embedded capsule network increases MIOU, OA and MPA by 3.37%, 0.63% and 1.6%, respectively, and increases IOUs of wheat1, wheat2, and wheat3 by 7.83%, 16.06%, and 2.6%, respectively. Comparing CV U-Net with L-CV-DeepLabv3+, their MIOUs, OAs and MPAs are very close to each other, which means that they have similar feature extraction ability on this dataset.

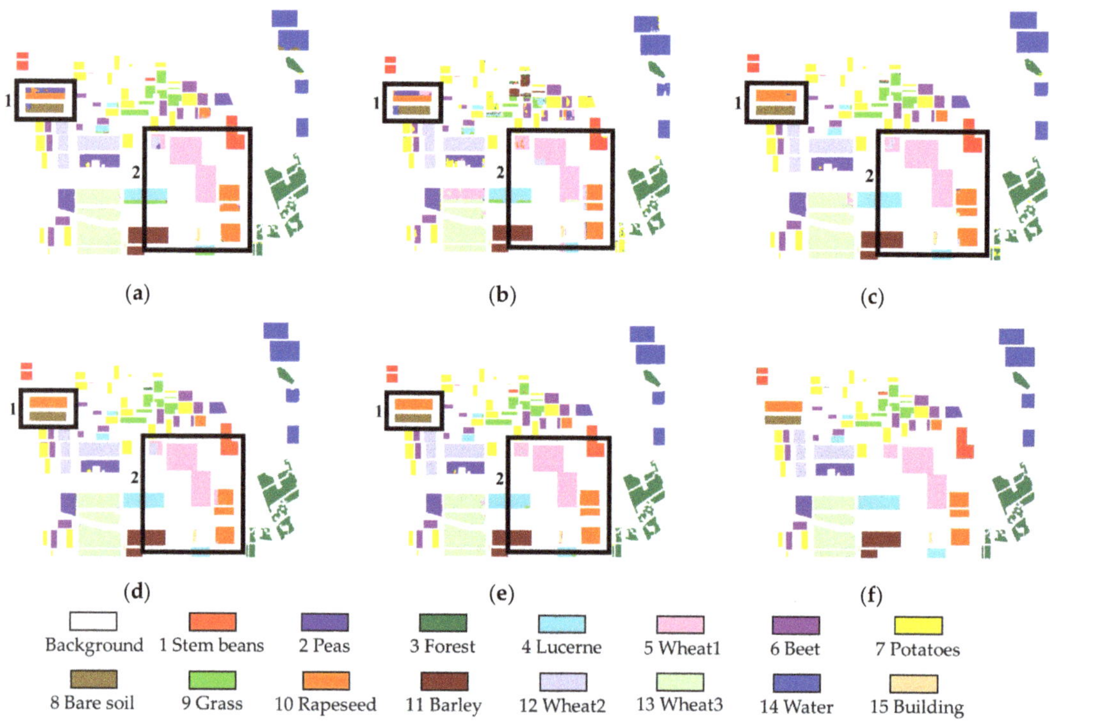

Figure 7. Segmentation results of Flevoland dataset. (**a**) U-Net; (**b**) DeepLabv3+; (**c**) CV U-Net; (**d**) L-CV-DeepLabv3+; (**e**) proposed network; (**f**) ground truth.

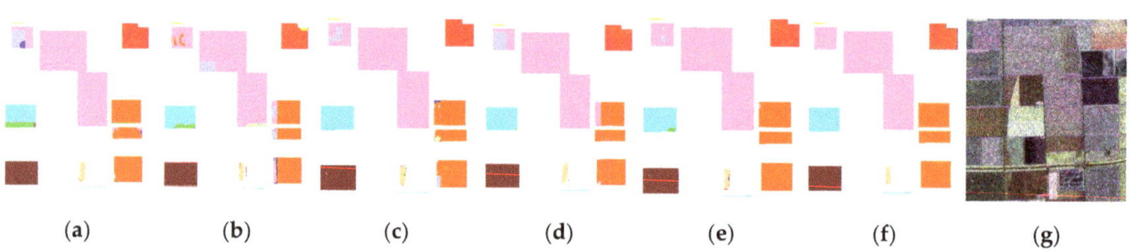

Figure 8. Enlarged views of black box area 2 in Figure 7. (**a**) U-Net; (**b**) DeepLabv3+; (**c**) CV U-Net; (**d**) L-CV-DeepLabv3+; (**e**) proposed network; (**f**) ground truth; (**g**) Pauli RGB image.

Table 2. Four indicators of Flevoland dataset.

Class	U-Net	DeepLabv3+	CV U-Net	L-CV-DeepLabv3+	Proposed
1	97.52	88.04	92.33	89.07	**98.58**
2	74.24	30.40	92.04	90.82	**92.97**
3	98.22	65.00	88.83	93.47	**99.47**
4	67.43	53.30	**97.99**	95.84	93.26
5	85.47	45.20	88.05	82.18	**95.88**
6	96.15	71.85	97.79	97.07	**98.14**
7	96.36	51.07	88.96	96.75	**98.51**
8	62.51	62.68	96.98	93.84	**99.70**
9	64.86	11.65	**97.05**	94.48	89.10
10	75.15	30.33	91.52	90.03	**98.02**
11	99.96	71.09	99.25	99.18	**99.42**
12	79.20	67.85	82.58	78.08	**98.64**
13	99.35	72.97	96.31	**98.98**	98.91
14	86.80	82.90	**97.76**	94.78	96.08
15	**92.70**	62.63	83.97	80.43	88.63
MIOU	85.99	60.32	93.20	92.13	**96.57**
OA	97.65	90.39	98.80	98.52	**99.43**
MPA	93.37	73.14	96.62	96.11	**98.22**

4.3.2. Experiments on San Francisco Dataset

The experimental results of the San Francisco dataset obtained by five networks are shown in Figure 9a–e. Figure 9f shows the ground truth. In the white box area marked with the number 1, the part of low density is misclassified into vegetation by two RV networks because of their similar scattering mechanism. However, the segmentation results obtained by three CV networks in this area are significantly better than those obtained by two RV networks. A similar situation occurs in the white box area marked with the number 2. The segmentation results of the developed urban area obtained by two RV networks are worse than those obtained by three CV networks.

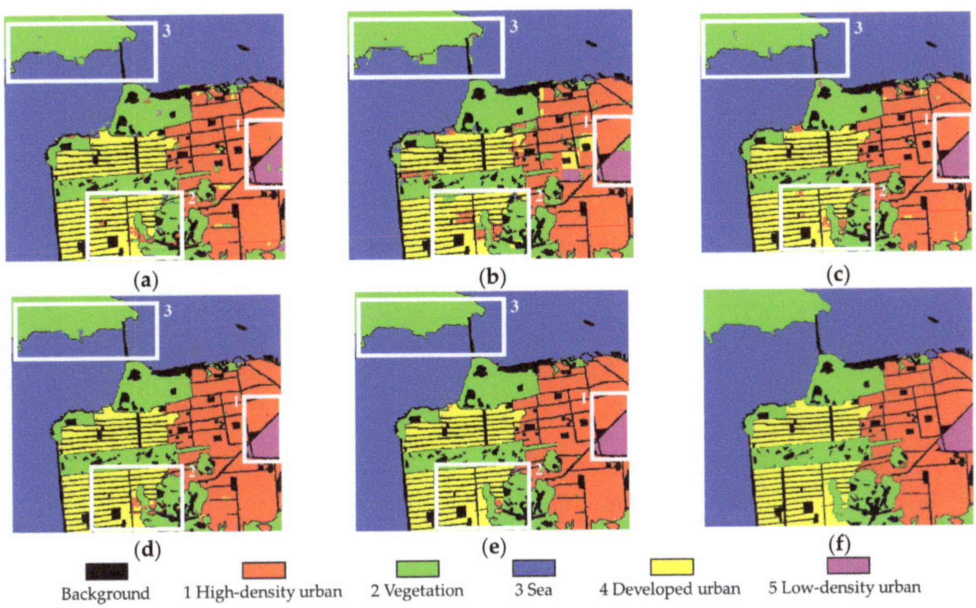

Figure 9. Segmentation results of San Francisco dataset. (**a**) U-Net; (**b**) DeepLabv3+; (**c**) CV U-Net; (**d**) L-CV-DeepLabv3+; (**e**) proposed network; (**f**) ground truth.

For the white box area marked with the number 3 shown in Figure 9a–e, they are magnified as Figure 10a–e, respectively. The enlarged ground truth of this area is given in Figure 10f. There are serious errors at the boundaries between vegetation and sea in Figure 10a,b. However, these boundaries are improved in Figure 10c,d. The boundaries shown in Figure 10e are very close to the ground truth. Thus, the proposed network has significant advantage in extracting the boundary information. The reason is that the embedded CV capsule network has a strong ability to extract features of land covers.

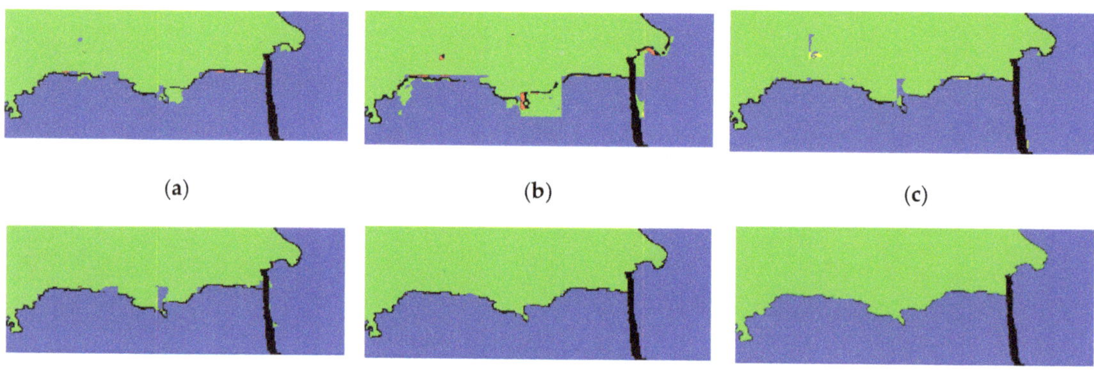

(a) (b) (c) (d) (e) (f)

Figure 10. Enlarged views of white box area 3 in Figure 9. (**a**) U-Net; (**b**) DeepLabv3+; (**c**) CV U-Net; (**d**) L-CV-DeepLabv3+; (**e**) proposed network; (**f**) ground truth.

Four indicators are also calculated to quantitatively analyze the segmentation performance, and they are shown in Table 3. It is obvious that three CV networks achieve higher IOUs, MIOUs, OAs and MPAs than two RV networks. Furthermore, the proposed network achieves the highest MIOU, OA and MPA among all the networks. Comparing the proposed network with CV U-Net, we can find that the embedded capsule network increases MIOU, OA and MPA by 3.13%, 1.25% and 1.91%, respectively, and increases IOUs of high-density urban and developed urban areas by 4.61% and 5.72%, respectively. Different from the previous dataset, L-CV-DeepLabv3+ achieves higher MIOU, OA and MPA than CV U-Net. This is because there are more boundaries between different categories in this dataset than in the previous dataset, and the boundary extraction ability of L-CV-DeepLabv3+ is better than that of CV U-Net.

Table 3. Four indicators of San Francisco dataset.

Class	U-Net	DeepLabv3+	CV U-Net	L-CV-DeepLabv3+	Proposed
1	88.95	72.24	92.15	94.13	**96.76**
2	87.55	77.13	94.21	94.23	**97.53**
3	98.18	96.14	99.08	99.05	**99.83**
4	88.75	68.63	90.34	93.43	**96.06**
5	70.35	45.25	87.06	**96.08**	91.46
MIOU	88.96	74.72	93.80	95.86	**96.93**
OA	96.57	90.89	97.93	98.24	**99.18**
MPA	93.60	85.17	96.83	97.80	**98.74**

4.3.3. Experiments on Hulunbuir Dataset

The experimental results of the Hulunbuir dataset obtained by five networks are shown in Figure 11a–e. Figure 11f shows the ground truth. In this dataset, the area ratio of land cover to the total area is small, and the intervals between different land covers are wide.

This enables all the networks except for DeepLabv3+ to achieve good segmentation results. In the white box area marked with the number 1, there is only wetland. We enlarged this area in Figure 11a–e to obtain Figure 12a–e, respectively. The enlarged ground truth of this area is shown in Figure 12f. It is easy to find that there are some errors in the boundaries of segmentation results obtained by two RV networks. However, three CV networks greatly improve this situation. In the white box area marked with the number 2, there are water and grasses. We enlarged this area in Figure 11a–e to obtain Figure 13a–e, respectively. The enlarged ground truth and Pauli RGB image of this area are shown in Figure 13f–g, respectively. We can see that grasses are seriously misclassified as water by DeepLabv3+ because of their similar scattering mechanisms. Furthermore, there are some errors at the boundaries obtained by CV U-Net and L-CV-DeepLabv3+. However, the proposed network achieves boundaries that are almost close to the ground truth.

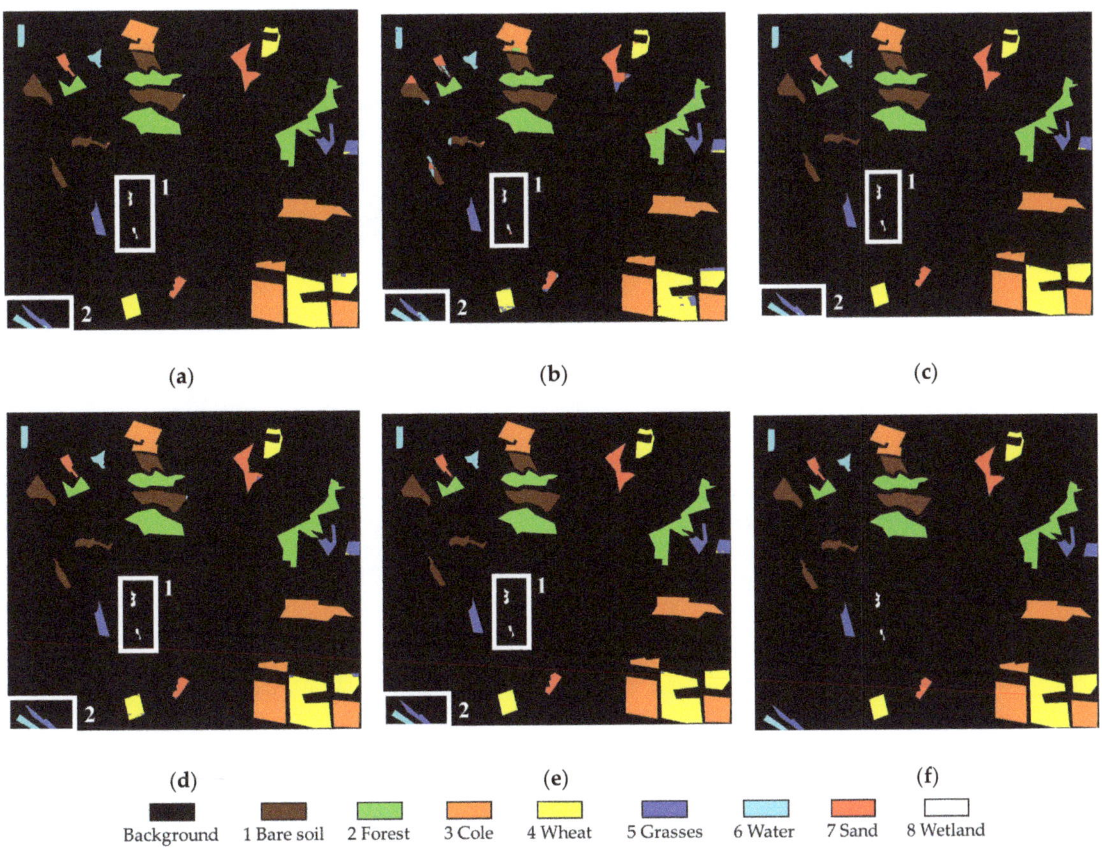

Figure 11. Segmentation results of Hulunbuir dataset. (**a**) U-Net; (**b**) DeepLabv3+; (**c**) CV U-Net; (**d**) L-CV-DeepLabv3+; (**e**) proposed network; (**f**) ground truth.

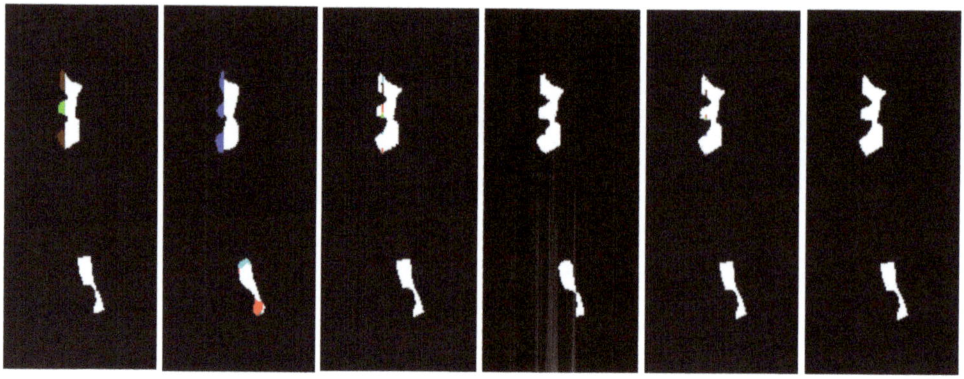

(a) (b) (c) (d) (e) (f)

Figure 12. Enlarged views of white box area 1 in Figure 11. (**a**) U-Net; (**b**) DeepLabv3+; (**c**) CV U-Net; (**d**) L-CV-DeepLabv3+; (**e**) proposed network; (**f**) ground truth.

(a) (b) (c) (d) (e) (f) (g)

Figure 13. Enlarged views of white box area 2 in Figure 11. (**a**) U-Net; (**b**) DeepLabv3+; (**c**) CV U-Net; (**d**) L-CV-DeepLabv3+; (**e**) proposed network; (**f**) ground truth; (**g**) Pauli RGB image.

Four indicators obtained by each network are shown in Table 4. The MIOUs, OAs and MPAs obtained by all the networks except for DeepLabv3+ are more than 90%. Furthermore, the proposed network achieves the highest MIOU, OA and MPA among the five networks. It also achieves the highest IOU of each type of land cover. Especially, for the small area of the wetland, the IOU obtained by the proposed network is 40.45%, 60.36%, 6.16% and 7.92% higher than that of U-Net, DeepLabv3+, CV U-Net and L-CV-DeepLabv3+, respectively. Comparing the proposed network with CV U-Net, we can find that the embedded capsule network increases MIOU, OA and MPA by 2.26%, 0.07% and 1.26%, respectively, and increases IOUs of grasses and water by 3.12% and 9%, respectively. In addition, CV U-Net achieves higher MIOU, OA and MPA than L-CV-DeepLabv3+. Therefore, based on the results of the three datasets, we can conclude that CV U-Net is more suitable for the simple dataset than L-CV-DeepLabv3+.

Table 4. Four indicators of Hulunbuir dataset.

Class	U-Net	DeepLabv3+	CV U-Net	L-CV-DeepLabv3+	Proposed
1	96.41	70.26	98.80	96.14	**99.16**
2	99.27	91.17	99.69	97.93	**99.94**
3	99.68	93.43	99.86	98.53	**99.97**
4	96.00	74.76	97.73	93.91	**97.97**
5	85.10	31.55	89.82	78.09	**92.94**
6	87.66	34.31	86.30	88.50	**95.30**
7	94.70	3.38	94.91	88.20	**95.97**
8	56.03	36.12	90.32	88.56	**96.48**
MIOU	90.53	59.29	95.27	92.15	**97.53**
OA	99.64	96.23	99.80	99.27	**99.87**
MPA	94.09	71.33	97.33	95.61	**98.59**

4.3.4. Parameters and Training Time

Since the network structure adopted by the three datasets is the same except for the number of categories, we take the Flevoland dataset as an example to analyze the parameters of five networks. The trainable, non-trainable and total parameters of the five networks are listed in Table 5. We can find that the parameters of CV U-Net are almost twice those of U-Net, because a CV number refers to both real and imagery parts. Similarly, the total parameters of L-CV-DeepLabv3+ are approximately 2.8 times of those of DeepLabv3+. Although the total parameters of the proposed network are more than those of CV U-Net, they are far less than those of L-CV-DeepLabv3+.

Table 5. Parameters of five networks for Flevoland dataset.

Parameter	U-Net	DeepLabv3+	CV U-Net	L-CV-DeepLabv3+	Proposed
Trainable	1,466,380	3,011,789	2,934,366	8,361,528	3,411,760
Non-trainable	3212	41,724	8030	144,620	8080
Total	1,469,592	3,053,513	2,942,396	8,506,148	3,419,840

Furthermore, we compare the training time of three CV networks and list the average training time of one epoch for the three datasets in Table 6. For each dataset, the proposed network only takes a little more time than CV U-Net, but much less time than L-CV-DeepLabv3+.

Table 6. Average training time of one epoch for three datasets (s).

Dataset	CV U-Net	L-CV-DeepLabv3+	Proposed
Flevoland dataset	9.28	29.81	9.68
San Francisco dataset	12.98	35.59	16.5
Hulunbuir dataset	9.44	32.68	10.76

4.3.5. Convergence Performance

We compared the convergence performance of the proposed network and CV U-Net. The training loss curves of the three datasets are shown in Figure 14a–c. When these two networks converge, the consumed epochs are listed in Table 7. Here, we define the convergence as the loss difference between two adjacent epochs that does not exceed 0.003 for five consecutive times. For each dataset, the proposed network consumes less epochs than CV U-Net. Therefore, it can be inferred from Tables 6 and 7 that the convergence speed of the proposed network is much faster than that of CV U-Net for each dataset.

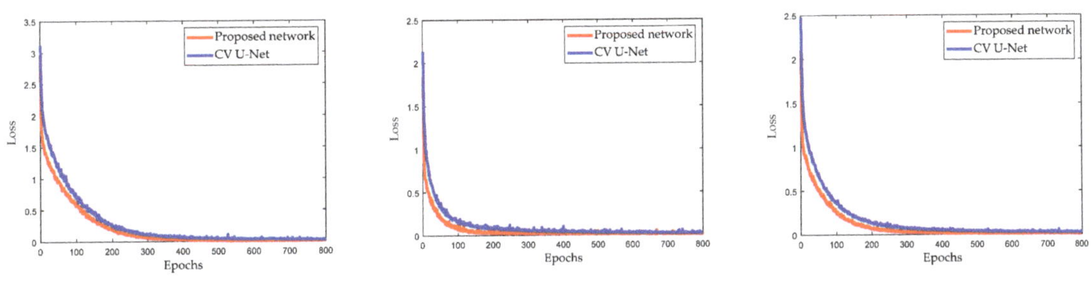

(a) (b) (c)

Figure 14. Curves of training loss. (**a**) Flevoland dataset; (**b**) San Francisco dataset; (**c**) Hulunbuir dataset.

Table 7. Epochs consumed by the convergence for three datasets (epochs).

Dataset	CV U-Net	Proposed
Flevoland dataset	540	445
San Francisco dataset	538	308
Hulunbuir dataset	527	397

5. Discussion

From the experimental results in Section 4, we know that the proposed network has significant advantages over the other four networks. In this section, we discuss the influence of training set size on segmentation performance of the proposed network and CV U-Net, and we analyze advantages of the embedded CV capsule network in feature extraction through the visualization of feature maps.

5.1. Influence of Training Set Size on Segmentation Performance

We compare the segmentation performance of the proposed network and CV U-Net when the expansion factor of the training samples changes. The curves of segmentation performance for the three datasets are shown in Figure 15a–c. When the training set is not expanded, the segmentation performance of the proposed network is far better than that of CV U-Net for each dataset. For the Flevoland dataset, the MIOU difference between these two networks is greater than 16%, and the MPA difference is greater than 10%. For the San Francisco dataset, these two differences are greater than 29% and 20%, respectively. For the Hulunbuir dataset, these two differences are greater than 25% and 17%, respectively. Furthermore, for the three datasets, the MIOUs obtained by the proposed network are equal to or greater than 90%. Therefore, the proposed network has significant advantages when the training sets are not expanded.

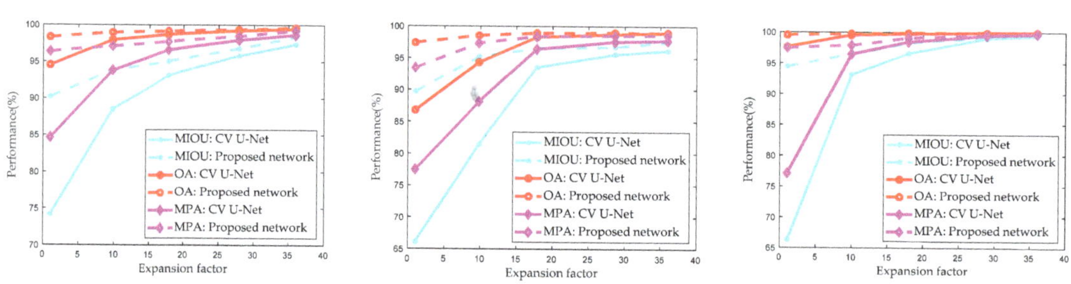

(a) (b) (c)

Figure 15. Segmentation performance changes with the expansion factor of training samples. (**a**) Flevoland dataset; (**b**) San Francisco dataset; (**c**) Hulunbuir dataset.

As the training samples increase, the segmentation performances obtained by these two networks are gradually improved for each dataset. This is because more features can be extracted from the expanded training samples. At the same time, we can find that the performance difference between these two networks gradually decreases for each dataset. However, the proposed network achieves higher MIOU, OA and MPA than CV U-Net, regardless of the expansion factor of training samples. In addition, we can also find that no matter the CV U-Net or proposed network, the OA obtained is the largest, followed by MPA, and finally MIOU. The reason is that the MIOU is stricter than the other two indicators.

5.2. Advantages of the Capsule Network in Feature Extraction

To demonstrate the feature extraction ability of the embedded capsule network, we visualize the feature maps obtained by the proposed network and CV U-Net. For the proposed network shown in Figure 4, feature maps (in the red box) obtained after the first upsampling and convolution operation are considered. For CV U-Net, feature maps obtained at the corresponding position are also considered. Taking the San Francisco dataset as an example, the ground truth of one test sample is shown in Figure 16a. When the number of training samples is not expanded, the visualization results of 128 feature maps obtained by the proposed network are shown in Figure 16c. The feature maps at the corresponding position obtained by CV U-Net are shown in Figure 16b. Comparing Figure 16b with Figure 16c, we can find that the texture features in Figure 16c are clearer than those in Figure 16b. The former is the global features. Therefore, the embedded CV capsule network enhances the feature extraction ability of CV U-Net.

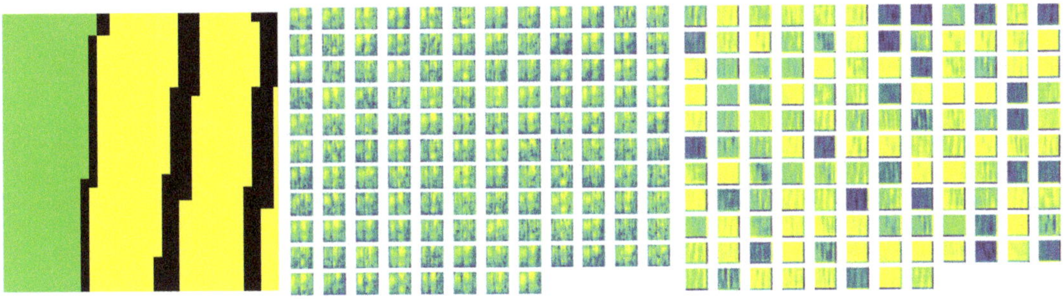

Figure 16. Test sample and visualization of feature maps. (**a**) Ground truth of the test sample; (**b**) feature maps obtained by CV U-Net; (**c**) feature maps obtained by the proposed network.

6. Conclusions

In this paper, we propose a CV U-Net with a capsule embedded for semantic segmentation of PolSAR images. The proposed network mainly includes two parts. One is the CV U-Net, and the other is the CV capsule network. The CV U-Net is obtained by extending the original U-Net to the CV domain and making its structure lightweight. Like the original U-Net, it also includes the encoder and the decoder. The CV capsule network is embedded between the CV encoder and the CV decoder. It is made up of the CV primary capsule and the segmented capsule. In order to accomplish the connection between these two types of capsules, CV dynamic routing is proposed. It can ensure the routing consistency of real and imaginary parts of CV capsules. From the experimental results for two airborne datasets and one spaceborne dataset, we can draw the following conclusions: (1) CV networks have better performance than RV networks because the former can make full use of both the amplitude and phase information of PolSAR data. (2) The structure of the network should match the dataset to achieve good performance. The reason is that overfitting can easily occur when a small number of samples are used for the training of a deep network. (3) The proposed network can not only effectively distinguish land covers with the similar scattering mechanism, but it can also obtain the accurate boundaries of land covers. These advantages are especially obvious when the number of training samples is small. The reason is that the embedded CV capsule network has a strong ability of feature extraction.

Author Contributions: Conceptualization, L.Y. and W.H.; Methodology, Q.S.; Software, Q.S. and Y.G.; Validation, M.L.; resources, X.X.; Writing, L.Y. and Q.S. All authors have read and agreed to the published version of the manuscript.

Funding: This research was funded by the National Natural Science Foundation of China (no. 62261027, no. 61901198, and no. 62266020), the Natural Science Foundation of Jiangxi Province (no. 20224BAB202002 and no. 20224BAB212008), the Science and Technology Project of Jiangxi Provincial Education Department (no. 211410), and the Special Innovation Project for Graduate Student of Jiangxi Province YC2021-S580.

Institutional Review Board Statement: Not applicable.

Informed Consent Statement: Not applicable.

Data Availability Statement: Not applicable.

Acknowledgments: The authors acknowledge Qiang Yin at Beijing University of Chemical Technology for providing the Gaofen-3 Hulunbuir dataset.

Conflicts of Interest: The authors declare no conflict of interest.

References

1. Lee, J.S.; Pottier, E. *Polarimetric Radar Imaging: From Basics to Applications*; CRC Press: Boca Raton, FL, USA, 2009.
2. Cloude, S.R.; Pottier, E. An entropy based classification scheme for land applications of polarimetric SAR. *IEEE Trans. Geosci. Remote Sens.* **1997**, *35*, 68–78. [CrossRef]
3. Lee, J.S.; Grunes, M.R.; Ainsworth, T.L.; Du, L.J.; Schuler, D.L.; Cloude, S.R. Unsupervised classification using polarimetric decomposition and the complex Wishart classifier. *IEEE Trans. Geosci. Remote Sens.* **1991**, *37*, 2249–2258.
4. Jiao, L.; Liu, F. Wishart deep stacking network for fast PolSAR image classification. *IEEE Trans. Image Process.* **2016**, *25*, 3273–3286. [CrossRef]
5. Zhang, Z.; Wang, H.; Xu, F.; Jin, Y.Q. Complex-valued convolutional neural network and its application in polarimetric SAR image classification. *IEEE Trans. Geosci. Remote Sens.* **2017**, *55*, 7177–7188. [CrossRef]
6. Cheng, J.; Zhang, F.; Xiang, D.; Yin, Q.; Zhou, Y. PolSAR image classification with multiscale superpixel-based graph convolutional network. *IEEE Trans. Geosci. Remote Sens.* **2022**, *60*, 1–14. [CrossRef]
7. Dong, H.; Zou, B.; Zhang, L.; Zhang, S. Automatic design of CNNs via differentiable neural architecture search for PolSAR image classification. *IEEE Trans. Geosci. Remote Sens.* **2020**, *58*, 6362–6375. [CrossRef]
8. Liu, H.; Yang, S.; Gou, S.; Chen, P.; Wang, Y.; Jiao, L. Fast classifiction for large polarimeteric SAR data based on refined spatial-anchor graph. *IEEE Geosci. Remote Sens. Lett.* **2017**, *14*, 1589–1593. [CrossRef]
9. Ding, L.; Zheng, K.; Lin, D.; Chen, Y.; Bruzzone, L. MP-ResNet: Multipath residual network for the semantic segmentation of high-resolution PolSAR images. *IEEE Geosci. Remote. Sens. Lett.* **2022**, *19*, 4014205. [CrossRef]
10. Xiao, D.; Wang, Z.; Wu, Y.; Gao, X.; Sun, X. Terrain segmentation in polarimetric SAR images using dual-attention fusion network. *IEEE Geosci. Remote Sens. Lett.* **2022**, *19*, 4006005. [CrossRef]
11. Garg, R.; Kumar, A.; Bansal, N.; Prateek, M.; Kumar, S. Semantic segmentation of PolSAR image data using advanced deep learning model. *Sci. Rep.* **2021**, *11*, 15365. [CrossRef]
12. Ren, S.; Zhou, F. Semi-supervised classification for PolSAR data with multi-scale evolving weighted graph convolutional network. *IEEE J. Sel. Top. Appl. Earth Observ. Remote Sens.* **2021**, *14*, 2911–2927. [CrossRef]
13. Chen, S.W.; Tao, C.S. PolSAR image classification using polarimetric-feature-driven deep convolutional neural network. *IEEE Geosci. Remote Sens. Lett.* **2018**, *15*, 627–631. [CrossRef]
14. Ni, J.; Zhang, F.; Yin, Q.; Zhou, Y.; Li, H.-C.; Hong, W. Random neighbor pixel-block-based deep recurrent learning for polarimetric SAR image classification. *IEEE Trans. Geosci. Remote Sens.* **2021**, *59*, 7557–7569. [CrossRef]
15. Liu, F.; Jiao, L.; Hou, B.; Yang, S. POL-SAR image classification based on Wishart DBN and local spatial information. *IEEE Trans. Geosci. Remote Sens.* **2016**, *54*, 3292–3308. [CrossRef]
16. Xie, W.; Jiao, L.; Hou, B.; Ma, W.; Zhao, J.; Zhang, S.; Liu, F. PolSAR image classification via Wishart-AE model or Wishart-CAE model. *IEEE J. Sel. Top. Appl. Earth Observ. Remote Sens.* **2017**, *10*, 3604–3615. [CrossRef]
17. Liu, F.; Jiao, L.; Tang, X. Task-oriented GAN for PolSAR image classification and clustering. *IEEE Trans. Neural Netw. Learn. Syst.* **2019**, *30*, 2707–2719. [CrossRef]
18. Fang, Z.; Zhang, G.; Dai, Q.; Kong, Y.; Wang, P. Semisupervised deep convolutional neural networks using pseudo labels for PolSAR image classification. *IEEE Geosci. Remote Sens. Lett.* **2022**, *19*, 4005605. [CrossRef]
19. Liu, H.; Xu, D.; Zhu, T.; Shang, F.; Yang, R. Graph convolutional networks by architecture search for PolSAR image classification. *Remote Sens.* **2021**, *13*, 1404. [CrossRef]
20. Bi, H.; Sun, J.; Xu, Z. A graph-based semisupervised deep learning model for PolSAR image classification. *IEEE Trans. Geosci. Remote Sens.* **2018**, *57*, 2116–2132. [CrossRef]
21. Liu, S.J.; Luo, H.; Shi, Q. Active ensemble deep learning for polarimetric synthetic apetrue radar image classification. *IEEE Geosci. Remote Sens. Lett.* **2021**, *18*, 1580–1584. [CrossRef]
22. Bi, H.; Xu, F.; Wei, Z.; Xue, Y.; Xu, Z. An active deep learning approach for minimally supervised PolSAR image classification. *IEEE Trans. Geosci. Remote Sens.* **2019**, *57*, 9378–9395. [CrossRef]

23. Zhou, Y.; Wang, H.; Xu, F.; Jin, Y.Q. Polarimetric SAR image classification using deep convolutional neural networks. *IEEE Geosci. Remote Sens. Lett.* **2016**, *13*, 1935–1939. [CrossRef]
24. Gao, F.; Huang, T.; Wang, J.; Sun, J.P.; Hussain, A.; Yang, E. Dual-branch deep convolution neural network for polarimetric SAR image classification. *Appl. Sci.* **2017**, *7*, 447. [CrossRef]
25. Wang, Y.; Chen, J.; Zhou, Y.; Zhang, F.; Yin, Q. A multi-channel fusion convolution neural network based on scattering mechanism for PolSAR image classification. *IEEE Geosci. Remote Sens. Lett.* **2022**, *19*, 4007805.
26. Dong, H.; Zhang, L.; Zou, A.B. PolSAR image classification with lightweight 3D convolutional networks. *Remote Sens.* **2020**, *12*, 396. [CrossRef]
27. Liu, X.; Jiao, L.; Tang, X.; Sun, Q.; Zhang, D. Polarimetric convolutional network for PolSAR image classification. *IEEE Trans. Geosc. Remote Sens.* **2019**, *57*, 3040–3054. [CrossRef]
28. Zhang, L.; Dong, H.; Zou, B. Efficently utilizing complex-valued PolSAR image data via a multi-task deep learning framework. *ISPRS J. Photogramm. Remote Sens.* **2019**, *157*, 59–72. [CrossRef]
29. Tan, X.; Li, M.; Zhang, P.; Wu, Y.; Song, W. Complex-valued 3D convolutional neural network for PolSAR image classification. *IEEE Geosci. Remote Sens. Lett.* **2020**, *17*, 1022–1026. [CrossRef]
30. Zhang, P.; Tan, X.; Li, B.; Jiang, Y.; Wu, Y. PolSAR image classification using hybrid conditional random fields model based on complex-valued 3D CNN. *IEEE Trans. Aerosp. Electron. Syst.* **2021**, *57*, 1713–1730. [CrossRef]
31. Xie, W.; Ma, G.; Zhao, F.; Zhang, L. PolSAR image classification via a novel semi-supervised recurrent complex-valued convolution neural network. *Neurocomputing* **2020**, *388*, 255–268. [CrossRef]
32. Long, J.; Shelhamer, E.; Darrell, T. Fully convolutional networks for semantic segmentation. In Proceedings of the 2015 IEEE Conference on Computer Vision and Pattern Recognition, Boston, MA, USA, 7–12 June 2015; pp. 3431–3440.
33. Ronneberger, O.; Fischer, P.; Brox, T. U-net: Convolutional networks for biomedical image segmentation. In Proceedings of the 18th International Conference on Medical Image Computing and Computer Assisted Interventions, Munich, Germany, 5–9 October 2015; pp. 234–241.
34. Badrinarayanan, V.; Kendall, A.; Cipolla, R. Segnet: A deep convolutional encoder-decoder architecture for image segmentation. *IEEE Trans. Pattern Anal. Mach. Intell.* **2017**, *39*, 2481–2495. [CrossRef] [PubMed]
35. Zhao, H.; Shi, J.; Qi, X.; Wang, X.; Jia, J. Pyramid scene parsing network. In Proceedings of the IEEE Conference on Computer Vision and Pattern Recognition, Honolulu, HI, USA, 21–26 July 2017; pp. 6230–6239.
36. Chen, L.C.; Zhu, Y.; Papandreou, G.; Schroff, F.; Adam, H. Encoder-decoder with atrous separable convolution for semantic image segmentation. In Proceedings of the 2018 European Conference on Computer Vision (ECCV), Munich, Germany, 8–14 September 2018; pp. 801–818.
37. Lin, G.; Milan, A.; Shen, C.; Reid, I. RefineNet: Multi-path refinement networks for high-resolution semantic segmentation. In Proceedings of the IEEE Conference on Computer Vision and Pattern Recognition, Honolulu, HI, USA, 21–26 July 2017; pp. 5168–5177.
38. Sun, X.; Shi, A.; Huang, H.; Mayer, H. BAS^{4}NET: Boundary-aware semi-supervised semantic segmentation network for very high resolution remote sensing images. *IEEE J. Sel. Top. Appl. Earth Observ. Remote Sens.* **2020**, *13*, 5398–5413. [CrossRef]
39. Shahzad, M.; Maurer, M.; Fraundorfer, F.; Wang, Y.; Zhu, X.X. Buildings detection in VHR SAR images using fully convolution neural networks. *IEEE Trans. Geosci. Remote Sens.* **2019**, *57*, 1100–1116. [CrossRef]
40. Shi, X.; Fu, S.; Chen, J.; Wang, F.; Xu, F. Object-level semantic segmentation on the high-resolution Gaofen-3 FUSAR-map dataset. *IEEE J. Sel. Top. Appl. Earth Observ. Remote Sens.* **2021**, *14*, 3107–3119. [CrossRef]
41. Bianchi, F.M.; Grahn, J.; Eckerstorfer, M.; Malnes, E.; Vickers, H. Snow avalanche segmentation in SAR images with fully convolutional neural networks. *IEEE J. Sel. Top. Appl. Earth Observ. Remote Sens.* **2021**, *14*, 75–82. [CrossRef]
42. Wang, Y.; He, C.; Liu, X.L.; Liao, M.S. A hierarchical fully convolutional network integrated with sparse and low-rank subspace representations for PolSAR imagery classification. *Remote Sens.* **2018**, *10*, 342. [CrossRef]
43. He, C.; He, B.; Tu, M.; Wang, Y.; Qu, T.; Wang, D.; Liao, M. Fully convolutional networks and a manifold graph embedding-based algorithm for PolSAR image classification. *Remote Sens.* **2020**, *12*, 1467. [CrossRef]
44. Li, Y.; Chen, Y.; Liu, G.; Jiao, L. A novel deep fully convolutional network for PolSAR image classification. *Remote Sens.* **2018**, *10*, 1984. [CrossRef]
45. Mohammadimanesh, F.; Salehi, B.; Mandianpari, M.; Gill, E.; Molinier, M. A new fully convolutional neural network for semantic segmentation of polarimetric SAR imagery in complex land cover ecosystem. *ISPRS J. Photogramm. Remote Sens.* **2019**, *151*, 223–236. [CrossRef]
46. Pham, M.T.; Lefèvre, S. Very high resolution airborne PolSAR image classification using convolutional neural networks. *arXiv* **2019**, arXiv:1910.14578.
47. Wu, W.; Li, H.; Li, X.; Guo, H.; Zhang, L. PolSAR image semantic segmentation based on deep transfer learning-realizing smooth classification with small training sets. *IEEE Geosci. Remote Sens. Lett.* **2019**, *19*, 977–981. [CrossRef]
48. Zhao, F.; Tian, M.; Wen, X.; Liu, H. A new parallel dual-channel fully convolutional network via semi-supervised fcm for PolSAR image classification. *IEEE J. Sel. Top. Appl. Earth Observ. Remote Sens.* **2020**, *13*, 4493–4505. [CrossRef]
49. Jing, H.; Wang, Z.; Sun, X.; Xiao, D.; Fu, K. PSRN: Polarimetric space reconstruction network for PolSAR image semantic segmentation. *IEEE J. Sel. Top. Appl. Earth Observ. Remote Sens.* **2021**, *14*, 10716–10732. [CrossRef]

50. Cao, Y.; Wu, Y.; Zhang, P.; Liang, W.; Li, M. Pixel-wise PolSAR image classification via a novel complex-valued deep fully convolutional network. *Remote Sens.* **2019**, *11*, 2653. [CrossRef]
51. Yu, L.; Zeng, Z.; Liu, A.; Xie, X.; Wang, H.; Xu, F.; Hong, W. A lightweight complex-valued DeepLabv3+ for semantic segmentation of PolSAR image. *IEEE J. Sel. Top. Appl. Earth Observ. Remote Sens.* **2022**, *15*, 930–943. [CrossRef]
52. Sabour, S.; Frosst, N.; Hinton, G.E. Dynamic routing between capsules. *arXiv* **2017**, arXiv:1710.09829.
53. Xiang, C.; Lu, Z.; Zou, W.; Yi, T.; Chen, X. Ms-CapsNet: A novel multi-scale capsule network. *IEEE Signal Process. Lett.* **2018**, *25*, 1850–1854. [CrossRef]
54. Jaiswal, A.; AbdAlmageed, W.; Wu, Y.; Natarajan, P. CapsuleGAN: Generative adversarial capsule network. *arXiv* **2018**, arXiv:1802.06167.
55. Mobiny, A.; Van Nguyen, H. Fast CapsNet for lung cancer screening. In Proceedings of the International Conference on Medical Image Computing and Computer-Assisted Intervention, Granada, Spain, 16–20 September 2018; Springer: Cham, Switzerland, 2018; pp. 741–749.
56. Afshar, P.; Plataniotis, K.N.; Mohammadi, A. Capsule networks for brain tumor classification based on MRI images and coarse tumor boundaries. In Proceedings of the ICASSP 2019 IEEE International Conference on Acoustics, Speech and Signal Processing (ICASSP), Brighton, UK, 12–17 May 2019; pp. 1368–1372.
57. Zhang, W.; Tang, P.; Zhao, L. Remote sensing image scene classification using CNN-CapsNet. *Remote Sens.* **2019**, *11*, 494. [CrossRef]
58. Yu, Y.; Liu, C.; Guan, H.; Wang, L.; Gao, S.; Zhang, H.; Zhang, Y.; Li, J. Land cover classification of multispectral lidar data with an efficient self-attention capsule network. *IEEE Geosci. Remote Sens. Lett.* **2022**, *19*, 6501505. [CrossRef]
59. Cheng, J.; Zhang, F.; Xiang, D.; Yin, Q.; Zhou, Y.; Wang, W. PolSAR image land cover classification based on hierarchical capsule network. *Remote Sens.* **2021**, *13*, 3132. [CrossRef]
60. LaLonde, R.; Bagci, U. Capsules for object segmentation. *arXiv* **2018**, arXiv:1804.04241.
61. Liu, A.; Yu, L.J.; Zeng, Z.X.; Xie, X.C.; Guo, Y.T.; Shao, Q.Q. Complex-valued U-Net for PolSAR image semantic segmentation. *IOP J. Phys. Conf. Ser.* **2021**, *2010*, 012102. [CrossRef]
62. Yu, L.; Hu, Y.; Xie, X.; Lin, Y.; Hong, W. Complex-valued full convolutional neural network for SAR target classification. *IEEE Geosci. Remote Sens. Lett.* **2020**, *17*, 1752–1756. [CrossRef]

remote sensing

MDPI

Article

Deep Network Architectures as Feature Extractors for Multi-Label Classification of Remote Sensing Images

Marjan Stoimchev [1,2,*], Dragi Kocev [1,2,3] and Sašo Džeroski [1,2]

1 Department of Knowledge Technologies, Jožef Stefan Institute, Jamova cesta 39, 1000 Ljubljana, Slovenia; dragi.kocev@ijs.si (D.K.); saso.dzeroski@ijs.si (S.D.)
2 Jožef Stefan International Postgraduate School, 1000 Ljubljana, Slovenia
3 Bias Variance Labs, 1000 Ljubljana, Slovenia
* Correspondence: marjan.stoimchev@ijs.si

Abstract: Data in the form of images are now generated at an unprecedented rate. A case in point is remote sensing images (RSI), now available in large-scale RSI archives, which have attracted a considerable amount of research on image classification within the remote sensing community. The basic task of single-target multi-class image classification considers the case where each image is assigned exactly one label from a predefined finite set of class labels. Recently, however, image annotations have become increasingly complex, with images labeled with several labels (instead of just one). In other words, the goal is to assign multiple semantic categories to an image, based on its high-level context. The corresponding machine learning tasks is called multi-label classification (MLC). The classification of RSI is currently predominantly addressed by deep neural network (DNN) approaches, especially convolutional neural networks (CNNs), which can be utilized as feature extractors as well as end-to-end methods. After only considering single-target classification for a long period, DNNs have recently emerged that address the task of MLC. On the other hand, trees and tree ensembles for MLC have a long tradition and are the best-performing class of MLC methods, but need predefined feature representations to operate on. In this work, we explore different strategies for model training based on the transfer learning paradigm, where we utilize different families of (pre-trained) CNN architectures, such as VGG, EfficientNet, and ResNet. The architectures are trained in an end-to-end manner and used in two different modes of operation, namely, as standalone models that directly perform the MLC task, and as feature extractors. In the latter case, the learned representations are used with tree ensemble methods for MLC, such as random forests and extremely randomized trees. We conduct an extensive experimental analysis of methods over several publicly available RSI datasets and evaluate their effectiveness in terms of standard MLC measures. Of these, ranking-based evaluation measures are most relevant, especially ranking loss. The results show that, for addressing the RSI-MLC task, it is favorable to use lightweight network architectures, such as EfficientNet-B2, which is the best performing end-to-end approach, as well as a feature extractor. Furthermore, in the datasets with a limited number of images, using traditional tree ensembles for MLC can yield better performance compared to end-to-end deep approaches.

Keywords: remote sensing; convolutional neural networks; tree ensemble methods; multi-label classification

Citation: Stoimchev, M.; Kocev, D.; Džeroski, S. Deep Network Architectures as Feature Extractors for Multi-Label Classification of Remote Sensing Images. *Remote Sens.* **2023**, *15*, 538. https://doi.org/10.3390/rs15020538

Academic Editor: Michael K. Ng

Received: 17 October 2022
Revised: 16 December 2022
Accepted: 5 January 2023
Published: 16 January 2023

1. Introduction

Remote sensing is the process of detecting and monitoring the physical characteristics of an area by measuring its reflected and emitted radiation at a distance. In the past few years, advances in satellite technology have resulted in large-scale remote sensing image (RSI) archives, which have attracted a considerable amount of research in various application areas. RSI can be used to monitor and predict various environmental phenomena, such as weather and climate change [1], land use and land cover changes at macro scale [2], deforestation [3], wildfires [4,5], and many others.

In machine learning terms, a wide range of applications stemming from RSI are approached using single-label classification, where the goal is to assign a single label/semantic category to an image [6]. However, real-world RSIs are typically complex and present more than a single semantic category within a single image. Hence, single-label classification is often insufficient to fully describe the presence of complex areas, which can carry semantically complex content [7].

To facilitate more realistic representation of the content of RSI, the analysis of RSI should be addressed through the task of multi-label classification (MLC), where a given image can be associated with multiple semantic concepts/labels taken from a predefined set of labels. In this way, the classification problem becomes more challenging as compared to single-label classification, as discussed in a recent large-scale comparative study and analysis of a wide range of MLC methods [8]. This study highlights the two major challenges that can limit the performance of MLC methods: the presence of complex label correlations and high-dimensional label spaces.

These challenges are attracting increasing amounts of attention from researchers focusing on deep learning (DL) methods, specifically on methods capable of automatically learning long-range dependencies (e.g., with the use of self-attention mechanisms in the vision transformer (ViT) base network architectures [9]), and handling the high-dimensional label spaces. Their internal mechanisms and structure, such as the hierarchical design and characteristics in convolutional neural networks (CNNs), with local connectivity and non-linearity, are capable of encoding information exceptionally well. Moreover, their flexible design offers knowledge to be extracted and discriminative representations to be learned from noisy data in an end-to-end manner, and achieves more accurate recognition performance in less-constrained environments as compared with traditional MLC approaches. Furthermore, the recent success of these methods can be associated with their ability to leverage large amounts of labeled data in order to learn meaningful knowledge.

Many of the existing approaches in computer vision try to address the challenges encountered by exploiting proven DL network architectures pre-trained on large-scale and diverse datasets such as ImageNet [10]. This is usually achieved by using the transfer learning paradigm [11], where specific parts of the model are fine-tuned in order to learn new features that generalize better to the new downstream task [12–14]. For example, Wang et al. [15] propose a framework based on a VGG-16 CNN model initialized with weights learned on ImageNet, which is used to extract semantic representations from images and is coupled with a recurrent neural network (RNN) network architecture with long short-term memory (LSTM) units to capture image/label relations and label dependency. Chen et al. [16] propose an end-to-end learning method based on Graph Convolutional Networks (GCN) to capture the label correlations, and a novel re-weighting scheme for creating the label correlation matrix that is used to guide the information propagation among the nodes in the GCN.

In addition, with the increased availability of RSI and the increased research interest in remote sensing applications, many interesting approaches have been proposed at the crossroads of remote sensing and computer vision [7,17–19]. For example, Hua et al. [20] propose a novel approach for MLC from aerial imagery—an attention aware label relational network, comprising a label-wise feature parcel module, an attention region extraction module, and a label relational inference module. Sumbul et al. [21] present a K-Branch CNN that uses a multi-attention strategy for bidirectional LSTM networks, which is specially developed to capture spatial and spectral contents from RS images of local image areas. Wang et al. [22], despite using local attention [21], are also able to maintain global context through global attention pooling, where the combination of both helps in modeling long-range dependencies among multiple objects and captures underlying relationships among multiple labels.

In this work, our main focus is on investigating the potential of several prominent deep learning architectures (variants of VGG [23], ResNet [24] and EfficientNet [25]) as feature extractors for the MLC of RSI as well as end-to-end approaches to MLC of RSI. To this end, we evaluate the performance of the architectures across seven MLC RSI datasets with different properties in terms of number of images, number of labels, and average number of labels per image (label cardinality). More specifically, we use pre-trained network architectures with weights learned on ImageNet as an initialization procedure. We further perform calibration of the pre-trained models by fine-tuning them on a small set of RSIs and compare the fine-tuned and pre-trained MLC performance. Finally, we use the learned models as feature extractors to describe the RSI, and use the resulting feature vectors to learn tree ensembles such as random forests and extra trees for MLC.

The main contributions of this paper can be summarized as follows:

- We present an experimental analysis of different approaches for MLC of RSI. More precisely, we investigate the performance of several deep learning network architectures by using pre-training and fine-tuning as the main learning strategies for the MLC task.
- We evaluate the effectiveness of the deep models used as end-to-end approaches to MLC, and used as feature extractors that provide feature representations of RSI, as inputs to tree ensemble methods for MLC. Moreover, we investigate which of the network architectures is the most suitable choice in terms of performance.
- We also investigate the performance of the considered methods in terms of the influence of the number of labeled training examples by providing the methods with different fractions of the data.

The remainder of this paper is organized as follows. Section 2 describes the MLC RSI data (Section 2.1) and the machine learning methods (Section 2.2) used to analyze them (deep learning architectures in Section 2.2.1 and tree ensembles in Section 2.2.2). Next, Section 3 gives the experimental questions and the specific experimental setup, including parameter instantiations for the methods (Section 3.1), the evaluation strategy (Section 3.2), and the different evaluation measures used to assess the performance of the models (Section 3.3). Furthermore, Section 4 discusses the results of our experiments, focusing on the outcomes of the different representation learning strategies (Section 4.1), deep architectures (Section 4.2), MLC approaches (Section 4.3), and number of images (Section 4.4). Finally, Section 5 concludes the paper by providing a summary of the presented work and directions for further work.

2. Materials and Methods

2.1. Datasets

We use seven publicly available MLC RSI datasets to assess the performance of the MLC methods, namely UC Merced (UCM) Land Use, AID Multilabel, Ankara HIS archive, DFC-15 Multilabel, MLRSNet, and two variants of the BigEarthNet dataset based on two Corine Land Cover (CLC) nomenclatures (Available at https://land.copernicus.eu/user-corner/technical-library/corine-land-cover-nomenclature-guidelines/html (accessed on 4 January 2023)), CLC with 43 labels (BigEarthNet-43), and CLC with 19 labels (BigEarthNet-19). When analysing the datasets that have hyperspectral RSI (meaning that they have several spectral bands), such as Ankara and BigEarthNet, we only use the RGB spectral band to train the models.

A summary of the properties of the seven datasets is given in Table 1. We can see that the selected datasets are diverse along several lines: number of images, locations, number of labels, image resolution, and number of labels per example image (label cardinality). This means that the trained predictive models are trained and evaluated in a challenging environments. The selection of typical images from the different datasets with their corresponding class labels is given in Figure 1.

Table 1. Description of the used RSI multi-label datasets. $|\mathcal{L}|$ denotes the number of possible labels; *Card* denotes label cardinality (i.e., average number of labels per image); *Dens* denotes label density (average proportion of images labeled with a given label); N is the number of images in the dataset, of which N_{train} are in the train and N_{test} in the test datasets; and $w \times h$ is the dimension of the images (in pixels).

Dataset	Image Type	$\|\mathcal{L}\|$	*Card*	*Dens*	N	N_{train}	N_{test}	$w \times h$
Ankara	Hyperspectral/Aerial RGB	29	9.120	0.536	216	171	45	64 × 64
UC Merced Land Use	Aerial RGB	17	3.334	0.476	2100	1667	433	256 × 256
AID Multilabel	Aerial RGB	17	5.152	0.468	3000	2400	600	600 × 600
DFC-15 Multilabel	Aerial RGB	8	2.795	0.465	3341	2672	669	600 × 600
MLRSNet	Aerial RGB	60	5.770	0.144	109,151	87,325	21,826	256 × 256
BigEarthNet	Hyperspectral/Aerial RGB	19	2.900	0.263	590,326	472,245	118,081	256 × 256
BigEarthNet	Hyperspectral/Aerial RGB	43	2.965	0.247	590,326	472,245	118,081	256 × 256

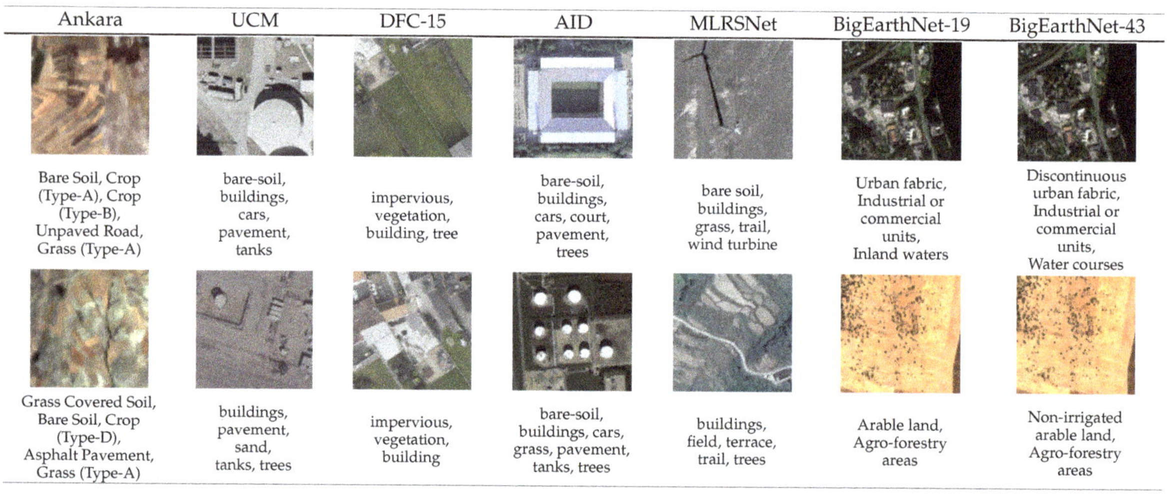

Figure 1. An illustration of the diversity of images from the different RSI datasets with the corresponding class labels.

2.1.1. UC Merced Land Use

The UC Merced data set contains 2100 images grouped into 21 broad categories at the scene level. There are a total of 100 images per category, with the size of 256 × 256 and a spatial resolution of 0.3 m. The initial version of this dataset was for single-label classification purposes [26]. Later, Chaudhuri et al. [27] relabeled the images with multiple labels. The total number of distinct object-level labels is 17: airplane, bare soil, buildings, chaparral, court, dock, field, grass, mobile home, pavement, sand, sea, ship, tanks, trees, and water. Each image is annotated with one or more (maximum 7) labels at the object level, containing 3.3 object-level labels per image on average.

2.1.2. AID Multilabel

The initial single-label AID dataset [28] was relabeled with multiple labels per image and became the AID multi-label dataset [20]. It is also a more challenging dataset than the U Merced dataset. It contains 3000 aerial images from 30 categories with manually assigned multiple-object labels. The resolution of the images is 600 × 600 pixels, where each image has 5.5 object-level labels on average (maximum 11). The spatial resolution varies from 0.3 to 8 m.

2.1.3. Ankara HIS Archive

This is a small hyperspectral dataset containing 216 image tiles with a size of 63 × 63 pixels [29]. The image patches are obtained by fragmenting large hyperspectral images, acquired by the NASA EO-1 satellite's Hyperion sensor from the area surrounding the city of Ankara in Turkey. Each image is associated with multiple object-level labels (land-cover-classes) and a single land-use scene-level label where the ground resolution is 30 m. There are 9 object-level labels per image on average, and a maximum of 17. It contains 119 channels of hyperspectral images and corresponding three-channel (RGB) images. We only use the RGB channels.

2.1.4. DFC-15 Multilabel

The DFC-15 Multilabel dataset [30] is built from a semantic segmentation dataset (DFC15), first used in the 2015 IEEE GRSS data fusion contest. This dataset is acquired over Zeebrugge with an airborne sensor (300 m off the ground). There are a total of 7 tiles, where each of the tiles is 10,000 × 10,000 pixels with a spatial resolution of 5 cm. The images are assigned pixel-level labels, where each pixel is categorized into 8 object classes: impervious, water, clutter, vegetation, building, tree, boat, and car. The final dataset contains 3342 image patches with 600 × 600 image resolution, where each image is associated with image-level multi-labels.

2.1.5. MLRSNet

The MLRSNet dataset is an RSI dataset containing optical satellite images with high spatial resolution [31]. It contains 109,161 RSI annotated with 60 predefined class labels, where the number of labels per image varies from 1 to 13. The images have a fixed resolution of 256 × 256 pixels, where the pixel resolution varies from ~10 m to 0.1 m. This dataset can be used for a wide range of learning tasks, such as multi-label classification, multi-class classification, multi-label image retrieval, and image segmentation. There are 10 images that do not have labels assigned, as shown in Figure 2; therefore, we exclude these images from the experiments. The final dataset contains 109,151 images.

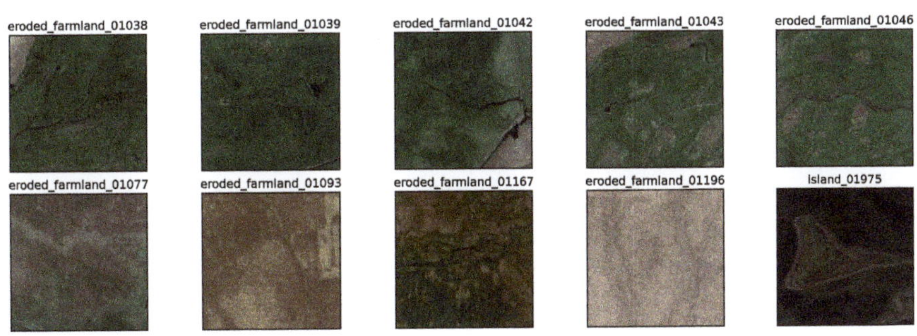

Figure 2. Images from the MLRSNet dataset that do not have labels assigned.

2.1.6. The BigEarthNet Archive

BigEarthNet [17] was constructed by the Remote Sensing image Analysis (RSiM) Group and the Database Systems and Information Management (DIMA) Group at the Technische Universität Berlin (TU Berlin). It is the largest dataset of image patches annotated with multiple labels available to date. There are 590,326 such patches for which Sentinel-2/S2 (and later also Sentinel-1/S1) images are available. For S2, twelve channels are available, and for S1, two channels are available. We only use three of the S2 channels (RGB images). Each image patch is annotated by multiple land-cover classes (i.e., multi-labels) taken from the CORINE Land Cover database of the year 2018 (CLC 2018). Originally, 43 labels were used. These were later merged into 19 labels [32].

2.2. *Overview of the Learning Methods for Multi-Label Classification*

We address the MLC task by applying a Convolutional Neural Network (CNN) as a deep learning approach and exploiting the the transfer learning paradigm. We use the weights learned on the ImageNet as the initialization procedure, and we further perform calibration by fine-tuning the models on the training sets of of RSI. The deep models are used in two different scenarios, namely as end-to-end approaches to directly perform the MLC task (Figure 3a,b), and as CNN-based feature extractors (Figure 3a) to generate feature representations, which are used as inputs to the tree-ensemble methods such as Random Forests and Extremely Randomized Trees for MLC (Figure 3c).

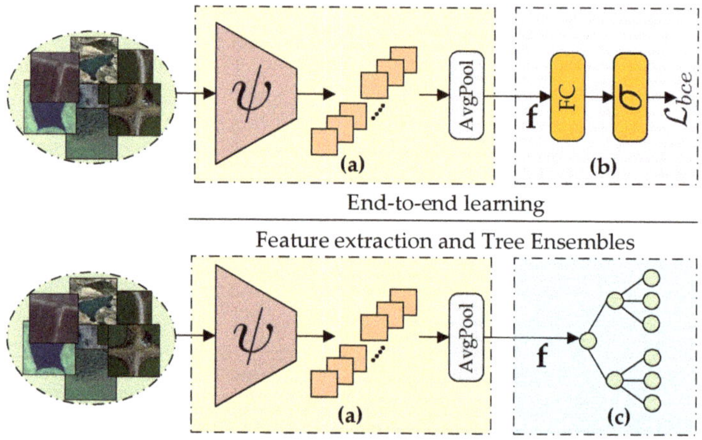

Figure 3. Overview of the proposed strategy for MLC of RSI. Part (**a**) of the strategy is the fully convolutional CNN backbone feature extractor. This part is same for both end-to-end learning, and feature extraction. The next part of the strategy, marked (**b**), is only used during end-to-end learning. The last part, marked with (**c**), is used for training tree ensemble methods from the extracted features, as a separate task and not a part of the end-to-end learning process. For this separate task the CNN feature extractor (**a**) is used in offline mode to generate the required feature representations, which serve as inputs to learn tree ensembles.

We make the following modifications to the used network architecture for the MLC task. To make it applicable to differently sized input images, we make sure the backbone CNN is fully convolutional. Next, to reduce the spatial dimension, the feature maps are aggregated through the average pooling layer, where d-dimensional feature representations $\mathbf{f}$ are extracted. The classification part is replaced with a single fully connected (FC) layer with the number of output units equal to the number of classes. This can also be seen as a layer providing non-normalized log probabilities z_i, i.e., scores (or also called $logits$), that a classification model generates. This $logit$ vector is then converted to a vector of probabilities by using the *sigmoid* activation function $\sigma(z_{i,c}) = 1/(1 + e^{-z_{i,c}})$, where $z_{i,c}$ is the $logit$ of the predicted class. The *sigmoid* function is suggested for MLC, instead of the $softmax$ function, as the activation of the last layer of the model, since the probabilities produced by *sigmoid* are independent and are not constrained to sum up to one. It follows a Bernoulli distribution and thus allows multiple label predictions [18]. It is important to mention that this part of the model (Figure 3b) is only used during training in the end-to-end mode to directly perform the MLC task.

The training process uses the standard binary cross-entropy learning objective, which takes the following form:

$$\mathcal{L}_{bce} = -\frac{1}{n}\sum_{i=1}^{n}[y_i log(\hat{y}_i) + (1 - y_i) log(1 - \hat{y}_i)], \quad (1)$$

where n is the number or samples in a given mini-batch, $y_i = [y_{i,1}, y_{i,2}, \ldots, y_{i,c}]$ is a binary vector representing the multi-labels of a given image, and $\hat{y}_i$ is the predicted output vector of probabilities from the *sigmoid* layer. This learning objective is most commonly used in MLC tasks [18].

2.2.1. Deep Learning Methods

To investigate the impact of the network architecture on the learning process, we consider several popular CNN configurations pre-trained on ImageNet. We use VGG (VGG-16, VGG-19) [23] , ResNet (ResNet-34, ResNet-50, ResNet-152) [24] and EfficientNet (EfficientNet-B0, EfficientNet-B1, EfficientNet-B2) [25] as backbone CNN models for the MLC task. Throughout the experiments, these methods are used either in an end-to-end manner or as CNN-based feature extractors.

- VGGs: VGG is a deep CNN network architecture developed by the Visual Geometry Group (VGG) team [23]. It is also the basis of ground-breaking object recognition models that surpass baselines on many tasks and datasets beyond ImageNet and is still one of the most popular image recognition architectures. Two variants of this family of architectures are intensively studied for their performance—VGG-16 and VGG-19. The VGG-16 model can be seen as an upgrade of AlexNet, while VGG-19 is similar to VGG-16 but contains more layers. They are modeled in such a way that convolutions would actually look simpler by replacing large AlexNet convolution filters with a 3×3 filter, while padding to maintain the same size before a 2×2 *MaxPool* layer down-samples the image size.
- ResNets: ResNets are a family of deep CNN architectures that follow the residual learning principle to ease the training of very deep networks [24]. Their design offers an efficient way to solve the issues related to the vanishing gradients. ResNet follows VGG's full 3×3 convolutional layer design. The residual block has two 3×3 convolutional layers with the same number of output channels. Each convolutional layer is followed by a batch normalization layer and a Rectified Linear Unit (*ReLU*) activation function. Then, there is a skip (or so-called skip connection) of those two convolution operations, where the input is directly added before the final *ReLU* activation function. This kind of design requires that the output of the two convolutional layers has to be the same shape as the input, so that they can be added together. By configuring different numbers of channels and residual blocks in the module, we can create different ResNet models, such as the deeper 152-layer ResNets, i.e., ResNet-152. For the experiments, we use three variants of ResNet: ResNet-34, ResNet-50, and ResNet-152.
- EfficientNets: Unlike conventional deep CNNs, which are often over-parameterized, and arbitrarily scale network dimensions, such as width, depth, and resolution, EfficientNets are methods that uniformly scale each dimension with a fixed set of scaling coefficients [25]. These models surpass state-of-the-art accuracy, with up to 10 times better efficiency (i.e., are both smaller and faster than competitors).

2.2.2. Tree Ensemble Methods

In the experiments, we consider two types of ensemble methods based on decision trees for MLC as a main learning model, namely, ensembles based on random forests for MLC [33] and extremely randomized trees for MLC [34], respectively.

- Random Forest: Random forest (RF) is an ensemble learning method for classification and regression, which creates a set of individual decision trees that operate as an ensemble. It uses bagging and feature randomness to create diversity among the predictors: At each node in the decision tree, a random subset of attributes is taken, and the best split is selected from this subset of attributes. Each individual tree in the random forest provides a class prediction, where the predictions can be aggregated by taking the average (for regression tasks) and the majority or probability distribution vote (for classification tasks). RFs were adapted for the task of MLC [33].
- Extremely Randomized Trees: Extremely Randomized Trees, or so-called Extra Trees (ET), is also an ensemble learning method similar to the Random Forest, which is based on extreme randomization of the tree construction algorithm. As compared

to the Random Forest ensemble, it operates with two key differences: it splits nodes by choosing the cut-points fully at random, and it uses the whole learning sample to grow the trees. The randomness in this method comes from the random splits of all observations, rather then bootstrapping the data as in RF. ETs were adapted for the task of MLC [34].

3. Experimental Design

This section presents the details of the experimental study design. It includes a detailed overview of the experimental setup describing the different learning settings in the end-to-end approaches and the tree ensemble methods. Next, we describe the evaluation strategy of the partitioning of the image datasets into disjoint splits used for training and testing MLC models. Finally, we describe the evaluation measures used to assess the predictive performance of the different methods as well as the statistical procedures used to analyze the results.

The experimental study is tailored to answer the following research questions:

(i) What is the influence of the learning strategy on the performance of end-to-end approaches: Is fine-tuning or pre-training only more suitable for solving the RSI-MLC task?
(ii) Which network architecture is the best choice for end-to-end MLC of RSI and for use as a feature extractor and further training of tree ensembles for MLC?
(iii) How do end-to-end learning and feature extraction plus tree ensembles compare on the task of RSI for MLC (assessed by using the best performing architecture from the previous analysis)? and
(iv) How does the number of training examples influence the predictive performance of the methods used?

3.1. Experimental Setup

3.1.1. End-to-End Learning Approaches

We train the deep network models by using two different learning strategies based on transfer learning. Both settings use the *ImageNet weights* for initialization, while fine-tuning different parts of the backbone CNN model. In the first learning setting, shown in Figure 4a, we fix all the model layers pre-trained on ImageNet and directly transfer the learned knowledge to the target domain. In the second learning setting, shown in Figure 4b, the entire network architecture is being fine-tuned to the new target domain. In both settings, we use a single application-dependent fully connected layer learned from scratch (i.e., the classification layer).

We use the binary cross entropy as an objective function (see Equation (1)) to optimize the model parameters, and apply the same training procedure and hyperparameters for 100 epochs across all datasets. The optimization of the model parameters is performed with the Adam optimizer with a learning rate of 1×10^{-4} and a mini-batch of 64.

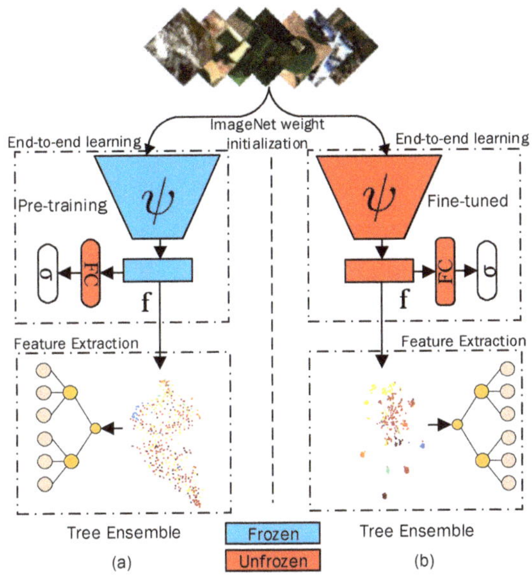

Figure 4. Two distinctive learning strategies for the MLC task. In the first learning setting, marked as (**a**), we rely on the ImageNet pre-training, where all the model layers are frozen, except the last fully connected layer, which is trained for the new target domain. In the second learning setting, marked as (**b**), the entire network architecture is fine-tuned along with the newly added fully connected layer. Moreover, we use both approaches either as feature extractors (in "offline" mode) jointly with tree ensemble methods, or as end-to-end learning machines to directly perform the MLC task. Note that the parts of the model highlighted with blue indicate that there is no update of the parameters in the model during the training process; on the other hand, with red, we highlight the opposite task, which means the parameters of the model are unfrozen and updated during training.

To prevent overfitting, we use data augmentation and modify data on the fly, so that our CNN models are transformation-invariant to the maximum possible extent. Moreover, we use data augmentation because we want to ensure that the predictive performance and generalization capability of the learned model is preserved to some extent, especially when training a predictive model on a dataset that contains a limited number of images (e.g., the Ankara dataset). Using the `Albumentations` library [35] (Available at: https://albumentations.ai (accessed on 4 January 2023)), we performed the following image transformations: horizontal flipping, random rotations in the range $\pm 10\%$, scaling by a factor in the range (0, 0.15), shifting by a factor in the range (0, 0.1), random crops with 50% of the original image size, random brightness within the range of $(-0.3, 0.3)$, and random contrast within the range of $(-0.3, 0.3)$. In addition, we consider the application of the following transformations to the input image: Contrast Limited Adaptive Histogram Equalization (CLAHE) to the input image, where the size of the grid for histogram equalization is set to 8×8 pixels; blurring with Gaussian kernel with σ value in the range (3, 7); median blur with aperture linear size value v randomly sampled once per image in the range (3, 7); motion blur with kernel size ω for blurring the input image, randomly chosen once per image in the range (3, 7); or Gaussian noise, where the variance range for noise is taken from the interval (10, 50).

The augmentations are performed in random order and with a 50% chance, which means that in such a setting there might be no augmentations applied over an image at all. This is important because, during the training process, the model needs to see the original image at least once in order to make reasonable predictions afterwards [36]. Some example augmentations are shown in Figure 5. Each row represents a specific dataset, where the left-most image marked with red is the original version, while the remaining images are the augmented versions of the same image. Note that multiple augmentations could be applied on the same image. Because of that, we are not mixing certain image augmentations, such as the blurring operations (i.e., Gaussian, median and motion blur) with color (i.e., CLAHE) and Gaussian noise corruption transformations, simultaneously. The main reason for this is that we want to avoid strong image degradation and loss of contextual information, which is crucial in remote sensing imagery.

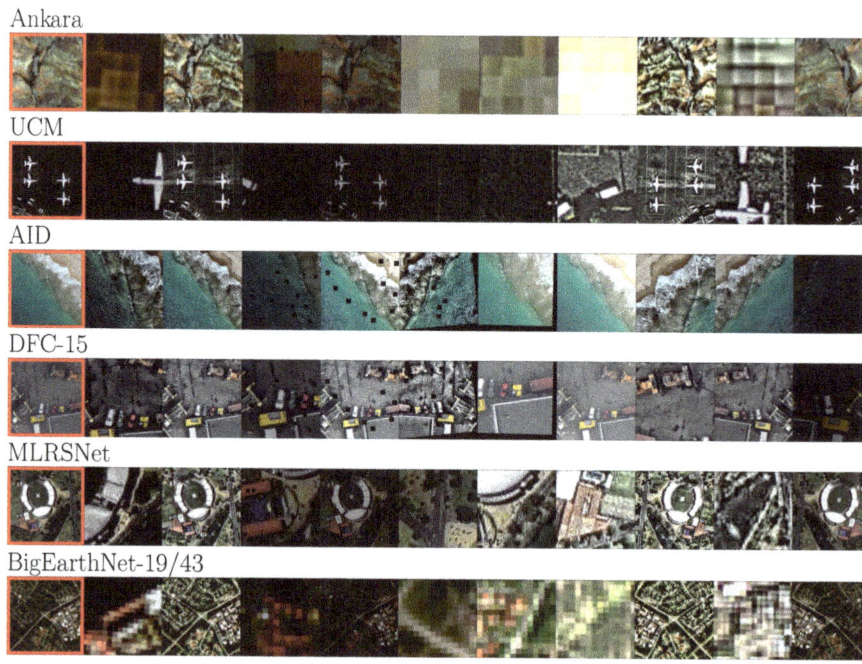

Figure 5. Augmentation examples. The left most image in each row marked with red is the original version, while the remaining images are the augmented versions. We can see the extent of variability added to the training datasets by the augmented images.

3.1.2. CNNs as Feature Extractors and Tree Ensembles

The deep learning methods presented in Section 2.2.1, are further used as feature extractors, for each of the two learning settings described in Section 3.1.1. Hence, we rely on two feature representations $\mathbf{f} \in \mathbb{R}^d$ extracted from the CNN feature extractors: representations based on pre-training (Figure 4a), and fine-tuning (Figure 4b). Furthermore, each feature representation is used as input to the tree ensemble methods, i.e., Random Forests (RF) and Extremely Randomized Trees (ERT)/Extra Trees (ET) for short. We use 150 base models in each of the ensembles and `sqrt` as the feature subset size.

3.2. Evaluation Strategy

We conducted experiments over the several publicly available RSI datasets described above and explored how the performance of the learning strategies is affected by data availability, image quality, and label dimensionality. To tackle this problem, we included datasets which are less challenging in terms of image resolution, such as DFC-15 and MLRSNet, and datasets such as Ankara, which is very limited in size and has poor image quality. To evaluate the effectiveness of the methods, we perform the splitting according to the datasets that already contain a predefined subset of images, such as AID and DFC-15, where the train and test ratios are approximately 80% and 20%. The remaining dataset is partitioned accordingly, by adopting an iterative stratified sampling strategy in order to preserve the relative frequency of the labels in the datasets to the maximum possible extent.

Moreover, we split a validation set of 10% from the overall training data, which is only used to monitor the learning progress and to provide an unbiased estimate of the model fit when tuning hyperparameters. A description of the datasets after the splitting procedure is given in Table 1.

3.3. Evaluation Measures and Statistical Analysis

Many evaluation measures are used to assess the predictive performance and effectiveness of MLC methods, offering different viewpoints on the performance of the methods. The evaluation measures can be grouped into two groups: measures based on bipartitions (example-based and label-based measures) and ranking-based evaluation measures [8]. The example-based evaluation measures compute the average difference between the true labels and the predicted labels for each data point, averaged across all the examples in the dataset. Unlike example-based measures, label-based measures evaluate each label separately and then average the performances across all labels. The ranking-based evaluation measures compare the predicted ranking of the labels with the ground truth ranking (where all present labels are ranked before all absent labels).

In this study, we present the results in terms of ranking-loss as an evaluation measure. Ranking loss evaluates the average fraction of label pairs that are misordered for a given example. Note, however, that we provide complete results in terms of all other evaluation measures for the MLC task in the Appendix A section (for the calculation of the evaluation measures, we use the `sckit-learn` implementation [37]). We focus on ranking loss, since we believe it is one of the best indicators for measuring the performance of methods for MLC. Moreover, this measure is threshold-independent, which means we do not rely on the use of techniques for threshold estimation to produce the predicted labels. Ranking loss is defined as follows:

$$rl = \frac{1}{N}\sum_{i=1}^{N}\sum_{(j,k):y_j>y_k}(I[r_i(j) < r_i(k)] + \frac{1}{2}I[r_i(j) = r_i(k)]), \quad (2)$$

where y_i and $\hat{y}_i$ are the true and the predicted labels, respectively, N is the number of examples, $r_i(j)$ is the ranking of label j for instance x_i, and I is an indicator function. The smaller the value of rl, the better the performance.

We use the Friedman test to assess whether the overall differences in performance of the used approaches evaluated across the RSI datasets are statistically significant and the post hoc Nemenyi test to detect between which methods the statistically significant differences occur. The obtained results are presented in the form of Nemenyi post hoc average rank diagrams [38] for the ranking loss measure. In the analysis, the significance level was set to $\alpha = 0.05$. The best-performing methods are on the left-most side of the diagram along the axis (average ranks closer to 1), and the methods whose predictive performance does not differ significantly at $\alpha = 0.05$, are connected with a red line.

3.4. Implementation Details

We implemented the deep learning models in the `PyTorch` framework (Available at: https://pytorch.org, accessed on 4 January 2023). We use the built-in implementations of VGGs, ResNets and EfficientNets, initialized with weights learned on ImageNet. We modify each of the backbone models in order to accept arbitrary-sized input images by replacing the last max pooling layer with the average pooling operation with a kernel size of 1, resulting in the following d-dimensional feature representations: $d = 4096$ for VGG-16 and VGG-19, $d = 512$ for ResNet-34, and $d = 2048$ for ResNet-50 and ResNet-152. We have 1280 output dimensions for EfficientNet-B0 and EfficientNet-B1, while EfficientNet-B2 has 1408 dimensions. Finally, for the tree ensemble methods, we use the `scikit-learn` [37] implementation of Random Forest tree ensembles for multi-label classification (MLC) [33] and Extra Tree ensembles for MLC [34]. The tree ensembles for MLC simultaneously predict the probability of each of the multiple class labels. The labels can then be ranked based on these predicted probabilities. The complete source code and the datasets used to execute the study are publicly and freely available at https://github.com/marjanstoimchev/RSMLC (accessed on 4 January 2023).

4. Results and Discussion

This section presents the results of our experimental study in MLC of RSI. It answers one of the experimental questions posed in Section 3 in each subsection. It thus discusses (1) the influence of the learning strategy, (2) the comparison of different network architectures used as end-to-end approaches, as well as feature extractors, (3) the comparison between end-to-end methods and tree ensembles, and (4) the influence of the number of available labeled images on predictive performance.

4.1. The Influence of the Learning Strategy

In the first experiment, we explored whether fine-tuning or pre-training only is the more appropriate learning strategy (as defined in Section 3.1.1). To provide the answer to this question, we conducted experiments by training different network architectures over the datasets, and using them in two different modes of operation: (1) as feature extractors providing the feature representations to the tree ensembles, and (2) as end-to-end learning methods. Moreover, for each learning approach, we present the difference in performance in the form of a heat map, which can be formally defined as follows:

$$\mathbf{H} = \begin{bmatrix} rl_{11} & rl_{12} & \cdots & rl_{1n} \\ rl_{21} & rl_{22} & \cdots & rl_{2n} \\ \vdots & \vdots & \ddots & \vdots \\ rl_{m1} & rl_{m2} & \cdots & rl_{mn} \end{bmatrix} \in \mathbb{R}^{m \times n}, \quad (3)$$

where $rl_{i,j} = rl_{i,j}^{fine-tune} - rl_{i,j}^{pre-train}$ is the difference between fine-tuning ($rl^{fine-tune}$) and pre-training ($rl^{pre-train}$) in terms of the ranking loss measure, calculated for the i-th dataset and the j-th method, respectively. This is performed for $i = 1, \ldots, m$ and $j = 1, \ldots, n$, where m denotes the number of datasets and n is the number of methods/CNN architectures.

These differences in performance are presented in Figure 6. We observe that for smaller datasets (e.g., Ankara, UCM, etc.), pre-training is the preferred choice, which can be mostly seen for the VGG architectures trained in an end-to-end manner over the DFC-15 dataset, where the largest differences in ranking loss are observed. These differences are slightly smaller for the DFC-15 dataset and the tree-ensemble methods. On the other hand, the fine-tuned versions of the models perform significantly better when data availability is not a problem. Furthermore, for the tree ensemble methods, the differences in performance are significantly increased (and more on the positive side), which points to the fact that the quality of the feature representations is of great importance for the tree ensemble methods to further boost their predictive performance. Overall, we can conclude that by solely relying on the ImageNet pre-training, we end up with worse model performance in almost all cases, because the content present in RSI is quite complex as compared to images present in the ImageNet dataset. More detailed results of the analysis are presented in Table A1 (Appendix A), where the performance figures for the ranking loss measure are given.

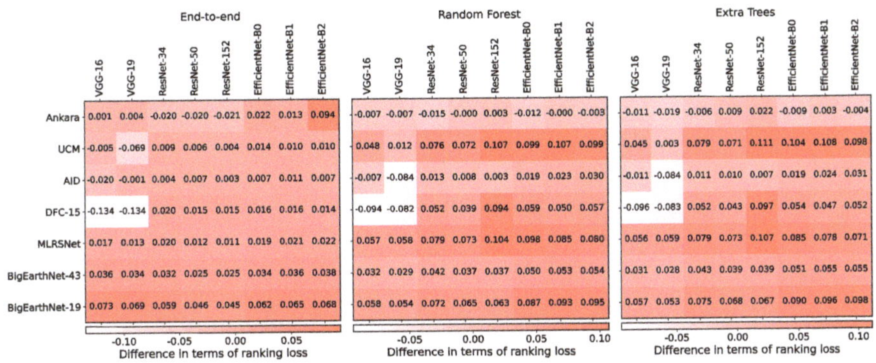

Figure 6. Comparison between fine-tuning and pre-training in terms of ranking loss. The results for the different architectures are presented in the form of a heat map showing the difference in performance between the learning approaches. Negative values indicate that pre-training is better than fine-tuning and positive when fine-tuning is better than pre-training. The intensities of the colors in the heat map are directly related to the difference in performance between the learning approaches (CNN architectures). The datasets on the y-axis are ordered by the number of images they contain.

4.2. Comparison of Different Network Architectures

Based on the analysis from Section 4.1, we used the fine-tuning approach to further explore which network architecture is the most suitable choice for RSI MLC tasks. The results are shown in terms of ranking loss and in the form of average rank diagrams. From Figure 7, we see that the EfficientNet variants, especially EfficientNet-B2, tend to produce the better results as compared to other network architectures. The differences in performance are statistically significant as compared to the VGG variants. The results also reveal that the EfficientNet-B2 model, used as base feature extractor in combination with tree ensemble methods, such as RF as in Figure 7b and ET in Figure 7c, is the clear winner, which indicates that this network architecture is the best choice for this task. Moreover, the ResNet-based network architectures are the closest competitors to EfficientNet variants (both when used with tree ensembles and in an end-to-end manner), i.e., the ResNet-152 model. Although the EfficientNet variants are not statistically significantly better than ResNet-152, they are by far more lightweight in terms of model parameters (e.g., EfficientNet-B0 is approximately 11× smaller, EfficientNet-B1 is 8×, and EfficientNet-B2 is approximately 6.5× smaller than ResNet-152, respectively). Comparison between different network architectures in terms of other MLC performance measures in the form of average rank diagrams is given in Figures A1–A6 in the Appendix A (Figures A1–A3 for fine-tuned features and label-based, example-based and ranking-based measures, respectively, Figures A4–A6 for pre-trained features and label-based, example-based and ranking-based measures, respectively).

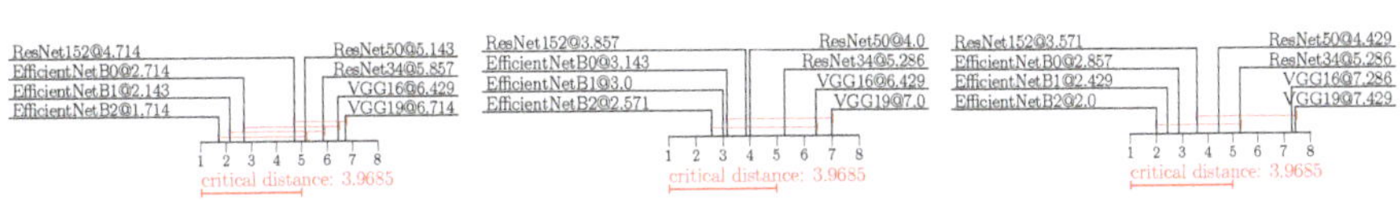

(**a**) End-to-end (**b**) Random forest (**c**) Extra trees

Figure 7. Comparison between different network architectures in terms of ranking loss. The results are presented in the form of average rank diagrams at 0.05 significance level for (**a**) End-to-end learning, (**b**) Random forests and (**c**) Extra trees. The best ranking methods are at the left-most side of the diagram. The difference in performance among the methods connected with a red line is not statistically significant.

Overall, we can conclude that in the fine-tuning setting, it is favorable to use lightweight network architectures in terms of model parameters for end-to-end learning, which are also capable of learning more discriminative feature representations when used as feature extractors. This is in direct relation to their capability of capturing high-level content present in RSI to a greater extent as compared to the other network architectures. Moreover, they are the preferred choice when addressing the problems encountered in challenging deployment scenarios, which means they can maintain good predictive performance while keeping the computational costs at a reasonable level.

4.3. Comparison of Different Learning Approaches

To answer the experimental question of how end-to-end learning and feature extraction plus the tree ensembles compare in the task of RSI MLC, we present the results in the form of average rank diagrams, where we use the pre-trained and fine-tuned versions of the EfficientNet-B2 model. Recall that EfficientNet-B2 produced the best results overall, either when used as a feature extractor where the extracted feature representations are further utilized in the tree ensembles, or as an end-to-end approach to directly address the MLC task. The results are shown in Figure 8a for fine-tuning and in Figure 8b for pre-training only.

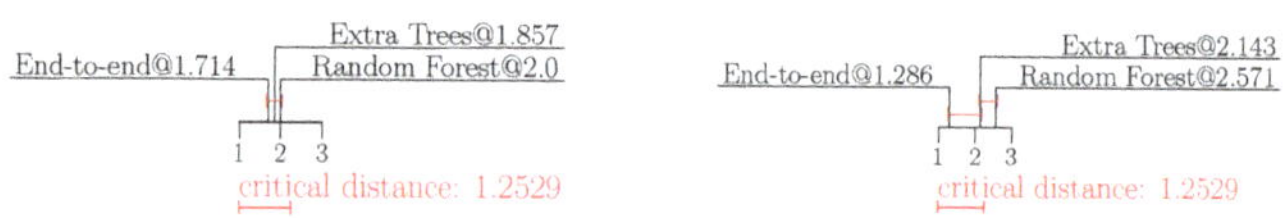

(**a**) Fine-tuning (**b**) Pre-training

Figure 8. Comparison between different learning methods in terms of ranking loss. The results are presented in the form of average rank diagrams at 0.05 significance level for (**a**) Fine-tuning and (**b**) Pre-training only. The best ranking methods are at the left-most side of the diagram. The differences among the methods connected with a red line is not statistically significant.

As seen from the diagrams, end-to-end learning is ranked best, followed by Extra trees and Random Forests. In the case of fine-tuning (Figure 8a), there is no statistically significant difference among the classification methods. In the pre-training setting, the end-to-end learning approach is significantly better than the random forest method. Overall, the differences in predictive performance depend on the specific deep neural network architecture and the learning setting. Comparison between different learning methods in terms of other MLC performance measures in the form of average rank diagrams is given in Figures A7 and A8 in the Appendix A. Figure A7 concerns the use of fine-tuned and Figure A8 the use of pre-trained features.

4.4. Influence of the Number of Available Labeled Images

To answer the fourth question about the relation of the number of available images and the performance of the different MLC models, we conducted additional experiments focusing on the biggest dataset available—BigEarthNet. We design an experimental protocol which partitions the BigEarthNet dataset into distinctive subsets of images, namely, fractions of: 0.1%, 0.5%, 1%, 5%, 10%, 25%, and 50% of 590,326 images, respectively. Each of the fractions is further split into disjoint subsets of training, validation, and testing images, with sizes 70%, 10%, and 20% of the fraction size, respectively. Moreover, the test sets are built in a cumulative manner, which means the test subset of images from the previous fractions are inherited into the test set of the next fraction. By doing this, we are evaluating the effectiveness of the models on new subsets of images, while taking the old ones into account, i.e., we simulate a scenario where we add new images. The sampling of the fractions and the subsets within the fractions can be done in two different ways: (i) sampling with stratification, and (ii) random sampling. Furthermore, we are also assessing the generalization capability of the model when exposed to different distribution shifts. To obtain a more reliable estimate of the predictive performance, we repeated the experiment five times and calculated the average performance and standard deviation ($\mu \pm \sigma$) in terms of ranking loss.

To carry out the experiment, we used the two versions of the BigEarthNet dataset, namely, BigEarthNet with 19 and 43 CLC nomenclatures. We selected this dataset because it is the largest one in terms of the number of images. For these experiments, we used EfficientNet-B2 as network architecture, since it produced the best results overall in the previous experiments. We fine-tuned the model parameters for 25 epochs and used the trained model as a feature extractor for the tree ensembles, as well as an end-to-end approach. We present the results in the form of learning curves in order to see how the number of training examples influences the predictive performance. The results of the experiment for the BigEarthNet-19 and BigEarthNet-43 are shown in Figure 9. We can see that the performance of the learning methods in all cases improves of the total number of training examples by up to 10%, after which it degrades. End-to-end learning methods perform worse than tree ensemble methods in all cases in the learning curve, but the differences in performance are only visible in the case of stratified sampling. The differences are more expressed in the BigEarthNet-19 dataset (Figure 9). This means that the different learning methods are affected differently by the choice of the sampling strategy.

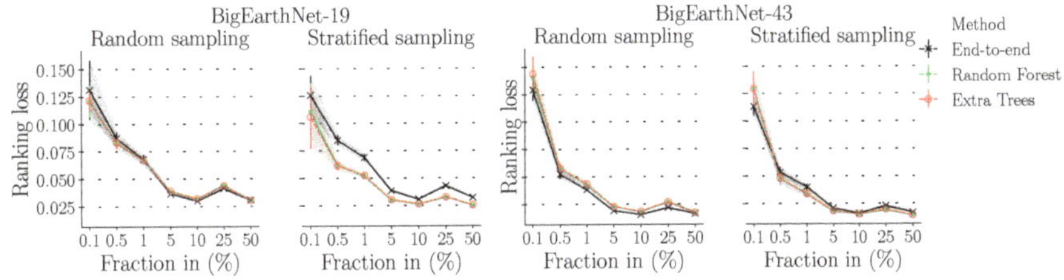

Figure 9. Comparison between different learning methods in terms of ranking loss when evaluated on different fractions of labeled examples from the BigEarthNet-19 and BigEarthNet-43 datasets,. The results are shown as learning curves depicting $\mu \pm \sigma$ for ranking loss across five repeats. The y-axis is shared across the figures.

5. Conclusions

In this study, we have presented a comparative analysis of methods for multi-label classification of remote sensing imagery. We have compared several popular deep learning methods based on two modes of operation, where they are (1) used as feature extractors in a combination with tree ensemble methods such as random forests and extra trees, and (2) used as end-to-end approaches to directly address the MLC task.

We focused on four aspects: (1) different transfer learning paradigms, namely, learning features based on ImageNet pre-training only, as well as learning features with fine-tuning, where the whole network architecture and the model parameters are trained on the new target domain of interest; (2) comparison between different network architectures; (3) comparison between tree ensembles and end-to-end approaches; and (4) investigating the influence of the number of labeled examples on the relative predictive performance of MLC methods. In the first dimension, we showed that it is beneficial to fine-tune the models on RSI to improve their performance. However, in certain cases, where the number of data is limited, it is better to extract features with ImageNet pre-training and only fine-tune the last layer for classification. Furthermore, we showed that it is very important to choose a proper network architecture: EfficientNets proved to be overall the most suitable choice for the task of MLC. They have significantly fewer parameters as compared to the ResNet variants, and no statistically significant difference in predictive performance is observed. We also showed that having the right feature extractor plays an important role in boosting the performance of tree ensemble methods, so that they outperform the end-to-end approaches in certain cases. In the last dimension, we investigated the influence of the amount of labeled data from the BigEarthNet dataset on the relative performance of MLC methods, where we applied two types of sampling strategies: random sampling and sampling with stratification. We showed that in such a setting, the tree ensemble methods outperform the end-to-end approaches, with the difference in performance clearly visible for stratified sampling.

Considering the findings from this study, further extension of this work should focus on several aspects. Firstly, we will incorporate even a wider range of deep learning models specially devised for the RS MLC task. Next, we will take into account different MLC loss functions to analyze whether they are more suitable for the RS MLC task. Moreover, since the label space of BigEarthNet dataset is organized in a hierarchical manner, we will further analyze the effect of using the hierarchical information in the tree ensemble methods and in the end-to-end approaches. Lastly, we will focus on the application of the methods in a semi-supervised learning setting, where we will exploit the abundance of the unlabeled data in RS domain and investigate if the semi-supervised counterparts can surpass the performance of the supervised learning methods.

Author Contributions: Conceptualization, S.D.; methodology, M.S., D.K., and S.D.; software, M.S.; validation, M.S. and D.K.; writing—original draft preparation, M.S.; writing—review and editing, D.K. and S.D.; supervision, S.D. All authors have read and agreed to the published version of the manuscript.

Funding: We acknowledge the support of the European Space Agency ESA through the activity of the AiTLAS - AI4EO rapid prototyping environment. This work was also partially supported by the Slovenian research agency through the knowledge technologies program P2-0103 and the project J2-2505.

Institutional Review Board Statement: Not applicable.

Informed Consent Statement: Not applicable.

Data Availability Statement: The AID Dataset in this study is openly and freely available at https://github.com/Hua-YS/AID-Multilabel-Dataset (accessed on 1 July 2022). The DFC-15 dataset in this study is openly and freely available at https://drive.google.com/drive/folders/1TKGS6TIRxQ6a7gdaj0cHs-mRCtv_J1HA (accessed on 1 July 2022). The MLRSNet dataset in this study is openly and freely available at https://data.mendeley.com/datasets/7j9bv9vwsx/2 (accessed on 1 July 2022). UCM dataset in this study is openly and freely available at https://drive.google.com/file/d/1DtKiauowCB0ykjFe8v0OVvT76rEfOk0v/view (accessed on 1 July 2022). The Ankara dataset in this study is openly and freely available at https://bigearth.eu/datasets (accessed on 1 July 2022). The BigEarthNet-19 and BigEarthNet-43 in this study are openly and freely available at https://bigearth.net/ (accessed on 1 July 2022).

Conflicts of Interest: The authors declare no conflict of interest.

Abbreviations

The following abbreviations are used in this manuscript:

RSI	Remote Sensing Images
MLC	Multi-Label-Classification
DNN	Deep Neural Network
CNN	Convolutional Neural Network
DL	Deep Learning
RNN	Recurrent Neural Network
LSTM	Long Short-Term Memory
GCN	Graph Convolutional Network
CLC	Corine Land Cover
ReLU	Rectified Linear Unit
RF	Random Forest
ET	Extra Trees

Appendix A. Complete Results from the Experimental Evaluation

Table A1. The performance in terms of ranking loss measure of fine-tuning and pre-training feature learning approaches with different network architectures and different MLC approaches. The best performing network architecture for each dataset is highlighted with green.

Approach	Datasets	VGG-16	VGG-19	ResNet-34	ResNet-50	ResNet-152	EfficientNet-B0	EfficientNet-B1	EfficientNet-B2
					Pre-Training				
End-to-end	Ankara	0.298	0.371	0.343	0.351	0.350	0.349	0.330	0.422
	UCM	0.186	0.180	0.154	0.149	0.135	0.194	0.185	0.184
	AID	0.215	0.208	0.171	0.181	0.179	0.198	0.194	0.188
	DFC-15	0.176	0.176	0.147	0.134	0.134	0.127	0.120	0.113
	MLRSNet	0.347	0.360	0.306	0.240	0.229	0.318	0.300	0.326
	BigEarthNet-19	0.557	0.550	0.461	0.399	0.391	0.460	0.476	0.478
	BigEarthNet-43	0.578	0.546	0.480	0.431	0.410	0.468	0.481	0.480
Random Forest	Ankara	0.322	0.320	0.324	0.329	0.388	0.317	0.337	0.345
	UCM	0.381	0.398	0.469	0.420	0.539	0.495	0.508	0.468
	AID	0.250	0.244	0.265	0.247	0.294	0.268	0.257	0.262
	DFC-15	0.297	0.337	0.235	0.201	0.342	0.259	0.222	0.242
	MLRSNet	0.529	0.545	0.566	0.549	0.615	0.587	0.567	0.548
	BigEarthNet-19	0.534	0.525	0.588	0.543	0.532	0.637	0.670	0.669
	BigEarthNet-43	0.547	0.537	0.600	0.557	0.545	0.654	0.686	0.683
Extra Trees	Ankara	0.318	0.312	0.331	0.325	0.370	0.328	0.320	0.330
	UCM	0.390	0.405	0.457	0.417	0.552	0.483	0.513	0.462
	AID	0.254	0.250	0.255	0.253	0.305	0.265	0.250	0.256
	DFC-15	0.297	0.338	0.235	0.204	0.351	0.235	0.218	0.229
	MLRSNet	0.530	0.545	0.567	0.549	0.620	0.573	0.556	0.534
	BigEarthNet-19	0.539	0.528	0.608	0.567	0.556	0.658	0.693	0.698
	BigEarthNet-43	0.550	0.540	0.622	0.581	0.570	0.676	0.709	0.713
					Fine-tuning				
End-to-end	Ankara	0.294	**0.285**	0.377	0.356	0.360	0.335	0.353	0.322
	UCM	0.224	0.508	0.101	0.097	0.112	0.088	0.096	**0.081**
	AID	0.265	0.202	0.152	0.143	0.147	0.137	**0.131**	0.137
	DFC-15	0.433	0.433	0.068	0.075	0.067	0.054	0.046	0.050
	MLRSNet	0.180	0.223	0.093	0.091	0.088	**0.082**	0.084	0.084
	BigEarthNet-19	0.276	0.282	0.235	0.236	0.210	0.207	0.203	**0.202**
	BigEarthNet-43	0.271	0.276	0.243	0.232	0.199	0.206	0.195	**0.194**
Random Forest	Ankara	0.318	0.319	0.344	0.338	0.345	0.304	0.335	0.320
	UCM	0.182	0.323	0.103	0.098	0.103	0.106	0.104	0.106
	AID	0.245	0.197	0.146	0.144	0.149	0.138	0.138	0.137
	DFC-15	0.433	0.433	0.050	0.047	0.050	0.046	**0.041**	0.044
	MLRSNet	0.185	0.221	0.104	0.103	0.102	0.093	0.095	0.090
	BigEarthNet-19	0.258	0.268	0.229	0.219	0.214	0.222	0.219	0.217
	BigEarthNet-43	0.255	0.267	0.235	0.230	0.219	0.228	0.221	0.224
Extra Trees	Ankara	0.364	0.348	0.342	0.322	0.362	0.313	0.343	0.321
	UCM	0.177	0.334	0.103	0.102	0.102	0.100	0.098	0.097
	AID	0.248	0.194	0.144	0.146	0.146	0.135	0.134	0.136
	DFC-15	0.433	0.433	0.049	0.050	0.046	0.046	0.041	0.043
	MLRSNet	0.184	0.221	0.106	0.104	0.103	0.097	0.100	0.094
	BigEarthNet-19	0.257	0.268	0.227	0.217	0.213	0.222	0.218	0.216
	BigEarthNet-43	0.255	0.267	0.233	0.228	0.217	0.227	0.219	0.224

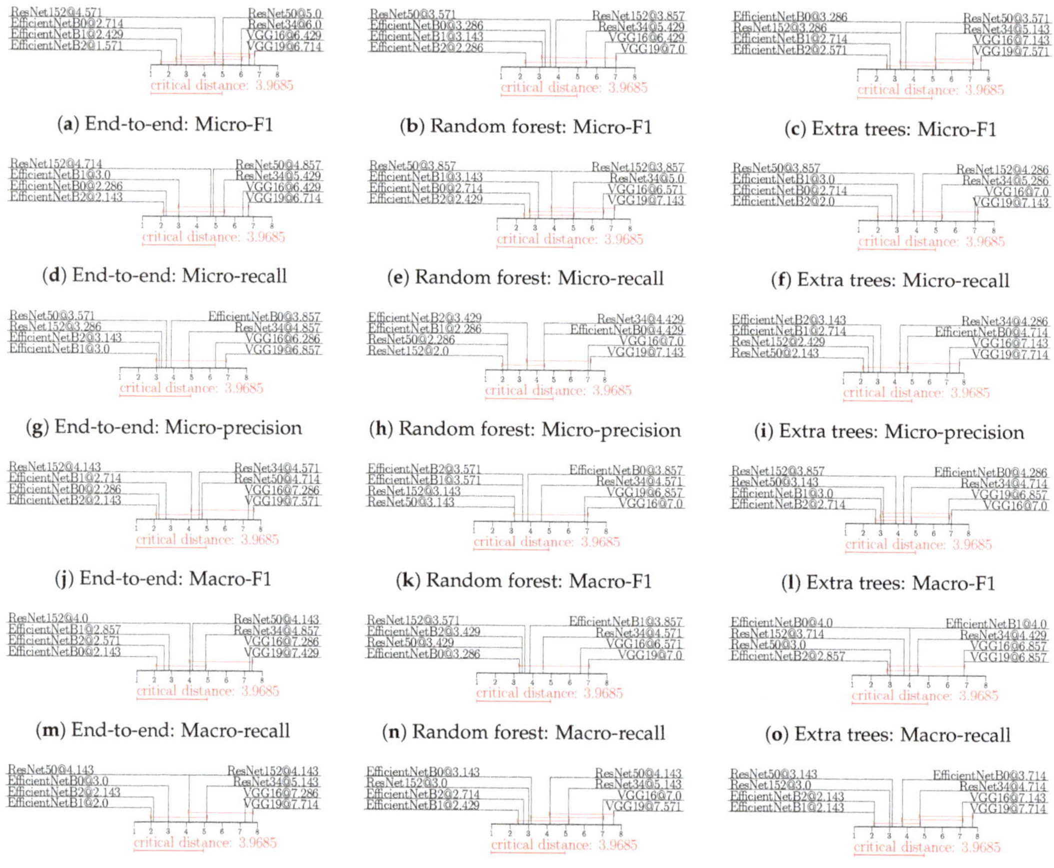

(**a**) End-to-end: Micro-F1 (**b**) Random forest: Micro-F1 (**c**) Extra trees: Micro-F1

(**d**) End-to-end: Micro-recall (**e**) Random forest: Micro-recall (**f**) Extra trees: Micro-recall

(**g**) End-to-end: Micro-precision (**h**) Random forest: Micro-precision (**i**) Extra trees: Micro-precision

(**j**) End-to-end: Macro-F1 (**k**) Random forest: Macro-F1 (**l**) Extra trees: Macro-F1

(**m**) End-to-end: Macro-recall (**n**) Random forest: Macro-recall (**o**) Extra trees: Macro-recall

(**p**) End-to-end: Macro-precision (**q**) Random forest: Macro-precision (**r**) Extra trees: Macro-precision

Figure A1. Performance of different network architectures in terms of label-based evaluation measures for fine-tuned features. The results are presented in the form of average rank diagrams at a 0.05 significance level. The best ranking methods are at the left-most side of the diagram. The difference among the methods connected with a red line is not statistically significant.

(**a**) End-to-end: Hamming loss (**b**) Random forest: Hamming loss (**c**) Extra trees: Hamming loss

(**d**) End-to-end: Subset Accuracy (**e**) Random forest: Subset Accuracy (**f**) Extra trees: Subset Accuracy

Figure A2. Performance of different network architectures in terms of example-based evaluation measures for fine-tuned features. The results are presented in the form of average rank diagrams at a 0.05 significance level. The best ranking methods are at the left-most side of the diagram. The difference among the methods connected with a red line is not statistically significant.

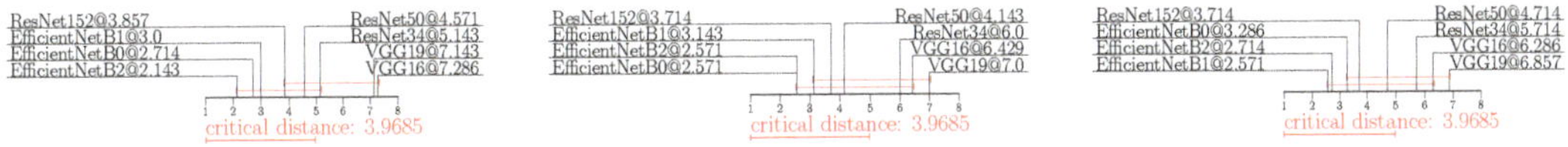

(**a**) End-to-end: Coverage (**b**) Random forest: Coverage (**c**) Extra trees: Coverage

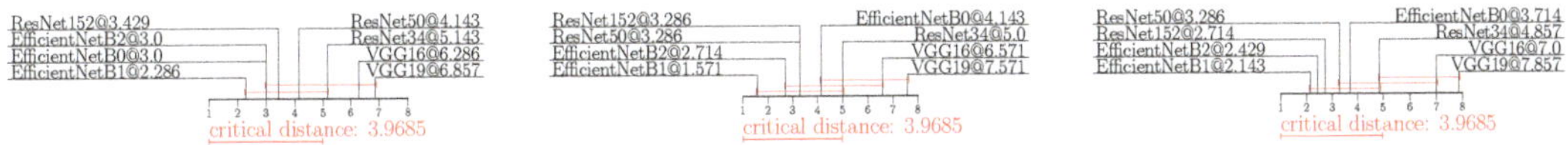

(**d**) End-to-end: Average Precision (**e**) Random forest: Average Precision (**f**) Extra trees: Average Precision

Figure A3. Performance of different network architectures in terms of ranking-based evaluation measures for fine-tuned features. The results are presented in the form of average rank diagrams at a 0.05 significance level. The best ranking methods are at the left-most side of the diagram. The difference among the methods connected with a red line is not statistically significant.

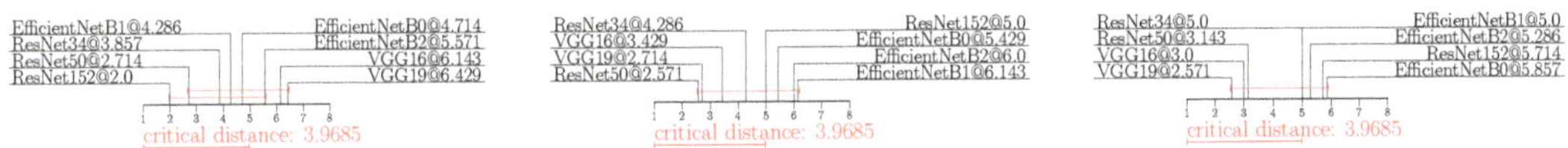

(**a**) End-to-end: Micro-F1 (**b**) Random forest: Micro-F1 (**c**) Extra trees: Micro-F1

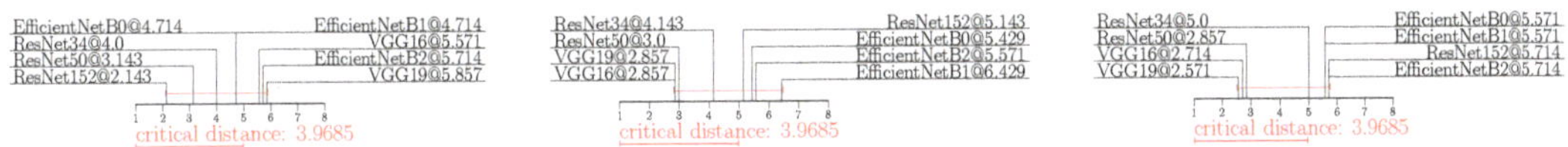

(**d**) End-to-end: Micro-recall (**e**) Random forest: Micro-recall (**f**) Extra trees: Micro-recall

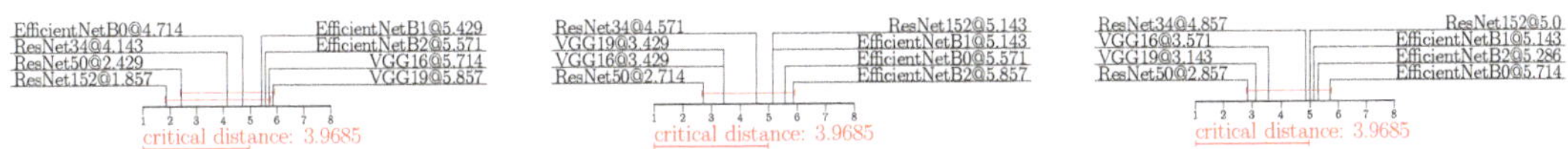

(**g**) End-to-end: Micro-precision (**h**) Random forest: Micro-precision (**i**) Extra trees: Micro-precision

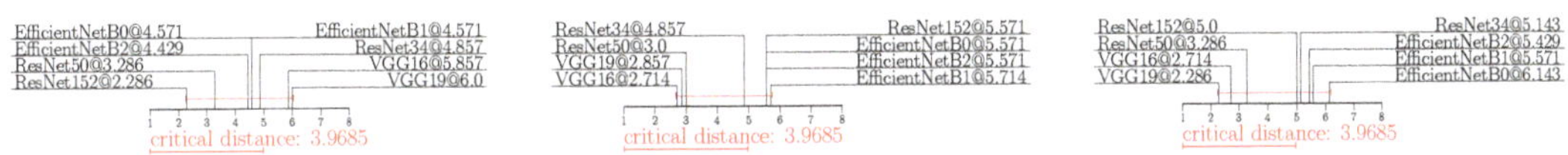

(**j**) End-to-end: Macro-F1 (**k**) Random forest: Macro-F1 (**l**) Extra trees: Macro-F1

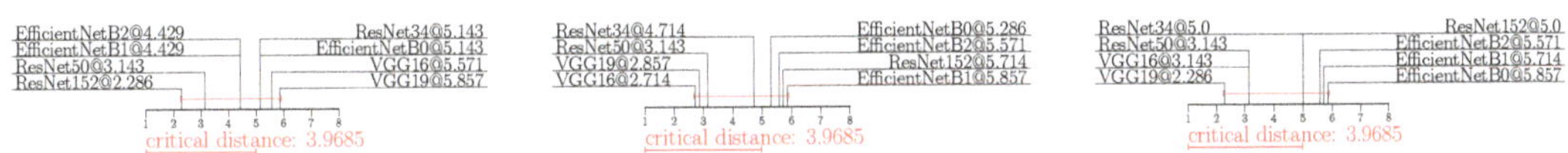

(**m**) End-to-end: Macro-recall (**n**) Random forest: Macro-recall (**o**) Extra trees: Macro-recall

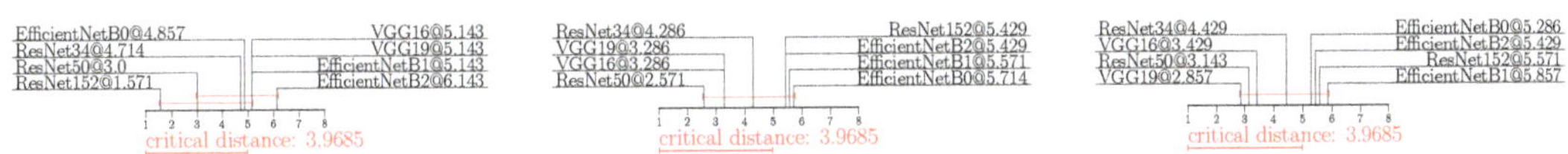

(**p**) End-to-end: Macro-precision (**q**) Random forest: Macro-precision (**r**) Extra trees: Macro-precision

Figure A4. Performance of different network architectures in terms of label-based evaluation measures for pre-trained features. The results are presented in the form of average rank diagrams at a 0.05 significance level. The best ranking methods are at the left-most side of the diagram. The difference among the methods connected with a red line is not statistically significant.

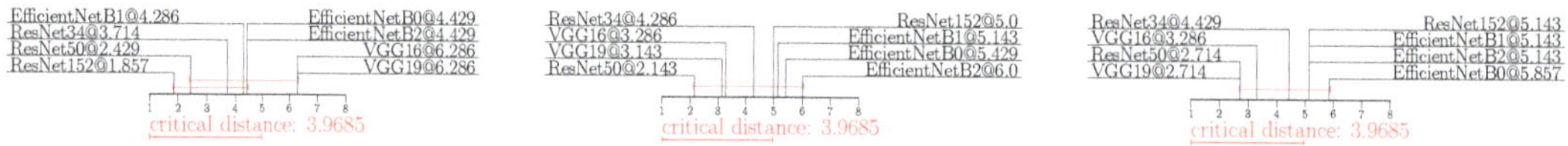

(**a**) End-to-end: Hamming loss (**b**) Random forest: Hamming loss (**c**) Extra trees: Hamming loss

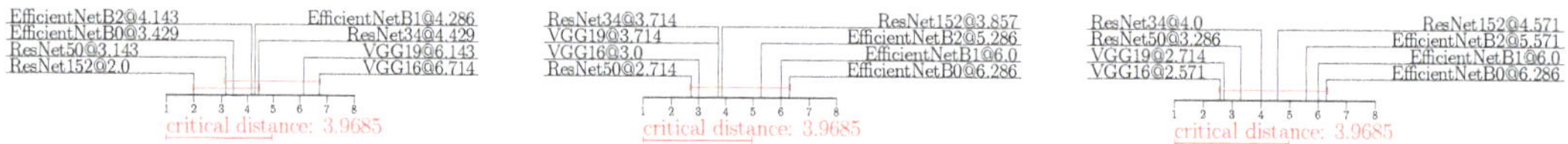

(**d**) End-to-end: Subset Accuracy (**e**) Random forest: Subset Accuracy (**f**) Extra trees: Subset Accuracy

Figure A5. Performance of different network architectures in terms of example-based evaluation measures for pre-trained features. The results are presented in the form of average rank diagrams at a 0.05 significance level. The best ranking methods are at the left-most side of the diagram. The difference among the methods connected with a red line is not statistically significant.

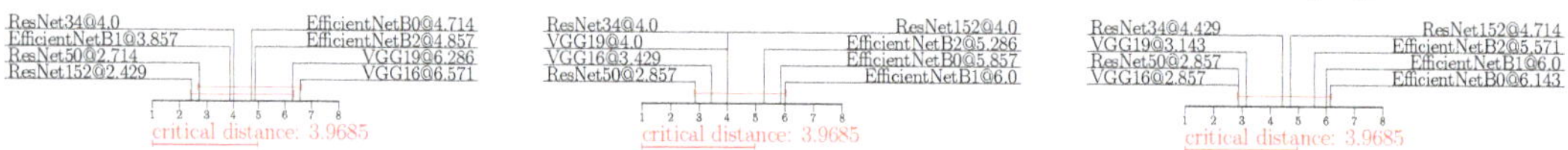

(**a**) End-to-end: Coverage (**b**) Random forest: Coverage (**c**) Extra trees: Coverage

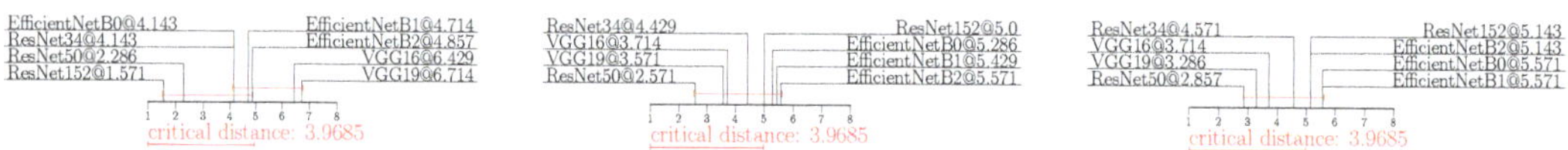

(**d**) End-to-end: Average Precision (**e**) Random forest: Average Precision (**f**) Extra trees: Average Precision

Figure A6. Performance of different network architectures in terms of ranking-based evaluation measures for pre-trained features. The results are presented in the form of average rank diagrams at a 0.05 significance level. The best ranking methods are at the left-most side of the diagram. The difference among the methods connected with a red line is not statistically significant.

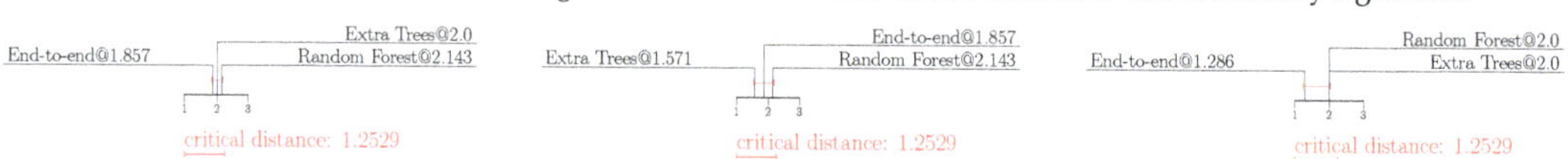

(**a**) Micro F_1 (**b**) Micro Precision (**c**) Micro Recall

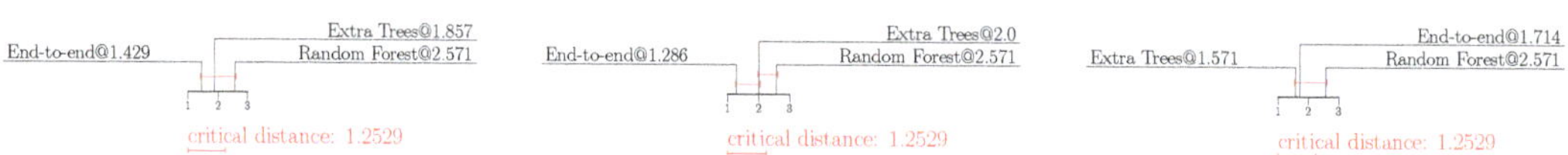

(**d**) Macro F_1 (**e**) Macro Recall (**f**) Macro Precision

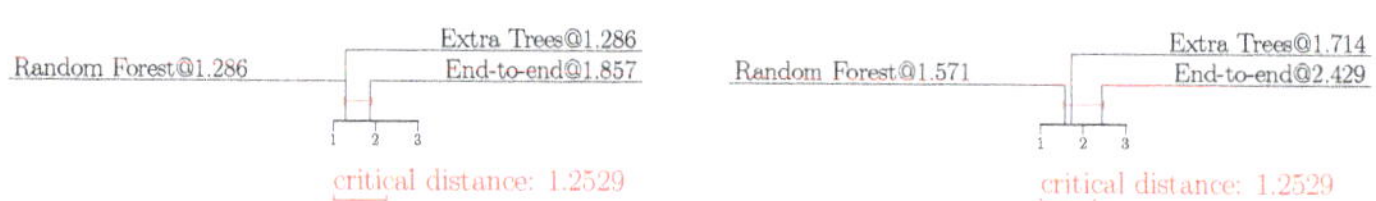

(**g**) Hamming loss (**h**) Subset Accuracy

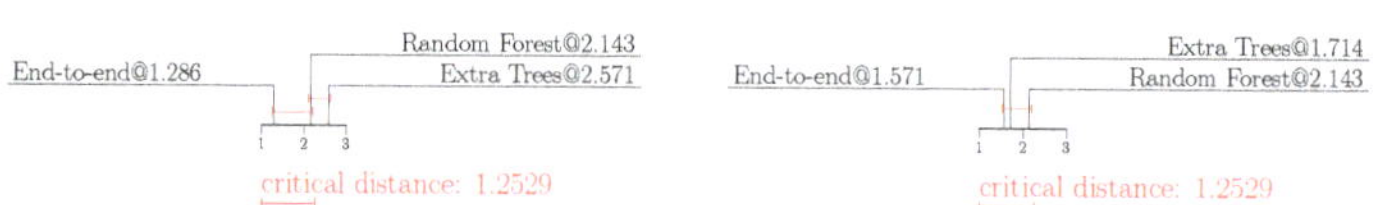

(**i**) Coverage (**j**) Average Precision

Figure A7. Performance of different learning methods in terms of all MLC evaluation measures for fine-tuned features. The results are presented in the form of average rank diagrams at a 0.05 significance level. The best ranking methods are at the left-most side of the diagram. The difference among the methods connected with a red line is not statistically significant.

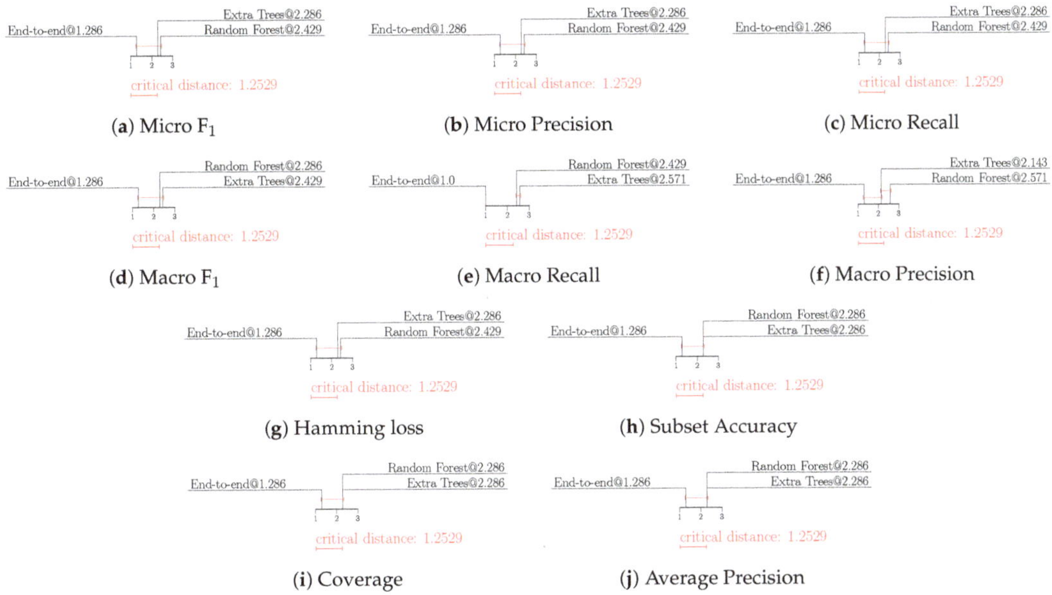

Figure A8. Performance of different learning methods in terms of all MLC evaluation measures for pre-trained features. The results are presented in the form of average rank diagrams at a 0.05 significance level. The best ranking methods are at the left-most side of the diagram. The difference among the methods connected with a red line is not statistically significant.

References

1. Ibrahim, S.K.; Ziedan, I.E.; Ahmed, A. Study of Climate Change Detection in North-East Africa Using Machine Learning and Satellite Data. *IEEE J. Sel. Top. Appl. Earth Obs. Remote Sens.* **2021**, *14*, 11080–11094. [CrossRef]
2. Chen, H.; Qi, Z.; Shi, Z. Remote Sensing Image Change Detection With Transformers. *IEEE Trans. Geosci. Remote Sens.* **2022**, *60*, 1–14. [CrossRef]
3. Ortega Adarme, M.; Queiroz Feitosa, R.; Nigri Happ, P.; Aparecido De Almeida, C.; Rodrigues Gomes, A. Evaluation of Deep Learning Techniques for Deforestation Detection in the Brazilian Amazon and Cerrado Biomes From Remote Sensing Imagery. *Remote Sens.* **2020**, *12*. [CrossRef]
4. Park, M.; Tran, D.Q.; Jung, D.; Park, S. Wildfire-Detection Method Using DenseNet and CycleGAN Data Augmentation-Based Remote Camera Imagery. *Remote Sens.* **2020**, *12*, 910. [CrossRef]
5. Zhang, Q.; Ge, L.; Zhang, R.; Metternicht, G.I.; Liu, C.; Du, Z. Towards a Deep-Learning-Based Framework of Sentinel-2 Imagery for Automated Active Fire Detection. *Remote Sens.* **2021**, *13*, 4790. [CrossRef]
6. Papoutsis, I.; Bountos, N.I.; Zavras, A.; Michail, D.; Tryfonopoulos, C. Benchmarking and scaling of deep learning models for land cover image classification. *ISPRS J. Photogramm. Remote Sens.* **2023**, *195*, 250–268. [CrossRef]
7. Yansheng, L.; Ruixian, C.; Yongjun, Z.; Mi, Z.; Ling, C. Multi-Label Remote Sensing Image Scene Classification by Combining a Convolutional Neural Network and a Graph Neural Network. *Remote Sens.* **2020**, *12*, 4003.
8. Bogatinovski, J.; Todorovski, L.; Džeroski, S.; Kocev, D. Comprehensive comparative study of multi-label classification methods. *Expert Syst. Appl.* **2022**, *203*, 117215. [CrossRef]
9. Dosovitskiy, A.; Beyer, L.; Kolesnikov, A.; Weissenborn, D.; Zhai, X.; Unterthiner, T.; Dehghani, M.; Minderer, M.; Heigold, G.; Gelly, S.; et al. An Image is Worth 16 × 16 Words: Transformers for Image Recognition at Scale. In Proceedings of the International Conference on Learning Representations (ICLR), virtual, 3–7 May 2021.
10. Deng, J.; Dong, W.; Socher, R.; Li, L.; Kai, L.; Fei-Fei, L. ImageNet: A Large-Scale Hierarchical Image Database. In Proceedings of the IEEE Conference on Computer Vision and Pattern Recognition (CVPR), Miami, FL, USA, 20–25 June 2009; pp. 248–255.
11. Dimitrovski, I.; Kitanovski, I.; Kocev, D.; Simidjievski, N. Current Trends in Deep Learning for Earth Observation: An Open-source Benchmark Arena for Image Classification. *arXiv* **2022**, arXiv:2207.07189.
12. Pires de Lima, R.; Marfurt, K. Convolutional Neural Network for Remote-Sensing Scene Classification: Transfer Learning Analysis. *Remote Sens.* **2020**, *12*, 86. [CrossRef]

13. Khaleghian, S.; Ullah, H.; Kræmer, T.; Hughes, N.; Eltoft, T.; Marinoni, A. Sea Ice Classification of SAR Imagery Based on Convolution Neural Networks. *Remote Sens.* **2021**, *13*, 1734. [CrossRef]
14. Wang, A.X.; Tran, C.; Desai, N.; Lobell, D.; Ermon, S. Deep Transfer Learning for Crop Yield Prediction with Remote Sensing Data. In Proceedings of the 1st ACM SIGCAS Conference on Computing and Sustainable Societies, San Jose, CA, USA, 20–22 June 2018; Association for Computing Machinery: New York, NY, USA, 2018; COMPASS'18.
15. Wang, J.; Yang, Y.; Mao, J.; Huang, Z.; Huang, C.; Xu, W. CNN-RNN: A Unified Framework for Multi-label Image Classification. In Proceedings of the IEEE Conference on Computer Vision and Pattern Recognition (CVPR), Las Vegas, NV, USA, 27–30 June 2016; pp. 2285–2294.
16. Chen, Z.; Wei, X.; Wang, P.; Guo, Y. Multi-Label Image Recognition with Graph Convolutional Networks. In Proceedings of the IEEE Conference on Computer Vision and Pattern Recognition (CVPR), Long Beach, CA, USA, 15–20 June 2019; pp. 5172–5181.
17. Sumbul, G.; Charfuelan, M.; Demir, B.; Markl, V. BigEarthNet: A Large-Scale Benchmark Archive for Remote Sensing Image Understanding. *IEEE Int. Geosci. Remote Sens. Symp.* **2019**, *12*, 5901–5904.
18. Yessou, H.; Sumbul, G.; Demir, B. A Comparative Study of Deep Learning Loss Functions for Multi-Label Remote Sensing Image Classification. In Proceedings of the IEEE International Geoscience and Remote Sensing Symposium (IGARSS), Waikoloa, HI, USA, 26 September–2 October 2020;
19. Sumbul, G.; Kang, J.; Demir, B. Deep Learning for Image Search and Retrieval in Large Remote Sensing Archives. *arXiv* **2020**, arXiv:2004.01613.
20. Hua, Y.; Mou, L.; Zhu, X. Relation Network for Multi-label Aerial Image Classification. *IEEE Trans. Geosci. Remote Sens.* **2019**, *58*, 4558–4572. [CrossRef]
21. Sumbul, G.; Demİr, B. A Deep Multi-Attention Driven Approach for Multi-Label Remote Sensing Image Classification. *IEEE Access* **2020**, *8*, 95934–95946. [CrossRef]
22. Wang, X.; Duan, L.; Ning, C. Global Context-Based Multilevel Feature Fusion Networks for Multilabel Remote Sensing Image Scene Classification. *IEEE J. Sel. Top. Appl. Earth Obs. Remote Sens.* **2021**, *14*, 11179–11196. [CrossRef]
23. Karen, S.; Andrew, Z. Very Deep Convolutional Networks for Large-Scale Image Recognition. In Proceedings of the International Conference on Learning Representations (ICLR), San Diego, CA, USA, 7–9 May 2015.
24. Kaiming, H.; Xiangyu, Z.; Shaoqing, R.; Jian, S. Deep Residual Learning for Image Recognition. In Proceedings of the 2016 IEEE Conference on Computer Vision and Pattern Recognition (CVPR), Las Vegas, NV, USA, 27–30 June 2016; pp. 770–778.
25. Tan, M.; Le, Q. EfficientNet: Rethinking Model Scaling for Convolutional Neural Networks. In Proceedings of the 36th International Conference on Machine Learning, Long Beach, CA, USA, 9–15 June 2019; Chaudhuri, K., Salakhutdinov, R., Eds.; Volume 97, pp. 6105–6114.
26. Yang, Y.; Newsam, S. Bag-of-visual-words and spatial extensions for land-use classification. In Proceedings of the 18th ACM SIGSPATIAL International Symposium on Advances in Geographic Information Systems, San Jose, CA, USA, 2–5 November 2010; pp. 270–279.
27. Chaudhuri, B.; Demir, B.; Chaudhuri, S.; Bruzzone, L. "Multilabel Remote Sensing Image Retrieval Using a Semisupervised Graph-Theoretic Method. *IEEE Trans. Geosci. Remote Sens.* **2018**, *56*, 1144–1158. [CrossRef]
28. Xia, G.S.; Hu, J.; Hu, F.; Shi, B.; Bai, X.; Zhong, Y.; Lu, X.; Zhang, L. AID: A Benchmark Data Set for Performance Evaluation of Aerial Scene Classification. *IEEE Trans. Geosci. Remote Sens.* **2017**, *55*, 3965–3981. [CrossRef]
29. Ömrüuzun, F.; Demir, B.; L. Bruzzone, L.; Çetin, Y. Content based hyperspectral image retrieval using bag of endmembers image descriptors. In Proceedings of the 2016 8th Workshop on Hyperspectral Image and Signal Processing: Evolution in Remote Sensing (WHISPERS), Los Angeles, CA, USA, 21–24 August 2016; pp. 1–4.
30. Hua, Y.; Mou, L.; Zhu, X. Recurrently exploring class-wise attention in a hybrid convolutional and bidirectional LSTM network for multi-label aerial image classification. *ISPRS J. Photogramm. Remote Sens.* **2019**, *149*, 188–199. [CrossRef]
31. Qi, Q.X.; Panpan, Z.; Yuebin, W.; Liqiang, Z.; Junhuan, P.; Mengfan, W.; Jialong, C.; Xudong, Z.; Ning, Z.; P.M.Takis. MLRSNet: A multi-label high spatial resolution remote sensing dataset for semantic scene understanding. *ISPRS J. Photogramm. Remote Sens.* **2020**, *169*, 337–350. [CrossRef]
32. Sumbul, G.; d. Wall, A.; Kreuziger, T.; Marcelino, F.; Costa, H.; Benevides, P.; Caetano, M.; Demir, B.; Markl, V. BigEarthNet-MM: A Large Scale Multi-Modal Multi-Label Benchmark Archive for Remote Sensing Image Classification and Retrieval. *IEEE Geosci. Remote Sens. Mag.* **2021**, *9*, 174–180. [CrossRef]
33. Kocev, D.; Vens, C.; Struyf, J.; Džeroski, S. Tree ensembles for predicting structured outputs. *Pattern Recognit.* **2013**, *46*, 817–833. [CrossRef]
34. Kocev, D.; Ceci, M.; Stepisnik, T. Ensembles of extremely randomized predictive clustering trees for predicting structured outputs. *Mach. Learn.* **2020**, *109*, 2213–2241. [CrossRef]
35. Buslaev, A.; Iglovikov, V.I.; Khvedchenya, E.; Parinov, A.; Druzhinin, M.; Kalinin, A.A. Albumentations: Fast and Flexible Image Augmentations. *Information* **2020**, *11*, 125. [CrossRef]
36. Xiao, Q.; Liu, B.; Li, Z.; Ni, W.; Yang, Z.; Li, L. Progressive Data Augmentation Method for Remote Sensing Ship Image Classification Based on Imaging Simulation System and Neural Style Transfer. *IEEE J. Sel. Top. Appl. Earth Obs. Remote Sens.* **2021**, *14*, 9176–9186. [CrossRef]

37. Pedregosa, F.; Varoquaux, G.; Gramfort, A.; Michel, V.; Thirion, B.; Grisel, O.; Blondel, M.; Prettenhofer, P.; Weiss, R.; Dubourg, V.; et al. Scikit-learn: Machine Learning in Python. *J. Mach. Learn. Res.* **2011**, *12*, 2825–2830.
38. Demšar, J. Statistical Comparisons of Classifiers over Multiple Data Sets. *J. Mach. Learn. Res.* **2006**, *7*, 1–30.

remote sensing

MDPI

Article

A Method of Fusing Probability-Form Knowledge into Object Detection in Remote Sensing Images

Kunlong Zheng [1,2], Yifan Dong [1,2,*], Wei Xu [1,2], Yun Su [1,2] and Pingping Huang [1,2]

1 College of Information Engineering, Inner Mongolia University of Technology, Hohhot 010051, China
2 Inner Mongolia Key Laboratory of Radar Technology and Application, Hohhot 010051, China
* Correspondence: yfdong@imut.edu.cn

Abstract: In recent years, dramatic progress in object detection in remote sensing images has been made due to the rapid development of convolutional neural networks (CNNs). However, most existing methods solely pay attention to training a suitable network model to extract more powerful features in order to solve the problem of false detections and missed detections caused by background complexity, various scales, and the appearance of the object. To open up new paths, we consider embedding knowledge into geospatial object detection. As a result, we put forward a method of digitizing knowledge and embedding knowledge into detection. Specifically, we first analyze the training set and then transform the probability into a knowledge factor according to an analysis using an improved version of the method used in existing work. With a knowledge matrix consisting of knowledge factors, the Knowledge Inference Module (KIM) optimizes the classification in which the residual structure is introduced to avoid performance degradation. Extensive experiments are conducted on two public remote sensing image data sets, namely DOTA and DIOR. The experimental results prove that the proposed method is able to reduce some false detections and missed detections and obtains a higher mean average precision (mAP) performance than the baseline method.

Keywords: convolutional neural networks (CNNs); remote sensing images; object detection; knowledge inference module

Citation: Zheng, K.; Dong, Y.; Xu, W.; Su, Y.; Huang, P. A Method of Fusing Probability-Form Knowledge into Object Detection in Remote Sensing Images. *Remote Sens.* **2022**, *14*, 6103. https://doi.org/10.3390/rs14236103

Academic Editor: Silvia Liberata Ullo

Received: 2 October 2022
Accepted: 17 November 2022
Published: 1 December 2022

1. Introduction

Object detection classifies and locates geospatial objects in aerial images and is a fundamental task in remote sensing. Recently, there have been great advances in object detection in aerial images (ODAI) [1,2] with the advent of deep convolutional neural networks. Because of this, applications in some fields, such as (UAVs) and wireless sensor networks (WSNs), have also benefited dramatically [3,4]. Although a large number of excellent methods [5–9] have been invented in the past few years, updating the previous state-of-the-art methods, there are still problems, such as missing detections and false detections caused by densely distributed objects, varied object sizes, and occluded objects. To illustrate, Figure 1a is a visualization of false harbor and helicopter detection, where a plane is detected as the harbor, and a piece of land is regarded as a helicopter. In Figure 1b, a piece of agricultural land is detected as a soccer ball field. In Figure 1c, a shaded basketball court is not detected. In the right of Figure 1d, two bridges are not detected, which damages the detection performance. The traditional object detection paradigm only takes the extracted feature into consideration. However, this is the bird eye view in aerial images. Therefore, detection algorithm inference only uses the roof information from aerial images. Moreover, due to its complex background features, such as the shape similarity and appearance similarity between objects and the background, the wrong labels are predicted. Because of the ambiguous appearance, those algorithms with an advantage in terms of extracting features do not have a significant effect. This is because when the instances are ambiguous, the extracted features are not that powerful, which leads to missed detection and false detection.

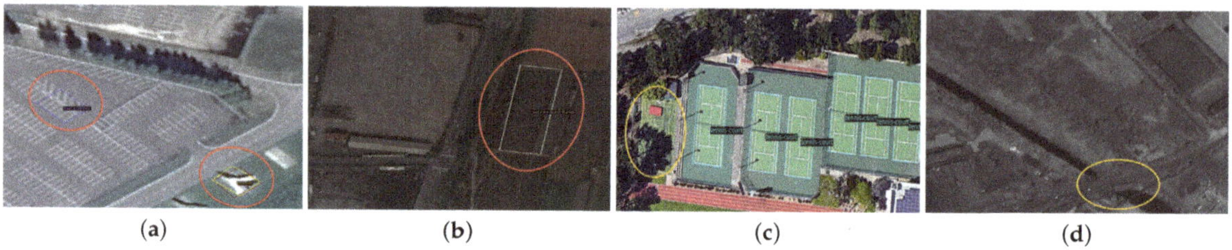

Figure 1. False detection in red circle and missed detection in yellow circle: (**a**) false detection of a helicopter and harbor; (**b**) false detection of a soccer ball field; (**c**) missed detection of a basketball court; (**d**) missed detection of a bridge.

However, the human visual recognition system informs us how to tackle this problem, because humans take their surroundings, as well as objects into account when conducting recognition tasks [10], which might be beneficial for geospatial object detection. When humans see one thing, it is common for them to think of another thing related to the seen one. Normally, for instance, when a harbor comes into view, there might be ship parking along the harbor. In addition, the ship and the harbor appear with water, which provides the link between the water and the object. Relationships between objects and relationships between a scene and objects can be assembled into a knowledge pool, which is then used in the object detection task, the process of which is shown in Figure 2. With the utilization of such a knowledge pool, the human visual recognition system is superior to the artificial intelligent algorithm in terms of its detection performance. The visual recognition method mentioned above benefits from the utilization of knowledge, which is worth adopting in the object detection algorithm.

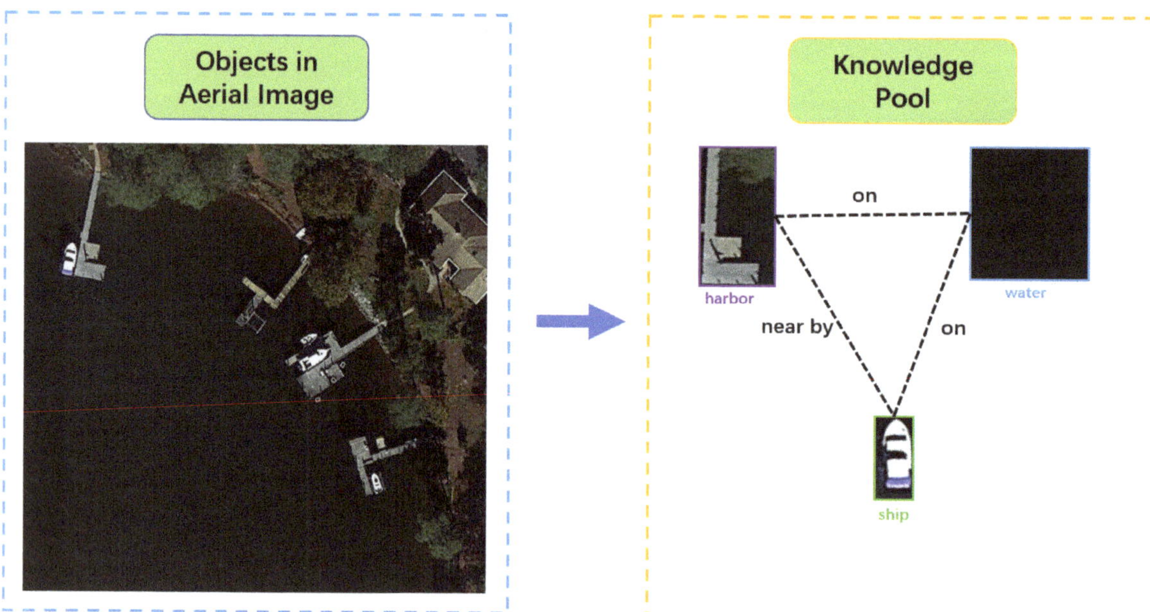

Figure 2. Relationships between the ship, harbor, and water assembled in the knowledge pool.

Obviously, it is by converting pictures into data that object detection algorithms process converted data and can detect objects. Thus, in order to improve the detection method added to work, such as the human visual system, the transformation of knowledge into data is essential. Refs. [11,12] were our sources of inspiration. Li et al. [11] presented statistics regarding the probability of objects appearing in each scene and assembled these data into a probability matrix with the aim of improving scene classification. Consequently, thinking in opposition, it is profitable to use the probability of an object appearing in the relevant scene to guide detection. In [12], Xu et al. utilized a class relation matrix to boost the object detection performance. Specifically, they integrated a class relation matrix with a high-dimensional feature to obtain an enhanced feature, which is not that intuitively explainable due to the high-dimensional space. As a result, to deal with false detections, such as a harbor being detected on land, in this work, we explored the co-relations between the water area and objects by analyzing a data set. Moreover, in order to utilize the implicit relation as knowledge, we analyzed the conditional co-occurrence probabilities of different categories, which is the expression of the implicit relation.

In this paper, an approach to utilizing knowledge is proposed. Our overall contributions are as follows:

- We extract two kinds of knowledge in the form of probabilities, namely the correlations between classes and the correlations between the water area and classes, by analyzing the DOTA [13] and DIOR [2] training sets. Then, we transform the extracted knowledge into a knowledge factor using a novel equation improved from [14].
- We propose a method, namely the Knowledge Inference Module (KIM), of integrating knowledge into object detection in remote sensing images. Through an evaluation of two public aerial data sets, our method obtains a higher mAP than the baseline model, Oriented R-CNN [15] with fewer false and missed detections.

2. Related Work

2.1. Object Detection in Remote Sensing Images

It is obvious that remote sensing images are very different from natural scene images due to their large aspect ratio, arbitrary orientation, variation in appearance and scale, densely distributed objects, etc., leading to difficulties in directly transferring object detection in natural scene images to object detection in aerial images. Therefore, references [9,16,17] proposed different ways to improve the detection performance. Han et al. [9] used the Feature Alignment Module (FAM) to generate anchors, encode orientation information, and obtain orientation-sensitive features using the Oriented Detection Module (ODM). Yang et al. [17] R^3Det refined features by re-encoding the positional information of the bounding box to the corresponding feature points through pixel-wise feature interpolation. Ming et al. [16] constructed critical features through the Polarization Refinement Module (PAM) and used the Rotation Anchor Refinement Module (R-ARM) to finally obtain a powerful semantic representation. Yang et al. [7] used a circular smooth label (CSL) to solve the problem of having discontinuous boundaries due to angular periodicity or corner ordering. Ding et al. [6] proposed the RoI-Transformer, which contains Rotated RoI Warping (RRoI Warping), which extracts rotation-invariant features, and the Rotated RoI Learner (RRoI Learner), which acquires objects' orientation information, to address detection misalignment. The aforementioned [6] achieved great success in boosting the detection performance. However, in this method, the amount of computation increases, the detection speed decreases, and the GPU memory is burdened. Consequently, some methods [15,18] with reduced computation have been proposed with the condition of maintaining accuracy. Xie et al. [15] built the Oriented-RCNN including the Oriented RPN to generate high-quality anchors and the Oriented R-CNN head, which has a faster detection speed than the RoI-Transformer [6]. Han et al. [18] proposed a novel backbone, the Rotation-equivariant ResNet (ReResNet), which has a reduction in parameters of over 60% compared to ResNet [19].

2.2. Utilization of Existing Knowledge

Works that integrate knowledge into object detection can be divided into two kinds. One involves learning relationships between objects and scenes or relationships between categories during the training process; another makes use of existing knowledge to improve the detection performance.

2.2.1. Utilization of Knowledge Learned in Training

Chen et al. [20] directly added a global image feature to an ROI feature, which is a simple and effective method. Liu et al. [21] designed a Structure Inference Network (SIN) with two parallel branches, one for global information extraction and another for global information extraction. Though this method is more sophisticated and useful, the GRU cell in each branch requires more GPU memory, making the process time-consuming. Siris et al. [22] applied an Attention [23] mechanism to combine global information and local information and produced a great performance in terms of salient object detection. Li et al. [24] obtained local features and contextual features from the Region of Interest (RoI), then integrated them into a Local-Contextual joint feature for geospatial object detection. Zhang et al. [25] proposed CAD-net, which can learn correlations between objects and scenes from global features and object features. Refs. [26–28] designed a module to enhance scene or global information in order to capture contextual information.

2.2.2. Utilization of Existing Knowledge

In [11,29], the prior scene-class graph was adopted to infer the relationship between a scene and an object through the Bayesian criterion. The adjacency matrix learned from the visual feature was adopted in the relation-reasoning module in [30,31]. Shu et al. [32] introduced the Graph Attention Network (GAT) and Graph Convolutional Network (GCN) to learn hidden knowledge from the obtained co-occurrence matrix and scene–object matrix. Fang [14] proposed a probability-based knowledge graph and graph-based knowledge graph with a cost function containing a knowledge graph to carry out knowledge-aware detection. In [33], an explicit knowledge module and an implicit knowledge module, containing an explicit knowledge graph and an implicit knowledge graph, respectively, were introduced, to enhance the RoI features.

3. Establishment of the Knowledge Matrix

3.1. Knowledge Matrix Establishment

In this section, the procedures used to create knowledge matrices, the category conditional co-occurrence knowledge matrix and the water area knowledge matrix from the data set are illustrated.

3.1.1. Conditional Co-Occurrence Knowledge Matrix

We first count the probability $n(l)$ that every category appears in images using Equation (1)

$$n(l) = \frac{N_{img}(l)}{N_{allimg}}. \tag{1}$$

where $N_{img}(l)$ denotes the number of images in which category l occur, and N_{allimg} denotes the number of images in the trainset. An analysis of the DOTA trainset is shown in Table 1, and an analysis of DIOR is shown in Appendix A .

Table 1. The number of images in which the category occurs $N_{img}(l)$ and the probabilities of the class occurrence $n(l)$.

Object Categories	$N_{img}(l)$	$n(l)$
Plane	197	0.1396
Baseball diamond	122	0.0864
Bridge	210	0.1488
Ground track field	177	0.1254
Small vehicle	486	0.3444
Large vehicle	380	0.2693
Ship	326	0.2310
Tennis court	302	0.2140
Basketball court	111	0.0786
Storage tank	161	0.1141
Soccer ball field	136	0.0963
Roundabout	170	0.1204
Harbor	339	0.2402
Swimming pool	144	0.1020
Helicopter	30	0.0212

Then, we determine the probability of conditional co-occurrence. In detail, we first count $N_{img}(l|l')$, the number of images in which class l and class l' appear together. Then, we use $N_{img}(l)$ diving $N_{img}(l|l')$.

Fang et al. [14] proposed Equation (2)

$$S_{l,l'} = max(log\frac{n(l|l')N_{allimg}}{n(l)n(l')},0) \tag{2}$$

where $n(l)$ denotes the probability of occurrence for category l, and N_{allimg} denotes the number of all images to transform the frequencies into a knowledge matrix. We applied this formula to the DOTA and DIOR data sets.

However, this knowledge matrix has some disadvantages. On one hand, the numerical dimension, which is over 10, is too large to be suitable for optimizing predicted class scores. The oversized numerical dimension has an excessive influence on the predicted class scores, causing the knowledge matrix to become the decisive element, whereas our original intention was to make use of the knowledge matrix to improve predictions. On the other hand, when two categories do not co-occur, the corresponding position in the knowledge matrix will be set as 0, which is a rigid way to deal with the aforementioned situation. Thus, we propose a novel processing approach that introduces a zero factor to replace 0. Additionally, simply setting zero could exert a negative impact on the generalization performance. This is because the none-conditional-co-occurrence of two categories in the data set only denotes that the two categories have little correlation, but it does not mean that the two categories never co-occur in the real world.

In order to address aforementioned problems, we modify Equation (2) and propose Equation (3).

$$S_{l,l'} = \begin{cases} \dfrac{log\frac{n(l,l')}{n(l)n(l')}(n(l)+n(l'))}{N_{img}(l)+N_{img}(l')} & n(l,l') \neq 0 \\ \dfrac{log\frac{\epsilon}{n(l)n(l')}(n(l)+n(l'))}{N_{img}(l)+N_{img}(l')} & n(l,l') = 0 \end{cases} \tag{3}$$

The modifications are as follows:

1. We abandon the $max()$ function, which means that the range of $S_{l,l'}$ extends to the negative axis, which is suitable for situations where category l and category l' do not co-occur or barely co-occur;

2. We regard the log part $log\frac{n(l,l')}{n(l)n(l')}$ as a conditional co-occurrence factor and abandon N_{allimg} and add $\frac{(n(l)+n(l'))}{N_{img}(l)+N_{img}(l')}$ as a scale factor into the equation in order to the scale knowledge factor into a proper numerical dimension and the make knowledge factor adaptive to the probability of category occurrence;
3. We split our novel equation into two branches, where the upper one is for situations where l and l' appear together in the training set and the lower one computes the knowledge factor when l and l' do not co-occur;
4. In the lower branch, we replace $n(l,l')$ with the zero factor ϵ in Equation (4), where N_{object} denotes the number of instances belonging to category l, making the equation more elegant when $n(l,l')$ equals 0.

$$\epsilon = \frac{1}{N_{object}(l)N_{object}(l')} \tag{4}$$

3.1.2. Water Area Knowledge Matrix

In this section, we first count the number of images containing water areas and the number not containing water areas and compute the occurrence probability of water appearing and not appearing as 0.5102 and 0.4898 in DOTA, respectively. Then, we determine the probability of a class appearing with water area or not using Equations (5) and (6)

$$n(l|w) = \frac{N_{img}(l|w)}{N_{img}(l)} \tag{5}$$

$$n(l|\overline{w}) = \frac{N_{img}(l|\overline{w})}{N_{img}(l)} \tag{6}$$

where $n(l|\overline{w})$ and $N_{img}(l|\overline{w})$ denote the probability of class l in a water area and the number of images in which class l does not occur in a water area. Those with a water area are denoted by $n(l|w)$ and $N_{img}(l|w)$. Probabilities of classes of DOTA appearing in a water area and not in a water area are demonstrated in Table 2. The probabilities of the classes of DIOR appearing in a water area and not in a water area are demonstrated in Table A2.

Table 2. Probabilities of classes appearing with water area and not with water area. Column $n(l|w)$ denotes the probability of category l appearing with water area; $n(l|\overline{w})$ is the probability of category l not appearing with water area.

Object Categories	$n(l\|w)$	$n(l\|\overline{w})$
plane	0.1748	0.8252
baseball-diamond	0.5543	0.4457
bridge	0.9755	0.0245
ground-track-field	0.6462	0.3538
small-vehicle	0.2837	0.7163
large-vehicle	0.1584	0.8416
ship	0.9994	0.0006
tennis-court	0.2945	0.7055
basketball-court	0.2544	0.7456
storage-tank	0.9402	0.0598
soccer-ball-field	0.5215	0.4785
roundabout	0.6166	0.3834
harbor	1	0
swimming-pool	1	0
helicopter	0.0015	0.9985

Similar to the conditional co-occurrence knowledge matrix, we also partly modify Equation (2) and propose Equation (7)

$$S_{l,*} = \begin{cases} \dfrac{log\frac{n(l|*)}{n(l)n(*)}n(l)}{N_{img}(l)} & n(l|*) \neq 0 \\ \dfrac{log\frac{\epsilon}{n(l)n(*)}n(l)}{N_{img}(l)} & n(l|*) = 0 \end{cases} \tag{7}$$

where ϵ, taken as the zero factor, equals $\frac{1}{N_{object}(l)}$, and $n(*)$ is the probability of a water area appearing, where $*$ is w or $\overline{w}$ meaning that there is a water area and that there is no water area, respectively. Equation (7) was also applied to DOTA and DIOR .

4. Methods

In this paper, we propose a knowledge-inferencing module that uses a residual structure to optimize the predicted class scores, which helps the detector to perform better. Our method can be applied to any two-stage object detection framework. In this work, we chose the Oriented R-CNN [15] as the framework and baseline model. The overall structure of the framework filled with the knowledge-aware bounding box head is presented in Figure 3.

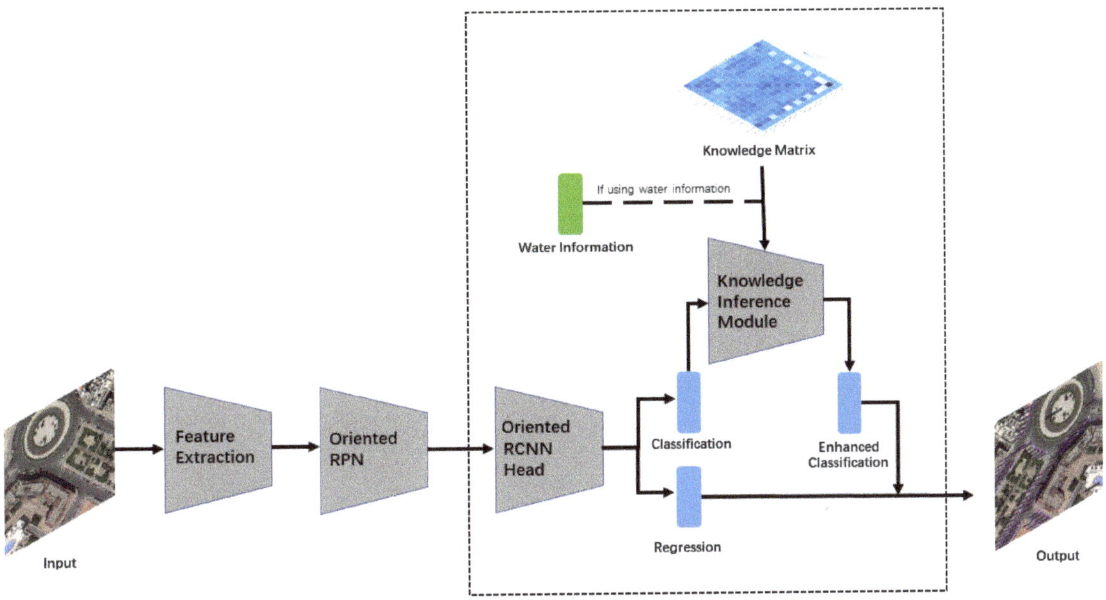

Figure 3. The overall structure of the framework filled with the Knowledge Inference module, a two-stage detector. Feature extraction module first extracts multi-scale features fed into the Oriented RPN. Region proposals are generated by the Oriented RPN and used for classifying and regression in the Oriented RCNN head. Finally, the classification scores are fed into the Knowledge Inference Module, resulting in knowledge-enhanced classification scores.

4.1. Feature Extraction

Resnet-50 [19] and FPN [34] are adopted into the Feature Extraction module to extract multi-scale features with which size-various objects can more easily be detected. Figure 4 shows the overall structure of the Feature Extraction module. In the bottom–up part, the input image is first input into a series of convolutional layers to generate low-semantic feature maps and high-semantic feature maps $\{C_1, C_2, C_3, C_4, C_5\}$. Moreover, the top–down part obtains more powerful features by fusing features. To be specific, M_4 is equal to the sum of the result of the upsampling of M_5 and the result of the 1×1 convolution of C_4. In this way, M_3 and M_2 are generated. P_2, P_3, P_4, and P_5 are obtained by feeding M_2, M_3, M_4, and M_5 into a 3×3 convolutional layer, and P_6 is obtained by maxpooling P_5. Finally, this module outputs the fused multi-scale features $\{P_2, P_3, P_4, P_5$, and $P_6\}$.

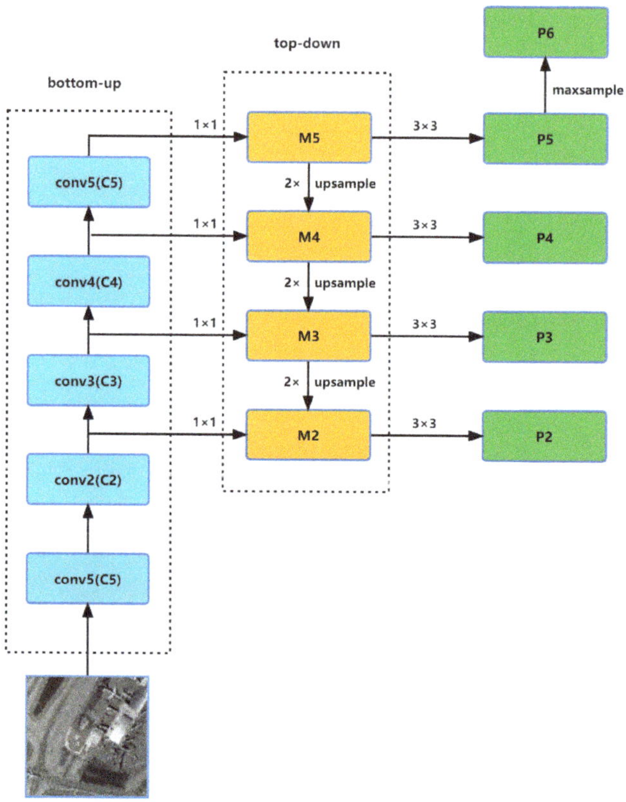

Figure 4. The overall structure of the Feature Extraction module.

4.2. Oriented RPN

The oriented RPN takes $\{P_2, P_3, P_4, P_5$, and $P_6\}$ as the input and output region proposals with location information and an objectness score. Figure 5 shows the structure of the Oriented RPN.

We set three horizontal anchors with aspect ratios of {1:2, 1:1, and 2:1} for every location in the features of all scales. The anchors correspond to $\{P_2, P_3, P_4, P_5$, and $P_6\}$ and have pixel areas of $\{32^2, 64^2, 128^2, 256^2$, and $512^2\}$, which are represented by a 4-dimensional vector $a = (a_x, a_y, a_w, a_h)$. a_x and a_y denote the horizontal and vertical locations of the anchor center; a_w and a_h correspond to the width and height of the anchor. The upper branch of Figure 5, i.e., the regression branch, outputs the offset $\delta = (\delta_x, \delta_y, \delta_w, \delta_h, \delta_\alpha, \delta_\beta)$ of proposals related to anchors. We decode the offset using Equation (8) to obtain oriented proposals $(x, y, w, h, \Delta\alpha, \Delta\beta)$, where (x, y) denotes the location of the proposed center coordinate, w and h correspond to the width and height of the external rectangle box of the proposal, $\Delta\alpha$ and $\Delta\beta$ denote the offsets of the proposal box vertex oriented to the midpoints of the top and right sides of the corresponding external rectangle. The lower branch of Figure 5, i.e., the classification branch, outputs objectness scores.

$$\begin{cases} \Delta\alpha = \delta_\alpha \cdot w, & \Delta\beta = \delta_\beta \cdot h \\ w = a_w \cdot e^{\delta_w}, & h = a_h \cdot e^{\delta_h} \\ x = \delta_x \cdot a_w + a_x, & y = \delta_y \cdot a_h + a_y \end{cases} \tag{8}$$

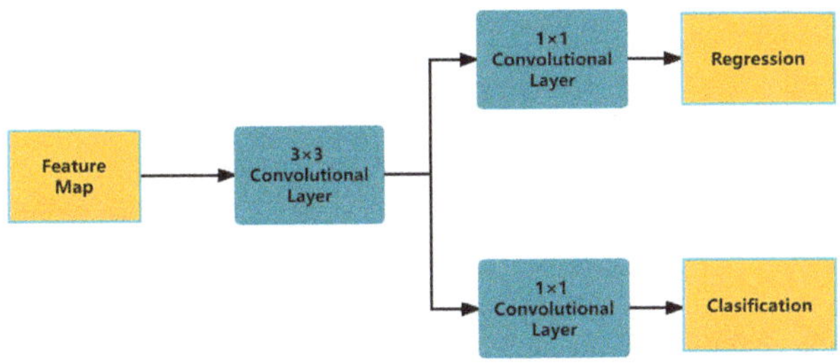

Figure 5. The structure of the Oriented RPN, which contains a 3 × 3 convolutional layer and two sibling 1 × 1 convolutional layers for classification and regression, respectively.

To represent the oriented object in elegant manner, the midpoint offset representation is introduced, the schematic of which is illustrated in Figure 6. In detail, the black horizontal box, i.e., the external rectangle of the blue one, is obtained from the anchor, where a_w and a_h are the width and height of the anchor, and the blue oriented box is the predicted oriented proposal box. The black dots and the light green dots are the midpoint of the external rectangle edges and the vertices of the oriented box, respectively. The predicted oriented proposal box can be represented as $O = (x, y, w, h, \Delta\alpha, \Delta\beta)$, which can be computed by Equation (8). Furthermore, the vertices of predicted oriented proposal are denoted by a set of coordinates v_1, v_2, v_3, v_4. Similarly, $\Delta\beta$ is the distance between v_2 and the midpoint $(x, y - \frac{h}{2})$ of the top side, and because of the symmetry, the distance between v_3 and the midpoint $(x, y + \frac{h}{2})$ of the bottom side equals $-\Delta\alpha$. It is noticing that $\Delta\alpha$ is distance between v_1 and the midpoint $(x, y - \frac{h}{2})$ of top side, and because of the symmetry distance between v_3 and the midpoint $(x, y + \frac{h}{2})$ of bottom side equals to $-\Delta\alpha$. Similarly, $\Delta\beta$ is the distance between v_2 and the midpoint $(x + \frac{w}{2}, y)$ of the right side, and the distance between v_4 and the midpoint $(x, y + \frac{h}{2})$ of the left side equals $-\Delta\beta$. As a result, the vertices of the oriented proposal $\{v_1, v_2, v_3, v_4\}$ can be computed using Equation (9).

$$\begin{cases} v_1 = (x, y - \frac{h}{2}) + (\Delta\alpha, 0) \\ v_2 = (x + \frac{w}{2}, y) + (0, \Delta\beta) \\ v_3 = (x, y + \frac{h}{2}) + (-\Delta\alpha, 0) \\ v_4 = (x - \frac{w}{2}, y) + (0, -\Delta\alpha) \end{cases} \tag{9}$$

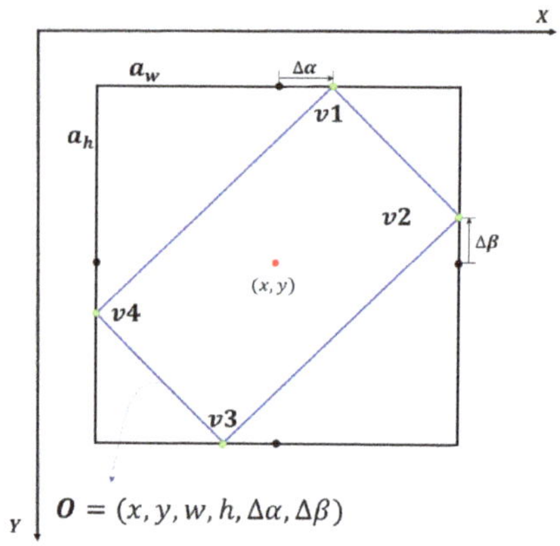

Figure 6. Schematic of the midpoint offset scheme.

In the training process, we assign positive and negative samples using the following rules:

1. An anchor that has an Intersection-over-Union(IoU) over 0.7 with any ground-truth box is regarded as a positive sample;
2. An anchor that has an IoU over 0.3 with a ground-truth box and the IoU is the highest;
3. An anchor that has an IoU lower than 0.3 is regarded as a negative sample;
4. Anchors that do not belong to the above cases are discarded during the training process.

4.3. Oriented RCNN Head with the Knowledge Inference Module

In this section, we apply our proposed Knowledge Inference module to the Oriented RCNN head in order to reduce missed and wrong detections by improving the predicted class scores. Specifically, the proposed module is applied on two kinds of knowledge: conditional co-occurrence knowledge and water area knowledge. Thus, the proposed module has two similar inferencing modes. The details can be seen in the middle part of Figure 3.

The oriented RCNN head first takes $\{P_2, P_3, P_4, P_5, \text{and } P_6\}$ and the oriented proposals from the oriented RPN as the input. In detail, feature vectors are obtained by rotating the RoI alignment to extract rotated RoI features according to the oriented proposals and transform them into fixed-length vectors. This is followed by two fully-connected layers. Then, we use two fully-connected sibling layers outputting classification scores and location predictions. For each image, we generate 512 predictions. Thus, the classification scores are denoted by the tensor of shape $[512, K+1]$, where $K+1$ denotes the number of classes plus the background, and the location predictions are denoted by the tensor of shape $[512, 5]$. With the aim of optimizing the predicted classification scores with the knowledge matrix, we propose the Knowledge Inference Module, the structure of which is illustrated in Figure 7. To be specific, Figure 7a shows the structure of the Knowledge Inference module applied to class conditional co-occurrence knowledge. The class scores $[512, 16]$ are first fed into the main-class-seeking-module to compute the major class in the image outputting the index of the main class. Then, the conditional co-occurrence matrix is sliced in terms of the main class index. The sliced matrix denotes the relationship between the main class and other classes, which is represented by the tensor of shape $[1, 16]$. Therefore, the Δ class score is a tensor with knowledge integrated. It is obtained by dot-multiplying class scores $[512, 16]$ and transposing the sliced matrix $[16, 1]$. However, our initial idea is to use knowledge to guide detection, so we apply a residual structure [19] into our proposed module, avoiding degradation brought about by using Δ class scores only. Enhanced class scores are the result of Δ class scores plus class scores. Additionally, the Knowledge

Inference module on water area knowledge is shown in Figure 7b and is similar to that of the category conditional co-occurrence matrix. The difference is that the Main Class Seeking module and the main class index are replaced by water information showing whether there is a water area in the image.

The structure of the Main-Class-Seeking module is illustrated in Figure 8. We first slice the tensor class scores [512, 16] into 512 small tensors [1, 16] and encode them into values of 1 to 512. Then, the sliced class scores are fed into the argmax() function outputting the classes with the highest classification scores. The function max check() is used to count the number of each category in the 512 predictions. This is performed sequentially, where the class with largest number is the main class.

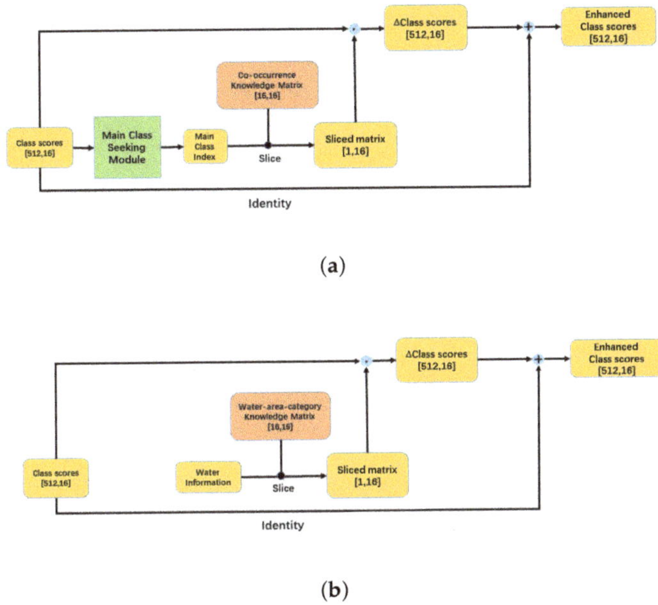

Figure 7. Structures of the Knowledge Inference module applied on two kinds of knowledge: (**a**) the structure of the Knowledge Inference module applied on conditional co-occurrence knowledge; (**b**) the structure of the Knowledge Inference module applied on water area knowledge.

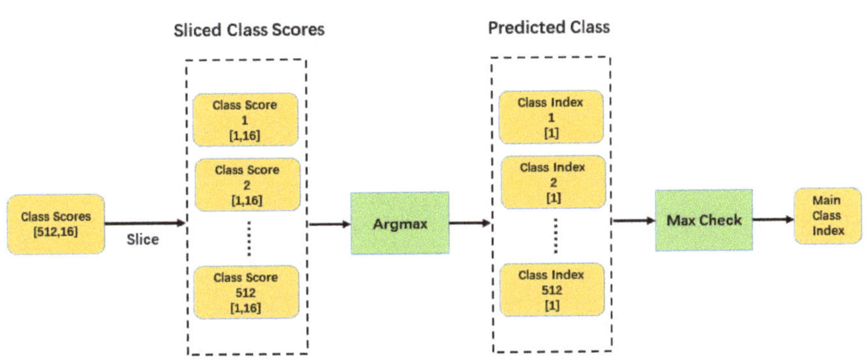

Figure 8. The Main Class Seeking module consists of a slice operation, argmax() function and a max check() function.

4.4. Loss Function

To train the Oriented RPN and Oriented RCNN head, we introduce Cross-Entropy Loss L_{cls} for the classification task and Smooth L1 Loss L_{reg} for the regression task. The whole loss function L is defined as follows:

$$L(p_i, t_i) = \frac{1}{N}\sum_{i}^{N} L_{cls}(p_i, p_i^*) + \frac{1}{N}\sum_{i}^{N} p_i^* L_{reg}(t_i, t_i^*) \quad (10)$$

$$L_{cls}(p_i, p_i^*) = -[p_i^* log(p_i) + (1 - p_i^*) log(1 - p_i)] \quad (11)$$

$$L_{reg}(t_i, t_i^*) = \begin{cases} 0.5(t_i - t_i^*)^2 & if\ |t_i - t_i^*| < 1 \\ |t_i - t_i^*| - 0.5 & otherwise \end{cases} \quad (12)$$

where N and i are, respectively, the number and index of the predicted anchors in a image, p_i is the probability of predicted anchors, p_i^* denotes the ground-truth label that belongs to $\{0, 1\}$, i.e., negative and positive, and t_i and t_i^* are the predicted box and the ground-truth box.

5. Experiments

In this section, we introduce the two geospatial object data sets used in this work. Then, evaluation metrics and implementation details are illustrated.

5.1. Data Sets

To evaluate the proposed method, we conduct experiments on two public aerial image data sets, i.e., DOTA and DIOR.

DOTA is the most popular large-scale data set for geospatial object detection, containing 2806 images and 188,282 instances with arbitrary-oriented objects. Moreover, there are 15 classes in the data set: bridge, harbor, ship, plane, helicopter, small vehicle, large vehicle, baseball diamond, ground track field, tennis court, basketball court, soccer ball field, roundabout, swimming pool , and storage tank. The image width ranges from 800 to 4000 pixels. In this work, the training set was used for training, and the validation set was used for evaluation.

DIOR is another data set that is widely used for geospatial object detection. It contains 23,463 optimal remote sensing images and 192,472 object instances annotated by a horizontal bounding box. There are 20 object classes in total, namely, airplane, airport, baseball field, basketball court, bridge, chimney, dam, expressway service area, expressway toll station, harbor, golf course, ground track field, overpass, ship, stadium, storage tank, tennis court, train station, vehicle, and windmill. The image size of the DIOR data set is 800 × 800 pixels. We trained the network on the training set and evaluated the method on the validation set.

5.2. Evaluation Metrics

To evaluate the performance of the proposed method, we utilized four popular evaluation metrics, i.e., precision, recall, average precision, and mean average precision, the calculation formulas of which are shown as follows:

$$Precision = \frac{TP}{TP + FP} \quad (13)$$

$$Recall = \frac{TP}{TP + FN} \quad (14)$$

TN, FN, and FP denote the number of true positives, the number of false negatives, and the number of false positives, respectively. *Precision* measures the number of correctly identified positive detections of the total number of positive detections and *Recall* measures the fraction of correctly identified positive detections of all positive samples.

AP is computed by calculating the average value of precision from recall = 0 to recall = 1.

$$AP = \int_0^1 P(R)dR \tag{15}$$

mAP is used to describe the multi-class object detection performance.

$$mAP = \frac{1}{N_{class}} \sum_{j=1}^{N_{class}} \int_0^1 P_j(R_j)dR_j \tag{16}$$

where N_{class} is the number of data set classes, j denotes the index of the class, and P_j and R_j are the precision rate and recall rate of the j-th class.

5.3. Implementation Details

The experiments were conducted on a single CPU, Intel Xeon CPU E5-2650 V4 at 2.20 GHz with a single GPU, NVIDIA Tesla P40 24 GB. The operating system was Ubuntu 18.04. The MMrotate [35] repository provided the training strategy. The size of the training image was 1024 × 1024, and the original DOTA images were split. All images in DIOR were 800 × 800 in size. Thus, there was no need to split the DIOR images. The objects in DIOR were annotated in the horizontal direction by the left-top vertex (x_1, y_1) and right-bottom vertex (x_1, y_1). Thus, we converted the annotations into a form suitable for the Oriented RCNN: $(x_1, y_1, x_2, y_1, x_2, y_2, x_1, y_2)$, corresponding to the vertices of oriented ground truth box in clockwise order. As for the hyperparameters, the optimizer was the stochastic gradient descent (SGD) with a learning rate of 0.005, a momentum of 0.9, and a weight decay of 0.0001. The batch size was 1, and the number of training epochs was 12.

6. Results

In this section, the results of the experiments on DOTA and DIOR are displayed.

6.1. DOTA Results

We applied the knowledge inference module to two kinds of knowledge: class co-occurrence knowledge and water area knowledge. The results show that our method achieved increases in mAP of 1.0% and 0.6%, respectively. Table 3 reports the comparison between the baseline model Oriented RCNN and our proposed method, in which the proposed method basically maintains the performance for both kinds of knowledge and improves the accuracy of several classes, for example 8.7% for the term helicopter with conditional co-occurrence knowledge, and 2.9% for the soccer ball field with water area knowledge.

Table 3. Comparison between the baseline model Oriented RCNN and our proposed method on DOTA data set.

METHOD	PL	BD	BR	GTF	SV	LV	SH	TC	BC	ST	SBF	RA	HA	SP	HC	mAP
R^3Det [17]	88.8	67.4	44.1	69.0	62.9	71.7	78.7	89.9	47.3	61.2	47.4	59.3	59.2	51.7	24.3	61.5
CSL [7]	88.1	72.2	39.8	63.8	64.3	71.9	78.5	89.6	52.4	61.0	50.5	66.0	56.6	50.1	27.5	62.2
S^2A-net [9]	89.1	72.0	45.6	64.8	65.0	74.8	79.5	90.1	60.2	67.3	49.3	62.2	60.6	53.4	37.6	64.8
FR-O [5]	89.3	76.0	49.3	74.7	68.1	75.5	87.1	90.7	64.2	62.3	57.0	65.8	66.6	59.6	38.2	68.3
RoI Trans [6]	89.9	76.5	48.1	73.1	68.7	78.2	88.7	90.8	73.6	62.7	62.0	63.4	73.7	57.2	47.9	70.3
Baseline [15]	89.8	75.7	50.2	77.3	69.4	84.8	89.3	90.8	69.2	62.6	63.1	65.0	75.3	57.5	45.3	71.0
Water area	89.6	75.6	50.3	76.4	68.4	84.3	89.4	90.7	72.9	62.6	66.0	67.2	75.6	56.5	48.8	71.6
Co-occurrence	89.6	76.0	50.7	77.0	68.3	84.4	89.3	90.7	73.6	62.4	63.8	66.8	75.1	57.6	54.0	72.0

The baseline is the Oriented RCNN [15], the Water area denotes the Knowledge Inference module on water area knowledge, and Co-occurrence is the Knowledge Inference module applied on conditional co-occurrence knowledge, where: PL: plane, BD: baseball diamond, BR: bridge, GFT: ground field track, SV: small vehicle, LV: large vehicle, SH: ship, TC: tennis court, BC: basketball court, ST: storage tank, SBF: soccer ball field, RA: roundabout, HA: harbor, SP: swimming pool, and HC: helicopter.

Moreover, to a certain degree, some missed detections and wrong detections were improved. Figures 9 and 10 display the reductions in missed detection and wrong detection using conditional co-occurrence knowledge and water area knowledge, respectively. In each subfigure, the left half is the result of the baseline model and the right half is the result of the proposed method. We use yellow circles to draw missed detections and red circles to draw false detections. Additionally, the first three subfigures shown in Figures 9 and 10 display missed detections, and the second three subfigures in Figures 9 and 10 display the false detections. The positive detections are shown in Figure 11.

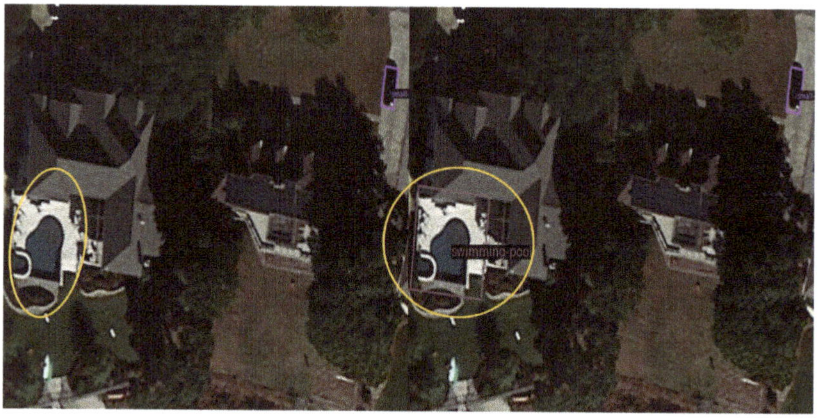

(a)

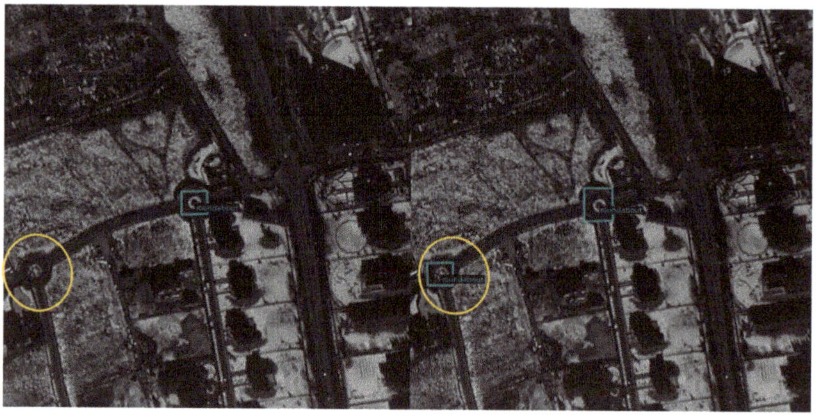

(b)

Figure 9. *Cont.*

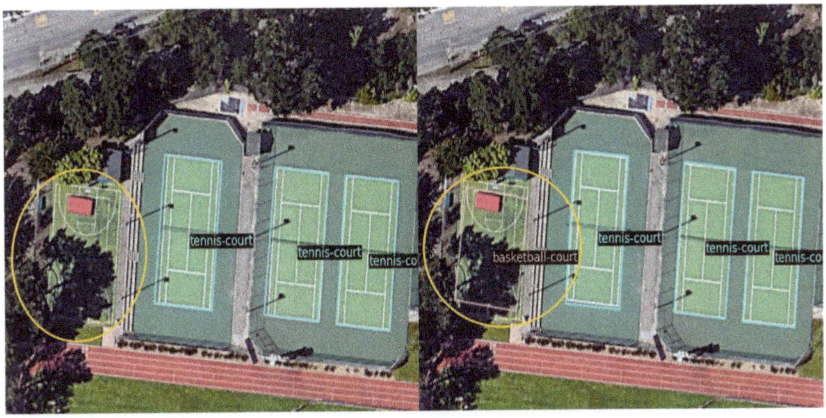

(**c**)

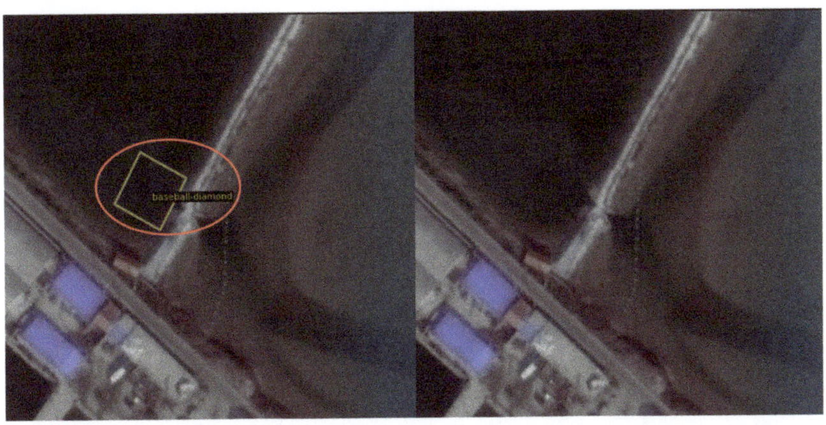

(**d**)

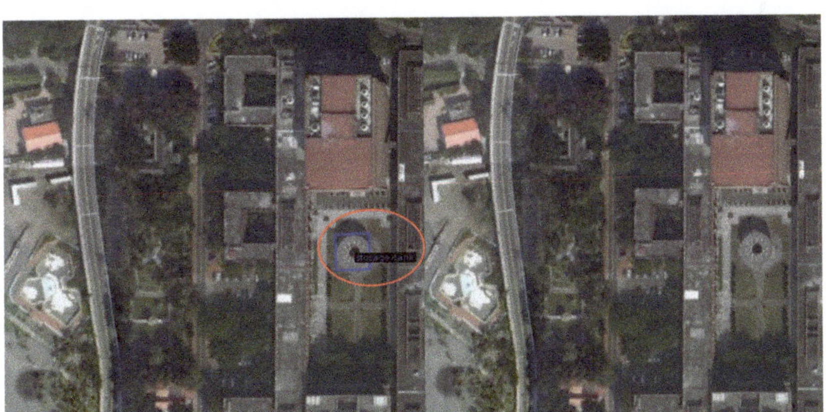

(**e**)

Figure 9. *Cont.*

(**f**)

Figure 9. Visualization of the results of the Knowledge Inference module applied to category occurrence knowledge. (**a**) missed detection of swimming pools; (**b**) missed detection of roundabouts; (**c**) missed detection of basketball courts; (**d**) false detection of baseball diamonds; (**e**) false detection of storage tanks; (**f**) false detection of basketball courts.

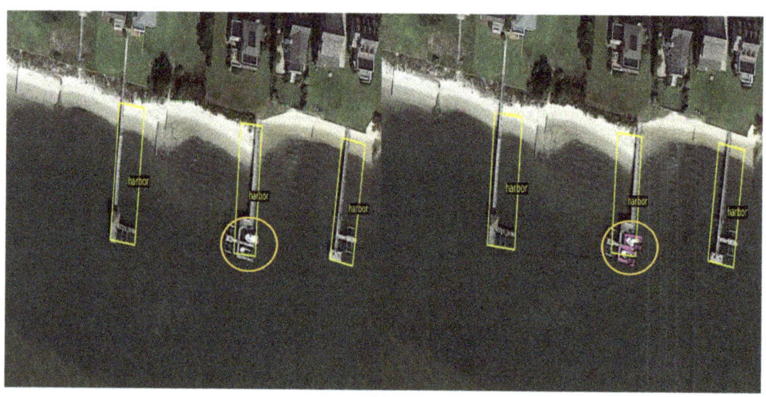

(**a**)

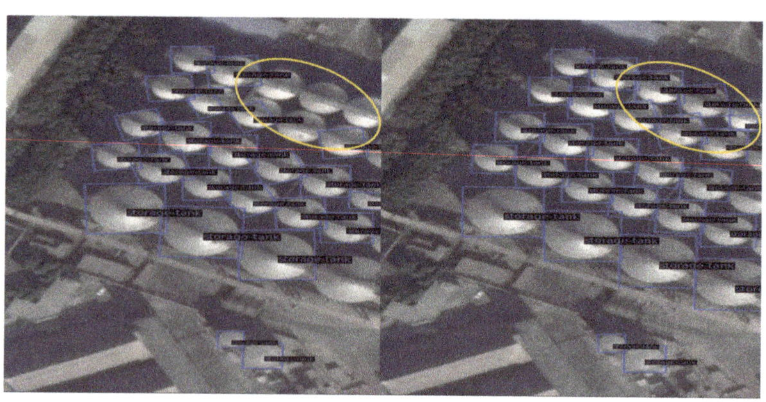

(**b**)

Figure 10. *Cont.*

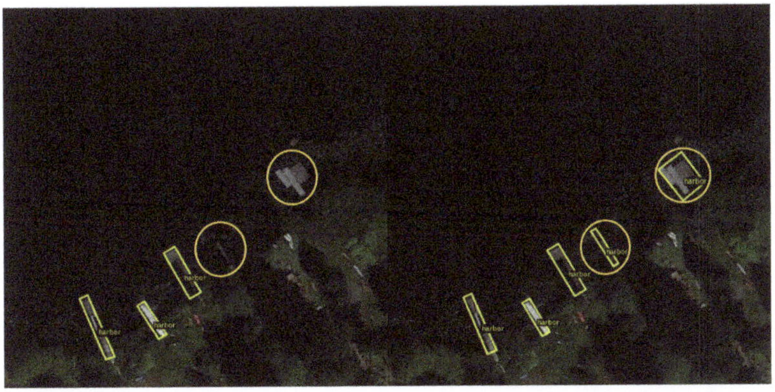

(c)

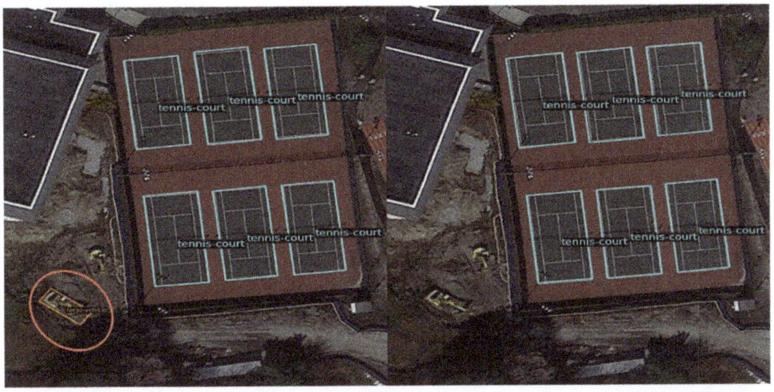

(d)

(e)

Figure 10. *Cont.*

(f)

Figure 10. Visualization of the results of the Knowledge Inference module applied to water area knowledge. (**a**) missed detections of ships in the middle of the image; (**b**) missed detection storage tanks; (**c**) missed detection of harbors; (**d**) false detection of large vehicles; (**e**) false detection of harbors; (**f**) false detection of baseball diamonds.

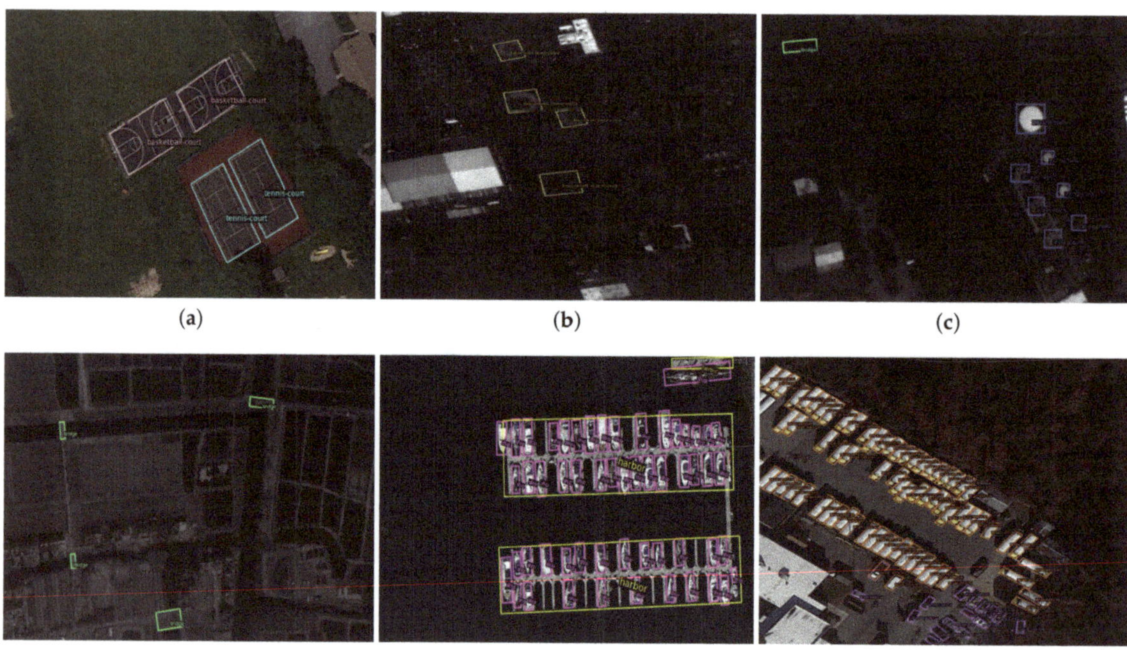

(a) (b) (c)

(d) (e) (f)

Figure 11. *Cont.*

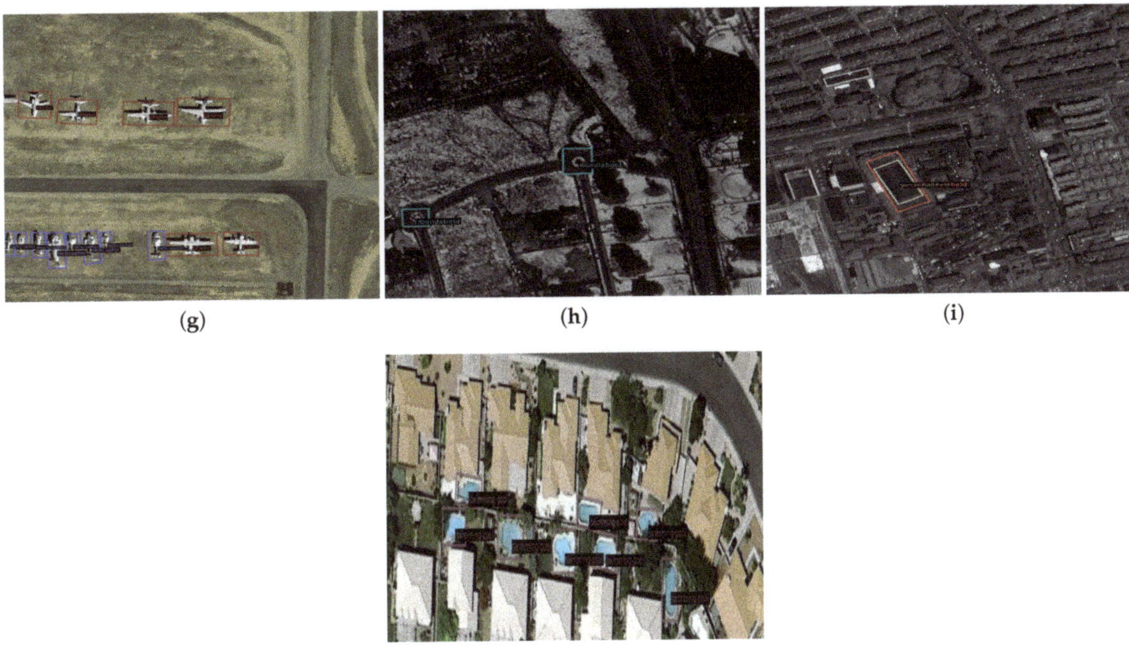

(g) (h) (i)

(j)

Figure 11. Visualization of the positive results. (**a**) basketball courts and tennis courts; (**b**) baseball diamonds; (**c**) bridge and storage tanks; (**d**) bridges; (**e**) harbors and ships; (**f**) large vehicles and small vehicles; (**g**) planes and helicopters; (**h**) roundabouts; (**i**) ground field tracks and soccer ball fields; (**j**) swimming pools.

Improvement occurs due to the utilization of knowledge. On the one hand, knowledge is used to optimize the predicted class scores. Thus, the performance of the classification is promoted; on the other hand, the class predictions optimized by knowledge can help the network iterate better during backpropagation. As a result, the more powerful features can be extracted by the network.

In terms of the inferencing speed, we compared the baseline and Knowledge Inference module for two kinds of knowledge, as shown in Table 4. With the Knowledge Inference module, there was no significant drop in speed.

Table 4. Comparison of the inferencing speed and accuracy between the baseline and proposed methods for two kinds of knowledge in the DOTA data set.

METHOD	FPS	mAP
Baseline	11.8	71.0
Water area	11.7	71.6
Conditional co-occurrence	11.5	72.0

6.2. DIOR Results

The experiments conducted on the DIOR data set achieved a better performance. The ap values and mAP values are shown in Table 5. As can be seen, both kinds of knowledge had beneficial impacts on the detection performance, and the mAP values increased by 0.5% and 3.9%, respectively. Similarly, missed detections and wrong detections were effectively eliminated.

Table 5. Comparison between the baseline model Oriented RCNN and our proposed method on DIOR data set .

METHOD	APL	APT	BF	BC	BR	CM	DA	ESA	EST	GF	GTF	HA	OPS	SP	STD	ST	TC	TS	VEH	WD	mAP
R^3Det [17]	89.6	6.40	89.5	71.2	14.4	81.7	8.90	26.5	48.1	31.8	65.6	8.20	33.4	69.2	51.9	72.9	81.1	21.4	54.2	44.7	48.5
CSL [7]	90.9	2.60	89.4	71.5	7.10	81.8	9.30	31.4	41.5	58.1	63.2	17.5	26.6	69.3	53.8	72.8	81.6	18.4	47.2	46.3	49.0
S^2A-Net [9]	90.8	14.0	89.9	72.7	17.6	81.7	9.50	32.9	50.1	50.9	70.9	16.3	43.6	80.1	52.5	75.7	81.7	23.9	59.0	45.6	53.0
RoI Trans [6]	90.8	12.1	90.8	79.8	22.9	81.8	8.20	51.3	54.1	60.7	77.2	30.6	40.5	89.9	88.2	79.5	81.8	20.6	67.9	55.1	59.2
FR-O [5]	90.9	13.1	90.7	79.9	22.0	81.8	10.4	49.8	53.1	58.5	75.4	35.5	41.6	89.1	85.4	79.3	81.8	32.7	66.4	55.5	59.6
Baseline	90.9	17.0	90.7	80.6	33.5	81.8	19.2	59.8	53.1	56.7	76.9	26.1	54.5	89.9	88.8	79.6	81.8	30.2	68.3	55.1	61.7
Water area	90.9	21.2	90.7	80.3	34.2	81.8	20.7	60.0	52.9	55.1	76.7	26.8	54.6	89.8	88.3	79.5	81.8	35.5	68.7	55.6	62.2
Co-occurrence	90.9	23.3	90.8	80.9	38.0	81.8	20.5	62.2	53.7	61.3	81.3	32.1	56.6	90.0	88.3	79.7	90.1	44.3	69.2	56.5	64.6

The baseline is the Oriented RCNN, the Water area denotes the Knowledge Inference module's knowledge on a water area, and conditional co-occurrence is the Knowledge Inference module for category conditional co-occurrence knowledge. APL: airplane, APT: airport, BF: baseball filed, BC: basketball court, BR: bridge, CM: chimney, DA: dam, ESA: expressway service area, ETS: expressway-toll-station, GF: golf field, GTF: ground track filed, HA: harbor, OPS: overpass, SP: ship, STM: stadium, ST: storage tank, TC: tennis court, TS: trainstation, VEH: vehicle, WD: windmill.

For the DIOR data set, we also visualized the impacts of two kinds of knowledge on the baseline, as shown in Figures 12 and 13. As for the visualization of the DOTA data set, the yellow circle and red circle denote missed detections and false detections, respectively. The positive detections are shown in Figure 14.

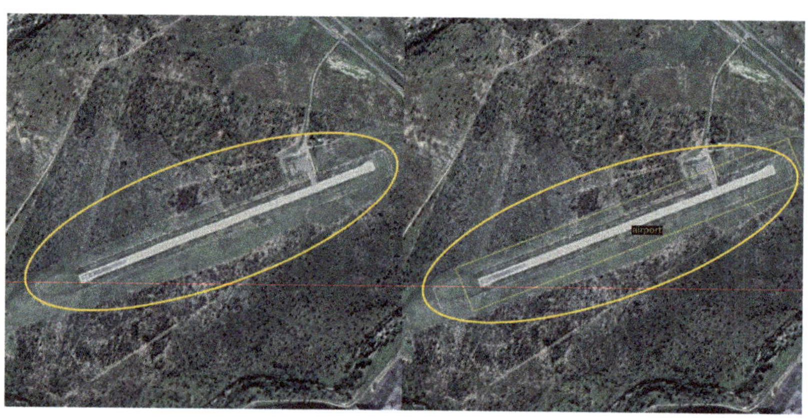

(a)

Figure 12. *Cont.*

(**b**)

(**c**)

(**d**)

Figure 12. Visualization of the results of the Knowledge Inference module applied on category occurrence knowledge: (**a**) missed detection of airports; (**b**) missed detection of expressway service areas; (**c**) false detection of bridges in the purple box and vehicles; (**d**) false detection of expressway toll stations.

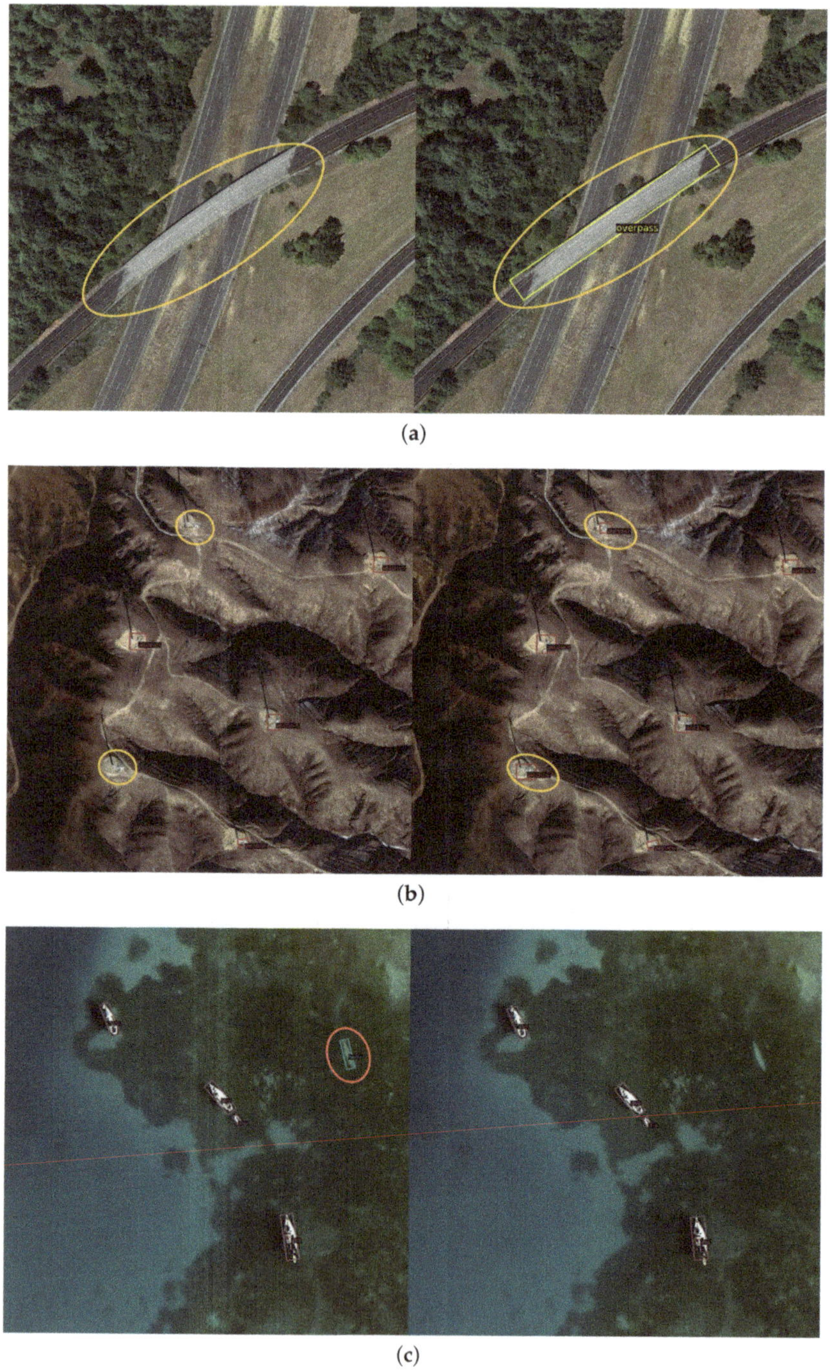

Figure 13. *Cont.*

(d)

Figure 13. Visualization of the results of the Knowledge Inference module applied on category occurrence knowledge. (**a**) missed detection of overpasses; (**b**) missed detection of windmills; (**c**) false detection of harbors; (**d**) false detection of storage tanks.

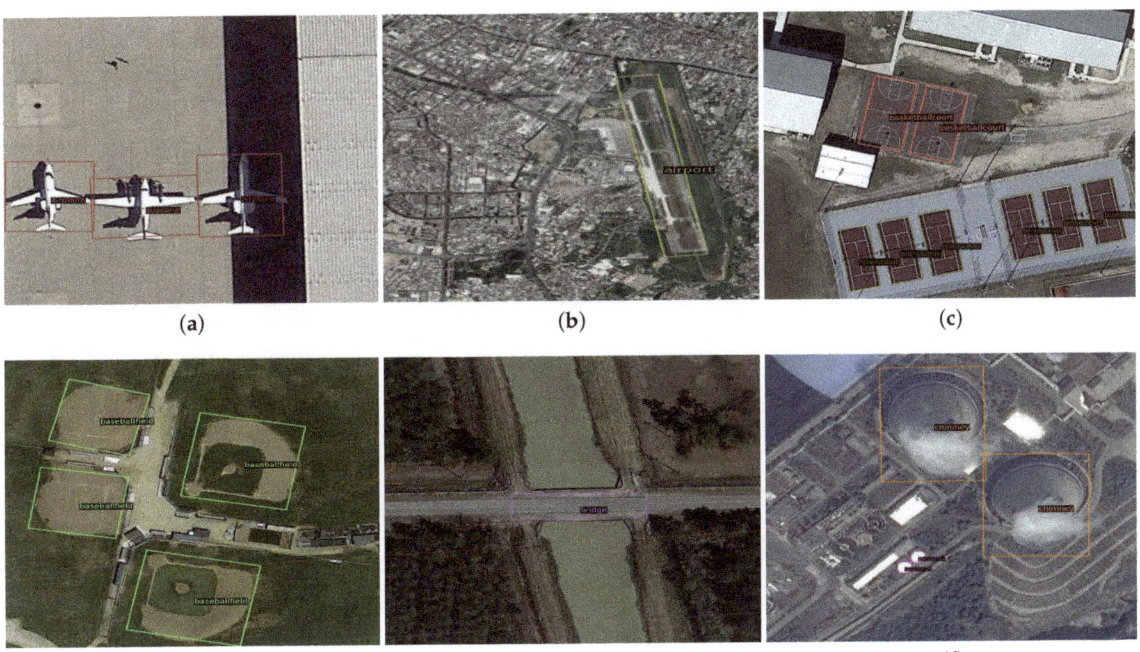

(a) (b) (c)

(d) (e) (f)

Figure 14. *Cont.*

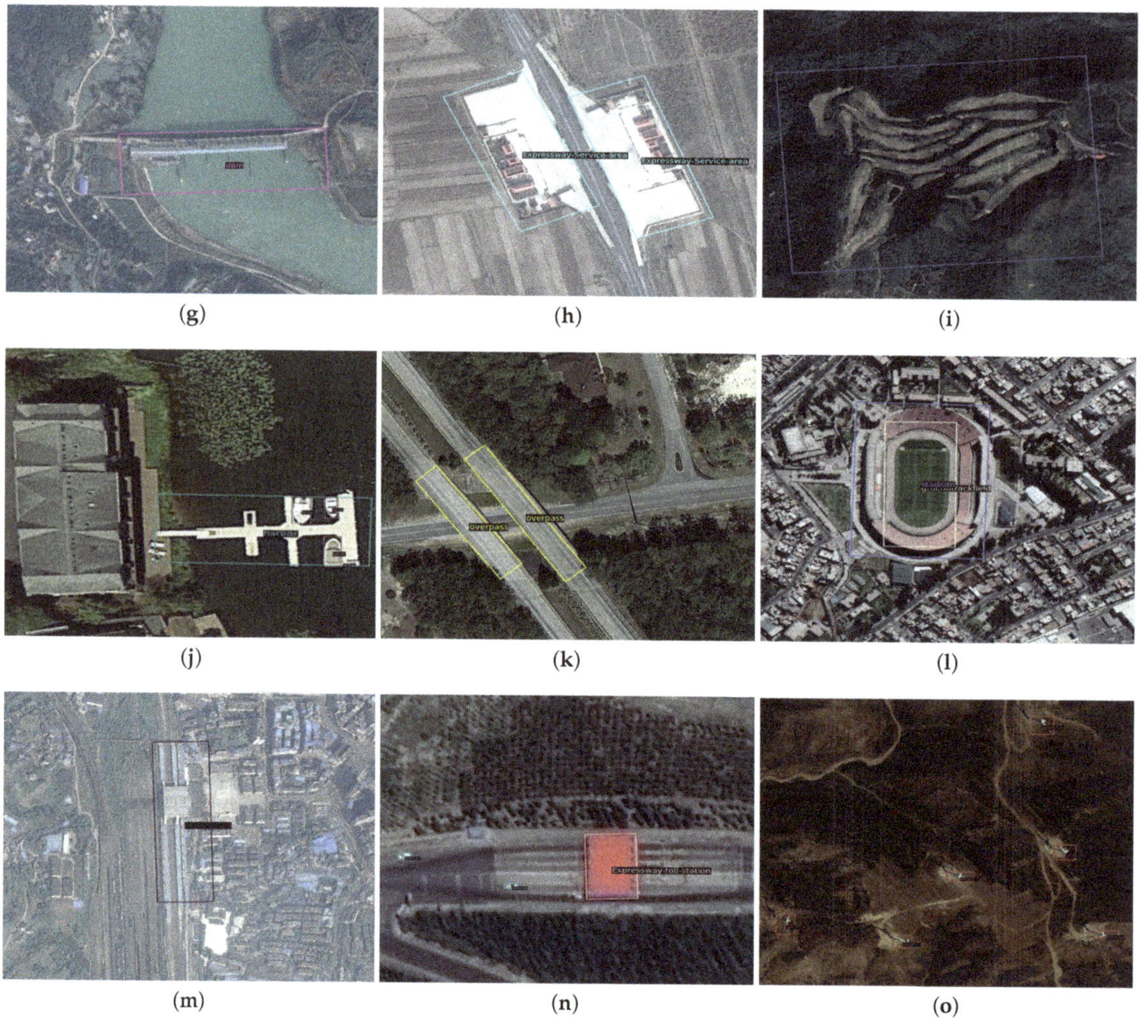

Figure 14. Visualization of the positive results. (**a**) airplanes; (**b**) airports; (**c**) basketball courts and tennis courts; (**d**) baseball fields; (**e**) bridges; (**f**) chimneys and storage tanks; (**g**) dams; (**h**) expressway service areas; (**i**) golf fields; (**j**) harbors and ships; (**k**) overpasses; (**l**) ground track fields and stadiums; (**m**) train stations; (**n**) expressway toll stations; (**o**) windmills.

A comparison of the inferencing speed and accuracy between the baseline and proposed methods for two kinds of knowledge is shown in Table 6. As can be seen, the Knowledge Inference module improved the detection accuracy with a negligible negative influence on the inferencing speed.

Table 6. Comparison of the inferencing speed and accuracy between the baseline and proposed methods for two kinds of knowledge in the DIOR data set.

METHOD	FPS	mAP
Baseline	9.3	61.7
Water area	9.1	62.2
Conditional co-occurrence	9.1	64.6

7. Conclusions

In this paper, in order to utilize knowledge to reduce false detections and missed detections caused by variation in the object appearance, varied object sizes, and complicated backgrounds, a series of steps were taken. We first established a knowledge matrix between the classes and a knowledge matrix between water areas and classes by analyzing the training set and proposed a novel equation, which can effectively avoid generalization degradation, to transform the relationship into form applicable for inferencing. Then, we proposed a method, the Knowledge Inference module, for integrating knowledge into object detection. The experiments were conducted on two public remote sensing data sets: DOTA and DIOR. The experimental results show that, compared to the baseline model, the proposed method achieved higher mAP values with fewer false detections and missed detections at an almost equal inferencing speed.

Author Contributions: Conceptualization, K.Z. and Y.D.; methodology, K.Z. and Y.D.; software, K.Z.; resources, W.X.; writing—original draft preparation, K.Z.; writing—review and editing, Y.D. and W.X.; visualization, K.Z. and Y.S.; supervision, P.H.; funding acquisition, Y.D. and W.X. All authors have read and agreed to the published version of the manuscript.

Funding: This research was funded by the Inner Mongolia Application Technology Research and Development Funding Project (2019GG138) and the Natural Science Foundation of the Inner Mongolia Autonomous Region (2020ZD18).

Data Availability Statement: Not applicable.

Conflicts of Interest: The authors declare no conflict of interest.

Appendix A

Table A1. The number of images where the category occurs $N_{img}(l)$ and the probabilities of class occurrences $n(l)$.

Object Categories	$N_{img}(l)$	$n(l)$
airplane	344	0.0586
airport	326	0.0556
baseball field	552	0.0941
basketball court	336	0.0573
bridge	378	0.0644
chimney	202	0.0344
dam	238	0.0406
expressway service area	279	0.0475
expressway toll station	285	0.0486
golf field	216	0.0368
ground track field	537	0.0916
harbor	329	0.0561
overpass	410	0.0699
ship	649	0.1107
stadium	289	0.0493
storage tank	390	0.0665
tennis court	605	0.1032
train station	244	0.0416
vehicle	1561	0.2662
windmill	404	0.0689

Table A2. Probabilities of classes of DIOR appearing with water area and not appearing with water area. Column $n(l|w)$ denotes the probability of category l appearing with water area; $n(l|\overline{w})$ is the probability that category l appears with no water area.

Object Categories	$n(l\|w)$	$n(l\|\overline{w})$
airplane	0.0412	0.9588
airport	0.4482	0.5518
baseball field	0.0714	0.9286
basketball court	0.0981	0.9019
bridge	0.9525	0.0475
chimney	0.1228	0.8772
dam	1	0
expressway service area	0.2719	0.7281
expressway toll station	0.1335	0.8665
golf field	0.9114	0.0886
ground track field	0.1293	0.8707
harbor	1	0
overpass	0.1179	0.8821
ship	0.9998	0.0002
stadium	0.1126	0.8874
storage tank	0.3126	0.6874
tennis court	0.1366	0.8634
train station	0.2398	0.7602
vehicle	0.2140	0.7860
windmill	0.0489	0.9511

References

1. Cheng, G.; Han, J. A survey on object detection in optical remote sensing images. *ISPRS J. Photogramm. Remote Sens.* **2016**, *117*, 11–28. [CrossRef]
2. Li, K.; Wan, G.; Cheng, G.; Meng, L.; Han, J. Object detection in optical remote sensing images: A survey and a new benchmark. *ISPRS J. Photogramm. Remote Sens.* **2020**, *159*, 296–307. [CrossRef]
3. Wu, X.; Li, W.; Hong, D.; Tao, R.; Du, Q. Deep learning for unmanned aerial vehicle-based object detection and tracking: A survey. *IEEE Geosci. Remote Sens. Mag.* **2021**, *10*, 91–124. [CrossRef]
4. Fascista, A. Toward Integrated Large-Scale Environmental Monitoring Using WSN/UAV/Crowdsensing: A Review of Applications, Signal Processing, and Future Perspectives. *Sensors* **2022**, *22*, 1824. [CrossRef] [PubMed]
5. Mo, N.; Yan, L. Improved faster RCNN based on feature amplification and oversampling data augmentation for oriented vehicle detection in aerial images. *Remote Sens.* **2020**, *12*, 2558. [CrossRef]
6. Ding, J.; Xue, N.; Long, Y.; Xia, G.S.; Lu, Q. Learning RoI transformer for oriented object detection in aerial images. In Proceedings of the IEEE/CVF Conference on Computer Vision and Pattern Recognition, Long Beach, CA, USA, 15–20 June 2019; pp. 2849–2858.
7. Yang, X.; Yan, J. Arbitrary-oriented object detection with circular smooth label. In Proceedings of the European Conference on Computer Vision, Glasgow, UK, 23–28 August 2020; Springer: Berlin/Heidelberg, Germany, 2020; pp. 677–694.
8. Guo, Z.; Liu, C.; Zhang, X.; Jiao, J.; Ji, X.; Ye, Q. Beyond bounding-box: Convex-hull feature adaptation for oriented and densely packed object detection. In Proceedings of the IEEE/CVF Conference on Computer Vision and Pattern Recognition, Nashville, TN, USA, 20–25 June 2021; pp. 8792–8801.
9. Han, J.; Ding, J.; Li, J.; Xia, G.S. Align deep features for oriented object detection. *IEEE Trans. Geosci. Remote Sens.* **2021**, *60*, 1–11. [CrossRef]
10. Torralba, A.; Oliva, A.; Castelhano, M.S.; Henderson, J.M. Contextual guidance of eye movements and attention in real-world scenes: The role of global features in object search. *Psychol. Rev.* **2006**, *113*, 766. [CrossRef] [PubMed]
11. Li, Z.; Wu, Q.; Cheng, B.; Cao, L.; Yang, H. Remote sensing image scene classification based on object relationship reasoning CNN. *IEEE Geosci. Remote Sens. Lett.* **2020** , *19*, 1–5. [CrossRef]
12. Xu, H.; Jiang, C.; Liang, X.; Lin, L.; Li, Z. Reasoning-rcnn: Unifying adaptive global reasoning into large-scale object detection. In Proceedings of the IEEE/CVF Conference on Computer Vision and Pattern Recognition, Long Beach, CA, USA, 15–20 June 2019; pp. 6419–6428.
13. Xia, G.S.; Bai, X.; Ding, J.; Zhu, Z.; Belongie, S.; Luo, J.; Datcu, M.; Pelillo, M.; Zhang, L. DOTA: A large-scale dataset for object detection in aerial images. In Proceedings of the IEEE Conference on Computer Vision and Pattern Recognition, Salt Lake City, UT, USA, 18–23 June 2018; pp. 3974–3983.
14. Fang, Y.; Kuan, K.; Lin, J.; Tan, C.; Chandrasekhar, V. Object detection meets knowledge graphs. In Proceedings of the International Joint Conferences on Artificial Intelligence, Melbourne, Australia, 19–25 August 2017.

15. Xie, X.; Cheng, G.; Wang, J.; Yao, X.; Han, J. Oriented R-CNN for object detection. In Proceedings of the IEEE/CVF International Conference on Computer Vision, Montreal, QC, Canada, 10–17 October 2021; pp. 3520–3529.
16. Ming, Q.; Miao, L.; Zhou, Z.; Dong, Y. CFC-Net: A critical feature capturing network for arbitrary-oriented object detection in remote-sensing images. *IEEE Trans. Geosci. Remote Sens.* **2021**, *60*, 1–14. [CrossRef]
17. Yang, X.; Yan, J.; Feng, Z.; He, T. R3det: Refined single-stage detector with feature refinement for rotating object. In Proceedings of the AAAI Conference on Artificial Intelligence, Vancouver, BC, Canada, 2–9 February 2021; Volume 35, pp. 3163–3171.
18. Han, J.; Ding, J.; Xue, N.; Xia, G.S. Redet: A rotation-equivariant detector for aerial object detection. In Proceedings of the IEEE/CVF Conference on Computer Vision and Pattern Recognition, Nashville, TN, USA, 19–15 June 2021; pp. 2786–2795.
19. He, K.; Zhang, X.; Ren, S.; Sun, J. Deep residual learning for image recognition. In Proceedings of the IEEE Conference on Computer Vision and Pattern Recognition, Las Vegas, NV, USA, 27–30 June 2016; pp. 770–778.
20. Chen, C.; Gong, W.; Chen, Y.; Li, W. Object detection in remote sensing images based on a scene-contextual feature pyramid network. *Remote Sens.* **2019**, *11*, 339. [CrossRef]
21. Liu, Y.; Wang, R.; Shan, S.; Chen, X. Structure inference net: Object detection using scene-level context and instance-level relationships. In Proceedings of the IEEE Conference on Computer Vision and Pattern Recognition, Salt Lake City, UT, USA, 18–23 June 2018; pp. 6985–6994.
22. Siris, A.; Jiao, J.; Tam, G.K.; Xie, X.; Lau, R.W. Scene context-aware salient object detection. In Proceedings of the IEEE/CVF International Conference on Computer Vision, Montreal, QC, Canada, 10–17 October 2021; pp. 4156–4166.
23. Vaswani, A.; Shazeer, N.; Parmar, N.; Uszkoreit, J.; Jones, L.; Gomez, A.N.; Kaiser, Ł.; Polosukhin, I. Attention is all you need. *Adv. Neural Inf. Processing Syst.* **2017**, *30*. [CrossRef]
24. Li, K.; Cheng, G.; Bu, S.; You, X. Rotation-insensitive and context-augmented object detection in remote sensing images. *IEEE Trans. Geosci. Remote Sens.* **2017**, *56*, 2337–2348. [CrossRef]
25. Zhang, G.; Lu, S.; Zhang, W. CAD-Net: A context-aware detection network for objects in remote sensing imagery. *IEEE Trans. Geosci. Remote Sens.* **2019**, *57*, 10015–10024. [CrossRef]
26. Zhang, K.; Wu, Y.; Wang, J.; Wang, Y.; Wang, Q. Semantic context-aware network for multiscale object detection in remote sensing images. *IEEE Geosci. Remote Sens. Lett.* **2021**, *19*, 1–5. [CrossRef]
27. Liu, J.; Li, S.; Zhou, C.; Cao, X.; Gao, Y.; Wang, B. SRAF-Net: A Scene-Relevant Anchor-Free Object Detection Network in Remote Sensing Images. *IEEE Trans. Geosci. Remote Sens.* **2021**, *60*, 1–14. [CrossRef]
28. Feng, X.; Han, J.; Yao, X.; Cheng, G. TCANet: Triple context-aware network for weakly supervised object detection in remote sensing images. *IEEE Trans. Geosci. Remote Sens.* **2020**, *59*, 6946–6955. [CrossRef]
29. Cheng, B.; Li, Z.; Xu, B.; Dang, C.; Deng, J. Target detection in remote sensing image based on object-and-scene context constrained CNN. *IEEE Geosci. Remote Sens. Lett.* **2021**, *19*, 1–5. [CrossRef]
30. Xu, H.; Jiang, C.; Liang, X.; Li, Z. Spatial-aware graph relation network for large-scale object detection. In Proceedings of the IEEE/CVF Conference on Computer Vision and Pattern Recognition, Long Beach, CA, USA, 15–20 June 2019; pp. 9298–9307.
31. Xu, H.; Fang, L.; Liang, X.; Kang, W.; Li, Z. Universal-rcnn: Universal object detector via transferable graph r-cnn. In Proceedings of the AAAI Conference on Artificial Intelligence, New York, NY, USA, 7–12 February 2020; Volume 34, pp. 12492–12499.
32. Shu, X.; Liu, R.; Xu, J. A Semantic Relation Graph Reasoning Network for Object Detection. In Proceedings of the 2021 IEEE 10th Data Driven Control and Learning Systems Conference (DDCLS), Suzhou, China, 14–16 May 2021; IEEE: Piscataway, NJ, USA, 2021; pp. 1309–1314.
33. Jiang, C.; Xu, H.; Liang, X.; Lin, L. Hybrid knowledge routed modules for large-scale object detection. *Adv. Neural Inf. Processing Syst.* **2018**, *31*. [CrossRef]
34. Lin, T.Y.; Dollár, P.; Girshick, R.; He, K.; Hariharan, B.; Belongie, S. Feature pyramid networks for object detection. In Proceedings of the IEEE Conference on Computer Vision and Pattern Recognition, Honolulu, HI, USA, 21–26 July 2017; pp. 2117–2125.
35. Zhou, Y.; Yang, X.; Zhang, G.; Wang, J.; Liu, Y.; Hou, L.; Jiang, X.; Liu, X.; Yan, J.; Lyu, C.; et al. MMRotate: A Rotated Object Detection Benchmark using PyTorch. *arXiv* **2022**, arXiv:2204.13317.

remote sensing

MDPI

Article

DSANet: A Deep Supervision-Based Simple Attention Network for Efficient Semantic Segmentation in Remote Sensing Imagery

Wenxu Shi [1,2], Qingyan Meng [1,2,3,*], Linlin Zhang [1,2,3], Maofan Zhao [1,2], Chen Su [1,2] and Tamás Jancsó [4]

1 Aerospace Information Research Institute, Chinese Academy of Sciences, Beijing 100049, China
2 University of Chinese Academy of Sciences, Beijing 100049, China
3 Key Laboratory of Earth Observation of Hainan Province, Hainan Research Institute, Aerospace Information Research Institute, Chinese Academy of Sciences, Sanya 572029, China
4 Alba Regia Technical Faculty, Obuda University, Budai ut 45, 8001 Szekesfehervar, Hungary
* Correspondence: mengqy@radi.ac.cn

Abstract: Semantic segmentation for remote sensing images (RSIs) plays an important role in many applications, such as urban planning, environmental protection, agricultural valuation, and military reconnaissance. With the boom in remote sensing technology, numerous RSIs are generated; this is difficult for current complex networks to handle. Efficient networks are the key to solving this challenge. Many previous works aimed at designing lightweight networks or utilizing pruning and knowledge distillation methods to obtain efficient networks, but these methods inevitably reduce the ability of the resulting models to characterize spatial and semantic features. We propose an effective deep supervision-based simple attention network (DSANet) with spatial and semantic enhancement losses to handle these problems. In the network, (1) a lightweight architecture is used as the backbone; (2) deep supervision modules with improved multiscale spatial detail (MSD) and hierarchical semantic enhancement (HSE) losses synergistically strengthen the obtained feature representations; and (3) a simple embedding attention module (EAM) with linear complexity performs long-range relationship modeling. Experiments conducted on two public RSI datasets (the ISPRS Potsdam dataset and Vaihingen dataset) exhibit the substantial advantages of the proposed approach. Our method achieves 79.19% mean intersection over union (mIoU) on the ISPRS Potsdam test set and 72.26% mIoU on the Vaihingen test set with speeds of 470.07 FPS on 512 × 512 images and 5.46 FPS on 6000 × 6000 images using an RTX 3090 GPU.

Keywords: convolutional neural network (CNN); deep supervision; lightweight model; remote sensing; semantic segmentation

Citation: DSANet: A Deep Supervision-Based Simple Attention Network for Efficient Semantic Segmentation in Remote Sensing Imagery. *Remote Sens.* **2022**, *14*, 5399. https://doi.org/10.3390/rs14215399

Academic Editors: Fahimeh Farahnakian, Jukka Heikkonen and Pouya Jafarzadeh

Received: 7 September 2022
Accepted: 20 October 2022
Published: 27 October 2022

1. Introduction

Remote sensing is a crucial technical tool for large-scale observations of the Earth's surface. With the rapid development of Earth observation and remote sensing imaging technology, remote sensing has entered the era of big data [1]. Big data qualities for remote sensing primarily involve three Vs: volume, velocity, and variety of data [2]. Every day, a massive volume of remote sensing data must be handled in the era of big data for remote sensing. Furthermore, increasingly diverse remote sensing data are playing important roles in several fields. Due to advances in imaging technology, very high-resolution (VHR) imagery has shown considerable potential in remote sensing images (RSIs) interpretation and has been the focus of semantic segmentation.

Semantic segmentation is a critical task in computer vision, and its special application to remote sensing is RSI interpretation. It requires pixelwise parsing of the input image to retrieve the predefined categories to which the elements belong. Semantic segmentation has broad and vital applications in a variety of fields. This is especially true in the realm of remote sensing, where subjects such as integrated land use and land cover mapping [3,4], town change detection [5,6], urban functional areas [7], building footprints [8], impervious surfaces [9], and water body [10] extraction. The majority of these applications and methodologies are based on VHR images and are constrained by the two issues listed below. (1) Information modeling with little detail. In comparison to prior low-resolution images, VHR images give unequal spatial and semantic information volume gains. The significant improvement in spatial resolution allows for the observation of previously unseen features. However, vital detail information is mixed in with a vast volume of redundant information, providing additional obstacles for information extraction. (2) Inefficient processing. On the data processing front, high-resolution imagery implies that the amount of data to be processed per unit of observation area for interpretation is rising dramatically, posing a considerable challenge for hardware and algorithms.

Researchers have proposed numerous ways to overcome the difficulties of semantic segmentation for VHR images in the age of big data. Deep learning algorithms are the primary techniques for semantic segmentation at the moment. Unlike classic machine learning algorithms based on prior knowledge and predetermined rules, deep learning algorithms are data-driven algorithms that perform poorly with tiny data samples but may be utilized to great advantage in the era of big data. Deep learning-based convolutional neural networks (CNNs) outperform classic machine learning methods in terms of performance. Fully convolutional networks (FCNs) [11] have been utilized to obtain outstanding results in the semantic segmentation of RSIs. Following study, numerous model variants based on the FCN architecture have been developed, making substantial advances in various aspects. UNet [12], which is based on an encoder-decoder architecture, enhances the FCN's capacity to represent the multiscale features of images through contraction paths and expansion paths for achieving high-precision road [13] and coastline recognition [14] in RSIs. The DeepLabv3 series [15,16] utilize parallelized atrous spatial pyramid pooling (ASPP) with varying ratios to expand the models' reception fields while obtaining multiscale features; these models are widely used in RSI semantic segmentation, cloud detection [17], etc. However, because to the poor inference speeds of these models and the high hardware needs placed on deployed devices, these approaches find it difficult to overcome the aforementioned two problems. Figure 1 depicts the problem of building segmentation models that take both efficiency and performance into account.

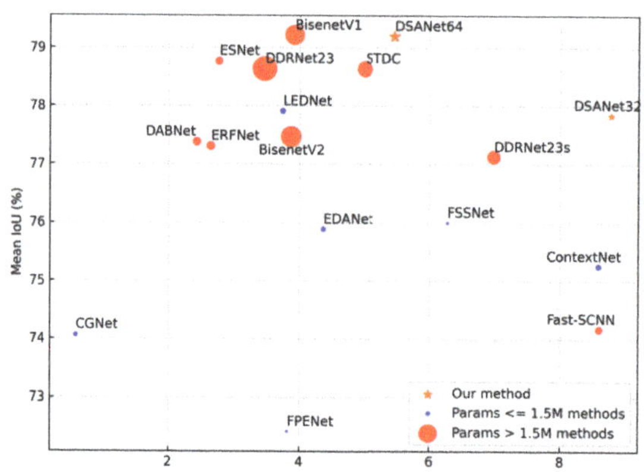

Figure 1. Speed-accuracy tradeoff yielded by different semantic segmentation methods on the ISPRS Potsdam dataset with a size of 6000 × 6000 pixels using an RTX 3090 GPU. Orange points: different versions of our proposed method. Red points: lightweight methods with more than 1.5 M parameters. Blue points: lightweight methods with less than 1.5 M parameters. Our proposed methods achieve the best speed-accuracy tradeoffs. It is worth noting that that the sizes of the corresponding points of the methods are positively correlated with their parameters.

In addition to investigating model segmentation performance, another approach is to optimize the efficiency and accelerate the inference speed of the utilized model. A conceivable way to accomplish lightweight model building is to reduce the number of model channels and add an attention mechanism to compensate for the loss in model performance [18]. In addition to incorporating an attention module, the introduction of a deep supervision [19] module can also enhance the segmentation performance of the model. By actively monitoring the body and edge characteristics of the object of interest, a lightweight semantic segmentation network was suggested to maximize the overall consistency and object details of semantic segmentation results [20]. Loss functions expressly designed for the semantic segmentation task can speed up the learning process of the resultant model for fundamental spatial information such as borders [21] and spatial correlations [22], as evidenced by higher performance with the same amount of training epochs. These lightweight networks struggle to capture the rich, detailed aspects of VHR images with fewer parameters, reducing accuracy significantly.

We investigate a solution for alleviating data interpretation burden in the era of large data for remote sensing that balances performance and inference speed. The functions of a lightweight network backbone, an attention mechanism, a deep supervision module, and a loss function in attaining effective semantic segmentation are thoroughly investigated in this paper. Our contributions are summarized here.

(1) To alleviate the VHR images interpretation mistake in the age of large data, an efficient deep-layer and shallow-channel network with spatial and semantic enhancement losses (DSANet) is developed.
(2) Without inference speed costs, two multiscale feature losses are proposed: improved multiscale spatial detail (MSD) and hierarchical semantic enhancement (HSE). The MSD loss is intended to improve the model's extraction of underlying spatial information, whilst the HSE loss assists the model in understanding the observed distribution of categories.
(3) The addition of the embedding attention module (EAM) decreases the attention module's complexity from quadratic (self-attention) to linear with equivalent accuracy, as well as increasing inference speed for large images.
(4) On the ISPRS Potsdam and Vaihingen benchmark data-sets, we attain outstanding results. Using an RTX 3090 GPU, we achieve mean intersection over unions (mIoUs) of 79.19% on the Potsdam test set and 72.26% on the Vaihingen test set, with a speed of 470.07 frames per second (FPS) on 512 × 512 images and 5.46 FPS on 6000 × 6000 images.

The rest of this paper is organized as follows. Section 2 reviews related works involving efficient network designs, efficient semantic segmentation approaches, information enhancement modules, and attention mechanisms. Section 3 presents the network structure of the proposed model and the detailed principles of its modules. Section 4 introduces the utilized datasets and demonstrates the implementation details of our experiments. The ablation experiments and a results comparison with state-of-the-art methods are also included in Section 5. Finally, Section 6 provides a summary of the paper.

2. Related Works

Many lightweight segmentation algorithms have obtained impressive results on many benchmarks in the domains of autonomous driving, video surveillance, and VHR remote sensing scene perception in the last 5–10 years. This section reviews efficient network designs and related works, categorizing them as follows: efficient network designs, efficient semantic segmentation approaches, information enhancement modules, and attention mechanisms.

2.1. *Efficient Network Designs*

Researchers are discovering that network design is becoming increasingly crucial as the Visual Geometry Group network (VGGNet) [23], the residual network (ResNet) [24], and DenseNet [25] models continue to be suggested. Because semantic segmentation is a dense prediction task, related models tend to have more parameters and slower inference speeds, which is harmful to model deployment and severely limits their application possibilities. An efficient network design paradigm lends itself well to the creation of efficient segmentation networks. By extensively replacing the 3 × 3 convolution in the model with a 1 × 1 convolution and reducing the number of channels in the 3 × 3 convolution, SqueezeNet [26] achieves comparable classification accuracy to AlexNet [27] with 2% of the total parameters. The MobileNet series [28–30] has steadily introduced new techniques to deep separable networks such as inverted residuals and neural architecture search (NAS). By integrating group convolution and channel shuffling operations and employing four recommendations, the ShuffleNet series [31,32] achieves a balance between accuracy and parameter number. 1. Equal channel widths minimize the memory access cost (MAC). 2. Excessive group convolution increases the MAC. 3. Network fragmentation reduces the degree of parallelism. 4. Elementwise operations are nonnegligible. Several outstanding and efficient semantic segmentation models have been presented as a result of these exploratory efforts on efficient network construction.

2.2. Efficient Semantic Segmentation Methods

Efficient semantic segmentation models strive for a balance between accuracy and speed, with considerable inference speed benefits at a low accuracy cost. They represent a significant development in the field of semantic segmentation in terms of efficiency, and they have created many good works based on the collaborative efforts of scholars. The two dominant approaches point the way to achieving high-accuracy and efficient semantic segmentation. 1. Light-weight backbones. ENet [33], a representative of earlier efficient segmentation models, greatly reduces the number of required parameters and floating point operations (FLOPs) by employing an asymmetric encoder-decoder structure and factorizing filters. Subsequent work has focused on asymmetric networks, with the goal of improving model performance by using deeply separable convolutions [34], dilated convolutions [35], factorized convolutions [36,37], dense connections [38], skip connections [39], pyramidal pooling [40] and channel splitting and shuffling [41]. The Fast-shallow CNN (SCNN) [42] adopts shared shallow network paths to encode details while learning contexts at low resolutions, saving computing costs. STDCNet [38] utilizes a lightweight backbone network from DenseNet with layer concatenation. Dual-resolution branch networks [43], exemplified by the bilateral segmentation network (BiSeNet) series [44,45], provide effective segmentation by modifying extraction branches for spatial and semantic information independently. 2. Feature aggregation. The deep feature aggregation network (DFANet) [46] recommends two deep branches where several bilateral fusions are conducted. By steering upper-level feature upsampling using low-level features, SFNet [47] achieves higher-resolution restoration and cross-layer feature aggregation. DDRNet [48] advises two deep branches between which multiple bilateral fusions are performed.

2.3. Information Enhancement Modules

The information in computer vision tasks can be divided into spatial and semantic information, both of which contribute significantly to accurate segmentation. (1) Enhancing spatial information. Typically, the shallow layer of the encoder may better describe spatial information. Ensuring that a branch has a high resolution preserves spatial information to the greatest extent possible. STDCNet adopts the Laplacian kernel of the pyramid hierarchy as an auxiliary loss function, which expedites the process of learning spatial edge features. Researchers suggest that the quantifications and statistics of spatial texture aspects are likewise of great significance due to quantization and counting operators. (2) Enhancing semantic information. PSPNet [49] adopts pyramid pooling to enhance the observed multiscale semantic features. The DeepLab series [15,16,50,51] utilizes parallel atrous convolutions with varying dilation rates; this approach is called ASPP, which can encode multiscale semantic information more effectively. DANet [52] models long-range dependencies in the channels and positions of sematic features using a dual self-attention module. OCRNet [53] explicitly turns the pixel classification problem into an object area classification problem, computes the relationship between each pixel and each object region, and augments the representation of each pixel with an object-contextual representation.

2.4. Attention Mechanisms

The selected attention mechanism is a crucial component of model design and is a key module for improving model performance. It is a descriptive weighting of the relationship between a particular attribute (from a small pixel value to an entire channel) and the data, so that it can be chosen to suppress or amplify that attribute at a particular location in order to achieve a selective representation of a particular feature for the model. The outstanding early approach is the squeeze-and-excitation network (SENet) [54], which squeezes the features on each channel by global maximum pooling and uses a fully connected layer to encode the features into a low-dimensional space before performing decoding. This makes the SENet an excellent attention module without imposing many additional parameters or a large computational burden on the subject network. The SENet's concept of squeezing and extracting channels and examining spatial attention inspired further research. Important follow-ups include the block attention module (BAM) [55] and convolutional BAM (CBAM) [56]. A BAM includes a two-branch parallel attention computation paradigm, with channel attention branches that adhere to the SENet's approach. Spatial features are squeezed in the channel dimension by a 1×1 convolution, key spatial features are extracted using a 3×3 convolution, and finally, a pixelwise summation operation is performed for both attention weights. A CBAM selects a multistep attention paradigm that combines channel attention and spatial attention simultaneously. The combination of spatial attention with gated mechanisms is another way to utilize attention mechanisms [57]. Unlike the idea of feature compression and extraction in the above work, self-attention [58] is a pixel-level attention mechanism. The computational complexity and resource needs of this method are an order of magnitude more than those of the preceding approaches, despite the fact that its performance is superior. Transformers [59], which outperform CNNs in many tasks, are excellent models based on self-attention; however, researchers are still designing optimizations for visual tasks such as patches [60] and hierarchical architectures [61,62] to overcome the fatal flaw of a computationally intensive attention mechanism. Fortunately, self-attention based on queries, keys and values can be optimized from O(n2) complexity to linear complexity by changing the order of computation [63], performing approximate computation [64], and conducting low-rank singular value decomposition [65].

It is typical practice for effective semantic segmentation networks [33,44] to utilize an attention module based on the SENet or linear simplified self attention due to its computational efficiency and inference speed.

3. Methodology

Our proposed segmentation model (DSANet) adheres to the original design concepts outlined below: (1) to adhere to Occam's razor: entities should not be multiplied beyond necessity; (2) to have the smallest possible number of parameters while obtaining acceptable accuracy; and (3) to avoid modules that improve the model's representation capabilities but consume an unacceptable amount of time during inference. Important aspects of the model include: (1) its low channel capacity and extra downsampling stages in the backbone to quickly obtain large perceptual fields, (2) the combined multiscale spatial detail loss and hierarchical semantic enhancement loss in the deep supervision module, and (3) a simple attention module with linear complexity. The details of DSANet can be seen in Figure 2.

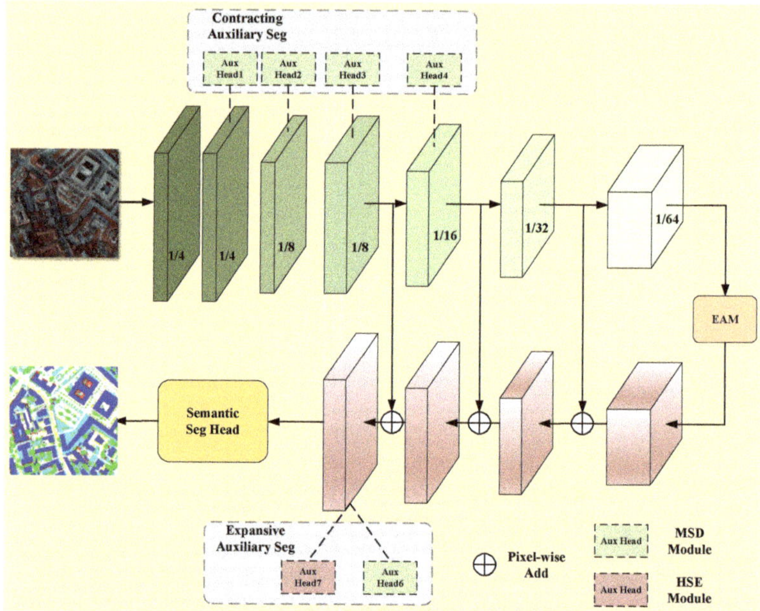

Figure 2. The network architecture of DSANet. Note that the green blocks are encoder components, whereas the red blocks and the semantic segmentation head are decoder components. All auxiliary segmentation heads within the dashed box are only engaged during model training, whereas only the remaining modules use inference time.

3.1. Network Architecture of the Proposed Method

DSANet is an asymmetric, U-shaped, basic network with an encoder for the contracting path and a decoder for the expansion path.

In contrast to prior lightweight semantic segmentation networks that employ two-branch designs, i.e., semantic and spatial branches, we employ a single backbone branch that is anticipated to extract both spatial and semantic information. For such single-branch networks, it is essential to improve spatial feature extraction. Good spatial texture information and color information are required for the model to sense semantics, and correct boundary detail information is essential for directing the high-resolution reconstruction of semantic components. Observing the inference time spent by BiSeNet (see details in Table 1) reveals that (1) the spatial path (SP) for extracting spatial information, the attention refinement module (ARM) for refining semantic features, and the feature fusion module (FFM) for feature interaction account for more than 30% of the model inference speed; (2) performing feature operations at the second-to-last scale (ARM16) is extremely time-consuming and unsatisfactory.

Table 1. Analysis of the Number of Parameters, Number of Computations and Inference Time of the BiSeNet Network.

Module	Params (M)	FLOPs (G)	Inference Time (ms)
SP	0.685	9.586	3.84
ARM32	4.521	1.023	0.74
ARM16	2.323	2.048	21.94
FFM	0.984	1.836	4.56
All	8.513	14.493	31.08

Note that all data are in percentages. ARM32 and ARM16 represent ARM applied to 1/32 and 1/16 scales, respectively.

To reduce the number of parameters in the model, including the number of layers and channel capacities, are redesigned. A typical semantic segmentation task only downsamples an image to 1/16 or 1/32 through the encoder and performs operations such as feature refinement and attention at this scale; this is totally insufficient for VHR images. Our method attempts to investigate the semantic content of VHR images at a more granular level. Semantic information extraction can benefit from increased channel capacity, but the resulting redundancy necessitates high model refinement and essential information discrimination. For this reason, a lightweight semantic segmentation job need to reduce the channel capacity of deep layers.

There are two suggested variants of DSANet, DSANet64 and DSANet32, with the numbers denoting the channel capacity of the model. Using DSANet64 as an example, the model encoder is briefly described in Table 2. At Stage 0, feature maps are subjected to continuous quick downsampling procedures to decrease the amount of computations performed from scratch. In stages 1–4, downsampling and feature extraction are alternated with a slower rate of channel capacity development. Another continuous quick downsampling procedure is done in the subsequent two steps. The final encoder extracts semantic information at a scale of 1/64 with a channel capacity of 256, which is quite low in comparison to other models' channel capacity of 1024. Finally, a self-attention module is used to simulate the most profound semantic information inside features over the long-range. Through skip connections, stages 7–9 merge the feature map with rich spatial information in the encoder with the upsampled semantic feature map and eventually restore the image's scale to 1/8 that of the original. The final result of semantic parsing is achieved via the segmentation head.

Table 2. Detailed Architecture of the DSANet Encoder.

Stages	Output Size	KSize	S	DSANet32		DSANet64	
				R	C	R	C
Image	512 × 512				3		3
Stage 0	256 × 256	3 × 3	2	1	32	1	64
	128 × 128		2	1		1	
Stage 1	128 × 128	3 × 3	1, 1	2	32	2	64
Stage 2	64 × 64	3 × 3	2, 1	1	32	1	64
	64 × 64		1, 1	1		1	
Stage 3	64 × 64	3 × 3	1, 1	2	64	2	128
Stage 4	32 ×32	3 × 3	2, 1	1	64	1	128
	32 × 32		1, 1	1		1	
Stage 5	16 × 16	3 × 3	2	1	64	1	128
Stage6	8 × 8	3 × 3	2	1	128	1	256
FLOPs				2.09G		7.46G	
Params				1.14M		4.58M	

Note that "Stage" in the table refers to the combination of a series of Conv-BN-rectified linear unit (ReLU). KSize, S, R, and C refer to the kernel size, stride, number of repetitions and number of out channels, respectively. Stages 1–4 use the basic DSA module, and the Stages 5 and 6 use the bottleneck version of the DSA module. FLOPs are calculated based on 512 × 512 images.

3.2. EAM

To compare and comprehend the features of the EAM and its advantages in terms of efficient semantic segmentation, we will first review the self-attention mechanism. As illustrated in Figure 3. A, the self-attention mechanism calculates the attention relations between various elements by the dot product operation, which allows for a more accurate representation of long-range information. Given a feature map $F \in \mathbb{R}^{C\times H\times W}$, where H, W, and C represent the length, width, and number of channels of the feature map F, respectively, the feature map F is reshaped to a sequence $X = \{x_1, x_2, \ldots, x_N\}$, where $x_i \in \mathbb{R}^C$ is the feature vector of element N and N (equal to $H \times W$) is the number of elements. Three linear transformations are performed on each of these feature vectors to encode the information into a high-dimensional space and to produce $Q \in \mathbb{R}^{N \times d_k}$, $K \in \mathbb{R}^{N \times d_k}$, and $V \in \mathbb{R}^{N \times d_v}$:

$$Q = W_Q(X), K = W_K(X), V = W_V(X) \tag{1}$$

where d_k and d_v are set to be equal to the same number for the simplicity of calculation in general.

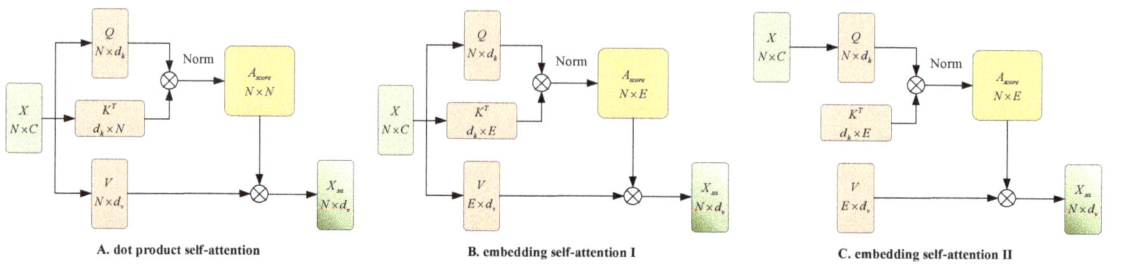

Figure 3. Dot product self-attention SA (**A**) and two kinds of embedding self-attention: EAM I (**B**) and EAM II (**C**).

The similarity measure between the i-th element and the j-th element can be calculated by the cosine similarity formula, expressed as $({q_i}^T k_j)$. The softmax function is chosen as the normalizing function because the attention given by the i-th element to the j-th element depends not only on their similarity but also on the attention paid by the i-th element to all other elements. The attention scores between elements and the outcomes of self-attention are computed by the following (2):

$$Attn(Q, K, V) = Norm(Similarity(Q, K)) \cdot V \tag{2}$$

where $Norm$ represents the softmax normalization function, and $Similarity(\cdot)$, which calculates the relationship between Q and K, is defined as:

$$Similarity(Q, K) = \frac{QK^T}{\sqrt{d_k}} \tag{3}$$

where $\sqrt{d_k}$ is the scaling factor that maintains the variance of $Similarity(Q, K)$ at 1, preventing the gradient from vanishing. $Similarity(Q, K)$ is abbreviated as A.

An intuitive approach for reducing the computational complexity of self-attention is that not every attention between a pair of elements is sufficiently useful, and so we may only need to obtain the attention relations between the i-th element and a set number of essential components. Two techniques are offered to accomplish the aforementioned concept.

(1) In accordance with the fundamental structure of self-attention, the feature vector X is linearly transformed to generate Q, K and V. The difference is that the dimensions of K and V are altered from $\mathbb{R}^{N \times d_k}$ to $\mathbb{R}^{E \times d_k}$, where E is the embedding dimensionality. The first dimension of K and V from N to E simulates the process of selecting the top E most important elements from N. Due to probable image size changes between training and test data, N cannot be predicted in advance for the semantic segmentation task; thus, adaptively pooling the feature vector X in advance is essential to achieve N with fixed dimensions. Theoretically, without considering adaptive pooling, the computational complexity of embedding self-attention I (Figure 3B) is $O(Ed_kN)$. In the real case, the computational complexity will be better than this value, satisfying a lower linear computational complexity.

(2) Unlike the first two attentional approaches, embedding self-attention II (Figure 2C) generates only Q using the feature vectors X, while the memory K and V are pre-

generated random matrices in $\mathbb{R}^{N \times d_k}$ and optimised during training phase. This strategy may successfully overcome the difficulty associated with the unpredictability of N and reduce calculation time for K and V. Due to the fact that K and V are fully independent of the feature vector X, the interactions between components are weak, making it difficult for EAM II to establish genuine attentional connections. We employ the approach in [65] to normalize the rows and columns of A independently, as it is possible that strengthening the connections between components using a single softmax function, which is often used in self-attention mechanisms, may not yield optimal results. L1 normalization is specifically applied following softmax activation. This method's computational complexity is also $O(Ed_kN)$. The following are the precise formulae for the softmax and L1 normalization functions.

$$\tilde{A}_{i,j} = softmax(Q,K)_{i,j} = \frac{\exp(A_{i,j})}{\sum_k^E \exp(A_{k,j})} \tag{4}$$

$$\hat{A}_{i,j} = L1\_Norm(\tilde{A}_{i,j}) = \frac{\tilde{A}_{i,j}}{\sum_k^{d_k} \tilde{A}_{i,k}} \tag{5}$$

3.3. MSD Loss

VHR images contain rich detail and texture information, necessitating lightweight models with strong spatial representation capabilities. The deep supervision module is an auxiliary segmentation head that helps mitigate problems such as gradient vanishing and slow network convergence during training and assists the intermediate layer in improving the model representation; this module is activated only during the model training phase. A novel deep supervision module based on the MSD loss was proposed in [38]. This module uses second-order differential operators to extract boundary and detail information from the labels at various scales to improve the spatial representation of the model. However, this method achieves suboptimal results on VHR images when applied to DSANet. Considering that the input of this module includes all feature maps from a shallow layer, it is difficult for the lightweight DSANet to effectively represent semantic features because VHR images are rich in semantic information and the deep spatial supervision process is too restrictive. It is recommended that our MSD loss with a selective kernel ratio will fix this issue. This kernel arbitrarily truncates portions of the feature maps so that the corresponding convolution kernels of the network layers may be less affected by spatial deep supervision. The selected feature maps enter the MSD module to improve the network's capacity to represent boundary details, while the other feature maps are transmitted to further layers to provide appropriate semantic representations. The particular MSD loss calculation procedure is as follows.

Constructing multiscale edge extraction pyramids. The most frequently used second-order differential operator is the Laplace operator in two dimensions, which is formulated as follows:

$$\Delta f = \frac{\partial^2 f}{\partial x^2} + \frac{\partial^2 f}{\partial y^2} \tag{6}$$

where f is a twice-differentiable real-valued function. For processing RSIs in the form of discrete data, the discrete Laplace operator O (see Equation (7)) is applied.

$$O = \begin{bmatrix} -1 & -1 & -1 \\ -1 & 8 & -1 \\ -1 & -1 & -1 \end{bmatrix} \tag{7}$$

Laplace convolution operators with varying strides are utilized to create multiscale detail maps $\mathcal{D}_0 \in \mathbb{R}^{H \times W}$, $\mathcal{D}_2 \in \mathbb{R}^{H \times W}$ and $\mathcal{D}_4 \in \mathbb{R}^{H \times W}$ in order to fully leverage the multiscale properties of the label maps. A pyramid detail map $P \in \mathbb{R}^{H \times W}$ is obtained by summing these multiscale detail maps.

$$\mathcal{P} = \mathcal{D}_0 \oplus \mathcal{D}_2 \oplus \mathcal{D}_4 \tag{8}$$

Given the output feature maps $\mathcal{F}_{in} \in \mathbb{R}^{C\times H\times W}$ of a shallow layer, the selected feature maps $\mathcal{F}_S \in \mathbb{R}^{C\times H\times W}$ are obtained through the selective kernel. Next, after a 3×3 convolution and a 1×1 convolution, the channel dimensionality of $\mathcal{F}_S \in \mathbb{R}^{H\times W}$ is reduced to 1, which matches the shape of the pyramid detail map.

Evaluation loss. For a sparse matrix with extremely unbalanced categories (such as the pyramid detail map), the percentage of pixels containing detailed information is very small (the pixels in red and black are compared in the pyramid detail map in Figure 4), so it is difficult to obtain better results with the binary cross-entropy (BCE) loss alone. A typically utilized strategy is to optimize the loss evaluation method by incorporating a category proportion-insensitive Dice loss that has a solid ability to distinguish between foreground and background information. The formulas for the BCE loss and Dice loss are as follows.

$$\mathcal{L}_{bce}(\mathcal{F},\mathcal{P}) = -\sum_{i=1}^{H\times W} \frac{f_i \cdot \log p_i + (1-f_i)\cdot \log(1-p_i)}{H\times W} \tag{9}$$

$$\mathcal{L}_{dice}(\mathcal{F},\mathcal{P}) = 1 - \frac{2\sum_{i=1}^{H\times W} f_i \cdot p_i + \varepsilon}{\sum_{i=1}^{H\times W}(f_i) + \sum_{i=1}^{H\times W}(p_i)^2 + \varepsilon} \tag{10}$$

where f_i and p_i represent the values of the i-th element derived from the feature map $\mathcal{F}$ and the pyramid detail map $\mathcal{P}$, respectively, and ε is a very small number used to smooth the gradient and is set to 1×10^{-8}.

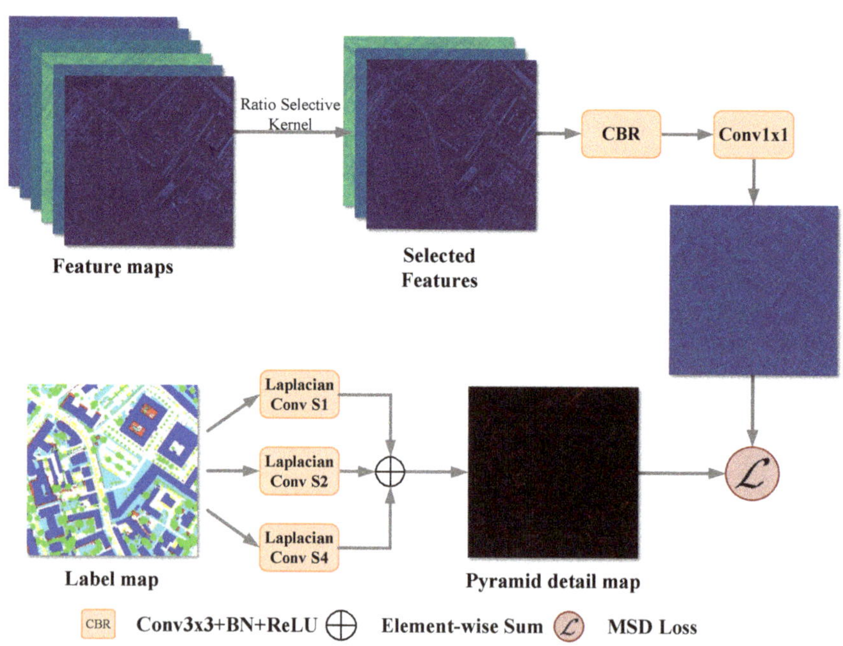

Figure 4. Schematic diagram of the MSD loss calculation process. Stridds 1, 2, and 4 in the Laplace convolution operator are denoted as S1, S2 and S4.

The mean squared error (MSE) combines these two losses, and the calculation formula is shown as follows.

$$\mathcal{L}_{\mathcal{MSD}}(\mathcal{F},\mathcal{P}) = \mathcal{L}_{bce}(\mathcal{F},\mathcal{P}) + \beta\mathcal{L}_{dice}(\mathcal{F},\mathcal{P}) \tag{11}$$

where β is a hyperparameter, which is 1.0 in this paper.

The specific calculation process of the MSE loss is shown in Figure 4.

3.4. HSE Loss

In contrast with the MSD loss, the HSE loss is proposed for enhancing the capacity of the model to discern the category distributions of images. Category parsing errors are frequently caused by the large number of categories in VHR images, which contain considerable intraclass spectral variations as well as moderate interclass spectral changes. Adding semantic information to the model can effectively reduce the impact of this issue on the segmentation results.

Our proposed HSE loss is embedded in the decoder without an inference cost. Figure 5 and Algorithm 1 provide detailed information. We denote the label map $y \in \mathbb{R}^{H \times W}$, which goes through the following process to obtain the HSE vector.

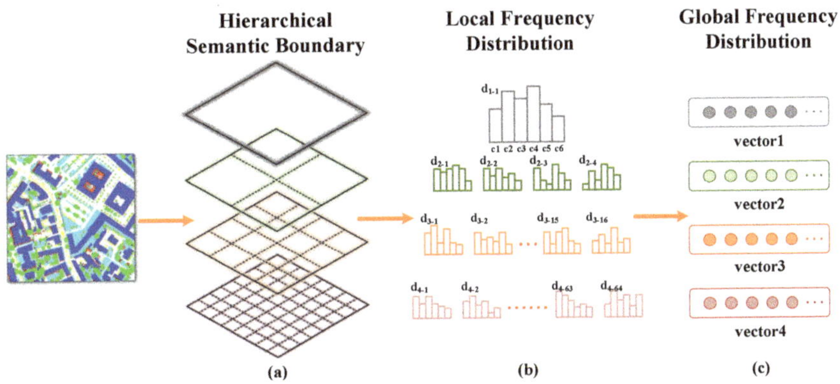

Figure 5. Schematic diagram of the HSE vector calculation process with 4 levels. (**a**) Boundaries in different colors indicate different segmentation scales. (**b**) Simplified diagram of the local frequency distribution. (**c**) Visual presentation of the HSE vectors.

(1) Set the hierarchical semantic boundary. Semantic boundaries at different locations can capture the distributions of categories in different local regions, and semantic boundaries at different scales can reflect the multiscale characteristics of category distribution features. Assume that N boundary levels and the boundary level of n slice the label map in 2^n patches along the length and width, respectively. The label patches are set as $y_n = \{y_1, y_2, \ldots, y_N\}$, where $y_n = \{y_n^{(j)}\}_{j=1}^{2^{2n}}$ is the set of feature patches in level n and $y_n^{(j)} \in \mathbb{R}^{\frac{H}{2^n} \times \frac{W}{2^n}}$.

(2) Calculate the local frequency distribution. The category distribution $d_n^{(j)}$ is calculated separately for each label patch $y_n^{(j)}$, where j is the sequence number of the label patch set.

(3) Aggregate the global distribution vector. The label patches at boundary level n are concatenated to generate the global frequency distribution vector $\hat{v}_n$. The same processing flow is applied to the prediction map output from network stage K in the decoder to obtain the vector v_n. The HSE vector of prediction is $v = \{v_n\}_{n=0}^{N}$, and that of the label map $\hat{v} = \{\hat{v}_n\}_{n=0}^{N}$. The HSE loss is obtained by computing the BCE between the HSE vector of the prediction v and the HSE vector of the label map $\hat{v}$.

The HSE algorithm is defined as follows:

Algorithm 1: HSE loss for the deep supervision module.

Input: Batch of examples with labels $\mathcal{X} = \{(x_i, y_i)\}_{i=1}^{\mathcal{B}}$
Output: HSE loss $\mathcal{L}_{\mathcal{HSE}}$
Result: Out stage K, Boundary levels N, Boundary weights $\{\alpha_i\}_{i=1}^{N}$, Numbers of classes c, Ignored label $c_{ignored}$, Length H and Width W of examples

for $b \leftarrow 1$ **to** $\mathcal{B}$ **do**
  $\mathcal{F}_{b,0} \leftarrow \text{Stage}_0(x_b)$
  for $k \leftarrow 0$ **to** $K-1$ **do**
    $\mathcal{F}_{b,k+1} \leftarrow \text{Stage}_k(\mathcal{F}_{b,k})$
  end
  $\mathcal{F}_b \leftarrow \mathit{Resize}(\arg\max \mathcal{F}_{b,\mathcal{K}}, (H, W))$
  for $n \leftarrow 0$ **to** N **do**
    Divide $\mathcal{F}_b$ and y_b equally into 2^{2n} patches `// along the length and width`
    Denote patches $\mathcal{F}_{b,n} \in \mathcal{R}^{2^{2n} \times \frac{H}{2^n} \times \frac{W}{2^n}}$ and $y_{b,n} \in R^{2^{2n} \times \frac{H}{2^n} \times \frac{W}{2^n}}$
    for $j \leftarrow 1$ **to** 2^{2n} **do**
      $\hat{d}_{b,n}^{(j)} \leftarrow \mathit{FrequencyDistribution}(y_{b,n}^{(j)}, c, c_{ignored})$ `// Ignore mask labels and compute frequency distribution`
      $d_{b,n}^{(j)} \leftarrow \mathit{FrequencyDistribution}(\mathcal{F}_{b,n}^{(j)}, c, c_{ignored})$
    end
    $\hat{v}_{b,n} \leftarrow \mathit{Concat}(\{\hat{d}_{b,n}^{(j)}\}_{j=1}^{2^{2n}})$ `// Aggregate the global distribution vector`
    $v_{b,n} \leftarrow \mathit{Concat}(\{d_{b,n}^{(j)}\}_{j=1}^{2^{2n}})$
    $\mathcal{L}_{b,n} \leftarrow \mathit{BCELoss}(\hat{v}_{b,n}, v_{b,n})$
  end
  $\mathcal{L}_{\mathcal{HSE},b} \leftarrow \frac{1}{n}\sum_n (ff_n \mathcal{L}_{b,n})$
end
$\mathcal{L}_{\mathcal{HSE}} \leftarrow \frac{1}{\mathcal{B}}\sum_b \mathcal{L}_{\mathcal{HSE},b}$

4. Data Set and Experimental Details

4.1. Benchmark Description

(1) ISPRS Potsdam Dataset (https://www.isprs.org/education/benchmarks/UrbanSemLab/2d-sem-label-potsdam.aspx (accessed on 2 September 2021).)

For the ISPRS competition, the Potsdam dataset serves as an urban modeling and semantic labeling baseline. Large building blocks, narrow streets, and dense settlement architecture may be seen in this typical old city. Data from DSM and nDSM orthophotography are available for each patch. For this dataset, there are 38 patches of the same size, all with the same ground sampling distance (GSD), and 24 of these patches are training data while the other 14 are validation data. Impervious surfaces, low-vegetation zones, trees, autos and the background are all manually determined categories. We used IRRG (near-infrared, red, and green bands) as the model's input data in order to compare it to other approaches.

(2) ISPRS Vaihingen Dataset (https://www.isprs.org/education/benchmarks/UrbanSemLab/2d-sem-label-vaihingen.aspx (accessed on 2 September 2021).)

Another benchmark from the ISPRS semantic labeling challenge is the Vaihingen dataset. It depicts a little community with a large number of single-story and small-scale multistory structures. The data types are configured in the same way that the Potsdam dataset was. With a GSD of 9 cm, it has 33 patches, 17 of which are set aside for validation. The patch sizes range from 1388 × 2555 to 3816 × 2550.

4.2. Evaluation Metrics

The mean of the classwise F1 score (mF1) and the mean of the classwise intersection over union are the most widely accepted metrics for evaluating model performance in semantic segmentation tasks (mIoU). The mF1 focuses on the evaluation of the outcomes predicted by the model at the pixel level, whereas the mIoU analyzes expected results in terms of the degree of overlap with the ground-truth labels.

The F1 score is the harmonic mean of the precision and recall, where the precision is the number of true-positive results divided by the number of all positive results, including those not identified correctly, and the recall is the number of true-positive results divided by the total number of samples that should have been identified as positive. Therefore, the precision, recall, and F1 score can be computed as

$$precision = \frac{TP}{TP + FP} \tag{12}$$

$$recall = \frac{TP}{TP + FN} \tag{13}$$

$$F1 = 2 \cdot \frac{precision \cdot recall}{precision + recall} \tag{14}$$

where TP, FP, FN, and TN represent true positives, false positives, false negatives and true negatives, respectively.

The IoU, also known as the Jaccard index, is a statistic used for gauging the similarity and the diversity between the predicted and the ground-truth labels. The IoU it can be represented as

$$IoU = \frac{|Sp \cap Sgt|}{|Sp \cup Sgt|} \tag{15}$$

where Sp and Sgt represent the set of predicted pixels and the set of ground-truth labels for the corresponding category, respectively, and $\cap$ and $\cup$ are the intersection and union operations defined on the set.

To compare the efficiency of the tested models, the study introduces the FLOPs and FPS as theoretical and practical measures of the model inference speed.

4.3. Data Preprocessing and Augmentation

The semantic interpretation of RSIs is characterized by fewer data samples but a higher size per picture than other standard computer vision tasks. In practice, we frequently confront two obstacles: (1) computing resources are constrained and GPUs struggle to enable direct input of full-frame RSI sample data; (2) a small sample size always results in overfitting and poor model generalizability. Through picture cropping, the image size may be lowered, and the data sample size can be raised proportionally. In addition, typical data augmentation techniques such as random cropping, random flipping, random rotation, and photometric distortion can successfully increase sample variability and enhance the generalizability of the used model. Given the considerable picture size variances in the Vaihingen dataset, it is required to standardize the image dimensions.

The specific strategy for data preprocessing and augmentation in this experiment is as follows. Preprocessing. (1) The raw images are cropped to a size of 500 $\times$ 500 pixels with a stride equal to half the size of the cropped image. (2) The images with lengths or widths that are less than a quarter of the cropped image size are discarded to ensure that enough valuable information exists within the images and to prevent images with excessive length and width differences, such as bars, from participating, as this can impair the model's ability to learn global features. Augmentation. (1) Multiscale resizing. A scale number is randomly selected from 0.5, 0.75, 1.0, 1.25, 1.5, 1.75 and 2.0 for the height and width. The current scale is equal to 512 multiplied by the scale number. Each cropped image is resampled to the set scale. (2) Random cropping. A 512 $\times$ 512 pixel block of data is cropped at a randomly selected location in the image; 512 is a common size in computer vision that satisfies an exponential multiple of 2, thereby avoiding the problem of indivisibility by 2 during operations such as pooling and downsampling. (3) Random flipping. (4) Photometric distortion. The brightness, contrast, saturation and hue levels of the images are randomly adjusted. (5) Normalization. The data distribution is adjusted to conform to a normal distribution.

Note that no data preprocessing and augmentation methods are used in the validation step except normalization, which is used to simulate the actual working flow of data processing.

4.4. Implementation Details

In all experiments, we establish a virtual Anaconda environment with Python 3.7 and PyTorch 1.8.2 as the standard. The specific graph computation platform contains CUDA 11.1, CUDNN 8.0.4 and TensorRT 7.2.3.4 on an NVIDIA RTX 3090 GPU. All latency benchmarks for our methods are computed by trtexec with a batch size of 1.

The specific parameter configuration is as follows. All experiments use a batch size of 16. To ignore the effect of the gradient descent algorithm on the experiments, stochastic gradient descent (SGD) is set as the standard optimizer. Following the optimizer parameter settings of most works, we choose a momentum of 0.9 and a weight decay of 5×10^{-4}. The learning rate (lr) is initially set to 0.001. All models are iterated 80,000 times with weights and evaluated for model performance. We utilize the "poly" policy as the learning rate update scheduler. The quantity lr can be calculated by the formula $lr = lr_0 \cdot (1 - \frac{iter}{iter_0})^{power}$, where lr_0 is the initialized learning rate, $iter_0$ is the maximum number of iterations and the power is 0.9. To prevent lr from being so small that the weights are almost negligible in the later iterations of the model training update process, we set the lower cutoff value for the learning rate to 1e-5. We employed the cross-entropy loss function, which is commonly used for semantic segmentation, to describe the difference between the final predictions and the labels. In the validation test session, we follow the settings in [66] and directly input the whole raw images, which benefits from the GPU parallelization of convolutions.

5. Experimental Results

5.1. Ablation Study

In this subsection, we design a series of ablation experiments to prove the effectiveness of our network. All the following experiments are evaluated on the ISPRS Potsdam and Vaihingen datasets.

5.1.1. Effectiveness of the EAM

Comparing various combinations of EAMs and normalizing techniques, Table 3 demonstrates that the optimal combination is EAM II + Softmax + L1 Norm. EAM I and EAM II with just softmax activation struggle to represent features accurately across resolutions. EAM II + Softmax + L1 Norm with varied encoding dimensions yields excellent performance, outperforming the backbone by 0.97 and 1.12 % based on the mIoU metric. To investigate the effect of the network stage in which the EAM is located on the results, we choose to insert the EAM in the deepest three layers for comparison experiments, taking the computational volume into account. Table 4 illustrates that the EAM is more efficient at deeper levels and larger image size. SA has the same level of inference speed as EAM for images of 512 size, but the drawback of quadratic computational complexity makes the inference speed much slower for images of size 6000, which is undesirable for large scale semantic segmentation applications. The stage-6 EAM is capable of boosting model performance by 1.12% mIoU, at the expense of only 6–7% of the inference speed. Comparatively, applying the EAM to stages 6/7 or 6/7/8 can significantly improve the model performance, but at the expense of a 20–60% reduction in inference speed, which is inefficient. Considering the increases in model performance and inference speed, the optimal placement of the EAM is in the network's deepest layer.

Table 3. Comparison Among Combinations of Different EAMs and Configurations on the Potsdam Dataset.

Method	Norm	E	mF1 (%)	mIoU (%)
DSANet64	-	-	86.90	77.05
+SA	Softmax	-	87.63	78.20
+EAM I	Softmax	64	85.98	75.54
+EAM II	Softmax	64	86.61	76.60
+EAM II	Softmax+Softmax	64	86.80	76.84
+EAM II	Softmax+L1	32	87.52	78.02
+EAM II	Softmax+L1	64	**87.61**	**78.17**

Table 4. Comparison of the EAM Results Obtained on the Potsdam Dataset in Different Stages.

Method	Stage	mF1 (%)	mIoU (%)	512 FPS	6000 FPS
DSANet64	-	86.90	77.05	503.77	5.83
+SA	6	87.63	77.20	478.57	4.35
+EAM	6	87.61	78.17	470.07	5.46
+EAM	6/7	87.75	78.40	405.00	4.15
+EAM	6/7/8	87.76	78.41	278.34	2.33

5.1.2. Effectiveness of the MSD loss and HSE loss

The selective kernel ratio is critical to the performance of the MSD loss. In Table 5 we can plainly see the model's performance at various ratios. Experiments conducted on DSANet32 and DSANet64 show that it is more effective to perform deep spatial supervision on parts of the feature maps, and the best ratio is 0.5; i.e., half of the feature maps need to be preserved for further learning of semantic information. This intuitive approach yields better results and reduces the required training time. Table 6 further explores the stages at which the insertion of the deep spatial supervision module is more effective in improving the model performance. The results are as expected: the deep spatial supervision module provides significant improvements for shallow-layer spatial representations but is not effective when applied to deep-layer semantic features. It is also found that imposing deep spatial supervision at each layer is not efficient enough. To select more effective MSD insertion locations and to be more intuitive, we finally choose to perform deep spatial supervision at stages 1–4 in the contracting path and at stage 9 in the expansion path. With multiscale spatial supervision and the spatial detail loss, the model is able to improve the mIoU by 0.91% on the Potsdam dataset without sacrificing inference speed. As shown in Table 7, the model performance improvement provided by the HSE loss is relatively small, and it can steadily improve the model performance by 0.28% mIoU. Using the HSE loss on the model with the EAM and MSD loss can still yield an mIoU improvement of 0.14%.

Table 5. Comparison of MSD Losses with Different Ratios on the Potsdam Dataset.

Method	MSD Ratio	mF1 (%)	mIoU (%)
DSANet32	0	86.09	75.80
	0.25	86.47	76.41
	0.5	**86.77**	**76.87**
	0.75	86.30	76.13
	1.0	86.51	76.46
DSANet64	0	86.89	77.05
	0.25	87.35	77.76
	0.5	**87.48**	**77.96**
	0.75	87.13	77.43
	1.0	86.98	77.17

Note: MSD Ratio 1.0 means the method in [38].

Table 6. Comparison of MSD Losses Achieved on the Potsdam Data Set in Different Stages.

Method	MSD Stage	mF1 (%)	mIoU (%)
DSANet64	no	86.89	77.05
	Stage 1	87.04	77.28
	Stage 2	87.21	77.53
	Stage 3	87.04	77.29
	Stage 4	87.31	77.70
	Stage 5	86.97	77.15
	Stage 6	87.07	77.32
	Stage 7	86.81	76.92
	Stage 8	86.93	77.12
	Stage 9	87.16	77.49
	Stages 1–9	87.32	77.73
	Stages 1–4 + Stage 9	87.48	77.96

Table 7. Results of the Ablation Study Conducted on the Potsdam Dataset.

Method	EAM	MSD	HSE	mIoU (%)	mF1 (%)
DSANet64				77.05	86.90
	✓			78.17	87.60
		✓		77.96	87.48
			✓	77.33	87.39
	✓	✓		79.06	88.17
	✓	✓	✓	79.20	88.25

5.1.3. Effectiveness of DSANet

All the ablation experiments conducted based on the Potsdam dataset are shown in Table 7. Introducing the EAM can produce 1.12% and 0.70% gains in the mIoU and mF1 scores of the model, respectively. The MSD loss effectively improves the mIoU and mF1 scores by 0.91% and 0.58%, respectively. Introducing the HSE loss in the decoder can modestly enhance the mIoU by 0.28%. The MSD loss brings a 0.90% mIoU increase and a 0.57% mF1 increase. The EAM is the most effective module, as it is accompanied by model mIoU and mF1 growths of 1.12% and 0.71%, respectively. By comparing the backbone of DSANet64 with DSANet64 for 80,000 iterations and 320,000 iterations (see details in Figure 6), we find that the improvement yielded by DSANet is significant; even if the number of training iterations for the backbone network is increased to 4 times the original amount, it is still difficult to obtain better model performance, while DSANet64 is able to continue achieving improved model performance up to an mIoU of 80% with the increase in the number of training iterations.

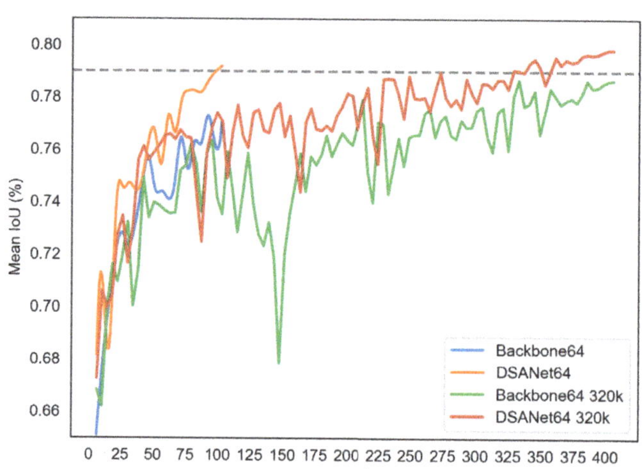

Figure 6. Comparisons between the mean IoU results of the DSANet64 backbone and DSANet64 with 80,000 iterations and 320,000 iterations on the Potsdam dataset.

5.2. Qualitative Analysis of Features

In order to examine the impact of various modules on the segmentation performance of the model, we visualize the obtained results in Figure 7. The visualization includes the original IRRG image, the labels, and the segmentation results of the backbone acquired after adding the feature enhancement modules separately and after adding all modules. Figure 7a–e are buildings, low-vegetation areas, trees, unmarked features, and buildings with complex boundaries, respectively. We observe that the results of segmentation based on the backbone frequently contain erroneous segmentation borders and even patch holes. Adding the EAM can effectively resolve the semantic discrimination issues, for example, by restoring the recognition results for the missing trees in Figure 7c. Adding MSD loss can help the segmentation process maintain better boundaries, but it cannot compensate for segmentation mistakes caused by semantics, such as identifying the connected buildings in Figure 7a while preserving the gaps in the buildings in Figure 7b,d. Adding the HSE loss can enhance the model's capacity for semantic perception, preventing the occurrence of the problem of missing semantics. DSANet64 with EAM, MSD loss, and HSE loss can combine the capabilities of each module and complement their benefits, and its segmentation results are more accurate than those of the backbone network.

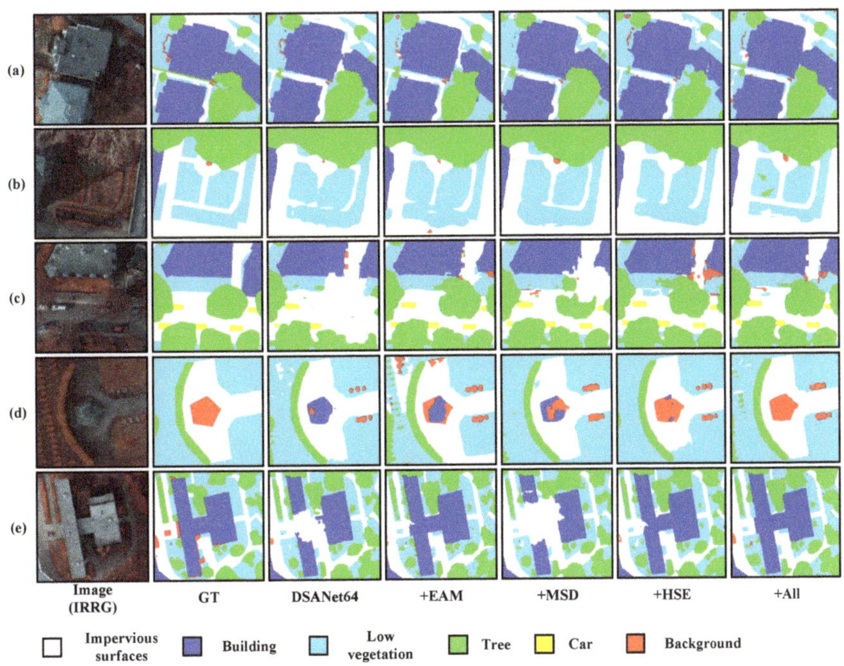

Figure 7. Comparison between the segmentation results of the DSANet64 backbone and the EAM, MSD, and HSE modules. (**a**–**e**) are derived from the Potsdam dataset. GT denotes the ground truth.

5.3. Quantitative Comparison with State-of-the-Art Methods

To measure the performance of our model, we compare DSANet with popular lightweight and efficient semantic segmentation networks whose numbers of parameters vary from 0.1 M to 21 M. We assess the performance of the models in terms of both accuracy and inference speed on both the Potsdam and Vaihingen datasets. To objectively evaluate the model performance, we fixed the cutoff threshold for the number of model parameters to 1.5 M. Table 8 reports the accuracy and inference speed results obtained on the Potsdam dataset.

Table 8. Comparison of DSANet with the State-of-the-Art Models on the Potsdam Dataset.

Method	Per-Class mIoU (%)					mIoU (%)	mF1 (%)	Params (M)
	Imperious Surface	Building	Low Vegetation	Tree	Car			
FPENet [40]	76.55	86.30	65.56	66.48	67.16	72.41	83.64	**0.11**
FSSNet [37]	79.90	86.83	68.69	69.40	75.20	76.00	86.20	0.17
CGNet [67]	78.08	84.88	66.86	68.32	72.17	74.06	84.93	0.48
EDANet [35]	79.83	87.50	69.24	70.73	72.16	75.89	86.13	0.67
ContextNet [43]	79.37	86.86	68.70	69.38	71.96	75.25	85.71	0.86
LEDNet [41]	**82.45**	**89.12**	**71.17**	**72.51**	74.28	**77.91**	**87.42**	0.89
Fast-SCNN [37]	78.15	83.29	68.76	69.74	70.89	74.17	85.05	1.45
DSANet32	82.04	88.79	70.70	72.09	**75.58**	77.84	87.38	1.28
ESNet [68]	82.31	88.16	**71.94**	73.37	78.09	78.77	88.00	**1.66**
DABNet [34]	81.30	88.23	70.95	73.24	73.20	77.38	87.10	1.96
ERFNet [36]	80.38	88.18	70.81	72.30	74.89	77.31	87.06	2.08
DDRNet23-slim [48]	81.27	89.09	69.91	72.37	72.99	77.13	86.91	5.81
STDCNet [38]	82.07	89.41	71.45	73.49	76.78	78.64	87.90	8.57
LinkNet [39]	80.71	88.08	70.75	72.13	76.11	77.56	87.22	11.54
BiSeNetV1 [44]	81.91	88.95	71.83	73.21	**80.18**	**79.22**	**88.27**	13.42
BiSeNetV2 [45]	81.23	89.21	71.03	72.6	73.29	77.47	87.14	14.77
SFNet [47]	80.52	84.97	71.37	72.92	79.94	77.94	87.51	13.31
DDRNet23 [48]	82.58	**90.07**	71.56	73.55	75.44	78.64	87.89	20.59
DSANet64	**83.02**	89.50	71.86	**74.26**	77.34	79.20	88.25	4.65

5.3.1. Segmentation Performances Achieved on the Potsdam Dataset

The comparison between the results produced by DSANet and the other state-of-the-art models on the Potsdam dataset are shown in Table 8. Among the models with fewer than 1.5 M model parameters, DSANet32 obtains the best mIoU result of 75.58% on the car segmentation task and achieves suboptimal performance. In terms of the accuracy-speed tradeoff, DSANet32 achieves a balance between accuracy and inference speed. DSANet32 is over 2.2 times more accurate than LEDNet, the most accuracy network, and is 2.59% more accurate than ContextNet, the fastest network. Among the models with more than 1.5 M model parameters, DSANet64 works best to segment impervious surfaces and trees and achieves comparable results to those of BiSeNet V1, yielding 79.20 % mIoU and 88.25 % mF1 scores with 35 % of the number of parameters in BiSeNet V1. Figure 8 provides a more intuitive comparison of the segmentation results obtained by DSANet and the other models on the Potsdam dataset under the small size settings. Figure 9 shows the whole-image segmentation results of DSANet and the other models.

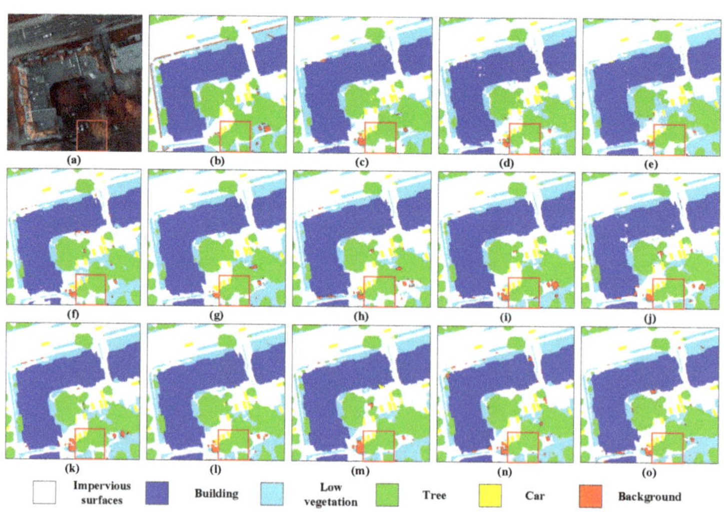

Figure 8. Examples of segmentation results derived from the Potsdam dataset under the small size setting. (**a**) IRRG image. (**b**) Ground truth. (**c**) FPENet. (**d**) FSSNet. (**e**) CGNet. (**f**) ContextNet. (**g**) Fast-SCNN. (**h**) ERFNet. (**i**) STDC1. (**j**) LinkNet. (**k**) ICNet34. (**l**) BiSeNet V1. (**m**) SFNet. (**n**) DDRNet23. (**o**) DSANet64.

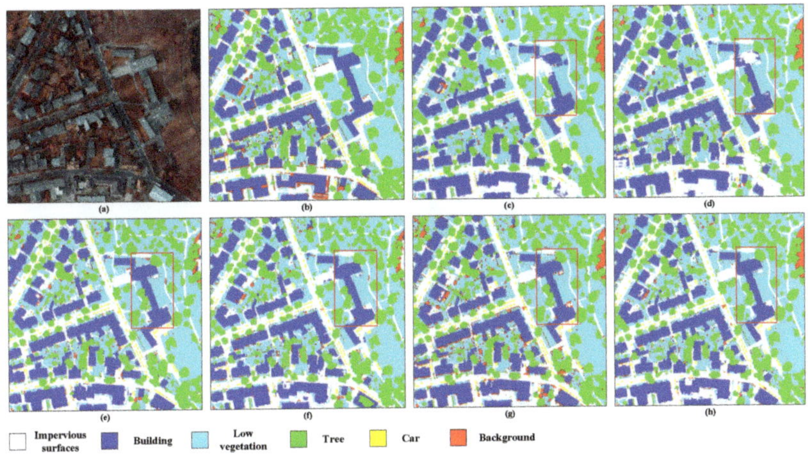

Figure 9. Examples of segmentation results derived from the Potsdam dataset with the whole image size setting. (**a**) IRRG image. (**b**) Ground truth. (**c**) FPENet. (**d**) ERFNet. (**f**) STDC1. (**g**) BiSeNet V1. (**h**) BiSeNet V2. (**i**) DSANet64.

5.3.2. Segmentation Performances Achieved on the Vaihingen Dataset

The comparison between the results produced by DSANet and the other state-of-the-art models on the Vaihingen dataset are shown in Table 9. Among the models with less than 1.5 M parameters, DSANet32 achieves the best results, with 85.30% and 53.74% mIoUs on the building and car segmentation tasks, respectively, and its overall 71.31% mIoU and 82.74% mF1 scores are impressive. In comparison with these other models, DSANet32 still obtains a better inference speed, although it has a disadvantage in terms of the number of required parameters. DSANet achieves the best car segmentation result, with an absolute 2.67% mIoU lead over the second-place method. DSANet64 achieves a 72.26% mIoU and a 83.49% mF1 on the Vaihingen dataset, which are also the best results. Figure 10 provides an intuitive comparison between the segmentation results obtained by DSANet and the other models on the Vaihingen dataset under the small size setting. Figure 11 shows the whole-image segmentation results of DSANet and the other models.

Table 9. Comparison of DSANet with the State-of-the-Art Models on the Vaihingen Dataset.

Method	Per-Class mIoU (%)					mIoU (%)	mF1 (%)
	Imperious Surface	Building	Low Vegetation	Tree	Car		
FPENet [40]	78.37	84.24	63.44	73.79	44.39	68.85	80.67
FSSNet [37]	76.88	83.75	62.96	73.03	45.74	68.47	80.51
CGNet [67]	77.86	84.63	64.88	**74.90**	47.80	70.01	81.61
EDANet [35]	78.76	84.56	64.51	74.32	51.65	70.76	82.36
ContextNet [43]	77.77	83.65	61.99	73.15	50.32	69.38	81.31
LEDNet [41]	**79.25**	85.00	**65.67**	74.72	50.73	71.07	82.48
Fast-SCNN [37]	76.21	82.08	61.06	71.47	44.45	67.05	79.48
DSANet32	79.17	**85.30**	64.30	74.05	**53.74**	**71.31**	**82.74**
ESNet [68]	79.74	**86.24**	64.35	74.47	53.77	71.71	82.99
DABNet [34]	78.48	84.42	63.92	73.90	54.16	70.98	82.55
ERFNet [36]	79.34	85.68	64.07	**74.51**	54.01	71.52	82.88
DDRNet23-slim [48]	78.81	84.53	64.55	73.96	52.92	70.95	82.49
STDC1 [38]	79.03	85.76	64.27	73.69	48.71	70.29	81.84
LinkNet [39]	**79.94**	85.94	**64.60**	74.29	54.32	71.82	83.09
BiSeNetV1 [44]	78.84	85.55	64.23	74.15	50.50	70.65	82.17
BiSeNetV2 [45]	79.14	84.91	64.26	74.09	55.59	71.60	83.00
DSANet64	79.50	85.98	63.86	73.60	**58.35**	**72.26**	**83.49**

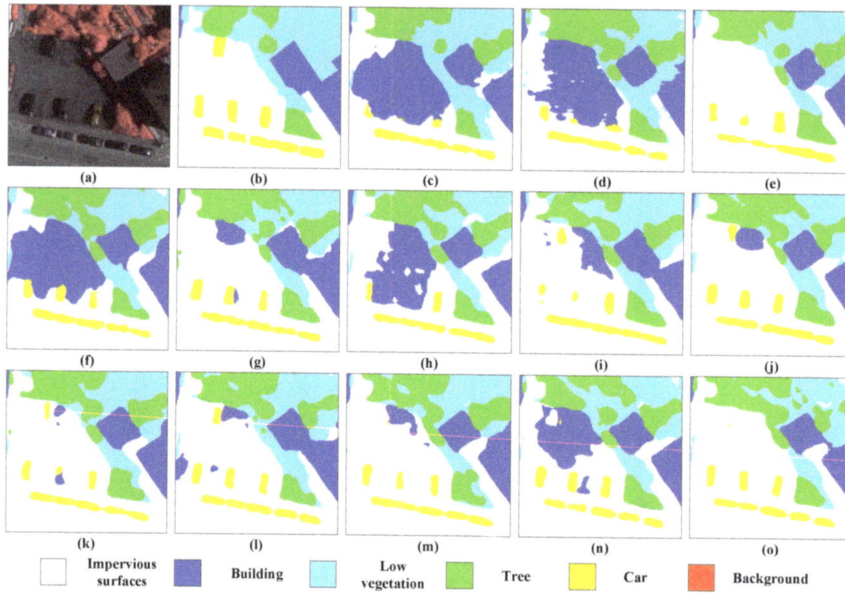

Figure 10. Examples of segmentation results derived from the Vaihingen dataset under the small size setting. (**a**) IRRG image. (**b**) Ground truth. (**c**) FPENet. (**d**) FSSNet. (**e**) CGNet. (**f**) ContextNet. (**g**) Fast-SCNN. (**h**) ESNet. (**i**) ERFNet. (**j**) DDRNet23-slim. (**k**) STDC1. (**l**) LinkNet. (**m**) BiSeNet V1. (**n**) BiSeNet V2. (**o**) DSANet64.

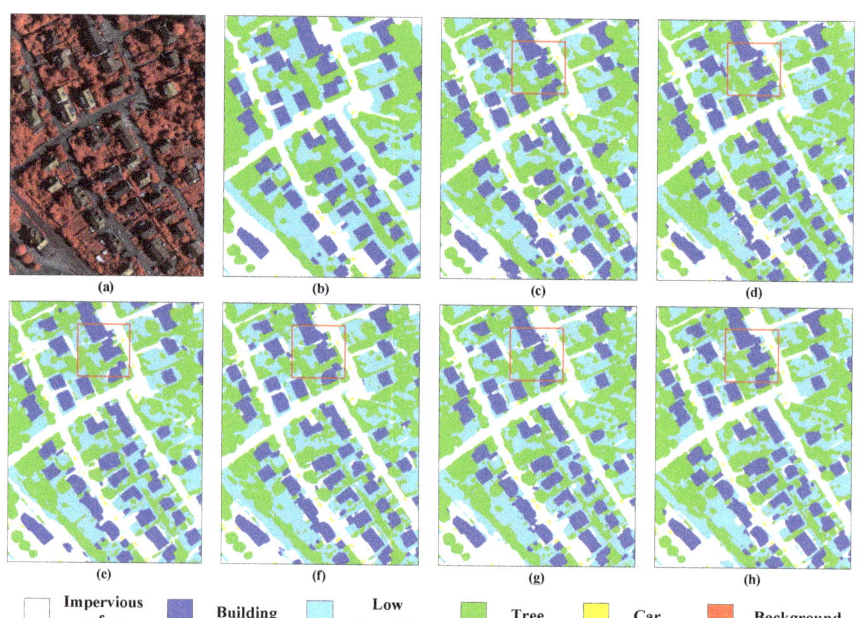

Figure 11. Examples of segmentation results derived from the Vaihingen dataset with the whole image size setting. (**a**) IRRG image. (**b**) Ground truth. (**c**) FPENet. (**d**) ERFNet. (**e**) DDRNet23-slim. (**f**) STDC1. (**g**) BiSeNet V1. (**h**) BiSeNet V2. (**i**) DSANet64.

5.3.3. Inference Speeds

The comparison between the inference speed results produced by DSANet and the other state-of-the-art models under different image sizes are shown in Table 10. Our proposed DSANet32 reaches an inference speed of 8.78 on the 6000 × 6000 images, which are derived from the Potsdam dataset. In comparison with the fastest inference model at sizes of 512 and 1024, DSANet32 is only 6-7% behind ContextNet, whose segmentation performance is far behind that of DSANet32. In a comparison with the corresponding models, DSANet64 achieves the best inference speed at a size of 512 with 470.07 FPS. At the 1024 and 6000 sizes, DSANet64 still achieves comparable results. Figure 1 gives a visualization of the segmentation speed-accuracy tradeoffs provided by all models. The closer the model's points are to the upper-right corner, the better that model performs in terms of the speed-accuracy tradeoff.

Table 10. FPS Results Obtained on the Potsdam Dataset Under Different Size Settings.

Method	mIoU (%)	FPS		
		512	1024	6000
FPENet [40]	72.41	173.47	73.13	2.44
FSSNet [37]	76.00	527.26	183.30	6.27
CGNet [67]	74.06	127.51	66.78	0.58
EDANet [35]	75.89	390.17	135.50	4.37
ContextNet [43]	75.25	**688.70**	**257.25**	8.59
LEDNet [41]	**77.91**	293.48	104.92	3.74
Fast-SCNN [37]	74.17	670.82	261.43	8.60
DSANet32	77.84	648.49	245.66	**8.78**
ESNet [68]	78.77	295.33	100.27	2.77
DABNet [34]	77.38	173.47	73.13	2.44
ERFNet [36]	77.31	282.66	96.00	2.65
DDRNet23-slim [48]	77.13	429.09	**208.38**	**6.98**
STDC1 [38]	78.64	437.41	147.07	5.00
BiSeNetV1 [44]	**79.22**	351.89	128.64	3.92
BiSeNetV2 [45]	77.47	242.27	114.15	3.87
DDRNet23 [48]	78.64	256.65	99.58	3.46
DSANet64	79.20	**470.07**	172.16	5.46

6. Conclusions

In this paper, we propose DSANet a deep supervision-based simple attention network, for large-scale RSI semantic segmentation; our network achieves an excellent balance between accuracy and inference speed. The main contributions of DSANet lie in three aspects: a simple attention module with linear complexity called the EAM, which is employed in the deepest network layer for long-range semantic information modeling; a improved deep supervision-based MSD loss for supervising portions of the feature map to directly learn the detailed spatial pyramid features; and a deep supervision-based HSE loss for supervising the network so that it learns the category frequency distribution of the training data.

Our DSANet provides consistently outstanding achievement on two benchmark datasets (i.e., the ISPRS Potsdam and Vaihingen datasets). On the ISPRS Potsdam test dataset, DSANet64 obtains a mean IoU of 79.20% at 5.46 FPS on 6000 × 6000 images and at 470.07 FPS on 512 × 512 images.

Author Contributions: Conceptualization, W.S. and Q.M.; methodology, W.S.; software, W.S.; validation, W.S., M.Z. and C.S.; formal analysis, W.S.; investigation, W.S., T.J. and Q.M.; resources, Q.M.; data curation, W.S.; writing—original draft preparation, W.S.; writing—review and editing, M.Z., C.S. and Q.M.; visualization, W.S.; supervision, L.Z. and Q.M.; project administration, Q.M.; funding acquisition, Q.M. All authors have read and agreed to the published version of the manuscript.

Funding: This research received no external funding.

Data Availability Statement: The ISPRS dataset can be found through International Society for Photogrammetry and Remote Sensing (https://www.isprs.org/education/benchmarks/UrbanSemLab/semantic-labeling.aspx (accessed on 2 September 2021)).

Conflicts of Interest: The authors declare no conflict of interest.

Abbreviations

The following abbreviations are used in this manuscript:

CNN	Convolutional Neural Network
RSIs	Remote Sensing Images
VHR	Very High-resolution
EAM	Embedding Attention Module
MSD	Multiscale Spatial Detail
HSE	Hierarchical Semantic Enhancement
mIoU	Mean Intersection over Union
FPS	Frames Per Second
FLOPs	Floating Point Operations

References

1. Liu, P. A survey of remote-sensing big data. *Front. Environ. Sci.* **2015**, *3*, 5. [CrossRef]
2. Laney, D. 3D data management: Controlling data volume, velocity and variety. *META Group Res. Note* **2001**, *6*, 1.
3. der Sande, C.V.; Jong, S.D.; Roo, A.D. A segmentation and classification approach of IKONOS-2 imagery for land cover mapping to assist flood risk and flood damage assessment. *Int. J. Appl. Earth Obs. Geoinf.* **2003**, *4*, 217–229. [CrossRef]
4. Costa, H.; Foody, G.M.; Boyd, D.S. Supervised methods of image segmentation accuracy assessment in land cover mapping. *Remote Sens. Environ.* **2018**, *205*, 338–351. [CrossRef]
5. Im, J.; Jensen, J.; Tullis, J. Object-based change detection using correlation image analysis and image segmentation. *Int. J. Remote Sens.* **2008**, *29*, 399–423. [CrossRef]
6. Chen, G.; Hay, G.J.; Carvalho, L.M.; Wulder, M.A. Object-based change detection. *Int. J. Remote Sens.* **2012**, *33*, 4434–4457. [CrossRef]
7. Du, S.; Du, S.; Liu, B.; Zhang, X. Mapping large-scale and fine-grained urban functional zones from VHR images using a multi-scale semantic segmentation network and object based approach. *Remote Sens. Environ.* **2021**, *261*, 112480. [CrossRef]
8. Wang, J.; Hu, X.; Meng, Q.; Zhang, L.; Wang, C.; Liu, X.; Zhao, M. Developing a method to extract building 3d information from GF-7 data. *Remote Sens.* **2021**, *13*, 4532. [CrossRef]
9. Li, P.; Guo, J.; Song, B.; Xiao, X. A multilevel hierarchical image segmentation method for urban impervious surface mapping using very high resolution imagery. *IEEE J. Sel. Top. Appl. Earth Obs. Remote Sens.* **2010**, *4*, 103–116. [CrossRef]
10. Miao, Z.; Fu, K.; Sun, H.; Sun, X.; Yan, M. Automatic water-body segmentation from high-resolution satellite images via deep networks. *IEEE Geosci. Remote Sens. Lett.* **2018**, *15*, 602–606. [CrossRef]
11. Long, J.; Shelhamer, E.; Darrell, T. Fully convolutional networks for semantic segmentation. In Proceedings of the IEEE Conference on Computer Vision and Pattern Recognition, Boston, MA, USA, 7–12 June 2015; pp. 3431–3440.
12. Ronneberger, O.; Fischer, P.; Brox, T. U-net: Convolutional networks for biomedical image segmentation. In Proceedings of the International Conference on Medical Image Computing and Computer-Assisted Intervention, Munich, Germany, 5–9 October 2015; pp. 234–241.
13. Zhang, Z.; Liu, Q.; Wang, Y. Road extraction by deep residual u-net. *IEEE Geosci. Remote Sens. Lett.* **2018**, *15*, 749–753. [CrossRef]
14. Heidler, K.; Mou, L.; Baumhoer, C.; Dietz, A.; Zhu, X.X. Hed-unet: Combined segmentation and edge detection for monitoring the antarctic coastline. *IEEE Trans. Geosci. Remote Sens.* **2021**, *60*, 1–14. [CrossRef]
15. Chen, L.-C.; Papandreou, G.; Schroff, F.; Adam, H. Rethinking atrous convolution for semantic image segmentation. *arXiv* **2017**, arXiv:1706.05587.
16. Chen, L.-C.; Zhu, Y.; Papandreou, G.; Schroff, F.; Adam, H. Encoder-decoder with atrous separable convolution for semantic image segmentation. In Proceedings of the European Conference on Computer Vision (ECCV), Munich, Germany, 8–14 September 2018; pp. 801–818.
17. Yang, J.; Guo, J.; Yue, H.; Liu, Z.; Hu, H.; Li, K. CDnet: CNN-based cloud detection for remote sensing imagery. *IEEE Trans. Geosci. Remote Sens.* **2019**, *57*, 6195–6211. [CrossRef]
18. Liu, S.; Cheng, J.; Liang, L.; Bai, H.; Dang, W. Light-weight semantic segmentation network for UAV remote sensing images. *IEEE J. Sel. Top. Appl. Earth Obs. Remote Sens.* **2021**, *14*, 8287–8296. [CrossRef]
19. Lee, C.-Y.; Xie, S.; Gallagher, P.; Zhang, Z.; Tu, Z. Deeply-Supervised Nets. In *Machine Learning Research, Proceedings of the Eighteenth International Conference on Artificial Intelligence and Statistics*; Lebanon, G., Vishwanathan, S.V.N., Eds.; PMLR: San Diego, CA, USA, 2015; Volume 38, pp. 562–570.
20. Deng, C.; Liang, L.; Su, Y.; He, C.; Cheng, J. Semantic segmentation for high-resolution remote sensing images by light-weight network. In Proceedings of the 2021 IEEE International Geoscience and Remote Sensing Symposium IGARSS, Brussels, Belgium, 11–16 July 2021; pp. 3456–3459.
21. Liu, S.; Ding, W.; Liu, C.; Liu, Y.; Wang, Y.; Li, H. ERN: Edge loss reinforced semantic segmentation network for remote sensing images. *Remote Sens.* **2018**, *10*, 1339. [CrossRef]
22. Yuan, W.; Xu, W. Neighborloss: A loss function considering spatial correlation for semantic segmentation of remote sensing image. *IEEE Access* **2021**, *9*, 641–675. [CrossRef]

23. Simonyan, K.; Zisserman, A. Very deep convolutional networks for large-scale image recognition. *arXiv* **2014**, arXiv:1409.1556.
24. He, K.; Zhang, X.; Ren, S.; Sun, J. Deep residual learning for image recognition. In Proceedings of the IEEE Conference on Computer Vision and Pattern Recognition, Las Vegas, NV, USA, 27–30 June 2016; pp. 770–778.
25. Huang, G.; Liu, Z.; Maaten, L.V.D.; Weinberger, K.Q. Densely connected convolutional networks. In Proceedings of the IEEE Conference on Computer Vision and Pattern Recognition, Honolulu, HI, USA, 21–26 July 2017; pp. 4700–4708.
26. Iandola, F.N.; Han, S.; Moskewicz, M.W.; Ashraf, K.; Dally, W.J.; Keutzer, K. SqueezeNet: AlexNet-level accuracy with 50x fewer parameters and <0.5 MB model size. *arXiv* **2016**, arXiv:1602.07360.
27. Krizhevsky, A.; Sutskever, I.; Hinton, G.E. Imagenet classification with deep convolutional neural networks. In Proceedings of the Advances in Neural Information Processing Systems, Lake Tahoe, NV, USA, 3–8 December 2012; pp. 1097–1105.
28. Howard, A.G.; Zhu, M.; Chen, B.; Kalenichenko, D.; Wang, W.; Weyand, T.; Andreetto, M.; Adam, H. Mobilenets: Efficient convolutional neural networks for mobile vision applications. *arXiv* **2017**, arXiv:1704.04861.
29. Sandler, M.; Howard, A.; Zhu, M.; Zhmoginov, A.; Chen, L.-C. Mobilenetv2: Inverted residuals and linear bottlenecks. In Proceedings of the IEEE Conference on Computer Vision and Pattern Recognition, Salt Lake City, UT, USA, 18–22 June 2018; pp. 4510–4520.
30. Howard, A.; Sandler, M.; Chu, G.; Chen, L.-C.; Chen, B.; Tan, M.; Wang, W.; Zhu, Y.; Pang, R.; Vasudevan, V.; et al. Searching for mobilenetv3. In Proceedings of the IEEE/CVF International Conference on Computer Vision, Long Beach, CA, USA, 16–20 June 2019; pp. 1314–1324.
31. Zhang, X.; Zhou, X.; Lin, M.; Sun, J. Shufflenet: An extremely efficient convolutional neural network for mobile devices. In Proceedings of the IEEE Conference on Computer Vision and Pattern Recognition, Salt Lake City, UT, USA, 18–22 June 2018; pp. 6848–6856.
32. Ma, N.; Zhang, X.; Zheng, H.-T.; Sun, J. Shufflenet v2: Practical guidelines for efficient CNN architecture design. In Proceedings of the European Conference on Computer Vision (ECCV), Munich, Germany, 14–18 September 2018; pp. 116–131.
33. Paszke, A.; Chaurasia, A.; Kim, S.; Culurciello, E. Enet: A deep neural network architecture for real-time semantic segmentation. *arXiv* **2016**, arXiv:1606.02147.
34. Li, G.; Yun, I.; Kim, J.; Kim, J. Dabnet: Depth-wise asymmetric bottleneck for real-time semantic segmentation. In Proceedings of the British Machine Vision Conference (BMVC), Cardiff, UK, 9–12 September 2019; pp. 186.1–186.12.
35. Lo, S.-Y.; Hang, H.-M.; Chan, S.-W.; Lin, J.-J. Efficient dense modules of asymmetric convolution for real-time semantic segmentation. In Proceedings of the ACM Multimedia Asia, Nice, France, 21–25 October 2019; pp. 1–6.
36. Romera, E.; Alvarez, J.M.; Bergasa, L.M.; Arroyo, R. Erfnet: Efficient residual factorized convnet for real-time semantic segmentation. *IEEE Trans. Intell. Transp. Syst.* **2017**, *19*, 263–272. [CrossRef]
37. Zhang, X.; Chen, Z.; Wu, Q.J.; Cai, L.; Lu, D.; Li, X. Fast semantic segmentation for scene perception. *IEEE Trans. Industr. Inform.* **2018**, *15*, 1183–1192. [CrossRef]
38. Fan, M.; Lai, S.; Huang, J.; Wei, X.; Chai, Z.; Luo, J.; Wei, X. Rethinking bisenet for real-time semantic segmentation. In Proceedings of the IEEE/CVF Conference on Computer Vision and Pattern Recognition, Montreal, QC, Canada, 11–17 October 2021; pp. 9716–9725.
39. Chaurasia, A.; Culurciello, E. Linknet: Exploiting encoder representations for efficient semantic segmentation. In Proceedings of the 2017 IEEE Visual Communications and Image Processing (VCIP), Venice, Italy, 22–29 October 2017; pp. 1–4.
40. Liu, M.; Yin, H. Feature pyramid encoding network for real-time semantic segmentation. *arXiv* **2019**, arXiv:1909.08599.
41. Wang, Y.; Zhou, Q.; Liu, J.; Xiong, J.; Gao, G.; Wu, X.; Latecki, L.J. Lednet: A lightweight encoder-decoder network for real-time semantic segmentation. In Proceedings of the 2019 IEEE International Conference on Image Processing (ICIP), Taipei, Taiwan, 22—25 September 2019; pp. 1860–1864.
42. Poudel, R.P.; Liwicki, S.; Cipolla, R. Fast-scnn: Fast semantic segmentation network. *arXiv* **2019**, arXiv:1902.04502.
43. Poudel, R.P.; Bonde, U.; Liwicki, S.; Zach, C. Contextnet: Exploring context and detail for semantic segmentation in real-time. *arXiv* **2018**, arXiv:1805.04554.
44. Yu, C.; Wang, J.; Peng, C.; Gao, C.; Yu, G.; Sang, N. Bisenet: Bilateral segmentation network for real-time semantic segmentation. In Proceedings of the European Conference on Computer Vision (ECCV), Munich, Germany, 8–14 September 2018; pp. 325–341.
45. Yu, C.; Gao, C.; Wang, J.; Yu, G.; Shen, C.; Sang, N. Bisenet v2: Bilateral network with guided aggregation for real-time semantic segmentation. *Int. J. Comput. Vis.* **2021**, *129*, 3051–3068. [CrossRef]
46. Li, H.; Xiong, P.; Fan, H.; Sun, J. Dfanet: Deep feature aggregation for real-time semantic segmentation. In Proceedings of the IEEE/CVF Conference on Computer Vision and Pattern Recognition, Long Beach, CA, USA, 15–20 June 2019; pp. 9522–9531.
47. Li, X.; You, A.; Zhu, Z.; Zhao, H.; Yang, M.; Yang, K.; Tan, S.; Tong, Y. Semantic flow for fast and accurate scene parsing. In *European Conference on Computer Vision*; Springer: Berlin/Heidelberg, Germany, 2020; pp. 775–793.
48. Hong, Y.; Pan, H.; Sun, W.; Jia, Y. Deep dual-resolution networks for real-time and accurate semantic segmentation of road scenes. *arXiv* **2021**, arXiv:2101.06085.
49. Zhao, H.; Shi, J.; Qi, X.; Wang, X.; Jia, J. Pyramid scene parsing network. In Proceedings of the IEEE Conference on Computer Vision and Pattern Recognition, Honolulu, HI, USA, 21–26 July 2017; pp. 2881–2890.
50. Chen, L.-C.; Papandreou, G.; Kokkinos, I.; Murphy, K.; Yuille, A.L. Semantic image segmentation with deep convolutional nets and fully connected crfs. *arXiv* **2014**, arXiv:1412.7062.

51. Chen, L.C.; Papreou, G.; Kokkinos, I.; Murphy, K.; Yuille, A.L. Deeplab: Semantic image segmentation with deep convolutional nets, atrous convolution, and fully connected crfs. *IEEE Trans. Pattern Anal. Mach. Intell.* **2017**, *40*, 834–848. [CrossRef]
52. Fu, J.; Liu, J.; Tian, H.; Li, Y.; Bao, Y.; Fang, Z.; Lu, H. Dual attention network for scene segmentation. In Proceedings of the IEEE/CVF Conference on Computer Vision and Pattern Recognition, Long Beach, CA, USA, 15–20 June 2019; pp. 3146–3154.
53. Yuan, Y.; Chen, X.; Wang, J. Object-contextual representations for semantic segmentation. In *European Conference on Computer Vision*; Springer: Berlin/Heidelberg, Germany, 2020; pp. 173–190.
54. Hu, J.; Shen, L.; Sun, G. Squeeze-and-excitation networks. In Proceedings of the IEEE Conference on Computer Vision and Pattern Recognition, Salt Lake City, UT, USA, 18–22 June 2018; pp. 7132–7141.
55. Park, J.; Woo, S.; Lee, J.-Y.; Kweon, I.S. Bam: Bottleneck attention module. *arXiv* **2018**, arXiv:1807.06514.
56. Woo, S.; Park, J.; Lee, J.-Y.; Kweon, I.S. Cbam: Convolutional block attention module. In Proceedings of the European Conference on Computer Vision (ECCV), Munich, Germany, 8–14 September 2018; pp. 3–19.
57. Meng, Q.; Zhao, M.; Zhang, L.; Shi, W.; Su, C.; Bruzzone, L. Multilayer feature fusion network with spatial attention and gated mechanism for remote sensing scene classification. *IEEE Geosci. Remote Sens. Lett.* **2022**, *19*, 1–5. [CrossRef]
58. Ambartsoumian, A.; Popowich, F. Self-attention: A better building block for sentiment analysis neural network classifiers. *arXiv* **2018**, arXiv:1812.07860.
59. Vaswani, A.; Shazeer, N.; Parmar, N.; Uszkoreit, J.; Jones, L.; Gomez, A.N.; Kaiser, L.; Polosukhin, I. Attention is all you need. In Proceedings of the Advances in Neural Information Processing Systems, Long Bench, CA, USA, 4–9 December 2017; pp. 5998–6008.
60. Dosovitskiy, A.; Beyer, L.; Kolesnikov, A.; Weissenborn, D.; Zhai, X.; Unterthiner, T.; Dehghani, M.; Minderer, M.; Heigold, G.; Gelly, S.; et al. An image is worth 16 × 16 words: Transformers for image recognition at scale. *arXiv* **2020**, arXiv:2010.11929.
61. Liu, Z.; Lin, Y.; Cao, Y.; Hu, H.; Wei, Y.; Zhang, Z.; Lin, S.; Guo, B. Swin transformer: Hierarchical vision transformer using shifted windows. In Proceedings of the IEEE/CVF International Conference on Computer Vision, Montreal, QC, Canada, 10–17 October 2021; pp. 10012–10022.
62. Wang, W.; Xie, E.; Li, X.; Fan, D.-P.; Song, K.; Liang, D.; Lu, T.; Luo, P.; Shao, L. Pyramid vision transformer: A versatile backbone for dense prediction without convolutions. In Proceedings of the IEEE/CVF International Conference on Computer Vision, Montreal, QC, Canada, 11–17 October 2021; pp. 568–578.
63. Li, R.; Zheng, S.; Zhang, C.; Duan, C.; Wang, L.; Atkinson, P.M. Abcnet: Attentive bilateral contextual network for efficient semantic segmentation of fine-resolution remotely sensed imagery. *ISPRS J. Photogramm. Remote Sens.* **2021**, *181*, 84–98.. [CrossRef]
64. Wang, S.; Li, B.Z.; Khabsa, M.; Fang, H.; Ma, H. Linformer: Self-attention with linear complexity. *arXiv* **2020**, arXiv:2006.04768.
65. Guo, M.-H.; Liu, Z.-N.; Mu, T.-J.; Hu, S.-M. Beyond self-attention: External attention using two linear layers for visual tasks. *arXiv* **2021**, arXiv:2105.02358.
66. Maggiori, E.; Tarabalka, Y.; Charpiat, G.; Alliez, P. Convolutional Neural Networks for Large-Scale Remote-Sensing Image Classification. *IEEE Trans. Geosci. Remote Sens.* **2017**, *55*, 645–657. [CrossRef]
67. Wu, T.; Tang, S.; Zhang, R.; Cao, J.; Zhang, Y. Cgnet: A light-weight context guided network for semantic segmentation. *IEEE Trans. Image Process.* **2020**, *30*, 1169–1179. [CrossRef]
68. Wang, Y.; Zhou, Q.; Xiong, J.; Wu, X.; Jin, X. Esnet: An efficient symmetric network for real-time semantic segmentation. In *Chinese Conference on Pattern Recognition and Computer Vision (PRCV)*; Springer: Berlin/Heidelberg, Germany, 2019; pp. 41–52.

remote sensing

MDPI

Article

Deep 1D Landmark Representation Learning for Space Target Pose Estimation

Shengli Liu, Xiaowen Zhu, Zewei Cao and Gang Wang *

College of Air and Missile Defense, Air Force Engineering University, Xi'an 710051, China
* Correspondence: sharesunny123@163.com

Abstract: Monocular vision-based pose estimation for known uncooperative space targets plays an increasingly important role in on-orbit operations. The existing state-of-the-art methods of space target pose estimation build the 2D-3D correspondences to recover the space target pose, where space target landmark regression is a key component of the methods. The 2D heatmap representation is the dominant descriptor in landmark regression. However, its quantization error grows dramatically under low-resolution input conditions, and extra post-processing is usually needed to compute the accurate 2D pixel coordinates of landmarks from heatmaps. To overcome the aforementioned problems, we propose a novel 1D landmark representation that encodes the horizontal and vertical pixel coordinates of a landmark as two independent 1D vectors. Furthermore, we also propose a space target landmark regression network to regress the locations of landmarks in the image using 1D landmark representations. Comprehensive experiments conducted on the SPEED dataset show that the proposed 1D landmark representation helps the proposed space target landmark regression network outperform existing state-of-the-art methods at various input resolutions, especially at low resolutions. Based on the 2D landmarks predicted by the proposed space target landmark regression network, the error of space target pose estimation is also smaller than existing state-of-the-art methods under all input resolution conditions.

Keywords: pose estimation; landmark regression; space target; 1D landmark representation; deep learning

Citation: Liu, S.; Zhu, X.; Cao, Z.; Wang, G. Deep 1D Landmark Representation Learning for Space Target Pose Estimation. *Remote Sens.* **2022**, *14*, 4035. https://doi.org/10.3390/rs14164035

Academic Editors: Jukka Heikkonen, Fahimeh Farahnakian and Pouya Jafarzadeh

Received: 13 July 2022
Accepted: 16 August 2022
Published: 18 August 2022

1. Introduction

Estimating the relative position and attitude (referred to as pose estimation) of a known non-cooperative space target is an essential capability for automated in-orbit space operations such as autonomous proximity, in-orbit maintenance, and debris removal. It is an attractive option that uses visual sensors such as camera for space target pose estimation because of small mass and low power consumption compared to active sensors such as Light Detection and Ranging (LIDAR) or Range Detection and Ranging (RADAR). Moreover, monocular cameras are more suitable for space missions than stereo vision systems due to spacecraft size, payload, and energy limitations. Therefore, it is necessary to design a space target pose estimation method, which can precisely estimate the pose of a known non-cooperative space target from a monocular image.

Deep Convolutional Neural Networks (DCNNs) have achieved state-of-the-art results in many computer vision tasks, such as image classification [1], object detection [2], and semantic segmentation [3]. In order to take advantage of DCNNs in computer vision tasks, Sharma et al. [4] pioneered the introduction of DCNNs into monocular vision-based pose estimation for a single known uncooperative space target. They treat the single space target pose estimation as a classification problem and use DCNNs to classify the target pose directly. Since then, several methods [5,6] have been put forward to use DCNNS to estimate pose-related parameters of the space target directly. Even though these methods have achieved success, they tend to overfit to the training set because they directly infer pose-related parameters, leading to the poor generalization ability and low robustness to background clutters and diverse poses. To tackle the above issues, other approaches first establish 2D-3D correspondences and then solve a Perspective-n-Point (PnP) problem to estimate the pose of a space target. Park et al. [7] are the first to propose using DCNNs to predict the 2D coordinates of pre-defined landmarks of the space target to build 2D-3D correspondences. Based on the pipeline as described in [7], several improved methods [8–10] were proposed to enhance the performance of landmark regression. In particular, the 2D heatmap representation plays an important role in regressing landmarks and significantly promotes the improvement of landmark regression performance. However, heatmap-based methods suffer from several drawbacks such as dramatic performance degradation for low-resolution images, expensive computational costs of upsampling operations (e.g., deconvolution operation [11]), and extra post-refinement for improving the precision of the coordinates of landmarks.

In this article, we propose a 2D-3D correspondences-based space target pose estimation method. To deal with the aforementioned shortcomings of 2D heatmap representation, we propose a 1D landmark representation for space target pose estimation, which owns the following advantages. First, it has less complexity than the representation of heatmap. Second, it also has less quantization error than heatmap. Third, it improves the accuracy of landmark regression at low input resolutions without post-processing. We propose a novel convolutional neural network based space target landmark regression model to predict 1D landmark representations. Experimental results demonstrate that our method is superior to the 2D heatmap representation based models at various input resolutions, especially at low resolutions. Based on the predicted landmarks predicted by the proposed landmark regression model, the pose of the space target is then recovered by an off-the-shelf PnP algorithm [12], which is more precise than the one estimated by existing methods. Our work has the following contributions:

- We propose a kind of 1D landmark representation, which describes the horizontal and vertical coordinates of a landmark as two independent fixed-length 1D vectors.
- We also propose a space target landmark regression network that predicts the 2D positions of landmarks in the input image using proposed 1D landmark representations.
- Comprehensive experiments are conducted on the SPEED dataset [13]. The proposed 1D landmark representation makes the space target landmark regression network achieve the competitive performance compared to 2D heatmap-based landmark regression methods and outperform them by a large margin at low input resolutions. Furthermore, predicted landmarks based on 1D landmark representations bring an improvement in the accuracy of space target pose estimation.

The rest of this article is organized as follows. We briefly review the related work in Section 2. In Section 3, we introduce the overall pipeline of space target pose estimation and propose the 1D landmark representation and space target landmark regression network in detail. In Section 4, we describe the dataset and evaluation metrics. Section 5 presents the settings and results of experiments and the comparison with the mainstream methods. Section 6 discusses the proposed 1D landmark representation. Finally, Section 7 summarizes the conclusions.

2. Related Work

We review closely-related deep learning-based methods developed mainly for monocular vision-based pose estimation of known non-cooperative space targets from two aspects: direct methods, and landmark-based methods.

2.1. Direct Methods

The most intuitive approach to predict the 6 degree of freedom (DoF) poses of space targets is to treat pose estimation as a classification or regression task and directly infer translation representations (e.g., Euclid coordinate) and rotation representations (e.g., Euler angle, axis-angle, and quaternion) of space targets from input monocular images.

Sharma et al. [4] treated single space target pose estimation as a classification task and pioneered the use of the AlexNet [14] based network to predict pose categories. Meanwhile, Proença et al. [5] treated single space target rotation estimation as a probabilistic soft classification task and proposed a ResNet [15] based network named UrsoNet to regress probabilities for each rotation category. To further improve the accuracy of rotation estimation, Spacecraft Pose Network (SPN) [6] exploits a hybrid classification-regression fashion to estimate space target rotations from coarse to fine.

These direct methods for monocular space target pose estimation use deep convolutional neural networks to directly learn a complex non-linear function that maps images to space target poses. Although such methods have made some achievements, they lack the ability of spatial generalization and have not acquired the same level of accuracy as landmark-based methods.

2.2. Landmark-Based Methods

Since the 3D model of a known space target is available, an indirect approach to predicting the 6 degree of freedom (DoF) pose of a space target is first to establish 2D-3D correspondences between 2D and 3D coordinates of the landmarks of the space target, then estimate the 6 DoF camera pose based on such 2D-3D correspondences using PnP algorithms (e.g., PST [16], RPnP [17], and EPnP [18]). The space target pose is also obtained by transforming the estimated camera pose.

Existing landmark-based methods commonly use DCNNs to detect 2D landmarks of the space target in the image. As the pioneer, Park et al. [7] proposed an improved landmark regression network based on the YOLO [19,20] framework in which the MobileNet [21,22] was used as the backbone to directly regress the 2D coordinates of landmarks. Based on [7], Hu et al. [10] introduced the Feature Pyramid Network (FPN) [23] into the landmark regression network to make the full use of multi-scale information. However, directly regressing 2D landmark coordinates by DCNNs still suffers from the poor accuracy of landmark detection. To solve this problem, Chen et al. [8] proposed exploiting 2D heatmaps to represent 2D landmarks. Compared with direct regression of 2D coordinates, predicting 2D heatmaps can explicitly preserve the spatial information of the space target. They proposed a top-down landmark regression method where the Faster R-CNN [24] was used to detect space targets and then HRNet [25] was used to regress the 2D heatmaps of landmarks for each space target. The reason for choosing HRNet [25] is that the high-resolution representation is maintained in the whole pipeline of the network, which enables the model to extract features with superior spatial relationships that are suitable for localization-related tasks. Based on [8], Xu et al. [9] applied dilated convolutions to fuse multi-scale features in the HRNet [25] to better mine global information, and presented an online hard landmark mining method to enhance the ability of the network to detect invisible landmarks. In addition, Wang et al. [26] proposed a set-based representation to fully explore the relationship among keypoints and the context between the keypoints and the satellite. They also constructed a transformer-based network to predict the set of keypoints, achieving better generalization ability.

Existing state-of-the-art landmark-based methods are more accurate and robust than direct methods. In particular, the methods based on heatmaps have achieved fairly good performance. However, to reduce the quantization error, heatmap-based methods usually need multiple upsampling layers to generate high-resolution heatmaps and require extra post-processing to obtain accurate 2D coordinates of landmarks, leading to heavy computational burden. Moreover, the performance degrades significantly for low-resolution images. To address these issues, we propose a 1D landmark representation to describe the horizontal and vertical coordinates of the landmark, respectively. Compared with the 2D heatmap, it has smaller complexity and quantization error as well as better representational ability in various input resolutions. In particular, the 1D landmark representation can increase the scales of the horizontal and vertical coordinates of the landmark, which improves the localization accuracy of the landmark in the image.

3. Methodology

The general problem statement for monocular space target pose estimation is to compute the relative position and attitude of the target frame T with respect to the camera frame C via a monocular image captured by the camera. The relative position is represented by a translation vector $\boldsymbol{t}_{TC}$, from the origin of C to the origin of T. The relative attitude is represented by a rotation matrix $\boldsymbol{R}_{TC}$, aligning the target frame with the camera frame. Figure 1 illustrates the target and camera reference frames to visualize the position and attitude variables.

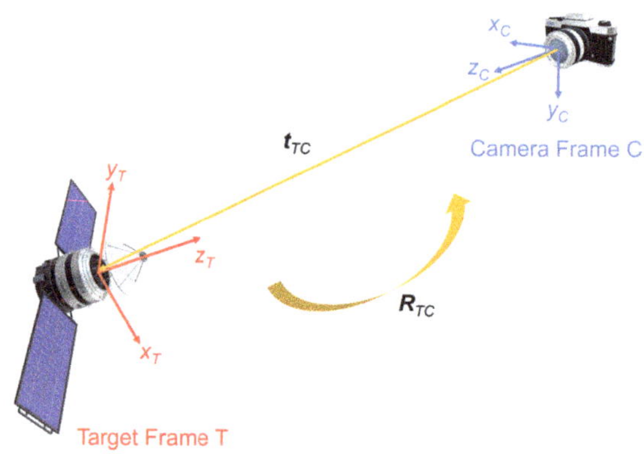

Figure 1. Definitions of the target reference frame (T), camera reference frame (C), relative position ($\boldsymbol{t}_{TC}$), and relative attitude ($\boldsymbol{R}_{TC}$).

Figure 2 illustrates the overall pipeline of our method for known non-cooperative space target pose estimation via monocular images. We use a top-down method for landmark regression. Specifically, to solve the issues of scale variations, the bounding box of a space target is first detected in an input image, and then a sub-image is cropped from the image based on the box followed by a resize operation to a fixed size. The resized sub-image is then used to regress the locations of landmarks of the space target. Therefore, the 2D bounding boxes of space targets need to be detected first. To this end, we employ an off-the-shelf object detection network (e.g., R-CNN series [24,27–32], YOLO series [19,20,33–36], FCOS [37], RepPoints [38], and DETR series [39–41]) to detect space targets in an input image. Next, we propose a space target landmark regression network to predict the landmarks of a space target in the resized sub-image. Based on the 3D model of the known space target, we build 2D-3D correspondences between the 2D pixel coordinates in the input image and 3D space coordinates in the 3D model for landmarks of the space target. Given the putative 2D-3D correspondences, we use a PnP solver in [12] within the RANdom SAmple Consensus (RANSAC) [42] framework to recover the pose of the space target.

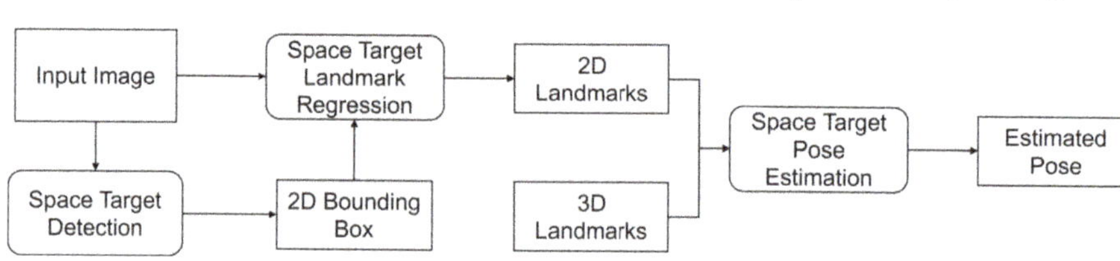

Figure 2. The overall pipeline of monocular vision-based known non-cooperative space target pose estimation.

In this pipeline, the accuracy of landmark regression significantly affects the performance of pose estimation. However, current 2D heatmap representation is redundant for the sparse landmarks. In this article, we present a 1D landmark representation for landmark regression, which performs well on low-resolution images and requires no post-refinement steps.

3.1. 1D Landmark Representation

As shown in Figure 3, the representation of a landmark contains two independent fixed-length 1D vectors, which represent the horizontal and vertical pixel coordinates of the landmark separately. The generation process of 1D landmark representation will be described in detail in this section.

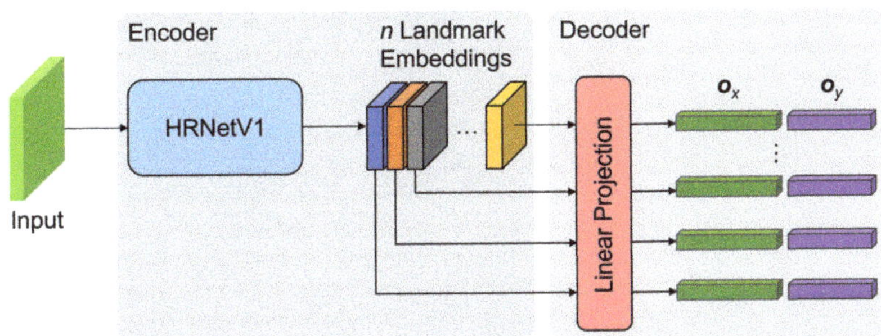

Figure 3. The overall structure of the proposed space target landmark regression network.

The input image has a size of $H \times W$, where H and W are the height and width, respectively. The ground-truth 2D coordinate of the k-th landmark is denoted as $\left(x^k, y^k\right)$, where x^k is the horizontal coordinate subjected to $0 \leq x^k \leq W$, and y^k is the vertical coordinate subjected to $0 \leq y^k \leq H$. To improve the accuracy of location description, we introduce an expansion factor $\mu \in \mathbb{R}^+$, to adjust the scale of the landmark coordinate. Thus, the ground-truth 2D coordinate of the k-th landmark is transformed into a new 2D coordinate:

$$p' = (x', y') = \left(\text{round}\left(\mu x^k\right), \text{round}\left(\mu y^k\right)\right) \tag{1}$$

where round$(\cdot)$ is the round function, and μ is an integer subjected to $\mu \geq 1$. The proposed expansion factor can improve the localization accuracy to the level of sub-pixel. Then, the 1D landmark representation of the k-th landmark is defined as follows:

$$\boldsymbol{p}_x^k = \left[x_0, x_1, \cdots, x_{\mu W-1}\right] \in \mathbb{R}^{\mu W}, x_i = \mathbb{I}(i = x') \tag{2}$$

$$\boldsymbol{p}_y^k = \left[y_0, y_1, \cdots, y_{\mu H-1}\right] \in \mathbb{R}^{\mu H}, y_j = \mathbb{I}(j = y') \tag{3}$$

where $i \in \{0, 1, \cdots, \mu W - 1\}$, $j \in \{0, 1, \cdots, \mu H - 1\}$, and $\mathbb{I}(\cdot)$ is the indicator function. Both $\boldsymbol{p}_x^k$ and $\boldsymbol{p}_y^k$ are 1D vectors. Such binary representation only encodes the 2D coordinate of a landmark while ignoring the adjacent point around the landmark. To describe the spatial relationships around landmarks, we introduce the Gaussian kernel to encode the 2D coordinates of landmarks and their surrounding points. For the k-th landmark, the Gaussian kernel-based 1D landmark representation is defined as follows:

$$\boldsymbol{p}_x^k = \left[x_0, x_1, \cdots, x_{\mu W-1}\right] \in \mathbb{R}^{\mu W}, x_i = \frac{1}{\sqrt{2\pi}\sigma} \exp\left(-\frac{(i - x')^2}{2\sigma^2}\right) \tag{4}$$

$$\boldsymbol{p}_y^k = \left[y_0, y_1, \cdots, y_{\mu H-1}\right] \in \mathbb{R}^{\mu H}, y_j = \frac{1}{\sqrt{2\pi}\sigma} \exp\left(-\frac{(j - y')^2}{2\sigma^2}\right) \tag{5}$$

where σ is the standard deviation.

In addition, the process of computing the 2D pixel coordinate of a landmark from 1D landmark representation is described as follows. Assuming the landmark regression network produces two 1D vectors $\boldsymbol{o}_x$ and $\boldsymbol{o}_y$ for a landmark, the final predicted 2D coordinate $(\hat{x}, \hat{y})$, of the landmark is computed by:

$$\hat{x} = \frac{1}{\mu} \arg\max_i (\boldsymbol{o}_x(i)) \quad (6)$$

$$\hat{y} = \frac{1}{\mu} \arg\max_j (\boldsymbol{o}_y(j)) \quad (7)$$

where $\boldsymbol{o}_x(i)$ denotes the i-th element of the $\boldsymbol{o}_x$, and $\boldsymbol{o}_y(j)$ denotes the j-th element of the $\boldsymbol{o}_y$.

3.2. Space Target Landmark Regression Network

We propose a top-down method to detect the landmarks of a space target in an input image. The input to the network is a fixed-size sub-image cropped from the input image using the detected bounding box. To enable the network to output 1D landmark representations, the network is designed to generate an embedding for each pre-defined landmark, and then use linear layers to project it into two 1D vectors with fixed lengths. As illustrated in Figure 3, we propose an encoder-decoder structured network to regress the locations of landmarks, where the encoder extracts landmark embeddings from the input, and the decoder predicts a 1D landmark representation for each landmark. The main modules of the proposed space target landmark regression network will be described in detail.

Encoder. To obtain high-quality landmark embeddings with rich spatial information, we use the HRNetV1 as described in [25,43] to extract embeddings for landmarks. Given n pre-defined landmarks, the output of HRNetV1 is a tensor of n embeddings corresponding one-to-one with landmarks only from the highest-resolution stream of the HRNetV1.

Decoder. The decoder comprises two shared linear projection layers that transform each embedding into a 1D landmark representation. Specifically, the embedding of a landmark is flattened and then fed into two shared fully-connected layers respectively to generate two independent 1D vectors $\boldsymbol{o}_x$ and $\boldsymbol{o}_y$, with the lengths of μW and μH, respectively.

Loss function. Because the 1D landmark representation is similar to the one-hot code, the proposed space target landmark regression network could be considered to perform a kind of classification task. Therefore, we use the cross-entropy loss function to train the proposed network and adopt the label smoothing strategy to help train the network. The equation of the cross-entropy loss function is as follows.

$$\mathcal{L}(p,t) = \frac{\sum_{n=1}^{N} l_n}{N} \quad (8)$$

$$l_n = -\sum_{c=1}^{C} \omega_c \log \frac{\exp(p_{n,c})}{\sum_{i=1}^{C} \exp(p_{n,i})} t_{n,c} \quad (9)$$

where p is the input tensor with the size of (N, C), t is the target tensor with the size of (N, C), C is the number of classes, N is the batch size, and ω is the weight that is a 1D tensor of size C assigning weight to each of the classes. In addition, when using the Gaussian kernel-based 1D landmark representation in the proposed space target landmark regression network, we exploit the Kullback–Leibler divergence as the loss function for network training. The equation of the Kullback–Leibler divergence loss function is as follows.

$$\mathcal{L}(p,t) = t \cdot \log \frac{t}{p} = t \cdot (\log t - \log p) \quad (10)$$

where p is the input tensor and t is the target tensor, they have the same shape.

4. Material

We briefly introduce the dataset and evaluation metrics used in the experiment in this article.

4.1. Dataset

The Spacecraft PosE Estimation Dataset (SPEED) [13] is the first publicly available dataset for space target pose estimation, mostly consisting of high-resolution synthetic grayscale images of the Tango satellite, as shown in Figure 4. There are 12,000 training images with ground truth pose labels and 2998 test images without ground truth labels. The ground truth pose label consists of a unit quaternion and a translation vector, describing the relative orientation and position of the Tango satellite with respect to the camera frame.

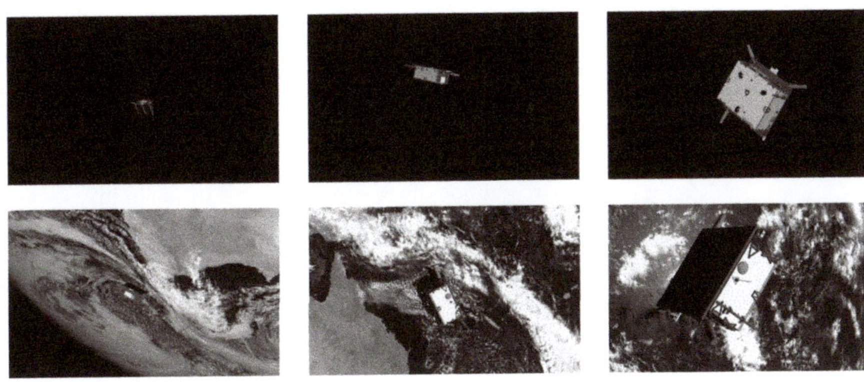

Figure 4. Sample images from the SPEED.

Since SPEED does not provide the ground truth pose labels for test images, we cannot conduct an in-depth analysis over them except for acquiring the total pose error score provided by the online server. Therefore, we conduct extensive experiments on the training images, where half of them have no background while the other half contain earth backgrounds. Notably, the size and orientation of the satellite, background, and illumination vary significantly in these images. For instance, the number of pixels of the satellite varies between 1 k and 500 k, as shown in Figure 5.

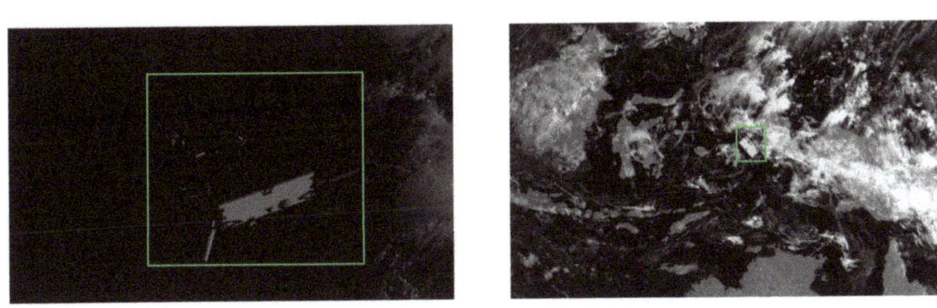

Figure 5. Large variation of the space target size in the SPEED dataset.

The intrinsic parameters of the camera used in SPEED are the horizontal focal length of 0.0176 m, the vertical focal length of 0.0176 m, the horizontal pixel pitch of 5.86×10^{-6} m/pixel, the vertical pixel pitch of 5.86×10^{-6} m/pixel, the image width of 1920 pixel, and the image height of 1200 pixel.

4.2. Evaluation Metrics

We introduce the evaluation metrics for space target landmark regression and pose estimation.

4.2.1. Space Target Landmark Regression Metrics

To evaluate the performance of space target landmark regression, predictions are considered to be true or false positives by measuring distances between predicted and ground truth landmarks. Therefore, the Object Keypoint Similarity (OKS) [44] is calculated as following:

$$\mathrm{OKS} = \frac{\sum_i \exp\left(-d_i^2/2s^2\kappa_i^2\right)\sigma(v_i > 0)}{\sum_i \sigma(v_i > 0)} \tag{11}$$

where d_i is the Euclidean distance between the i-th landmark's predicted and corresponding ground truth coordinates, κ_i is a constant, v_i is the visibility flag, and s is the scale defined as the square root of the 2D bounding box area of the space target. For each landmark, this yields a similarity that ranges between 0 and 1. By setting a threshold for OKS, we can distinguish whether a prediction is a true positive or a false positive. Specifically, for each prediction, if the value of OKS is greater than the threshold, it is considered as a true positive, and vice versa. Then, we can compute precision as following:

$$\mathrm{precision} = \frac{\mathrm{TruePositives}}{\mathrm{TruePositives} + \mathrm{FalsePositives}} = \frac{\mathrm{TruePositives}}{\mathrm{AllPredictions}} \tag{12}$$

In addition, there may be ground truth landmarks with no matching predictions, so they are called false negatives. Thus, we also need to compute recall as following:

$$\mathrm{recall} = \frac{\mathrm{TruePositives}}{\mathrm{TruePositives} + \mathrm{FalseNegatives}} = \frac{\mathrm{TruePositives}}{\mathrm{AllGroundTruths}} \tag{13}$$

Then, the precision-recall curve is calculated to compute the average precision value, as described in [45].

The Average Precision (AP) [45] is used as the space target landmark regression metric. For better measuring the performances of methods, it is averaged over multiple OKS values [44]. Specifically, AP is computed at 10 OKS thresholds from 0.50 to 0.95 with a step size of 0.05, AP_{50} is computed at a single OKS threshold of 0.50, and AP_{75} is computed at a single OKS threshold of 0.75.

4.2.2. Space Target Pose Estimation Metrics

To evaluate the estimated pose of each space target, a rotation error e_r, a translation error e_t, and a pose error e_p, are calculated.

The rotation error e_r, is calculated as the angular distance between the estimated and ground-truth rotation quaternions $\hat{\boldsymbol{q}}$ and $\boldsymbol{q}$, of a space target with respect to the camera. The rotation error e_r, is defined as

$$e_r = 2 \cdot \arccos(|\langle \hat{\boldsymbol{q}}, \boldsymbol{q} \rangle|) \tag{14}$$

where $\langle \cdot \rangle$ denotes the inner product operation.

The translation error e_t, is calculated as the 2-norm of the difference of the estimated and ground-truth translation vectors $\hat{\boldsymbol{t}}$ and $\boldsymbol{t}$, from the camera frame to the space target frame. The translation error e_t, is defined as

$$e_t = ||\hat{\boldsymbol{t}} - \boldsymbol{t}||_2 \tag{15}$$

The normalized translation error $\bar{e}_t$, is also defined as

$$\bar{e}_t = \frac{e_t}{||\boldsymbol{t}||_2} \tag{16}$$

which penalizes the translation error more heavily when the space target is closer to the camera.

The pose error for a single space target e_p, is the sum of the rotation error e_r, and the normalized translation error $\bar{e}_t$,

$$e_p = e_r + \bar{e}_t \tag{17}$$

Finally, the total error E is the average of the pose errors for all space targets,

$$E = \frac{1}{N}\sum_{i=1}^{N} e_p^i \tag{18}$$

where N is the number of space targets.

5. Experimental Results

In this section, we first give the details of the experiment, then we evaluate the performance of our method and compare it with several state-of-the-art methods.

5.1. Experiment Details

As a preliminary, since the 3D model of the Tango satellite is unavailable in SPEED, we first reconstruct the 3D landmarks of the Tango satellite via multi-view triangulation using a small set of manually selected training images with pose labels. Moreover, to obtain ground-truth 2D bounding boxes and landmarks, we first project the reconstructed 3D landmarks of the Tango satellite into the image to obtain the 2D landmark labels and then extract the closest bounding box of the reprojected points as the label for 2D object detection. Finally, we briefly introduce the implementation details of the experiment.

5.1.1. 3D Landmark Reconstruction

For Tango satellite, we select its 8 corners and the tips of its 3 antennas as landmarks. To reconstruct the 3D coordinates of the 11 landmarks using multi-view triangulation, we manually annotate 2D points corresponding to each landmark over a small set of hand-picked close-up training images in which the space target is well-illuminated and has various poses. Assuming $\boldsymbol{z}_{i,j}$ is the 2D coordinate vector of the i-th landmark in the j-th image, the following optimization problem is solved to obtain the 3D coordinate vector of the i-th landmark $\boldsymbol{x}_i$, in the space target frame,

$$\min_{\boldsymbol{x}_i} \sum_j \left\| \gamma_{i,j}\boldsymbol{z}_{i,j} - \boldsymbol{K}\left(\boldsymbol{R}_j\boldsymbol{x}_i + \boldsymbol{t}_j\right) \right\|_2^2 \tag{19}$$

where, for i-th landmark, the sum of the reprojection error is minimized over a set of images where the i-th landmark is visible. In Equation (19), $\gamma_{i,j}$ is a scale factor denoting the depth of the i-th landmark in the j-th image, $\boldsymbol{K}$ is a known camera intrinsic matrix, and $\left(\boldsymbol{R}_j, \boldsymbol{t}_j\right)$ is a ground truth camera pose of the j-th image. The optimization variable in Equation (19) is the 3D coordinate vector $\boldsymbol{x}_i$. The routine of the triangulateMultiview function in MATLAB is used to solve this optimization problem. Figure 6 visualizes the 11 selected 3D landmarks and the reconstructed wireframe model of the Tango satellite.

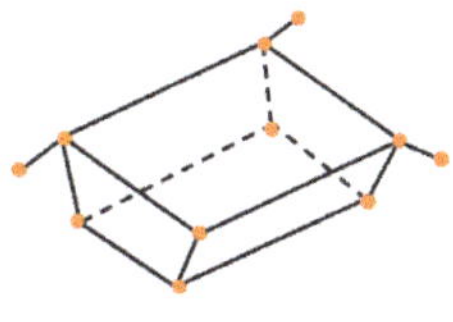

Figure 6. The reconstructed 3D wireframe model with 11 landmarks.

5.1.2. 2D Landmark and Bounding Box Label

To obtain ground-truth landmarks of the Tango satellite in the image, we reproject the aforementioned set of reconstructed 3D landmarks to the image plane using the ground-truth camera pose. Since the convex hull of such reprojected 2D landmarks almost covers the whole target in any image, we slightly relax the horizontal rectangle which encloses all 2D landmarks, and take it as the ground-truth bounding box of the Tango satellite. Figure 7 shows several examples of obtained 2D landmark and bounding box labels.

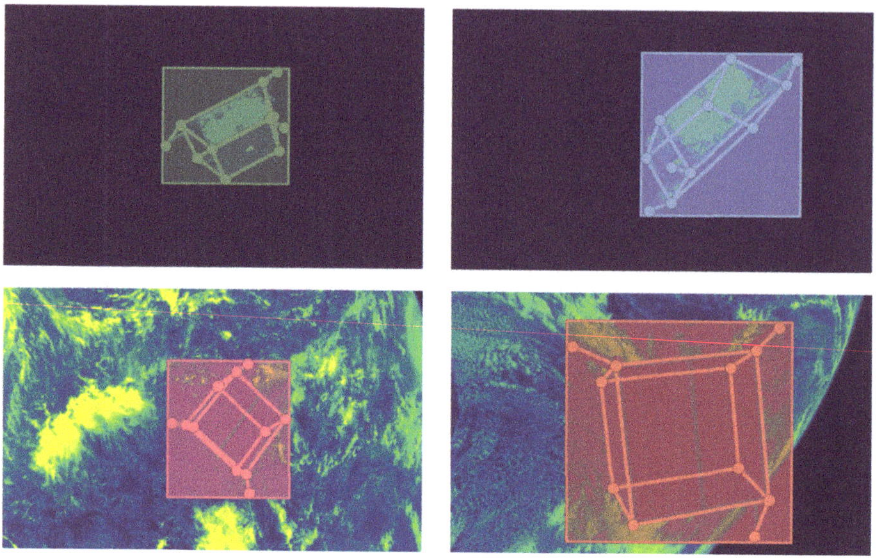

Figure 7. Examples of 2D landmark and bounding box labels.

5.1.3. Implementation Details

We conduct experiments using 6-fold cross-validation over the training images of the SPEED dataset. Specifically, we split the 12,000 training images into 6 groups, and then for each group, we test a space target landmark regression network trained with the remaining 5 groups. In addition, we use the off-the-shelf object detection network Sparse R-CNN [32] to detect 2D bounding boxes. The neural network architecture is implemented with PyTorch 1.11.0 and CUDA 11.3 and runs on an Intel Core i9-10900X CPU @ 3.70 GHz with an NVIDIA Geforce RTX 3090. The pose estimation algorithm is implemented with MATLAB.

Data augmentation. To increase the diversity of training samples and alleviate the accuracy error of the detected 2D bounding boxes, we introduce a bounding box augmentation strategy which random shifts bounding box centers slightly while the targets are still within bounding boxes. What is more, we also use data augmentation strategies such as random rotation ($[-80^\circ, 80^\circ]$) and random scaling ($[0.5, 1.5]$) for training images. We do not apply a random flipping strategy to the training images because it might generate unrealistic geometric relationships between 2D landmarks, which may result in performance degradation of networks.

Training setting. We use the Adam optimizer [46] for network training. The base learning rate is set to 1×10^{-3}, and is dropped to 1×10^{-4} and 1×10^{-5} at the 16th and 22nd epochs, respectively. The total training process is terminated within 24 epochs. The batch contains 8 randomly sampled samples per iteration. The HRNetV1 backbone in the proposed space target landmark regression network is initialized with a pre-trained model HRNetV1-W32 [43] which is pretrained with the COCO dataset [44].

5.2. Results of Space Target Landmark Regression

As shown in Table 1, we evaluate the performance of our proposed space target landmark regression network in terms of the AP, AP_{50}, and AP_{75} metrics. In this experiment, we set up the expansion factor as 2 and the standard deviation of the Gaussian kernel σ, as 2. For each fold, we test the performances of the proposed space target landmark regression network using 1D landmark representations with or without Gaussian kernel at four input resolutions including 512 × 512, 256 × 256, 128 × 128, and 64 × 64. The AP is improved as the input resolution increases, no matter whether the Gaussian kernel is used in the 1D landmark representation. When the Gaussian kernel is introduced into the 1D landmark representation, the AP is commonly superior to the one without the Gaussian kernel at any input resolution. It empirically demonstrates that spatial relationships play an important role in describing the coordinates of landmarks and improve the performances of landmark regression networks. Figure 8 shows qualitative results where the orange points are the landmarks predicted by the proposed landmark regression network using Gaussian kernel-based 1D landmark representations at the input resolution of 512 × 512.

Table 1. Quantitative results of space target landmark regression and pose estimation using the proposed 1D landmark representation.

Fold	Gaussian Kernel	Input Size	AP_{50} (↑)	AP_{75} (↑)	AP (↑)	E_r (↓)	E_t (↓)	E (↓)
1	w/o	512 × 512	100.0	98.7	94.6	0.0125	0.0036	0.0161
		256 × 256	99.0	98.9	94.0	0.0164	0.0048	0.0212
		128 × 128	99.0	97.6	90.6	0.0278	0.0083	0.0361
		64 × 64	98.9	81.2	70.2	0.0816	0.0204	0.1021
	w/	512 × 512	100.0	98.9	96.4	0.0112	0.0030	**0.0141**
		256 × 256	100.0	98.8	**97.0**	0.0132	0.0040	0.0172
		128 × 128	100.0	97.3	90.8	0.0347	0.0094	0.0442
		64 × 64	97.7	85.1	74.3	0.0758	0.0195	0.0953
2	w/o	512 × 512	100.0	98.6	93.2	0.0130	0.0038	0.0168
		256 × 256	100.0	98.7	94.6	0.0165	0.0049	0.0214
		128 × 128	99.0	97.3	89.5	0.0325	0.0090	0.0415
		64 × 64	97.8	79.9	69.9	0.0924	0.0228	0.1153
	w/	512 × 512	100.0	98.8	95.3	0.0103	0.0030	**0.0133**
		256 × 256	100.0	98.9	**96.6**	0.0158	0.0043	0.0201
		128 × 128	99.0	97.5	91.4	0.0314	0.0086	0.0400
		64 × 64	97.6	82.8	72.9	0.0862	0.0203	0.1065
3	w/o	512 × 512	100.0	99.0	95.0	0.0121	0.0036	0.0157
		256 × 256	100.0	98.8	94.7	0.0149	0.0045	0.0194
		128 × 128	99.0	97.7	90.8	0.0282	0.0085	0.0367
		64 × 64	98.9	80.9	70.1	0.0801	0.0220	0.1021
	w/	512 × 512	100.0	98.9	95.2	0.0104	0.0031	**0.0135**
		256 × 256	100.0	99.0	**96.9**	0.0149	0.0042	0.0191
		128 × 128	100.0	97.5	91.8	0.0284	0.0080	0.0364
		64 × 64	98.8	88.0	76.0	0.0736	0.0195	0.0930
4	w/o	512 × 512	99.0	98.8	93.7	0.0119	0.0034	0.0153
		256 × 256	100.0	98.8	93.9	0.0164	0.0048	0.0212
		128 × 128	99.0	96.0	88.8	0.0329	0.0092	0.0420
		64 × 64	98.8	80.7	71.1	0.0842	0.0219	0.1061
	w/	512 × 512	100.0	99.0	96.1	0.0100	0.0030	**0.0130**
		256 × 256	100.0	99.0	**96.3**	0.0139	0.0041	0.0180
		128 × 128	100.0	97.6	90.8	0.0296	0.0083	0.0380
		64 × 64	98.8	83.0	72.8	0.0795	0.0202	0.0997
5	w/o	512 × 512	100.0	98.9	95.1	0.0128	0.0037	0.0165
		256 × 256	100.0	98.9	94.0	0.0162	0.0048	0.0210
		128 × 128	100.0	97.7	90.2	0.0284	0.0085	0.0369
		64 × 64	98.9	82.1	71.4	0.0908	0.0223	0.1131
	w/	512 × 512	100.0	98.7	94.9	0.0100	0.0030	**0.0131**
		256 × 256	100.0	98.9	**96.7**	0.0143	0.0042	0.0185
		128 × 128	100.0	97.5	90.9	0.0314	0.0086	0.0399
		64 × 64	98.8	84.7	73.6	0.0805	0.0198	0.1003

Table 1. *Cont.*

Fold	Gaussian Kernel	Input Size	AP50 (↑)	AP75 (↑)	AP (↑)	E_r (↓)	E_t (↓)	E (↓)
6	w/o	512 × 512	100.0	98.6	93.6	0.0138	0.0036	0.0174
		256 × 256	100.0	98.8	93.9	0.0162	0.0049	0.0211
		128 × 128	100.0	97.5	89.6	0.0318	0.0087	0.0405
		64 × 64	97.8	79.8	69.1	0.0876	0.0226	0.1102
	w/	512 × 512	100.0	98.8	96.0	0.0112	0.0030	**0.0142**
		256 × 256	100.0	98.9	**96.5**	0.0142	0.0043	0.0184
		128 × 128	99.0	97.5	90.4	0.0330	0.0089	0.0419
		64 × 64	98.7	85.3	74.5	0.0828	0.0201	0.1029

Figure 8. Qualitative results for the proposed space target landmark regression network using Gaussian kernel-based 1D landmark representations at the input resolution of 512 × 512.

5.3. Results of Space Target Pose Estimation

As shown in Table 1, we evaluate the performance of space target pose estimation that uses 2D landmarks predicted by our proposed space target landmark regression network. The evaluation metrics for space target pose estimation are composed as follows. E_r is the average of the rotation errors for all space targets. E_t is the average of the normalized translation errors for all space targets. E is the total error which is the sum of E_r and E_t. For each fold, all kinds of errors decrease as the input resolution increases. What is more, estimating spatial target pose using predicted landmarks described by the Gaussian kernel-based 1D landmark representation generally leads to a performance boost. Figure 9 shows qualitative results of space target pose estimation using predicted landmarks based on the Gaussian kernel-based 1D landmark representation under the input resolution condition of 512 × 512. The satellite body frame's correspondence between colors and directions is red—x, green—y, and blue—z.

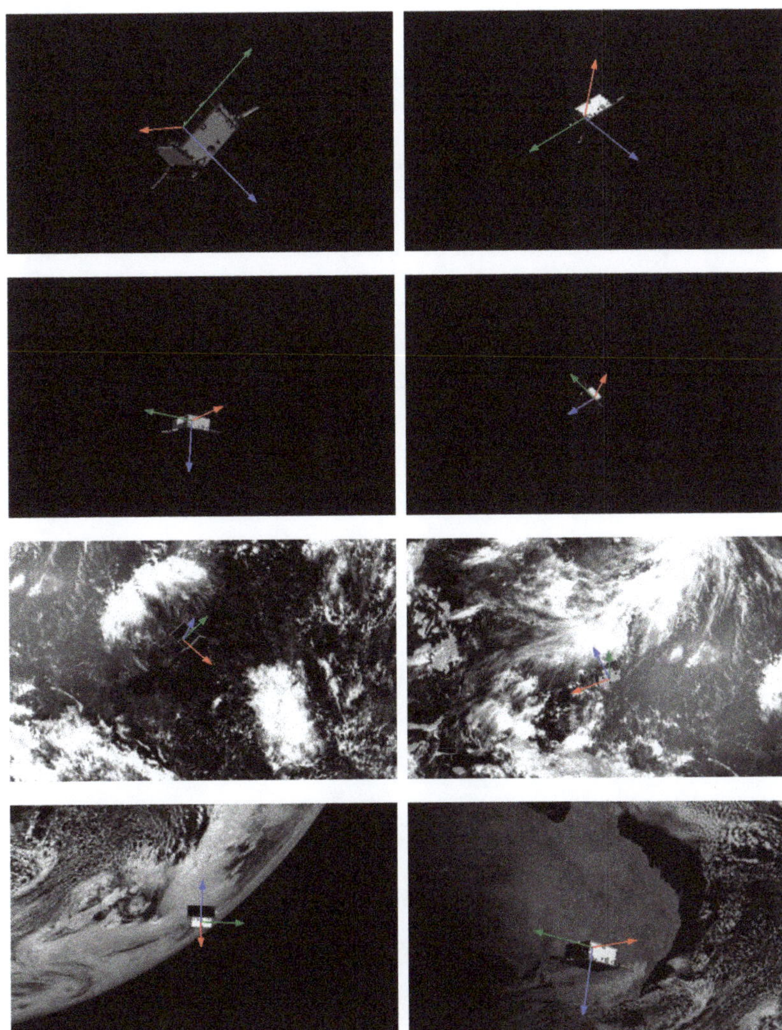

Figure 9. Qualitative results for space target pose estimation using 2D landmarks predicted by the proposed space target landmark regression network using Gaussian kernel-based 1D landmark representations at the input resolution of 512 × 512.

5.4. Comparisons with State-of-the-Art Methods

We compare our method with the state-of-the-art methods in space target landmark regression and pose estimation. To compare with them, we re-implement the same pipeline as [7,8], both of which estimate the poses of space targets by regressing the pre-defined landmarks of space targets in images.

5.4.1. Performance of Space Target Landmark Regression

To evaluate the performance of space target landmark regression, we compare our method with [7,8] in terms of the AP, AP_{50}, and AP_{75} metrics. In [7], the coordinates of landmarks are directly regressed by the network using Darknet53 [20] as the backbone. In [8], HRNetV1-W32 [43] is used to produce heatmaps for landmarks, and then the coordinates of landmarks are calculated by post-processing on these heatmaps. It should be pointed out that we use pre-trained models which are pretrained on the COCO dataset [44] to initialize the weights of Darknet53 and HRNetV1-W32 during training re-implemented networks. In our proposed method, both the expansion factor and standard deviation of the Gaussian kernel are defined as 2. Table 2 presents the comparison between our method and the state-of-the-art methods, in which each of the values of a metric is the average over the values for the metric from all over the folds. As shown in Table 2, our methods outperform the direct regression method [7] at any input size condition. Our methods achieve the same level of AP as the heatmap regression method [8] for the input sizes of 512×512 and 256×256, respectively. When the input size decreases to 128×128 and 64×64, our method without Gaussian kernel gets 89.9 and 70.3 AP, respectively, which have 6.8 and 33.5 improvements compared to the heatmap regression method [8], while the one with Gaussian kernel gets 91.0 and 74.0 AP, respectively, which achieve 7.9 and 37.2 improvements in contrast to the heatmap regression method [8]. Meanwhile, our methods have similar model sizes and GFLOPs to the heatmap regression method [8].

Table 2. Comparison with the state-of-the-art methods in terms of the evaluation metrics for space target landmark regression.

Method	Input Size	#Params	GFLOPs	AP_{50} (↑)	AP_{75} (↑)	AP (↑)
Direct Regression [7]	512 × 512	23.55 M	21.52	96.4	71.7	62.8
	256 × 256	23.55 M	5.38	97.7	74.0	63.7
	128 × 128	23.55 M	1.35	91.9	42.5	46.9
	64 × 64	23.55 M	0.34	32.0	1.4	8.1
Heatmap Regression [8]	512 × 512	28.54 M	41.08	99.8	**99.0**	**96.1**
	256 × 256	28.54 M	10.27	100.0	**99.0**	96.3
	128 × 128	28.54 M	2.57	99.0	95.2	83.1
	64 × 64	28.54 M	0.64	89.6	21.1	36.8
Ours (w/o) [1]	512 × 512	62.09 M	41.45	99.8	98.8	94.2
	256 × 256	32.73 M	10.32	99.8	98.8	94.2
	128 × 128	29.06 M	2.57	99.3	97.3	89.9
	64 × 64	28.6 M	0.64	**98.5**	80.8	70.3
Ours (w/) [2]	512 × 512	62.09 M	41.45	**100.0**	98.9	95.7
	256 × 256	32.73 M	10.32	**100.0**	98.9	**96.7**
	128 × 128	29.06 M	2.57	**99.7**	**97.5**	**91.0**
	64 × 64	28.6 M	0.64	98.4	**84.8**	**74.0**

[1] This method uses the 1D landmark representation without Gaussian kernel. [2] This method uses the Gaussian kernel-based 1D landmark representation.

Figure 10 illustrates that our methods consistently provide significant gains for the performance of space target landmark regression, especially in low-resolution input conditions. Whereas, both the direct regression method [7] and the heatmap regression method [8] suffer from drastic performance drop as the input size degrades.

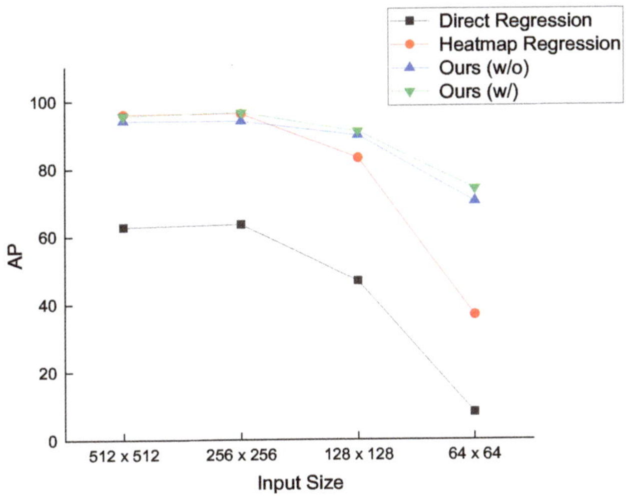

Figure 10. Comparison between our methods and the state-of-the-art methods in terms of the AP metric at all input size conditions.

5.4.2. Performance of Space Target Pose Estimation

To evaluate the performance of space target pose estimation that uses predicted 2D landmarks to recover the pose of a space target, we compare our method with [7,8] in terms of E_r, E_t, and E metrics. Furthermore, in order to investigate the impact of 2D landmarks predicted by various methods on space target pose estimation, we universally use the RANSAC-based PnP solver [12] instead of the improved PnP algorithm as described in [8] to compute space target poses. Table 3 presents the comparison between our method and the state-of-the-art methods, where the values of each metric are all the averages of the values of the metric from all folds. As shown in Table 3, our methods achieve smaller pose error than the direct regression method [7] and the heatmap regression method [8] under any input size condition. Compared with the heatmap regression method [8], our methods without and with Gaussian kernel get little gains for the input size of 512 × 512 while attaining increasing gains as the input size decreases. Especially, at the input size of 64 × 64, the total errors of our methods without and with Gaussian kernel are lower by 0.1571 and 0.1657, respectively.

Table 3. Comparison with the state-of-the-art methods for space target pose estimation.

Method	Input Size	E_r (↓)	E_t (↓)	E (↓)
Direct Regression [7]	512 × 512	0.0753	0.0371	0.1124
	256 × 256	0.0809	0.0319	0.1128
	128 × 128	0.1088	0.0383	0.1471
	64 × 64	0.2083	0.0818	0.2901
Heatmap Regression [8]	512 × 512	0.0139	0.0042	0.0181
	256 × 256	0.0274	0.0083	0.0358
	128 × 128	0.0669	0.0192	0.0861
	64 × 64	0.2141	0.0512	0.2653
Ours (w/o) [1]	512 × 512	0.0127	0.0036	0.0163
	256 × 256	0.0161	0.0048	0.0209
	128 × 128	**0.0303**	0.0087	**0.0390**
	64 × 64	0.0861	0.0220	0.1082
Ours (w/) [2]	512 × 512	**0.0105**	**0.0030**	**0.0135**
	256 × 256	**0.0144**	**0.0042**	**0.0186**
	128 × 128	0.0314	**0.0086**	0.0401
	64 × 64	**0.0797**	**0.0199**	**0.0996**

[1] This method uses the 1D landmark representation without Gaussian kernel. [2] This method uses the Gaussian kernel-based 1D landmark representation.

Figure 11 illustrates that our proposed methods maintain a low error level of space target pose estimation for any input size. However, the total errors of the direct regression method [7] and the heatmap regression method [8] increase sharply when the input size decreases. In particular, the total error of the direct regression method [7] is the largest among all methods at any input size.

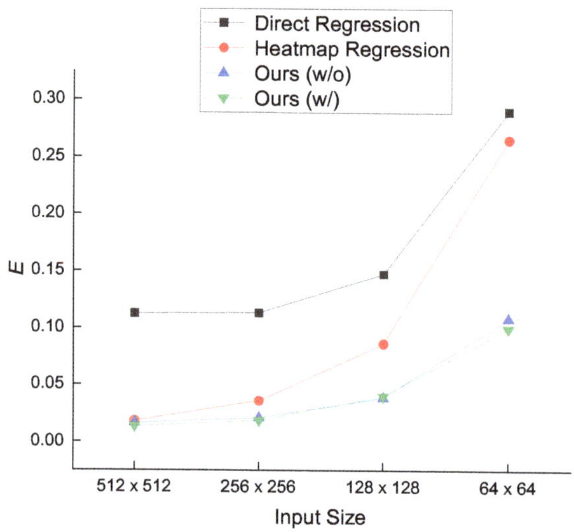

Figure 11. Comparison between our methods and the state-of-the-art methods in terms of the E metric at all input size conditions.

6. Discussion

We give a discussion on the property of 1D landmark representation compared to the 2D heatmap representation. The discussion is conducted from the perspectives of complexity and quantization error. In addition, we conduct an analysis on the expansion factor μ.

6.1. Representation Complexity

Given an input image with the resolution of $H \times W$, heatmap regression methods aim to generate a 2D heatmap with the resolution of $\frac{H}{\lambda} \times \frac{W}{\lambda}$, where λ is the downsampling factor. Since λ is a constant, the spatial complexity of the heatmap representation is $O(H \times W)$. Whereas, our method aims to yield two 1D vectors with the length of $H \cdot \mu$ and $W \cdot \mu$ respectively, where μ is the expansion factor. Considering μ is a constant, the spatial complexity of the proposed 1D landmark representation is $O(H + W)$. Therefore, we can find that the 1D landmark representation is more efficient than the heatmap representation. Particularly, it enables some methods to remove extra independent upsampling operations (e.g., deconvolution [11]) used to obtain high-resolution landmark embeddings directly so that the computational costs of networks can be reduced.

6.2. Quantization Error

Given an input image with the resolution of $H \times W$ and the ground-truth 2D coordinate of a landmark (x, y), we take the horizontal coordinate x, as an example to analyze the quantization errors of 1D landmark representation and 2D heatmap representation, respectively.

For 1D landmark representation, x can be rewritten as:

$$x = \frac{n_v + z_v}{\mu} = \frac{n_v}{\mu} + \sigma_v \tag{20}$$

where μ is the expansion factor subjected to $\mu \geq 1$, $n_v \in \mathbb{N}$, $0 \leq z_v < 1$, and $0 \leq \sigma_v < \frac{1}{\mu}$. The ground-truth x is then rescaled by μ into a new one:

$$x_v = \text{round}(x \cdot \mu) = n_v + \text{round}(\sigma_v \cdot \mu) = \begin{cases} n_v & 0 \leq \sigma_v < \frac{1}{2\mu} \\ n_v + 1 & \frac{1}{2\mu} \leq \sigma_v < \frac{1}{\mu} \end{cases} \tag{21}$$

The quantization error of 1D landmark representation $|\Delta_v|$, is calculated as:

$$|\Delta_v| = \left| \frac{x_v}{\mu} - x \right| = \begin{cases} \left| \frac{n_v}{\mu} - \frac{n_v}{\mu} - \sigma_v \right| & 0 \leq \sigma_v < \frac{1}{2\mu} \\ \left| \frac{n_v+1}{\mu} - \frac{n_v}{\mu} - \sigma_v \right| & \frac{1}{2\mu} \leq \sigma_v < \frac{1}{\mu} \end{cases} \tag{22}$$

Therefore, the quantization error of 1D landmark representation $|\Delta_v|$, satisfies $0 \leq |\Delta_v| < \frac{1}{2\mu}$.

For heatmap representation, x can be rewritten as:

$$x = n_h \cdot \lambda + \sigma_h \tag{23}$$

where λ is the downsampling factor subjected to $\lambda \geq 1$, $n_h \in \mathbb{N}$, $0 \leq \sigma_h < \lambda$. Due to that the heatmap resolution is usually downsampled from the original input image resolution, the ground-truth x is transformed as:

$$x_h = \text{round}\left(\frac{x}{\lambda}\right) = n_h + \text{round}\left(\frac{\sigma_h}{\lambda}\right) = \begin{cases} n_h & 0 \leq \sigma_h < \frac{\lambda}{2} \\ n_h + 1 & \frac{\lambda}{2} \leq \sigma_h < \lambda \end{cases} \tag{24}$$

The quantization error of heatmap representation $|\Delta_h|$, is calculated as:

$$|\Delta_h| = |x_h \cdot \lambda - x| = \begin{cases} |\sigma_h| & 0 \leq \sigma_h < \frac{\lambda}{2} \\ |\lambda - \sigma_h| & \frac{\lambda}{2} \leq \sigma_h < \lambda \end{cases} \tag{25}$$

Therefore, the quantization error of heatmap representation $|\Delta_h|$, satisfies $0 \leq |\Delta_h| < \frac{\lambda}{2}$.

This reduces the quantization error from the level of $\left[0, \frac{\lambda}{2}\right)$ to the level of $\left[0, \frac{1}{2\mu}\right)$. In order to alleviate the quantization error of heatmap representation, extra time-consuming refinements are always employed. In contrast, 1D landmark representation does not require extra coordinate post-processing and reduce the quantization error, which promote the improvement of space target landmark regression.

6.3. Analysis on Expansion Factor

The expansion factor, μ, is the only hyperparameter in the 1D landmark representation, which controls the sub-pixel accuracy level of the landmark location. Specifically, the larger μ is, the lower the quantization error of 1D landmark representation is. Nevertheless, the increase in the expansion factor brings more computational burden to network training. Therefore, there is a trade-off between the quantization error and the network performance. We test $\mu \in \{1, 2, 3, 4\}$ in the 1D landmark representation without Gaussian kernel based on the proposed space target landmark regression network at various input resolutions. As shown in Figure 12, the performance of the proposed network tends to decrease as μ increases. For the input sizes of 64×64 and 128×128, the best choice to μ is 1 so that the proposed network achieves the AP of 72.4 and 90.5, respectively. For the input sizes of 256×256 and 512×512, the value of μ affects the network performance indistinctively. Specifically, the proposed network achieves the best AP of 94.4 and 94.6 when μ is set up as 1.

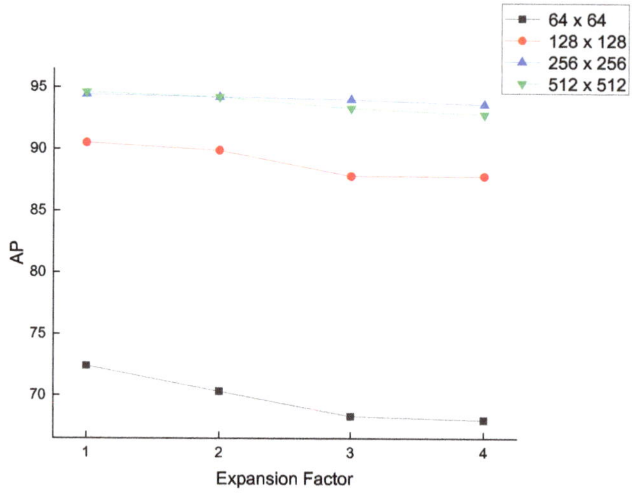

Figure 12. Analysis of the value of expansion factor μ, under various input resolution conditions.

7. Conclusions

In this article, we study the task of the space target pose estimation from monocular images and propose a novel approach. Based on the 2D-3D correspondences-based pose estimation pipeline, we present a 1D landmark representation to describe the 2D coordinates of pre-defined landmarks in images, to deal with several drawbacks of the 2D heatmap representation that dominates the landmark regression task. Next, we propose a space target landmark regression network to predict the horizontal and vertical coordinates of landmarks respectively using 1D landmark representations. Experimental results on the SPEED dataset show that our method achieves superior performance to the existing state-of-the-art methods in space target landmark regression and pose estimation, especially when the input images are low in resolution. In the space target landmark regression task, our method using the 1D landmark representation without Gaussian kernel achieves 70.3, 89.9, 94.2, and 94.2 AP respectively at the input sizes of 64×64, 128×128, 256×256, and 512×512, while using the Gaussian kernel-based 1D landmark representation, our method achieves 74.0, 91.0, 96.7, and 95.7 AP respectively. In the space target pose estimation task, the total errors of our method using the 1D landmark representation without Gaussian kernel are 0.1082, 0.0390, 0.0209, and 0.0163 respectively at the input sizes of 64×64, 128×128, 256×256, and 512×512, while using the Gaussian kernel-based 1D landmark representation, the total errors of our method are 0.0996, 0.0401, 0.0186, and 0.0135 respectively. Furthermore, we discuss the property of the proposed 1D landmark representation, which has less computational complexity and quantization error than the 2D heatmap. We also analyze the settings of the value of expansion factor at various input resolutions. In particular, for the input sizes of 64×64 and 128×128, the best choice for the value of the expansion factor is 1, while for the input sizes of 256×256 and 512×512, the value of the expansion factor affects the network performance indistinctively. In the future, we would like to adopt more geometric information such as the edges, parts, and structures of space targets to our method and extend our approach to other targets, such as aircraft and vehicles.

Author Contributions: Conceptualization, S.L. and G.W.; methodology, S.L.; software, S.L.; validation, S.L., X.Z., Z.C. and G.W.; formal analysis, S.L.; investigation, S.L.; resources, S.L.; data curation, S.L.; writing—original draft preparation, S.L.; writing—review and editing, X.Z. and Z.C.; visualization, S.L.; supervision, G.W.; project administration, G.W.; funding acquisition, G.W. All authors have read and agreed to the published version of the manuscript.

Funding: This research was funded by the National Natural Science Foundation of China (No. 62106283) and the Natural Science Foundation of Shaanxi Province (No. 2021JM-226).

Data Availability Statement: The SPEED dataset is available at https://kelvins.esa.int/satellite-pose-estimation-challenge/ (accessed on 12 July 2022).

Acknowledgments: We thank the Advanced Concepts Team (ACT) of the European Space Agency and the Space Rendezvous Laboratory (SLAB) of Stanford University for providing the Spacecraft PosE Estimation Dataset (SPEED) online.

Conflicts of Interest: The authors declare no conflict of interest.

References

1. Chen, L.; Li, S.; Bai, Q.; Yang, J.; Jiang, S.; Miao, Y. Review of Image Classification Algorithms Based on Convolutional Neural Networks. *Remote Sens.* **2021**, *13*, 4712. [CrossRef]
2. Zaidi, S.S.A.; Ansari, M.S.; Aslam, A.; Kanwal, N.; Asghar, M.N.; Lee, B. A survey of modern deep learning based object detection models. *Digit. Signal Process.* **2022**, *126*, 103514. [CrossRef]
3. Mo, Y.; Wu, Y.; Yang, X.; Liu, F.; Liao, Y. Review the state-of-the-art technologies of semantic segmentation based on deep learning. *Neurocomputing* **2022**, *493*, 626–646. [CrossRef]
4. Sharma, S.; Beierle, C.; D'Amico, S. Pose estimation for non-cooperative spacecraft rendezvous using convolutional neural networks. In Proceedings of the 2018 IEEE Aerospace Conference, Big Sky, MT, USA, 3–10 March 2018; pp. 1–12.
5. Proença, P.F.; Gao, Y. Deep Learning for Spacecraft Pose Estimation from Photorealistic Rendering. In Proceedings of the 2020 IEEE International Conference on Robotics and Automation (ICRA 2020), Paris, France, 31 May–31 August 2020; pp. 6007–6013.
6. Sharma, S.; D'Amico, S. Neural Network-Based Pose Estimation for Noncooperative Spacecraft Rendezvous. *IEEE Trans. Aerosp. Electron. Syst.* **2020**, *56*, 4638–4658. [CrossRef]
7. Park, T.H.; Sharma, S.; D'Amico, S. Towards Robust Learning-Based Pose Estimation of Noncooperative Spacecraft. *arXiv* **2019**, arXiv:1909.00392.
8. Chen, B.; Cao, J.; Bustos, Á.P.; Chin, T. Satellite Pose Estimation with Deep Landmark Regression and Nonlinear Pose Refinement. In Proceedings of the 2019 IEEE/CVF International Conference on Computer Vision Workshops (ICCV Workshops 2019), Seoul, Korea, 27–28 October 2019; pp. 2816–2824.
9. Xu, J.; Song, B.; Yang, X.; Nan, X. An Improved Deep Keypoint Detection Network for Space Targets Pose Estimation. *Remote Sens.* **2020**, *12*, 3857. [CrossRef]
10. Hu, Y.; Speierer, S.; Jakob, W.; Fua, P.; Salzmann, M. Wide-Depth-Range 6D Object Pose Estimation in Space. In Proceedings of the IEEE Conference on Computer Vision and Pattern Recognition (CVPR 2021), Virtual, 19–25 June 2021; pp. 15870–15879.
11. Noh, H.; Hong, S.; Han, B. Learning Deconvolution Network for Semantic Segmentation. In Proceedings of the 2015 IEEE International Conference on Computer Vision (ICCV 2015), Santiago, Chile, 7–13 December 2015; pp. 1520–1528.
12. Gao, X.; Hou, X.; Tang, J.; Cheng, H. Complete Solution Classification for the Perspective-Three-Point Problem. *IEEE Trans. Pattern Anal. Mach. Intell.* **2003**, *25*, 930–943. [CrossRef]
13. Kisantal, M.; Sharma, S.; Park, T.H.; Izzo, D.; Märtens, M.; D'Amico, S. Satellite Pose Estimation Challenge: Dataset, Competition Design, and Results. *IEEE Trans. Aerosp. Electron. Syst.* **2020**, *56*, 4083–4098. [CrossRef]
14. Krizhevsky, A.; Sutskever, I.; Hinton, G.E. ImageNet Classification with Deep Convolutional Neural Networks. In Proceedings of the 26th Annual Conference on Neural Information Processing Systems, Lake Tahoe, NV, USA, 3–6 December 2012; pp. 1106–1114.
15. He, K.; Zhang, X.; Ren, S.; Sun, J. Deep Residual Learning for Image Recognition. In Proceedings of the 2016 IEEE Conference on Computer Vision and Pattern Recognition (CVPR 2016), Las Vegas, NV, USA, 27–30 June 2016; pp. 770–778.
16. Li, S.; Xu, C. A Stable Direct Solution of Perspective-Three-Point Problem. *Int. J. Pattern Recognit. Artif. Intell.* **2011**, *25*, 627–642. [CrossRef]
17. Li, S.; Xu, C.; Xie, M. A Robust O(n) Solution to the Perspective-n-Point Problem. *IEEE Trans. Pattern Anal. Mach. Intell.* **2012**, *34*, 1444–1450. [CrossRef] [PubMed]
18. Lepetit, V.; Moreno-Noguer, F.; Fua, P. EPnP: An Accurate $O(n)$ Solution to the PnP Problem. *Int. J. Comput. Vis.* **2009**, *81*, 155–166. [CrossRef]
19. Redmon, J.; Farhadi, A. YOLO9000: Better, Faster, Stronger. In Proceedings of the 2017 IEEE Conference on Computer Vision and Pattern Recognition (CVPR 2017), Honolulu, HI, USA, 21–26 July 2017; pp. 6517–6525.
20. Redmon, J.; Farhadi, A. YOLOv3: An Incremental Improvement. *arXiv* **2018**, arXiv:1804.02767.
21. Howard, A.G.; Zhu, M.; Chen, B.; Kalenichenko, D.; Wang, W.; Weyand, T.; Andreetto, M.; Adam, H. MobileNets: Efficient Convolutional Neural Networks for Mobile Vision Applications. *arXiv* **2017**, arXiv:1704.04861.
22. Sandler, M.; Howard, A.G.; Zhu, M.; Zhmoginov, A.; Chen, L. MobileNetV2: Inverted Residuals and Linear Bottlenecks. In Proceedings of the 2018 IEEE Conference on Computer Vision and Pattern Recognition (CVPR 2018), Salt Lake City, UT, USA, 18–22 June 2018; pp. 4510–4520.

23. Lin, T.; Dollár, P.; Girshick, R.B.; He, K.; Hariharan, B.; Belongie, S.J. Feature Pyramid Networks for Object Detection. In Proceedings of the 2017 IEEE Conference on Computer Vision and Pattern Recognition (CVPR 2017), Honolulu, HI, USA, 21–26 July 2017; pp. 936–944.
24. Ren, S.; He, K.; Girshick, R.B.; Sun, J. Faster R-CNN: Towards Real-Time Object Detection with Region Proposal Networks. *IEEE Trans. Pattern Anal. Mach. Intell.* **2017**, *39*, 1137–1149. [CrossRef] [PubMed]
25. Sun, K.; Xiao, B.; Liu, D.; Wang, J. Deep High-Resolution Representation Learning for Human Pose Estimation. In Proceedings of the IEEE Conference on Computer Vision and Pattern Recognition (CVPR 2019), Long Beach, CA, USA, 16–20 June 2019; pp. 5693–5703.
26. Wang, Z.; Zhang, Z.; Sun, X.; Li, Z.; Yu, Q. Revisiting Monocular Satellite Pose Estimation with Transformer. *IEEE Trans. Aerosp. Electron. Syst.* **2022**. [CrossRef]
27. He, K.; Gkioxari, G.; Dollár, P.; Girshick, R.B. Mask R-CNN. In Proceedings of the IEEE International Conference on Computer Vision (ICCV 2017), Venice, Italy, 22–29 October 2017; pp. 2980–2988.
28. Cai, Z.; Vasconcelos, N. Cascade R-CNN: Delving Into High Quality Object Detection. In Proceedings of the 2018 IEEE Conference on Computer Vision and Pattern Recognition (CVPR 2018), Salt Lake City, UT, USA, 18–22 June 2018; pp. 6154–6162.
29. Pang, J.; Chen, K.; Shi, J.; Feng, H.; Ouyang, W.; Lin, D. Libra R-CNN: Towards Balanced Learning for Object Detection. In Proceedings of the IEEE Conference on Computer Vision and Pattern Recognition (CVPR 2019), Long Beach, CA, USA, 16–20 June 2019; pp. 821–830.
30. Wu, Y.; Chen, Y.; Yuan, L.; Liu, Z.; Wang, L.; Li, H.; Fu, Y. Rethinking Classification and Localization for Object Detection. In Proceedings of the 2020 IEEE/CVF Conference on Computer Vision and Pattern Recognition (CVPR 2020), Seattle, WA, USA, 13–19 June 2020; pp. 10183–10192.
31. Zhang, H.; Chang, H.; Ma, B.; Wang, N.; Chen, X. Dynamic R-CNN: Towards High Quality Object Detection via Dynamic Training. In Proceedings of the Computer Vision—ECCV 2020—16th European Conference, Glasgow, UK, 23–28 August 2020; pp. 260–275.
32. Sun, P.; Zhang, R.; Jiang, Y.; Kong, T.; Xu, C.; Zhan, W.; Tomizuka, M.; Li, L.; Yuan, Z.; Wang, C.; et al. Sparse R-CNN: End-to-End Object Detection With Learnable Proposals. In Proceedings of the IEEE Conference on Computer Vision and Pattern Recognition (CVPR 2021), Virtual, 19–25 June 2021; pp. 14454–14463.
33. Redmon, J.; Divvala, S.K.; Girshick, R.B.; Farhadi, A. You Only Look Once: Unified, Real-Time Object Detection. In Proceedings of the 2016 IEEE Conference on Computer Vision and Pattern Recognition (CVPR 2016), Las Vegas, NV, USA, 27–30 June 2016; pp. 779–788.
34. Bochkovskiy, A.; Wang, C.; Liao, H.M. YOLOv4: Optimal Speed and Accuracy of Object Detection. *arXiv* **2020**, arXiv:2004.10934.
35. Chen, Q.; Wang, Y.; Yang, T.; Zhang, X.; Cheng, J.; Sun, J. You Only Look One-Level Feature. In Proceedings of the IEEE Conference on Computer Vision and Pattern Recognition (CVPR 2021), Virtual, 19–25 June 2021; pp. 13039–13048.
36. Ge, Z.; Liu, S.; Wang, F.; Li, Z.; Sun, J. YOLOX: Exceeding YOLO Series in 2021. *arXiv* **2021**, arXiv:2107.08430.
37. Tian, Z.; Shen, C.; Chen, H.; He, T. FCOS: Fully Convolutional One-Stage Object Detection. In Proceedings of the 2019 IEEE/CVF International Conference on Computer Vision (ICCV 2019), Seoul, Korea, 27 October–2 November 2019; pp. 9626–9635.
38. Yang, Z.; Liu, S.; Hu, H.; Wang, L.; Lin, S. RepPoints: Point Set Representation for Object Detection. In Proceedings of the 2019 IEEE/CVF International Conference on Computer Vision (ICCV 2019), Seoul, Korea, 27 October–2 November 2019; pp. 9656–9665.
39. Carion, N.; Massa, F.; Synnaeve, G.; Usunier, N.; Kirillov, A.; Zagoruyko, S. End-to-End Object Detection with Transformers. In Proceedings of the Computer Vision—ECCV 2020—16th European Conference, Glasgow, UK, 23–28 August 2020; pp. 213–229.
40. Zhu, X.; Su, W.; Lu, L.; Li, B.; Wang, X.; Dai, J. Deformable DETR: Deformable Transformers for End-to-End Object Detection. In Proceedings of the 9th International Conference on Learning Representations (ICLR 2021), Virtual Event, 3–7 May 2021.
41. Dai, X.; Chen, Y.; Yang, J.; Zhang, P.; Yuan, L.; Zhang, L. Dynamic DETR: End-to-End Object Detection with Dynamic Attention. In Proceedings of the 2021 IEEE/CVF International Conference on Computer Vision (ICCV 2021), Montreal, QC, Canada, 10–17 October 2021; pp. 2968–2977.
42. Fischler, M.A.; Bolles, R.C. Random Sample Consensus: A Paradigm for Model Fitting with Applications to Image Analysis and Automated Cartography. *Commun. ACM* **1981**, *24*, 381–395. [CrossRef]
43. Wang, J.; Sun, K.; Cheng, T.; Jiang, B.; Deng, C.; Zhao, Y.; Liu, D.; Mu, Y.; Tan, M.; Wang, X.; et al. Deep High-Resolution Representation Learning for Visual Recognition. *IEEE Trans. Pattern Anal. Mach. Intell.* **2021**, *43*, 3349–3364. [CrossRef] [PubMed]
44. Lin, T.; Maire, M.; Belongie, S.J.; Hays, J.; Perona, P.; Ramanan, D.; Dollár, P.; Zitnick, C.L. Microsoft COCO: Common Objects in Context. In Proceedings of the Computer Vision—ECCV 2014—13th European Conference, Zurich, Switzerland, 6–12 September 2014; pp. 740–755.
45. Everingham, M.; Eslami, S.M.A.; Gool, L.V.; Williams, C.K.I.; Winn, J.M.; Zisserman, A. The Pascal Visual Object Classes Challenge: A Retrospective. *Int. J. Comput. Vis.* **2015**, *111*, 98–136. [CrossRef]
46. Kingma, D.P.; Ba, J. Adam: A Method for Stochastic Optimization. In Proceedings of the 3rd International Conference on Learning Representations (ICLR 2015), San Diego, CA, USA, 7–9 May 2015.

remote sensing

MDPI

Article

Generalizing Spacecraft Recognition via Diversifying Few-Shot Datasets in a Joint Trained Likelihood

Xi Yang [1], Dechen Kong [1], Ren Lin [1] and Dong Yang [2,*]

[1] State Key Laboratory of Integrated Services Networks, School of Telecommunications Engineering, Xidian University, Xi'an 710071, China; yangx@xidian.edu.cn (X.Y.); kong_dc@stu.xidian.edu.cn (D.K.); linriversluv@stu.xidian.edu.cn (R.L.)

[2] Xi'an Institute of Space Radio Technology, Xi'an 710100, China

* Correspondence: yangd504@126.com

Abstract: With the exploration of outer space, the number of space targets has increased dramatically, while the pressures of space situational awareness have also increased. Among them, spacecraft recognition is the foundation and a critical step in space situational awareness. However, unlike natural images that can be easily captured using low-cost devices, space targets can suffer from motion blurring, overexposure, and excessive dragging at the time of capture, which greatly affects the quality of the images and reduces the number of effective images. To this end, specialized or sufficiently versatile techniques are required, with dataset diversity playing a key role in enabling algorithms to categorize previously unseen spacecraft and perform multiple tasks. In this paper, we propose a joint dataset formulation to increase diversity. Our approach involves reformulating two local processes to condition the Conditional Neural Adaptive Processes, which results in global feature resampling schemes to adapt a pre-trained embedding function to be task-specific. Specifically, we employ variational resampling to category-wise auxiliary features, adding a generative constraint to amortize task-specific parameters. We also develop a neural process variational inference to encode representation, using grid density for conditioning. Our evaluation of the BUAA dataset shows promising results, with no-training performance close to a specifically designed learner and an accuracy rate of 98.2% on unseen categories during the joint training session. Further experiments on the Meta-dataset benchmark demonstrate at least a 4.6% out-of-distribution improvement compared to the baseline conditional models. Both dataset evaluations indicate the effectiveness of exploiting dataset diversity in few-shot feature adaptation. Our proposal offers a versatile solution for tasks across domains.

Keywords: spacecraft recognition; few-shot feature adaptation; generative family; neural processes

Citation: Yang, X.; Kong, D.; Lin, R.; Yang, D. Generalizing Spacecraft Recognition via Diversifying Few-Shot Datasets in a Joint Trained Likelihood. *Remote Sens.* **2023**, *15*, 4321. https://doi.org/10.3390/rs15174321

Academic Editors: Jukka Heikkonen, Fahimeh Farahnakian and Pouya Jafarzadeh

Received: 1 July 2023
Revised: 28 August 2023
Accepted: 30 August 2023
Published: 1 September 2023

1. Introduction

The growing tension over resource constraints has directly accelerated the need to explore space beyond Earth. However, the large number of space targets with different shapes and forms increases the difficulty of space situational awareness, and misjudgment of space targets will directly affect the space order and delay the popularization of space knowledge, which requires the recognition of spacecraft. Still, space target images suffer from distortion and blurring due to target attitude instability, sensor performance and image channel transmission [1], which results in a limited number of available space target images.

Traditional spacecraft recognition methods extract features from images by manual design. The construction of these methods requires the knowledge and experience of domain experts to ensure that appropriate features and algorithms are selected. For example, SIFT methods focus on detecting key points at different scales and rotation angles, while HOG methods focus on the edge and texture information of the image. However, traditional methods are greatly limited by the problems of complex scene modeling and limited data samples for space targets. A realistic scenario like this presents challenges in few-shot formulation. Deep learning methods [2], especially Convolutional Neural Networks (CNNs), have achieved significant advantages in image classification tasks for space targets by automatically learning abstract features through multilevel neural networks. Among them, DCNN [3] achieves space target recognition through an end-to-end approach and copes with the few-shot problem by means of data augmentation and data simulation. Discriminative Deep Nearest Neighbor Neural Network (D2N4) [4] overcomes the significant intra-class differences of space targets by introducing center loss. On the other hand, global pooling information is introduced for each depth-local descriptor to reduce the interference of local background noise and thus enhance the robustness of the model.

However, as can be seen from the rise of the cross-domain few-shot learning field, poor performance outside the target task domain (which we call out-of-distribution data) is a crucial constraint on few-sample tasks. Furthermore, the design of D2N4 on discriminating spacecraft is an approximate overfitting problem, which contradicts further generalization to the unseen category. Each of these data domains is distributed differently, maintains large domain gaps between each of them, and shows significant deviations from the target data domain (which we call in-distribution data). Solutions dedicated to one area tend to fail the corner or generally less common observations. Thus, they better have long-term support [5]. The alternative can choose to be multi-functional and sufficiently capable of current concerns, meaning a successful algorithm should address its majors well and easily generalize to the rare rest [6]. Human exploration of the planet still suffices as a good example. Compared with natural images, space target images have a single background and are greatly affected by illumination; in addition, space target images have the problem of sizable intra-class gaps and small inter-class gaps. The introduction of multi-domain natural images covers the target domain's data features by increasing the data's diversity. Therefore, it is more reasonable to take the different domains of natural images as the main task of adapting the features of space target images while placing the space target images precisely in the "rest of the domain".

Similar to the paradigm that learns the major features and evaluates the rest, the few-shot learning model uses a handful of examples to categorize previously unseen observations into known labels. A simple few-shot learner achieves matching of the extracted features to the distribution of the dataset via fine-tuning a small classifier [7–9] or calibrating target distribution [10]. Another popular alternative, meta-learning approaches [11], takes "learning to learn from diverse few-shot examples and evaluate the unseen one, even the unseen domain" to generalize to wild datasets. Practically, the meta-learner considers optimally measured feature distances and, therefore, can be characterized by constructing a universal representation with great computation [12,13], or elegantly adapting models from a good initialization [14,15]. However, the rare context examples still matter if moving towards unseen categories of spacecraft images.

To cover the corner, the researcher presents grayscale spacecraft images, the BUAA dataset [16], to simulate the contexts for recognition. With Figure 1, when inspecting the dataset content, it is at that early deep era when much analysis [4,17] measuring fine-grained properties and intra-class variance from scratch score well in those offline archives. However, less diversity in such learning procedures potentially under-fit future generalizations [18]. Fortunately, these years of milestones in AI research make publicly accessible assets handy, which could help allow any meta-learner to converge in the range of the large-scale dataset instead. Even in the few-shot setting, much of the meta-learner now utilizes a large labeled dataset, episodically simulating few-shot constraints [19]. For instance, Triantafillou et al. [20] present a large-scale Meta-dataset (ten labeled datasets in composition, with eight for training, e.g., ILSVRC-2012 [21] and two for the testing, e.g., MSCOCO [22]) that makes few-shot classifications in the cross-dataset setting and catches up on real-world events. Therefore, the recognition of spacecraft dataset raises the problem of whether to use a single space target image for training or to use a large-scale dataset to aid in training. When the protocol is defined with large-scale Meta-dataset, it samples random tasks into episodes, and the solutions must consider (1) being adapted to an indicated domain by using a few task-specific examples, and (2) exploring shared knowledge between each task. Using shared structures to adapt models with task-specific formulation is a critical factorization for such an algorithm. Furthermore, limited overhead in the whole process would be preferred.

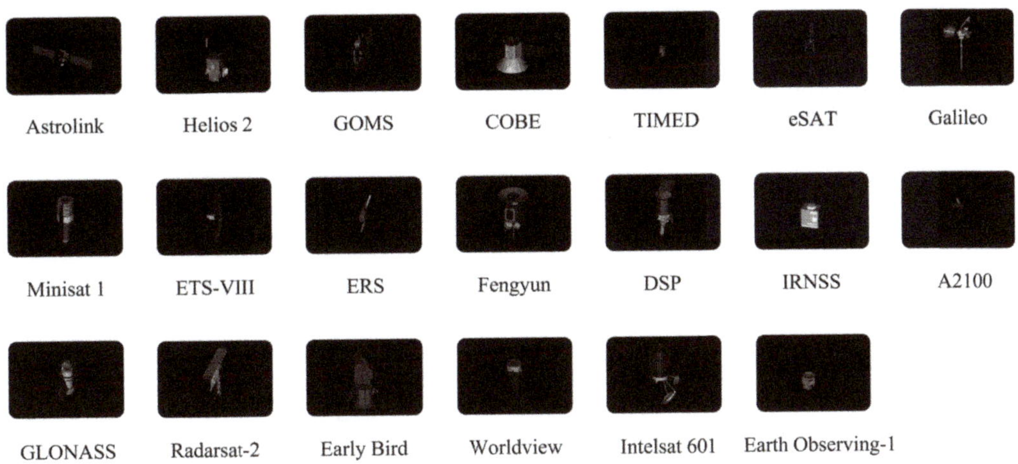

Figure 1. Inspection of the BUAA dataset. The data generation uses 3D triangulated models with dark backgrounds, simulating aircraft in deep space. Images of each category are uniformly rendered from 230 viewpoints on a default view port in 3Ds MAX software.

Members in the Neural Processes family (NPF) [23] meta-learn a mapping directly from observations to a distribution over functions, known as the stochastic process, exploiting prior assumptions to quickly infer a new task-specific predictor at the test time. Conditional Neural Adaptive Processes (CNAPs) [24], the conditional model [25] for adaptable few-shot classification, introduce an amortization of FiLM layers [26] in distribution modeling, offering fast and scaleable realizations from a pre-trained template to predict unseen multi-task datasets. Observations from [27] also suggest that this is one of the cases where training images from diverse domains benefit the distribution approximation.

The pipeline in Figure 2 highlights an adaptation of a conditional embedding function when solving each few-shot classification. CNAPs model amortizes the computational cost of the model by learning a functional approximator in the meta-training phase, which generates most of the parameters in Resnet-18 [28] by evaluating the sample. Further, with no explicit adaptation on the classifier (contrasting with CNAPs), Simple CNAPs conclude multi-task likelihood estimation in formulating each non-parametric Mahalanobis distance measurement. By its mathematical definition, two participants in acquiring the distance and proper conditional embedding function are responsible for such a design. Figure 2 also shows similarities to another specific design for cross-domain few-shot learning: Task-Specific Adapters (TSA). As the latest method, TSA attaches its task-specific adapters to a single universal network distilled from a cross-domain dataset and learns those adapters on a few labeled image examples. This means the task-specific adapters can be plugged into a pre-trained feature extractor to adapt it to multiple datasets, similar to the FiLM layers of the Conditional Neural Adaptive Processes.

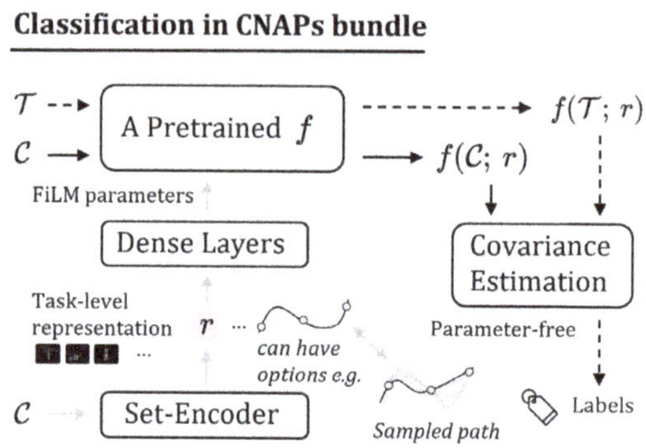

Figure 2. The "Simple" variant of Conditional Neural Adaptive Processes in few-shot classifications [29]. The model also follows the rules of a meta-learning algorithm that learns from labeled context-set ($\mathcal{C}$) images and predicts unlabeled target-set images ($\mathcal{T}$).

Back to the highlighted feature adaptation from Figure 2, a solid fact is that generating task-specific embedding functions always constitutes their task-specific parameters [30], also known as the amortization parameter, according to a task-level feature representation. The process connects to the capacity of scaling to complex functions for given datasets. Since the deterministic representation cannot match the diversity in data domains, a direct likelihood estimation for the model would fail. It also could be interpreted as an underfitting phenomenon (or amortization gap [6]) that the task-level aggregate mathematically captures a distribution over task-specific parameters but potentially under-fits small-scale examples [31]. Simply observing sufficient data would, in turn, violate the few-shot setting.

In this work, we specify a meta-objective to diversify the conditional feature adaptation, generalizing the large in-distribution datasets [20] to perform spacecraft recognition, particularly in the few-shot setting. The idea is to resample the task-level representation explicitly (see comments in Figure 2). To achieve this, we adopt variational learning to parameterize generative density, from which we can directly sample the reformulated features. We further assume the neural process variational inference to approximate a distribution over tensor values that encoder the context features, allowing us to sample a collection of embedding schemes for each latent feature. Our resampling formulation has two implementations: (1) A conditional variational auto-encoder pipeline that provides controllable constraints in directly estimating the generative density of task-level representative; and (2) A latent Neural Process [23,32] that reformulates the meta-learned embedding function to encode task-level representation. Overall, we improve the robustness of the model by reformulating the local progressions to obtain more representative class prototypes that are resistant to data bias in the few-shot setting. Furthermore, the adaptation scheme is applied on a single backbone that is only pre-trained on ImageNet dataset [21], yet it achieves comparable performance against methods in universal representation [12,13]. The evaluation on the Meta-dataset also shows that the extended version of the latest few-shot classification algorithm has the highest average rank, particularly with a large margin on out-of-distribution tasks. A summary of our contributions to solving few-shot classification are:

(1) We investigate a generative re-sampling scheme for representation learning. The sampled representation is used to condition an amortized strategy and universal backbone in adapting the embedding extractor function to multiple datasets;
(2) From a self-supervised perspective, we propose a data encoding function based on neural process variational inference;
(3) We present a comparable out-of-distribution performance against methods with a specific design on BUAA dataset and with universal representation on the Meta-dataset.

The remainder of the paper is organized as follows: Section 2 briefly reviews the related background to our approach. Our approach is studied and detailed in Section 3. The introduction to the dataset, model ablation studies and experimental results are analyzed and presented in Section 4. We conclude our work in Section 5.

2. Background

Our preliminary insight into solving a visual classification problem is related to meta-learning formulations in classification and Neural Processes Family and Generative distribution modeling.

2.1. Meta-Learning Approaches in a Few-Shot Setting

The meta-learning [33] technique applies the "learning to learn" concept to reduce the data required for the model. During the meta-training phase, task samples are extracted from large-scale labeled datasets based on the task distribution $P(\mathcal{D})$. Tasks are randomly divided into few-shot context and target sets. That is, per iteration, the classification dataset $\mathcal{D}$, also named a task sample, has x images to be classified as known labels y, and can further subdivide into two subsets: $\mathcal{C}$ that stores contextual supervision and $\mathcal{T}$ that requires generalization from those few contexts.

Over all accessible tasks, a meta-algorithm updates its parametric modeling to minimize a general object function:

$$\mathcal{L} := \sum_{\mathcal{C},\mathcal{T}} l\,(\text{one-hot} \cdot d(f(x_{\mathcal{T}}); \mathcal{C}),\, y_{\mathcal{T}}), \tag{1}$$

in which l is the cross-entropy loss. The distance function d measures how close a categorical distribution (think about what softmax function outputs) is to the true layout $y_{\mathcal{T}}$. In contrast, the embedding function f ideally defines a feature space where the pre-defined metric d is assumed to be optimal for drawing boundaries. Then, there are two main streams on the table to conclude one meta-learner: one is based on distance (also known as metric learning), and the other shares the core of adaptation. Compared to earlier explorations [34,35] on a generic metric space, classic examples for the latter can be found in optimization-based approaches to few-shot classifications [17]. The algorithms follow particular rules for learning to adapt from a few observations and addressing unseen bunches in a specific task. However, what does "adapt for a specific classification task" mean for an algorithm? Is there a unified formulation, and is it optimal?

If we view model adaptation as solving an optimization with a set of parameters Φ that updates a set of weights Θ of a classifier $f(x;\Theta)$ to a set of adapted weights Θ' in predicting $\mathcal{C}$, then optimal parameters are defined to have

$$\Theta' = \underset{\Theta}{\text{argmax}}\, \mathbb{E}_{\mathcal{C}\sim P(\mathcal{C})}\left[\prod_{c=1}^{|\mathcal{C}|} p(y = y^{(c)} |\, f(x^{(c)}; \Theta)\,; \Phi\right]. \tag{2}$$

The classifier under Θ' is responsible for maximizing the expected density estimation in predicting $\mathcal{T}$:

$$\mathop{\mathbb{E}}_{\mathcal{T}\sim P(\mathcal{T})} [\prod_{t=1}^{|\mathcal{T}|} p(y = y^{(t)} |\, f(x^{(t)}; \Theta')]. \tag{3}$$

From this formulation, performing just full gradient descent learning (see MAML, [14]) can be one of the adaptation rules. Still, in most cases, they tend to overfit few-shot data with expensive computation of the second derivatives. Cheaper implementations are to ignore the second derivative (fo-MAML) [36], and to incorporate inductive bias from the prototypes into an initialization scheme (Proto-MAML) [20]. A more efficient way is to adopt amortized inference [6,24,29] on the contextual information and to enable sharing global parameters for a learned distribution over embedding functions. Such an inference allows us to rapidly instantiate a function $f \sim p(f)$ to participate in a specific task, and further reduces the cost of adaptation. After iteratively refining their performance across various instances of few-shot tasks drawn from the distribution $P(\mathcal{D})$ [19], meta-learners that acquire the ability to adapt ensure, in theory, that their current parameter set θ' becomes optimal for the specific task $\mathcal{D}$ [33]:

$$\theta' = \underset{\theta}{\operatorname{argmin}} \, \mathbb{E}_{\mathcal{D} \sim P(\mathcal{D})} \left[\mathcal{L}(\mathcal{D}; \theta) \right]. \tag{4}$$

It shows a natural capacity to adapt multi-task scenarios, despite a task formulation underexposed or unseen settings. The paradigm hence builds our solution.

2.2. *Neural Processes Family*

Models in the Neural Process Family meta-learn a distribution over random functions and can be distinguished between two assumptions.

2.2.1. The Conditional Neural Process Family

Members in this family employ a factorization assumption [25]: a conditional model first explores the entire context set $\mathcal{C}$ for a global representation r, using a parameterized (sub-) encoder ϕ and aggregator ρ to compute representation:

$$r = \rho\Big(\sum_{c=1}^{|\mathcal{C}|} \phi(x^{(c)}, y^{(c)})\Big). \tag{5}$$

The two modules define an encoder architecture in the family members. Then, the predictive distribution at any set of target inputs $x_{\mathcal{T}}$ is factorized and conditioned on the global representation r:

$$p_\theta(y_{\mathcal{T}} | x_{\mathcal{T}}; \mathcal{C}) = \prod_{t=1}^{|\mathcal{T}|} p_\theta(y^{(t)} | x^{(t)}, r). \tag{6}$$

The factorization assumption in the conditional models allows for directly maximizing the log-likelihood $\log p_\theta(y_{\mathcal{T}} | x_{\mathcal{T}}; \mathcal{C})$ on the target set to train the parameters. Furthermore, this log-likelihood formulation builds our overall framework.

2.2.2. Latent Neural Process Family

The latent models [23] instead introduce a stochastic latent variable into the parameterization of predictive distribution, formulated as:

$$p_\theta(y_{\mathcal{T}} | x_{\mathcal{T}}; \mathcal{C}) = \int \prod_{t=1}^{|\mathcal{T}|} p_\theta\left(y^{(t)} | x^{(t)}, z\right) p_\theta(z|r) dz. \tag{7}$$

This is conditioned on a sampled z, a latent representation from posterior distribution $p(z; \mathcal{C}, \mathcal{T})$, and an analytically intractable posterior, which lies in approximation $p_\theta(z|r)$.

In practice, Garnelo et al. [23] propose to map all informative samples in $\mathcal{D}$ to the distribution over z as a sampling distribution in approximating the true posterior $p(z; \mathcal{C}, \mathcal{T}) \approx p_\theta(z|\mathcal{D})$. A latent model thus has different choices of encoder architecture, for example, to first have a deterministic representation after having observed both the context-set and target-set, and then use it to parameterize a distribution over z. Therefore, the encoder, i.e., an inference network as in the methods [37] for performing approximate inference and learning probabilistic global latents, and the decoder, which is the same as the conditional models except for using sampled latent representation $z \sim p_\theta(z|\mathcal{D})$, are jointly trained to compute an approximation of likelihood objective $p_\theta(y_{\mathcal{T}} | x_{\mathcal{T}}; \mathcal{C})$ at the target inputs $x_{\mathcal{T}}$ in an amortized variational inference:

$$\log p_\theta(y_\mathcal{T}|x_\mathcal{T};\mathcal{C}) \geq \int p_\theta(z|\mathcal{D}) \cdot [\log \prod_{t=1}^{|\mathcal{T}|} p_\theta(y^{(t)}|\, x^{(t)}, z) + \log p_\theta(z\,|\,\mathcal{C}) - \log p_\theta(z\,|\,\mathcal{D})\,] = \mathbb{E}_z\Big[\log \prod_{t=1}^{|\mathcal{T}|} p_\theta(y^{(t)}|\, x^{(t)}, z)\Big] - KL(\,p_\theta(z\,|\,\mathcal{D})\,\|\,p_\theta(z\,|\,\mathcal{C})\,), \tag{8}$$

by first placing $p_\theta(z|\mathcal{D})$ in an identity trick and using Jensen's inequality to derive a lower bound targeting the intractable integral problem [37], with an expectation and a Kullback–Leibler divergence.

Members of the Neural Process Family highlight the capacity to interpolate the context information to produce predictions on unseen in-distribution data, but cannot deal with a distribution shift [27] from simulated to real-world data at test time. Current feature weighting on universal representation [12,13] or task-specific parameters combined on dataset generalization [15] span a large-scale feature space and promote out-of-distribution adaptation.

2.3. The Generative Family in Learning Representation

Machine learning models refer to different probabilistic frameworks [38]. If involved in vision tasks, discriminative models learn a probability distribution $p(y|x)$ that predicts the probability of true y when given an image x. The evaluation then determines whether categorical distribution $p(y|x)$ would match the ground truth. In contrast, a member in the generative family formulates density function $p(x)$ over all possible inputs x. Conditional generative modules take further steps to learn $p(x|y)$, conditioning on y in every pair. The variational generative method performs the density estimation by maximizing a logarithm lower bound to achieve the true parametric density$p(x;\vartheta^2)$. That is, by first assuming $x{\sim}p(x|z;\vartheta^2)$ holds for all possible input, and the introduced $p(x|z)$ is subject to a latent $z := \mu + \sigma \cdot \epsilon$ commonly sampled from a Gaussian prior $p(z)$. Then, in a joint optimization, a parametric posterior $q(z|x;\vartheta^1)$ approximates to capture the prior $p(z)$ (in place of the true posterior $p(x|z)$).

Being subject to the below normalization constraint of probabilistic finite integrals, for the generative model, all possible inputs x compete benignly for the probability mass of their generation.

$$\int_x p(x)dx = 1. \tag{9}$$

Generative models, thereby, can represent seen data and explore more on its latent pattern. Likewise, in a conditional generative model, every companion y induces a separate competition among all x, but properties derived from the normalization still hold in the conditional generative formulation.

3. Methods

Classification algorithms can be broken down into two parts: representation learning (Section 3.1) and classifier building (Section 3.2). We use the Simple CNAPs model as an example method, and later we will explore its extension. Straightforwardly, the method is based on learning from a public large-scale dataset and maximizing the log-likelihood presented in Equation (15). First, learning the representations of both labeled/unlabeled images and building a classifier on those representations solve the regular classification problem. An overall learning object sees Section 3.3.

3.1. Reformulated Representation Learning

A versatile model not only handles spacecraft recognition well but also the rest. Learning to adapt the future data satisfies the purpose. Here, we first introduce the extension of the Conditional Neural Process Family into adaptable multi-task classification.

Likewise to approaches including MAML [14] and its variants that explore a well-behaved initialization in modeling, the predictive distribution $p_\theta(y = y^{(t)}|f(x^{(t)}); \mathcal{C})$ in Conditional Neural Adaptive Processes (CNAPs) [24] has a specific form in parametric modeling:

$$f(;\mathcal{C}) = f(;_{\vartheta}, \phi), \phi = g(r) \text{ and } r = \rho(\sum w(\mathcal{C})). \tag{10}$$

Behind the symbol, a fixed-cost adaptation mechanism (as the set of parameters ϕ in part of θ) shares a dataset encoder–decoder structure and amortizes $\mathcal{C}$ and its feature encodings into parameters γ and β, a channel-wise function scales and biases input feature x via $\gamma(\mathcal{C}) \odot x \oplus \beta(\mathcal{C})$ to condition pre-trained f on context information. For few-shot classifications, it retains every property seen among the family members to infer unseen patterns and, more importantly, protect against over-fitting by requiring a single forward pass rather than multiple gradient back props with that optimization-based approaches. However, the point here is that if going back to the pipeline in Figure 2, task representation r that indicates the current dataset plays a leading role in the conditional neural process models. Allowing the diversity of the data domain to broaden the selectivity of the target task can satisfy the variability of the random function f distribution realization. For this reason, following formulations, reconsider the process of specifying an encoder–decoder structure to map handful examples $\mathcal{C}$ into FiLM parameters [26] that adapt $f(x;_{\vartheta})$ to have a set of weights $[\vartheta, \phi]$, and practically extend the former statistical deterministic aggregation into: (1) $r \sim p(r;\mathcal{C})$, but the adapted version of ϑ can still be $[\vartheta, \phi = g(r)]$; or (2) $w \sim p(w;\mathcal{C})$ and $r = \rho \cdot (\sum \phi(\mathcal{C}))$, such that $[\vartheta, \phi]$ would have the result of $g(r)$ as well. Then, to sum up, this is a representation learning problem, and Conditional Neural Processes make such representations the condition.

Following the global encoder–decoder structure defined in the Conditional Neural Processes collection [25], the function approximator g would relate to the so-called decoder that generates the desired function set. This results in $f' := f(;\vartheta, \phi = g(r))$, which is adapted for each task.

3.1.1. Formulation (1), Directly Resampling Task-Representation from Generative Density

Consider w as a deterministic function. Our first formulation adds a global sampling on the aggregate r. With this in mind, the procedure in Figure 3 indirectly forces a generative constraint to estimate the density $p(r)$ with a local variational inference. Here, we amortize a portion of ϕ (as a reminder, $f(;\vartheta, \phi)$) using a conditional variational auto-encoder [39], where an encoder–decoder pair is jointly trained to approximate the conditional density function $p(r|r^c)$, using estimated $r^c := 1/|\mathcal{C}^c| \cdot \sum_{x \in \mathcal{C}^c} w(x)$ and, likewise, $r := \rho \cdot \sum w(x_{\mathcal{C}})$. That is, q_ϑ is the encoder that takes duplicated and concatenated input to reparameterize a Gaussian distribution over the introduced latent z with its associated means μ_z and variance Σ_z; and $p_{\vartheta 2}$ is the decoder to resample a r' from the mean of each conditional distribution $p(r|r^c, z)$.

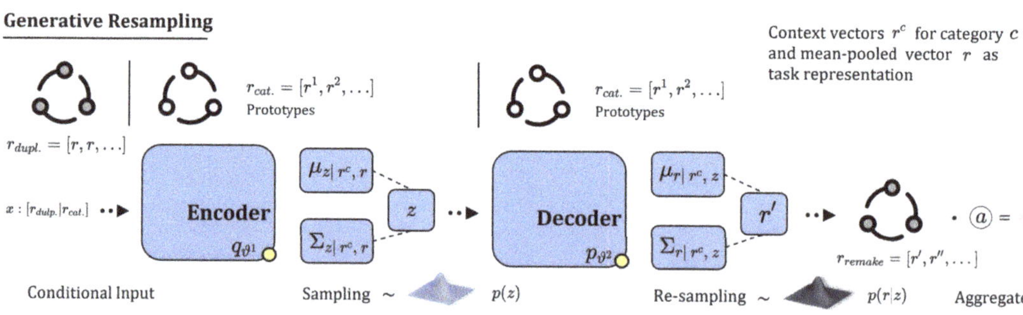

Figure 3. Overview of Formulation (1): the resampling task representative of a latent distribution posterior reparamertized by a conditional variational inference.

In a latent variable perspective, we always sample from the generative density $p(r|\,r^c)$ in the highest probability, making the variables r easy to populate. This satisfies our representative purpose. In later optimization, we follow the variational inference to draw a log-likelihood lower bound to capture conditional density $p(r|r^c)$. It assumes variable z to have a relaxed Gaussian prior $p(z) = \mathcal{N}(0, \mathcal{I})$.

3.1.2. Formulation (2), a Resample Embedding Function from Grid Density

The second formulation considers a distribution over functions and places $w \sim p(w)$ into aggregating $r := \rho \cdot (\sum w(x_C, y_C))$. Considering, in 2D image regression, that a colored image corresponds to a mapping from an actual 2D grid location x_i to its RGB pixel intensity $y_i \in \mathbb{R}^3$, each latent feature map $\nu(x^{(c)}) \in \mathbb{R}^{d \times h \times e}$ can be equivalently interpreted as an instance function from a stochastic process. Here, we apply the same manipulation as in Section 3.1.1 (but instead denoted as ν) to first include the latent version of $\mathcal{C}$.

Then, we can specify a function instance w on a fixed $h \times e$ grid [23,40]. Through reconstruction on each feature map, the target representation r can be bound to all the realizations $p(w)$. For illustration, we refer $m := \nu(x), x \in \mathcal{C}$ to latent feature maps and $n \in \mathbb{R}^{h \times e}$, the absolute position of the entries that constitutes m, to each 2D array. Further, we randomly select 2D indices to gather a (sub-)context-set $\mathcal{M}_c$ and target-set $\mathcal{M}$ from all input pairs (n, m), and approximating $p(w)$ through predicting $\mathcal{M}$ by given $\mathcal{M}_c$. $\mathcal{M}_c \subseteq \mathcal{M}$ is of note. The following parameterization is left for the Attentive Neural Processes (ANPs) [32], a latent collection of the Neural Process Family, with multi-head cross-attention [41–43] to predict the target-set feature maps.

The introduced model of ANPs has two branches featuring deterministic and latent properties, respectively. From Figure 4, a well-developed attention mechanism reformulates local encodings (i.e., the values) into representation r_d in the deterministic path. It allows a given target location $n^{(t)}$ (query) to attend the location of the relevant context (i.e., the keys) in $\mathcal{M}_c$ and to encode the dot-product relations. The latent path instead samples a variational latent z that captures distribution properties for subsequent prediction on grid values. We denote the overall parameterization as ϑ. These fixed dimension representations model a global structure of stochastic process realization, whereas r_d in the deterministic path models is a fine-grained structure. Finally, a decoder takes representations $[z; r_d]$ from the two paths generating grid density $p_\vartheta(m|\,n, z;\, r_d)$ to estimate the final $\mathcal{M}$. Aggregation of $r := \rho(\sum m \cdot p_\vartheta(m|\,n, z;\, r_d))$ makes the representative vector of $\mathbb{R}^d$.

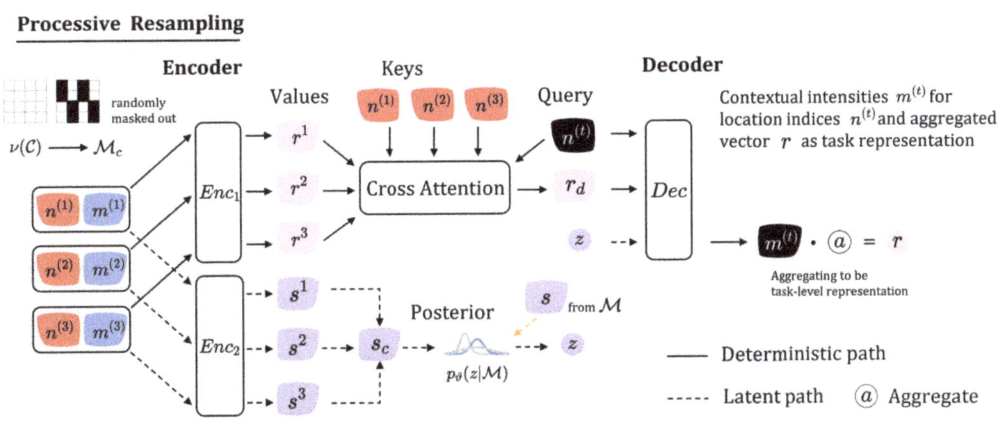

Figure 4. Overview of Formulation (2): resampling a latent embedding function in grid density. The pipeline is adapted from [32], except that we manipulate grid samples.

Theoretically, when $p_\theta(m^{(t)}|\, n^{(t)}, z; r_d)$ specifies a Gaussian density characterized by respective $[\mu^{(t)}, \sigma^{2(t)}]$ to present each grid value, the factorization (see Section 2.2) goes into an infinite mixture of Gaussians:

$$p_\theta(y_\mathcal{T}|x_\mathcal{T};\mathcal{C}) = \int p_\theta(z|\,\mathcal{C}) \prod_{t=1}^{|\mathcal{T}|} \mathcal{N}(y^{(t)}; \mu^{(t)}, \sigma^{2(t)}) dz, \tag{11}$$

meaning that the predictive distribution $p(w)$ conceptually allows us to scale the complex likelihood function to diversify the global features. Our implementation additionally considers an experimental manipulation to reduce test-time complexity. That is, aggregating r solely on prototypes $\bar{m}$ and their reconstruction, while interpolation on 2D grid features is left for regularization. For each class k, the prototype is simply:

$$\bar{m}^k = 1/|\mathcal{C}^k| \cdot \sum_{c=1}^{|\mathcal{C}^k|} \nu(x^{(c)}). \tag{12}$$

3.2. Building Estimated Classifier

To classify the adapted features of unlabeled targets, a non-parametric version of CNAPs [29] introduces a convex combination $\lambda^k \cdot \Sigma^k + \left(1-\lambda^k\right) \cdot \Sigma$ into estimating covariance matrix Q^k in the squared Mahalanobis distance for label k:

$$d_k\left(f(x), \bar{x}^k\right) = \left(f(x) - \bar{x}^k\right)^T \left(Q^k\right)^{-1} \left(f(x) - \bar{x}^k\right). \tag{13}$$

Mathematically, each covariance estimation in the combination is dealt with using a sample covariance matrix; an unbiased and efficient estimator of the covariance matrix in this case. We first derive $\bar{x}^k := 1/|\mathcal{C}^k| \cdot \sum_{x\in\mathcal{C}^k} f(x)$ for class k and task-level prototype $\bar{x} := 1/|\mathcal{C}| \cdot \sum_{x\in\mathcal{C}} f(x)$. Then, by definition, each sample covariance relies on the difference between each observation and the sample mean, i.e., for a class-related covariance matrix:

$$\Sigma^k = \frac{1}{\left|\mathcal{C}^k\right| - 1} \sum_{x\in\mathcal{C}^k} \left(f(x) - \bar{x}^k\right) \left(f(x) - \bar{x}^k\right)^T. \tag{14}$$

Then, use observations $x \in \mathcal{C}$ that cover whole context set for task-related covariance matrix Σ. Combining two matrices and considering all individual distributions in the task, the full covariance estimation presents a hierarchical regularization scheme [44]. Finally, the conditional predictive model finalizes its factorization assumption with an adapted embedding function f in a probabilistic mixture model [45], directly estimating the below likelihood function:

$$p(y = k\,|\, f(x); \mathcal{C}) = \frac{exp\left(-d_k(\, f(x),\ \bar{x}^k\right)\,)}{\sum_{k'} exp\left(-d_{k'}(f(x),\ \bar{x}^{k'})\right)}. \tag{15}$$

And, maximizing the correct likelihood helps to classify image x in $\mathcal{T}$. Notice that Equation (15) is deterministic and exclusively dependent on the distribution over embedding function f, which is the key to our formulation below. Furthermore, the parameter-free structure with the squared Mahalanobis distance explores the theoretical properties of Bregman divergences well. The family of distance functions suggests a minimal distance in a softmax classifier from the sample as prototypes of all assigned data points.

There remains one downside to note: the above formulations and their classifier structure see deficiencies in generalizing out-of-distribution regimes, since true $P(\mathcal{D})$ is still hard to cover. One of the strongest recommendations is to let $(\mathcal{C}, \mathcal{T}) \sim P(\mathcal{D})$ be a diverse dataset preparation for maximum-likelihood training. There then lies the top priority to digest diverse data, e.g., all training sources from BUAA [16] and the Meta-dataset [20] in joint training sessions.

3.3. Training Objects

Let $p_\theta(y_\mathcal{T}|\, x_\mathcal{T}; \mathcal{C})$ correspond to use of the parametric model θ to obtain a joint categorical distribution. In general, the target is to sample f for each task $(\mathcal{C}, \mathcal{T})$ and participate in classifying target images $\mathcal{T}$ into known labels from $\mathcal{C}$. Now that the classifier is deterministic, Equation (15) can be directly maximized through episodic training with the associated classification dataset. Furthermore, we simultaneously summarize such episodic training into solving below local and global optimization.

In the local part, our first formulation instantiates the variational lower bound with an approximate *KL divergence* $\mathcal{D}_{KL}$, as well as an expectation that forms a log-likelihood lower bound $\mathcal{L}_1(\vartheta; \mathcal{C})$ to have $\log p(r|\, r^c)$ maximized:

$$\mathcal{L}_1(\vartheta; \mathcal{C}) = \mathcal{D}_{KL} + \mathbb{E}_{z\sim q_{\vartheta^1}(z|\, r^c, r)}[\log p_{\vartheta^2}(r|\, r^c, z)] \qquad (16)$$
$$\text{s.t.} \quad \mathcal{D}_{KL} = -KL(q_{\vartheta^1}(z|r^c, r)||\, p(z)).$$

And, by definition, an encoder–decoder pair (parameterized by ϑ) is used for approximating the true posterior $q(z|\, r^c, r)$ and conditional density $p(r|\, r^c, z)$. Alternatively, our second inference principle is neural process variational inference [23]. The target objective evaluates learning the task-level representation r by applying Equation (8) to the Evidence Lower Bound (ELBO), specifically for embeddings of contexts $\mathcal{C}$. During training, we infer latent variable z on a posterior sampling from $p_\vartheta(z|\, \mathcal{M})$, maximizing the lower-bound $\mathcal{L}_{VI}(\vartheta; \mathcal{M})$ to predict latent feature maps:

$$\mathcal{L}_{VI}(\vartheta; \mathcal{M}) = \mathbb{E}_{z\sim p_\vartheta(z\,|\,\mathcal{M})}\left[\, \log \prod_{t=1}^{|\mathcal{M}|} p_\vartheta(m^{(t)}\,|\, n^{(t)}, z)\,\right] \qquad (17)$$
$$- KL(\, p_\vartheta(z\,|\,\mathcal{M})\, \|\, p_\vartheta(z\,|\,\mathcal{M}_c)\,),$$

where $\mathcal{M}$ and $\mathcal{M}_c$ denote context-set and target-set. Not to confuse image–label pairs, here we use n to refer to 2D grid coordinates and m to refer to feature maps. To the best of our knowledge, we consider Equation (16), which takes advantage of whole $\mathcal{C}$ in a feasible objective, to meta-learn representations for the downstream task of adapting f. Essentially, we expect the model to reconstruct targets while being regularized by a fine-grained exploration within their structures.

Globally, in the maximum likelihood part, we have log-likelihood function specified to each classification dataset $(\mathcal{C}, \mathcal{T})$ instead. Furthermore, eventually, we evaluate the expected log-likelihood $\hat{\mathcal{L}}(\theta; \mathcal{C}, \mathcal{T})$ on all those accessible tasks, approximately over distribution $P(\mathcal{C}, \mathcal{T})$:

$$\hat{\mathcal{L}}(\theta; \mathcal{C}, \mathcal{T}) = \mathbb{E}_{(\mathcal{C},\mathcal{T})\sim P(\mathcal{C},\mathcal{T})}[\log p_\theta(y_\mathcal{T}|x_\mathcal{T}; \mathcal{C})]. \qquad (18)$$

With reference to the provided pseudocode (see Algorithm 1), we repeat sampling $(\mathcal{C}, \mathcal{T})$ with image–label pairs and maximize Equation (18) in three steps: (1) learning from the given observation set $\mathcal{C}$, (2) evaluating targets $\mathcal{T}$, and (3) applying a gradient step until training converges.

3.4. Architecture with Formulation

In this paper, we first summarize the overall architecture into a local encoder–decoder pair and global encoder–decoder structure of Simple CNAPs, such that the encoder–decoder definition is consistent with the framework defined in the Neural Processes Family [23].

Specific choices of the local auxiliary architecture highly connect to the training objects. Behind the symbol ϕ, the first introduced is a conditional auto-encoder structure. Basically, the encoder part involves 4 convolutional layers, each followed by their own batch normalization and ReLU function that locally encodes category-wise sample means r^c and the raw

r into a predictive mean and a diagonal variance, implementing the reparameterization trick (with two individual linear layers) under a latent variable assumption [39] of a multivariate Gaussian distribution over each latent variable z, while the decoder part organizes subsequent 3 transposed convolutional blocks, with each LeakyReLU and a final Sigmoid activation on the top, to encourage our global resampling from the generative distribution $p(r|\, z, r^c)$. Likewise, the second parameterization indicated with a latent neural process structure [23] involves 3 convolutional layers and their ReLU activations in its deterministic encoder Enc_1, which locally encodes each concatenation $[n^{(c)}, m^{(c)}]$ into $r^{(c)}$. A two-head cross-attention layer with linear embedding functions follows behind. The latent encoder Enc_2 comprises the same amount of convolutional blocks, but is followed by 2 linear layers to reparameterize a latent posterior as in a VAE model [37]. Instead, the conditional decoder Dec takes 3 transposed convolutional layers and ReLUs together with linear layers (familiarly, a predictive mean and a diagonal variance function) to predict multivariate Gaussians, which each instance function w can be resampled from. A mean aggregator follows to take all realizations $w \sim p(w)$ into task-level representation r. Note that all the hidden representations will be of 128 dimensions.

Algorithm 1: Example Maximum Likelihood Training for Simple CNAPs

Given a distribution over meta-training tasks $P(\mathcal{D})$;
Given a pre-trained template $f(;\ \varphi)$;
Freeze φ and initialize ϑ, w, g randomly;
while *not converged* **do**
  Uniformly sample tasks $\mathcal{D} = (\mathcal{C} \cup \mathcal{T}) \sim P(\mathcal{D})$;
  if *formulation 1 stay true* **then**
    Evaluate lower bound $\mathcal{L}_1$ (as Equation (16));
    Let $\phi = g(r)$;
  else if *formulation 2 stay true* **then**
    Evaluate lower bound $\mathcal{L}_2$ (as Equation (16)) $\setminus$ by using $\mathcal{M}$ and *subset* $\mathcal{M}_c$ according to $\mathcal{C}$;
    Let $r = \rho(\sum w(x_C, y_C))$ and $\phi = g(r)$;
  Let $f' \leftarrow f(;\ \varphi, \phi)$;
  foreach k **in** *unique label set of* C **do**
    Estimate Q^k by sample covariance matrix;
  Evaluate a joint categorical distribution on $\mathcal{T}$ $\setminus$ by using Equation (15);
  Update $\theta = [\vartheta, w, g]$ to maximize Equation (18);

Regarding the global structure, the encoder only requires us to produce task-level representation r and can be one of the two formulations above; the decoder is simply the ResNet-18 and a few amortization steps to generate plug-in FiLM layers [26]. In those steps, stacks of linear blocks map the aggregate representation r into those parameters of the channel-wise transformation, specifying the final adapted version f'.

We apply the formulation to a new model: Task-Specific Adapters (TSAs) [46]. From Figure 5, TSAs are used for cross-domain few-shot classification that aims to learn a classifier from previously unseen classes and domains with few labeled samples. TSAs are commonly used with Universal Representation Learning (URL) [47], a single universal network learned and distilled from the labeled context set C. Briefly speaking, a TSA adapts the feature extractor created by URL using task-specific adapters. In this paper, we consider methods (Formulations (1) and (2)) in Section 3.1 as the pre-processed formulation and post-processed formulation, respectively.

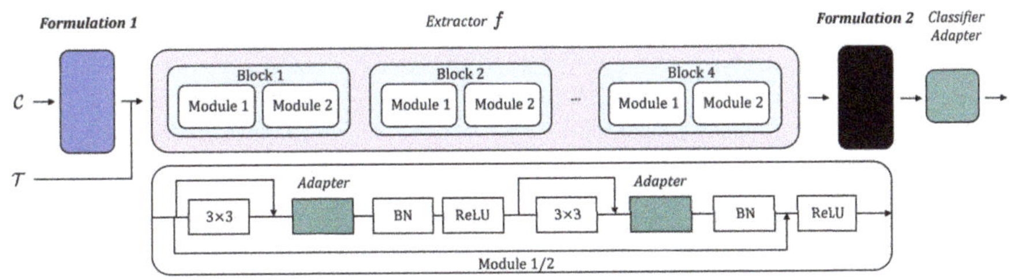

Figure 5. Overview of reformulated Task-specific Adapters (TSAs). The TSA model adapts extractor distilled from the Universal Representation Learning (URL) technique. Here, we reformulate its preprocessing part.

4. Experiments

We first detail the Experimental setups in Sections 4.1 and 4.2. Following comparisons in the BUAA dataset (see Section 4.3.1), we choose a simple distance-based learner D2N4 [4] that features adding image global pooling information into each feature descriptor, and our baseline Simple CNAPs [29], and the optimization-based learner, as the selected models.

4.1. Dataset Format

We evaluated our approach on the joint dataset, which is composed of the BUAA and the Meta-Dataset [20]. The Meta-Dataset is a large few-shot learning benchmark and consists of multiple datasets of different data distributions that feature 10 existing classification problems to help with anxiety. From Figure 6, the benchmark collects labeled data on diverse domains, varying from natural images with 1000 categories in ImageNet (ILSVRC-2012), FGVC-Aircraft (aircraft), QuickDraw (hand-drawn sketches), VGG Flower (flower images), FGVCx Fungi (mushroom), Omniglot (hand-written characters), CUB-200-2011 (birds), Describable Textures (texture), Traffic Signs, and MSCOCO (nature images). For any algorithm, samples from the first eight and the name of in-distribution tasks should not have overlapped during training, validation, and final testing. The unseen out-of-distribution split, instead, holds the combination of Traffic Signs, MSCOCO and held-out MNIST, CIFAR10, and CIFAR100. The algorithm should take them only for testing. BUAA, a space target dataset, has collected 20 classes of satellite models of different types, shapes, and functions. It is based on a space target 3-D model, using 3Ds MAX software to generate a space target full viewpoint simulation image. The data set consists of 4600 gray images from 230 viewpoints sampled on a viewing sphere, so each class has 230 examples. We adopt average accuracy and rank as the evaluation metric in our experiments. Considering the model's rank makes it possible to obtain a more complete picture of its generalization performance, ensuring that the model remains stable and performs well under different data distributions and characteristics.

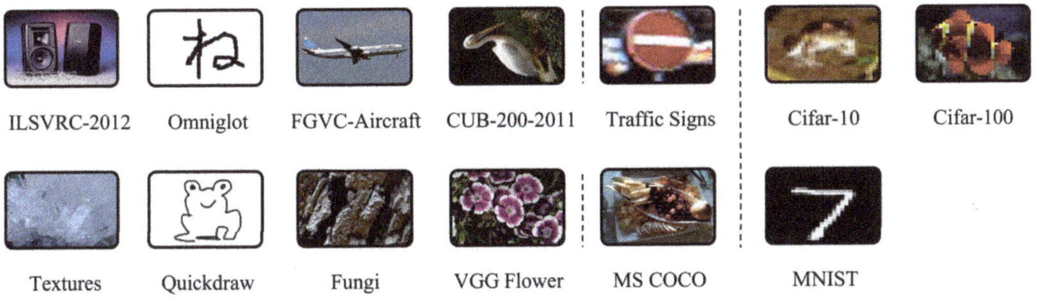

Figure 6. The datasets of the dataset, including the Meta-dataset's 10 datasets and the left three datasets Mnist and Cifar10/100 for additional tests.

4.2. Implementation Details

4.2.1. Dataset Setting

To investigate the applicability of our scheme, we partitioned the BUAA dataset by placing 25% to 70% of the data categories (with a 5% gradient increase) inside the distribution and placing the remaining data categories outside the distribution. In our experiments, we only trained on the in-distribution data and used the out-of-distribution data for testing. In below sections, performance at each percentage accuracy will be reported by averaging over 600 proxy classification tasks with a 95% confidence interval. The benchmark setting does not restrict few-shot tasks to have fixed ways and shots, thus representing a more realistic scenario.

4.2.2. Training

In Section 3.1.2, we highlighted global sampling in latent neural process models, where the encoder, as the inference network, plays a dual role [23] of being an approximate posterior $p_\theta(z|\mathcal{M})$, and also of defining the prior, having observed $\mathcal{M}_c$, suggesting different behaviors in between stages: during training, we sample function instances from the approximate posterior when we have whole $\mathcal{M}$ in observation; during inference, instead, the sampling can only turn to the prior $p_\theta(z|\mathcal{M}_c)$ given a subset of entries each feature map. We train an overall θ with 110,000 sampled tasks, using a task batch of 16 on all training splits from the Meta-dataset [20]. Our maximum likelihood training remains identical to [29] as we use episodic context/target splits $\mathcal{C}, \mathcal{T}$ and set the size of all x within to be 84×84. A step learning rate (from 1×10^{-3}) scheduler is set to configure the Adam optimizer. The whole procedure is loaded on a single NVIDIA RTX 3090 GPU.

4.3. Results and Extendable Discussion

We first report selected modelings on the BUAA dataset. However, before we discuss the benchmark part, we would like to demonstrate the potential of our proposal formulations in Section 3.1. As Section 2.2 explains, the task-level representation r plays an important role in the Conditional Neural Processes model. The proposal Formulations (1) and (2) are designed for this representation. However, we conclude that these formulations are generally applicable to the neural feature learning problem.

Notice the "test-only" setting means Table 2 evaluating spacecraft images as an out-of-distribution dataset. It can set a vital criterion to examine the model's capacity to generalize unseen domains with public in-distribution examples. Then, we present a leaderboard on the Meta-Dataset to read how generalizable the models are in close-to-realistic applications.

4.3.1. Benchmarking Spacecraft Dataset

The validation configures two different settings, and we would suggest a "single-domain training session" and "multi-domain joint training session".

The setting of the first part adopts BUAA dataset as the only training data. Table 1 is more likely to represent the basic capacity of the baseline models when we exclude large-scale dataset and train models with our target domain data only. This can also help to ablate our later max likelihood training. As shown, an average rank via aggregating across each column of Table 1 both suggests baseline TSA and Simple CNAPs outperform the simple distance learner D2N4; the modification on TSA encourages its capacity when having nearly at least half percent available training data. A more intuitive demonstration can refer to Figure 7a, where all the conditional models maintain an upward performance if expanding the training splits, but a similar conclusion does not hold for D2N4.

Table 1. Results reported on unseen in-distribution tasks of BUAA dataset using models trained on a single domain.

	Avg. Rank	25%.	30%.	35%.	40%.	45%.	50%.	55%.	60%.	65%.	70% Categories
D2N4 [4]	5.7	88.6 ± 0.7	89.9 ± 0.6	92.1 ± 0.5	90.8 ± 0.6	91.1 ± 0.6	92.1 ± 0.6	93.8 ± 0.5	94.5 ± 0.5	94.9 ± 0.4	93.9 ± 0.4
Simple CNAPs [29]	**2.0**	92.9 ± 0.6	93.8 ± 0.5	**94.4 ± 0.6**	94.2 ± 0.6	93.9 ± 0.4	94.7 ± 0.6	95.0 ± 0.5	95.2 ± 0.6	96.3 ± 0.5	97.0 ± 0.4
TSA [46]	3.0	**93.6 ± 0.5**	**93.9 ± 0.5**	**94.4 ± 0.4**	94.1 ± 0.5	94.3 ± 0.5	94.4 ± 0.4	94.7 ± 0.5	94.9 ± 0.3	96.1 ± 0.4	96.4 ± 0.3
Sim. CNAPs∼fo. 1	5.9	85.8 ± 0.7	88.7 ± 0.6	91.6 ± 0.5	91.5 ± 0.6	91.9 ± 0.7	90.2 ± 0.7	92.2 ± 0.6	94.7 ± 0.5	94.0 ± 0.5	97.0 ± 0.3
Sim. CNAPs∼fo. 2	5.1	88.0 ± 0.6	92.0 ± 0.5	91.1 ± 0.6	93.5 ± 0.6	91.6 ± 0.7	93.2 ± 0.5	93.8 ± 0.5	94.5 ± 0.6	94.5 ± 0.4	96.3 ± 0.3
TSA∼fo. 1	4.1	88.1 ± 0.4	89.6 ± 0.5	91.7 ± 0.6	92.4 ± 0.4	93.5 ± 0.4	94.6 ± 0.5	94.9 ± 0.4	95.0 ± 0.5	94.5 ± 0.5	96.1 ± 0.4
TSA∼fo. 2	2.1	90.4 ± 0.4	91.7 ± 0.6	93.0 ± 0.5	**94.6 ± 0.4**	**95.4 ± 0.6**	**95.2 ± 0.5**	**95.7 ± 0.3**	**95.6 ± 0.4**	**96.8 ± 0.5**	**97.5 ± 0.4**

The setting of the second part adopts all accessible datasets as the training data instead. Furthermore, what Table 2 conveys is that both of our formulations turn more competitive in a joint maximum likelihood modeling with all training splits of the Meta-Dataset included; particularly, the second formulation prominently leads the result. From a direct comparison between Figure 7a,b, a wide range of cross-domain training also benefits the distance-based learner, while some cases see declines for Simple CNAPs. An interesting result can also be found in the test-only case (Table 2) and a proportion of 70% categories used in training (Table 1). The result of our formulation using only BUAA for testing in Table 2 is even comparable to the result of D2N4 in Table 1 using the 70% of samples used in training. Modifications on TSA convey such improvement, too, making them the highest average rank among the listed methods. However, such an advantage becomes less when Simple CNAPs∼fo. 2 makes a well-matched competition with at least 60% of available training categories.

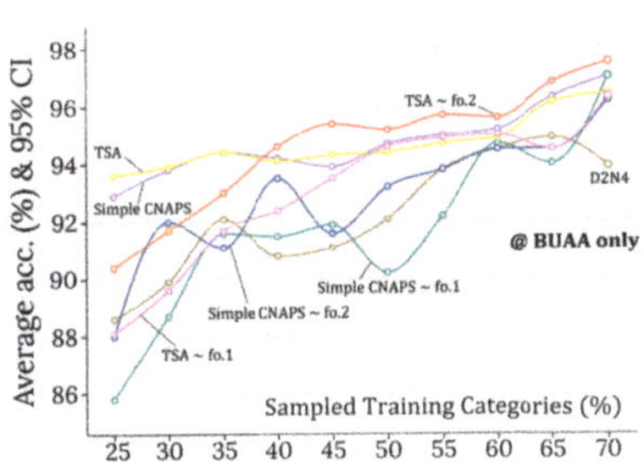

(**a**) Single-domain training result

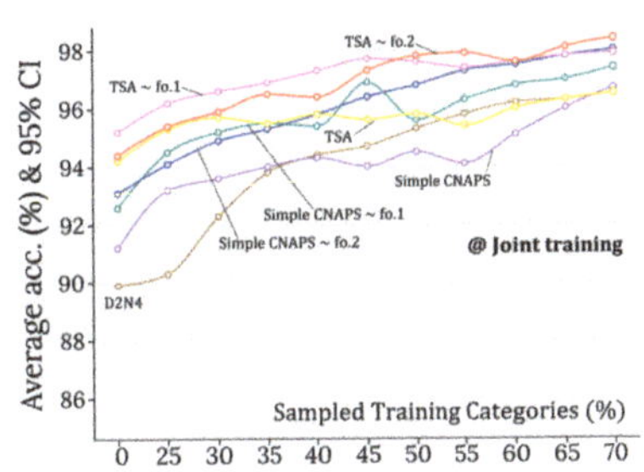

(**b**) Multi-domain training result

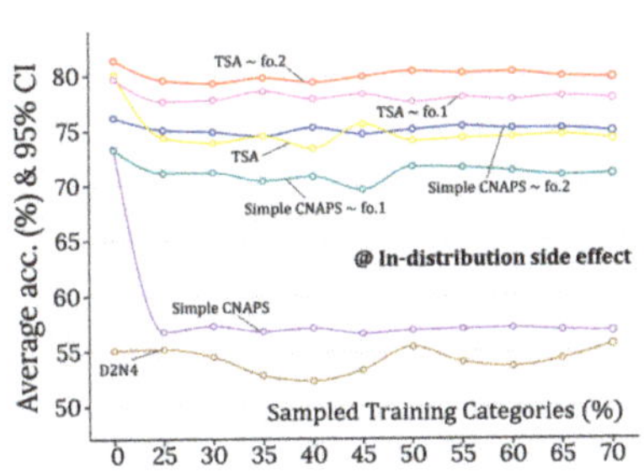

(**c**) Unseen in-distribution result

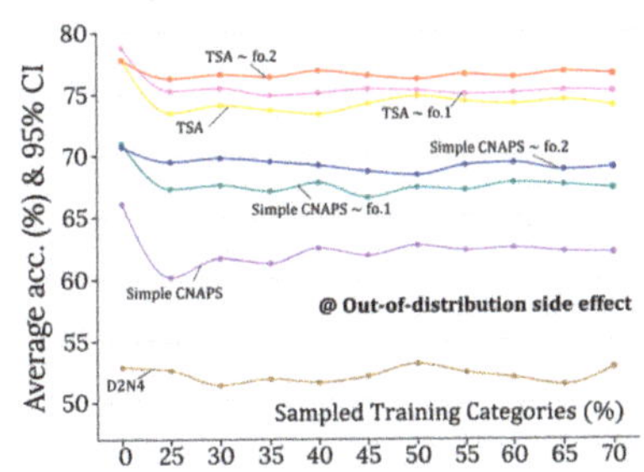

(**d**) Out-of-distribution performance

Figure 7. Comparisons from (**a**,**b**) evaluate models trained on BUAA dataset only, and in similar settings, but instead with joint maximum likelihood training. (**c**,**d**) show performance on Meta-Dataset in-distribution/out-of-distribution testing using models trained with (none-zero x axis) or without (zero x axis) feeding spacecraft images.

Table 2. Results reported on joint In-distribution tasks of BUAA dataset using models trained on multiple domains.

	Avg. Rank	Test Only	25%.	30%.	35%.	40%.	45%.	50%.	55%.	60%.	65%.	70% Categories
D2N4 [4]	6.1	89.9 ± 0.6	90.3 ± 0.6	92.3 ± 0.5	93.8 ± 0.5	94.4 ± 0.5	94.7 ± 0.5	95.3 ± 0.4	95.8 ± 0.5	96.2 ± 0.4	96.3 ± 0.4	96.5 ± 0.4
Simple CNAPs [29]	6.4	91.2 ± 0.7	93.2 ± 0.6	93.6 ± 0.6	94.0 ± 0.6	94.3 ± 0.6	94.0 ± 0.5	94.5 ± 0.6	94.1 ± 0.5	95.1 ± 0.6	96.0 ± 0.5	96.7 ± 0.5
TSA [46]	4.1	94.2 ± 1.0	95.3 ± 0.7	95.7 ± 0.7	95.5 ± 0.6	95.8 ± 0.6	95.6 ± 0.7	95.8 ± 0.6	95.4 ± 0.7	96.0 ± 0.5	96.3 ± 0.6	96.5 ± 0.6
Sim. CNAPs∼fo. 1	4.3	92.6 ± 0.5	94.5 ± 0.4	95.2 ± 0.4	95.5 ± 0.5	95.4 ± 0.4	96.9 ± 0.4	95.6 ± 0.4	96.3 ± 0.4	96.8 ± 0.3	97.0 ± 0.3	97.4 ± 0.3
Sim. CNAPs∼fo. 2	3.3	93.1 ± 0.6	94.1 ± 0.5	94.9 ± 0.5	95.3 ± 0.4	95.8 ± 0.4	96.4 ± 0.3	96.8 ± 0.4	97.3 ± 0.4	97.5 ± 0.3	97.8 ± 0.4	98.2 ± 0.3
TSA∼fo. 1	**1.7**	**95.2 ± 0.8**	**96.2 ± 0.6**	**96.6 ± 0.7**	**96.9 ± 0.6**	**97.3 ± 0.5**	**97.7 ± 0.5**	97.6 ± 0.6	97.4 ± 0.6	**97.6 ± 0.5**	97.8 ± 0.4	97.9 ± 0.4
TSA∼fo. 2	2.0	94.4 ± 1.0	95.4 ± 0.5	95.9 ± 0.4	96.5 ± 0.5	96.4 ± 0.5	97.3 ± 0.3	**97.8 ± 0.4**	**97.9 ± 0.4**	**97.6 ± 0.5**	**98.1 ± 0.4**	**98.4 ± 0.3**

4.3.2. Benchmarking Meta-Dataset

The validation in this subsection goes into two different settings, and we would suggest a "leaderboard version of unseen tasks performance" and "joint-training version of unseen tasks performance".

The first part sees models trained at all available datasets of the Meta-Dataset and tested on it. Tables 3 and 4 display the in/out-of-distribution statuses for benchmark models. This setting, however, excludes the BUAA dataset and makes spacecraft images test-only (the source of that column in Table 2). In Table 3, we also compare methods with the unseen parts of training sources. The universal representation approach, i.e., SUR [12] and URT [13], achieves classification by training a respective embedding function for each intra-distributed dataset and then linearly combining each embedding function to form a specific embedding function based on the task query set. However, a considerable domain gap against training sources in out-of-distribution settings explains their need to be more generalizable from the eight extractors. Instead, we are motivated to approximate a distribution over dataset-specified embedding functions for Simple CNAPs and to sample the proper one for each test-time task, which is more efficient [24]. Modifications of TSAs convey such an idea too. Further evidence of a comparable average rank among listed models shows our meta-learned formulations promote feature adaptation over the same embedding function as in Simple CNAPs. Another promising result would be the highest average rank for TSA∼fo. 1 in Table 4, where we generalize all models to the out-of-distribution splits of Meta-Dataset. Generative formulation 1 extends the embedding extractor function of Simple CNAPs and TSAs from encoding static training datasets only to having generative density, such that resampling schemes can efficiently encourage adapting out-of-distribution tasks.

Table 3. Results reported on in-distribution tasks using models trained on all training datasets.

	Avg. Rank	ILSVRC	Omniglot	Aircraft	Birds	Textures	QuickDraw	Fungi	Flower
D2N4 [4]	15.3	26.1 ± 0.8	82.8 ± 0.9	72.8 ± 0.9	34.6 ± 1.0	52.7 ± 0.7	66.6 ± 0.9	32.3 ± 0.9	72.8 ± 0.8
fo-MAML [20]	14.6	37.8 ± 1.0	83.9 ± 0.9	76.4 ± 0.7	62.4 ± 1.1	64.2 ± 0.8	59.7 ± 1.1	33.5 ± 1.1	80.0 ± 0.8
ProtoNet [34]	14.3	44.5 ± 1.0	79.6 ± 1.1	71.1 ± 0.9	67.0 ± 1.0	65.2 ± 0.8	64.9 ± 0.9	40.3 ± 1.1	86.8 ± 0.7
Proto-MAML [20]	12.6	46.5 ± 1.0	82.7 ± 1.0	75.2 ± 0.8	69.9 ± 1.0	68.2 ± 0.8	66.8 ± 0.9	42.0 ± 1.1	88.7 ± 0.7
CNAPs [24]	11.1	51.0 ± 1.0	90.7 ± 0.6	72.3 ± 0.8	73.0 ± 0.8	54.8 ± 0.7	74.2 ± 0.6	50.2 ± 1.0	88.5 ± 0.6
Simple CNAPs [29]	9.0	56.5 ± 1.0	91.7 ± 0.6	82.4 ± 0.7	74.9 ± 0.9	67.8 ± 0.7	77.5 ± 0.8	46.9 ± 1.0	89.7 ± 0.6
SUR [12]	8.3	56.1 ± 1.1	93.1 ± 0.5	84.6 ± 0.7	70.6 ± 1.0	71.0 ± 0.8	81.3 ± 0.6	64.2 ± 1.1	82.8 ± 0.8
FLUTE [15]	7.5	51.8 ± 1.0	93.2 ± 0.5	87.2 ± 0.5	79.2 ± 0.8	68.8 ± 0.8	79.5 ± 0.7	58.1 ± 1.1	91.6 ± 0.6
Transductive CNAPs [48]	6.9	**57.9 ± 1.1**	94.3 ± 0.4	84.7 ± 0.5	78.8 ± 0.7	66.2 ± 0.8	77.9 ± 0.6	48.9 ± 1.2	92.3 ± 0.4
URT [13]	6.7	55.7 ± 1.0	94.4 ± 0.4	85.8 ± 0.6	76.3 ± 0.8	71.8 ± 0.7	82.5 ± 0.6	63.5 ± 1.0	88.2 ± 0.6
URL [47]	3.1	57.5 ± 1.1	94.5 ± 0.4	88.6 ± 0.5	80.5 ± 0.7	76.2 ± 0.7	81.8 ± 0.6	68.7 ± 1.0	92.1 ± 0.5
TSA [46]	2.8	57.3 ± 1.0	95.0 ± 0.4	**89.3 ± 0.4**	81.4 ± 0.7	76.7 ± 0.7	82.0 ± 0.6	67.4 ± 1.0	92.2 ± 0.5
Simple CNAPs∼fo. 1	9.5	52.5 ± 1.1	88.2 ± 0.8	74.5 ± 0.8	73.2 ± 0.9	74.0 ± 0.8	80.5 ± 0.7	53.4 ± 1.1	90.2 ± 0.6
Simple CNAPs∼fo. 2	7.8	55.1 ± 1.1	92.2 ± 0.6	81.4 ± 0.6	78.1 ± 0.8	72.9 ± 0.9	80.4 ± 0.7	59.4 ± 1.0	89.7 ± 0.6
TSA∼fo. 1	4.0	54.1 ± 0.8	93.8 ± 0.6	85.8 ± 1.0	78.4 ± 1.0	**80.1 ± 0.8**	84.2 ± 0.8	68.7 ± 0.7	**93.1 ± 0.8**
TSA∼fo. 2	**2.3**	56.6 ± 0.6	**96.4 ± 0.8**	88.4 ± 0.5	**83.1 ± 0.8**	78.6 ± 1.0	**84.4 ± 0.8**	**69.5 ± 1.0**	92.8 ± 0.7

Table 4. Results reported on out-of-distribution tasks using models trained on all training datasets.

	Avg. Rank	Traffic Signs	MSCOCO	Mnist	Cifar10	Cifar100
fo-MAML [20]	15.8	42.9 ± 1.3	29.4 ± 1.1	-	-	-
ProtoNet [34]	14.3	46.5 ± 1.0	39.9 ± 1.0	-	-	-
D2N4 [4]	11.5	60.7 ± 1.1	28.2 ± 0.9	92.9 ± 0.5	44.0 ± 0.7	39.0 ± 1.0
Proto-MAML [20]	11.2	52.4 ± 1.1	41.7 ± 1.1	-	-	-
CNAPs [24]	10.8	56.5 ± 1.1	39.4 ± 1.0	92.7 ± 0.4	61.5 ± 0.7	50.1 ± 1.0
SUR [12]	10.0	53.4 ± 1.0	50.1 ± 1.0	94.3 ± 0.4	66.8 ± 0.9	56.6 ± 1.0
URT [13]	9.6	51.1 ± 1.1	52.2 ± 1.1	94.8 ± 0.4	67.3 ± 0.8	56.9 ± 1.0
Simple CNAPs [29]	9.2	59.2 ± 1.0	42.4 ± 1.1	93.9 ± 0.4	74.3 ± 0.7	60.5 ± 1.0
Transductive CNAPs [48]	7.1	59.7 ± 1.1	42.5 ± 1.1	95.7 ± 0.3	75.7 ± 0.7	62.9 ± 1.0
FLUTE [15]	6.6	58.4 ± 1.1	50.0 ± 1.0	95.6 ± 0.5	78.6 ± 0.7	67.1 ± 1.0
URL [47]	6.7	63.3 ± 1.2	54.0 ± 1.0	94.7 ± 0.4	74.2 ± 0.8	63.5 ± 1.0
TSA [46]	2.3	83.5 ± 0.9	55.7 ± 1.1	**96.7 ± 0.4**	**82.9 ± 0.7**	70.4 ± 0.9
Simple CNAPs∼fo. 1	6.4	69.5 ± 0.8	52.6 ± 0.7	93.6 ± 0.4	70.5 ± 0.8	69.0 ± 1.0
Simple CNAPs∼fo. 2	6.5	67.4 ± 1.0	55.3 ± 0.7	92.5 ± 0.5	68.4 ± 0.8	69.8 ± 0.9
TSA∼fo. 1	**1.8**	**85.4 ± 0.8**	**58.7 ± 1.0**	95.1 ± 0.6	81.5 ± 0.6	**73.3 ± 0.8**
TSA∼fo. 2	2.4	84.1 ± 0.6	56.6 ± 0.8	96.4 ± 0.6	80.3 ± 0.8	71.7 ± 0.7

The second part is attached to the "multi-domain joint training session" of Section 4.3.1, where the setting instead evaluates the incorporation of spacecraft images into Meta-Dataset performance. Figure 7c,d represent in- and out-of-distribution splits of the Meta-Dataset, respectively. Similar to spacecraft images, cases of Simple CNAPs on the auxiliary Meta-Dataset suggest a sharp drop, indicating a deterministic representation in the conditional model is less likely to adapt more domains than the stochastic one. Fewer side effects can be found with all our modifications, suggesting a potential capability to continue to learn data from a new domain (at least for spacecraft images).

4.4. Ablation Study

By analyzing Tables 1 and 2, it can be seen that the recognition accuracy of the target domain can be improved dramatically when using the joint data as the auxiliary data of the target domain. The effect is significant, even if the target domain is not used in the training phase.

Since space target images have large intra-class gaps and small inter-class gaps and are close to fine-grained classification settings, we use a classifier based on Mahalanobis distance to improve classification performance by analyzing the overall and local variance. Table 5 shows that the Mahalanobis distance outperforms the Euclidean distance when only the space target image is used for testing. However, when the amount of space target data is increased during training, the performance of both is almost equal. This is because as the number of training samples in the target domain increases, the learnable feature extractor can achieve the optimization of the intra-class distribution, which reduces the effect of the Mahalanobis distance.

Table 5. Results of different metrics in the classifier on BUAA.

	Simple CNAPs [29]	Simple CNAPs∼fo. 1	Simple CNAPs∼fo. 2	TSA [46]	TSA∼fo. 1	TSA∼fo. 2
Mahalanobis distance (test only)	**91.2 ± 0.7**	**92.6 ± 0.5**	**93.1 ± 0.6**	**94.2 ± 1.0**	**95.2 ± 0.5**	94.4 ± 1.0
European distance (test only)	90.7 ± 0.8	91.2 ± 0.4	92.4 ± 0.6	93.4 ± 0.8	94.5 ± 0.3	**94.6 ± 0.9**
Mahalanobis distance (70% for training)	**96.7 ± 0.5**	97.4 ± 0.3	**98.2 ± 0.3**	96.5 ± 0.6	**97.9 ± 0.4**	**98.4 ± 0.3**
European distance (70% for training)	96.6 ± 0.5	**97.6 ± 0.5**	97.9 ± 0.8	**96.8 ± 0.3**	97.7 ± 0.7	98.2 ± 0.6

5. Conclusions and Limitations

This paper uses conditional variational inference and latent neural processes [32] to learn diversified representations over multiple datasets while generalizing spacecraft recognition to a generalization problem to improve space target recognition performance with the aid of multi-domain datasets. On optimizing Equation (4), we propose to condition such meta-learned task-level representations on feature adaptation; as a result, the promoted adaptation of Simple CNAPs and TSAs improves their performance for benchmark classification for the spacecraft images both in-distribution and out-of-distribution. The algorithm also shows competitive results on larger benchmark settings for multitasking or versatility purposes. Similar conclusions still hold when we extend the feature learning to a more general stage, where the proposed formulation preprocesses and post-processes. The algorithm also shows competitive results on a larger benchmark setting for multi-task purposes. Our approach starts from representation and solves the few-shot problem by generalizing the prototypical representation of the target data, so it has a strong generalization ability for the domains where data acquisition is expensive, such as medical engineering and ocean observations [49]. However, BUAA dataset lacks light intensity variations as well as stellar interference compared to real space target images, and further validation of the model will be performed when the real data is complete. Our experiments also suggest a catastrophic interference, as all the modifications end up forgetting the Meta-dataset training after including spacecraft images with the joint training. A possible future work approach is extending our rough applications of neural processes to support future learning.

Author Contributions: Methodology, X.Y. and D.Y.; Software, D.K. and R.L.; Formal analysis, D.K.; Data curation, D.K.; Writing—original draft, X.Y. and R.L.; Writing—review & editing, D.Y.; Supervision, D.Y. All authors have read and agreed to the published version of the manuscript.

Funding: National Natural Science Foundation of China: 62372348, 61976166; Shaanxi Outstanding Youth Science Fund Project: 2023-JC-JQ-53; Fundamental Research Funds for the Central Universities: QTZX23057; Open Research Projects of Laboratory of Pinghu.

Conflicts of Interest: The authors declare no conflict of interest.

References

1. Zhao, Z.; Xu, G.; Zhang, N.; Zhang, Q. Performance analysis of the hybrid satellite-terrestrial relay network with opportunistic scheduling over generalized fading channels. *IEEE Trans. Veh. Technol.* **2022**, *71*, 2914–2924. [CrossRef]
2. Heidari, A.; Jafari Navimipour, N.; Unal, M.; Zhang, G. Machine learning applications in internet-of-drones: Systematic review, recent deployments, and open issues. *ACM Comput. Surv.* **2023**, *55*, 1–45. [CrossRef]
3. Zeng, H.; Xia, Y. Space target recognition based on deep learning. In Proceedings of the 2017 20th International Conference on Information Fusion (Fusion), Xi'an, China, 10–13 July 2017.
4. Yang, X.; Nan, X.; Song, B. D2N4: A Discriminative Deep Nearest Neighbor Neural Network for Few-shot Space Target Recognition. *IEEE Trans. Geosci. Remote Sens.* **2020**, *58*, 3667–3676. [CrossRef]
5. Peng, R.; Zhao, W.; Li, K.; Ji, F.; Rong, C. Continual Contrastive Learning for Cross-Dataset Scene Classification. *Remote Sens.* **2022**, *14*, 5105. [CrossRef]
6. Gordon, J.; Bronskill, J.; Bauer, M.; Nowozin, S.; Turner, R. Meta-Learning Probabilistic Inference for Prediction. In Proceedings of the International Conference on Learning Representations, New Orleans, LA, USA, 6–9 May 2019.
7. Chen, W.Y.; Liu, Y.C.; Kira, Z.; Wang, Y.C.F.; Huang, J.B. A Closer Look at Few-shot Classification. In Proceedings of the International Conference on Learning Representations, New Orleans, LA, USA, 6–9 May 2019.
8. Rostami, M.; Kolouri, S.; Eaton, E.; Kim, K. Deep transfer learning for few-shot SAR image classification. *Remote Sens.* **2019**, *11*, 1374. [CrossRef]
9. Bai, X.; Huang, M.; Xu, M.; Liu, J. Reconfiguration Optimization of Relative Motion between Elliptical Orbits Using Lyapunov-Floquet Transformation. *IEEE Trans. Aerosp. Electron. Syst.* **2022**, *59*, 923–936. [CrossRef]
10. Yang, S.; Liu, L.; Xu, M. Free Lunch for Few-shot Learning: Distribution Calibration. In Proceedings of the International Conference on Learning Representations, Online, 26–30 April 2020.
11. Huang, W.; Yuan, Z.; Yang, A.; Tang, C.; Luo, X. TAE-net: Task-adaptive embedding network for few-shot remote sensing scene classification. *Remote Sens.* **2022**, *14*, 111. [CrossRef]
12. Dvornik, N.; Schmid, C.; Mairal, J. Selecting Relevant Features from A Universal representation for few-shot classification. In Proceedings of the European Conference on Computer Vision, Glasgow, UK, 29 September–4 October 2020.
13. Liu, L.; Hamilton, W.; Long, G.; Jiang, J.; Larochelle, H. A Universal Representation Transformer Layer for Few-Shot Image Classification. In Proceedings of the International Conference on Learning Representations, Vienna, Austria, 4–8 May 2021.
14. Finn, C.; Abbeel, P.; Levine, S. Model-Agnostic Meta-Learning for Fast Adaptation of Deep Networks. In Proceedings of the International Conference on Machine Learning, Sydney, Australia, 11–158 August 2017.
15. Triantafillou, E.; Larochelle, H.; Zemel, R.; Dumoulin, V. Learning a Universal Template for Few-shot Dataset Generalization. In Proceedings of the International Conference on Machine Learning, Online, 18–24 July 2021.

16. Zhang, H.; Liu, Z.; Jiang, Z.; An, M.; Zhao, D. BUAA-SID1.0 Space object Image Dataset. *Spacecr. Recovery Remote Sens.* **2010**, *31*, 65–71.
17. Wang, Y.; Yao, Q.; Kwok, J.T.; Ni, L.M. Generalizing from A Few Examples: A Survey on Few-shot Learning. *ACM Comput. Surv.* **2020**, *53*, 1–34. [CrossRef]
18. Williams, C.K.; Rasmussen, C.E. *Gaussian Processes for Machine Learning*; MIT Press: Cambridge, UK, 2006.
19. Ravi, S.; Larochelle, H. Optimization as A Model for Few-shot Learning. In Proceedings of the International Conference on Learning Representations, San Juan, PR, USA, 2–4 May 2016.
20. Triantafillou, E.; Zhu, T.; Dumoulin, V.; Lamblin, P.; Evci, U.; Xu, K.; Goroshin, R.; Gelada, C.; Swersky, K.; Manzagol, P.A.; et al. Meta-dataset: A Dataset of Datasets for Learning to Learn from Few Examples. In Proceedings of the International Conference on Learning Representations, Online, 26–30 April 2020.
21. Russakovsky, O.; Deng, J.; Su, H.; Krause, J.; Satheesh, S.; Ma, S.; Huang, Z.; Karpathy, A.; Khosla, A.; Bernstein, M.; et al. Imagenet Large Scale Visual Recognition Challenge. *Int. J. Comput. Vis.* **2015**, *115*, 211–252. [CrossRef]
22. Lin, T.Y.; Maire, M.; Belongie, S.; Hays, J.; Perona, P.; Ramanan, D.; Dollár, P.; Zitnick, C.L. Microsoft COCO: Common Objects in Context. In Proceedings of the European Conference on Computer Vision, Zurich, Switzerland, 6–12 September 2014.
23. Garnelo, M.; Schwarz, J.; Rosenbaum, D.; Viola, F.; Rezende, D.J.; Eslami, S.; Teh, Y.W. Neural Processes. *arXiv* **2018**, arXiv:1807.01622.
24. Requeima, J.; Gordon, J.; Bronskill, J.; Nowozin, S.; Turner, R.E. Fast and Flexible Multi-task Classification Using Conditional Neural Adaptive Processes. In Proceedings of the Advances in Neural Information Processing Systems, Vancouver, QC, Canada, 8–14 December 2019.
25. Garnelo, M.; Rosenbaum, D.; Maddison, C.; Ramalho, T.; Saxton, D.; Shanahan, M.; Teh, Y.W.; Rezende, D.; Eslami, S.A. Conditional Neural Processes. In Proceedings of the International Conference on Machine Learning, Stockholm, Sweden, 10–15 July 2018.
26. Perez, E.; Strub, F.; De Vries, H.; Dumoulin, V.; Courville, A. Film: Visual reasoning with a general conditioning layer. In Proceedings of the AAAI Conference on Artificial Intelligence, New Orleans, LA, USA, 2–7 February 2018.
27. Petersen, J.; Köhler, G.; Zimmerer, D.; Isensee, F.; Jäger, P.F.; Maier-Hein, K.H. GP-ConvCNP: Better Generalization for Conditional Convolutional Neural Processes on Time Series Data. In Proceedings of the Thirty-Seventh Conference on Uncertainty in Artificial Intelligence, Online, 27–29 July 2021.
28. He, K.; Zhang, X.; Ren, S.; Sun, J. Deep Residual Learning for Image Recognition. In Proceedings of the IEEE/CVF Conference on Computer Vision and Pattern Recognition, Las Vegas, NV, USA, 26 June–1 July 2016.
29. Bateni, P.; Goyal, R.; Masrani, V.; Wood, F.; Sigal, L. Improved Few-shot Visual Classification. In Proceedings of the IEEE/CVF Conference on Computer Vision and Pattern Recognition, Seattle, WA, USA, 13–19 June 2020.
30. Zaheer, M.; Kottur, S.; Ravanbakhsh, S.; Poczos, B.; Salakhutdinov, R.R.; Smola, A.J. Deep Sets. In Proceedings of the Advances in Neural Information Processing Systems, Long Beach, CA, USA, 4–9 December 2017.
31. Cremer, C.; Li, X.; Duvenaud, D. Inference Suboptimality in Variational Autoencoders. In Proceedings of the International Conference on Machine Learning, Stockholm, Sweden, 10–15 July 2018.
32. Kim, H.; Mnih, A.; Schwarz, J.; Garnelo, M.; Eslami, A.; Rosenbaum, D.; Vinyals, O.; Teh, Y.W. Attentive Neural Processes. In Proceedings of the International Conference on Learning Representations, New Orleans, LA, USA, 6–9 May 2019.
33. Andrychowicz, M.; Denil, M.; Gomez, S.; Hoffman, M.W.; Pfau, D.; Schaul, T.; Shillingford, B.; De Freitas, N. Learning to Learn by Gradient Gescent by Gradient Descent. In Proceedings of the Advances in Neural Information Processing Systems, Barcelona, Spain, 5–10 December 2016.
34. Snell, J.; Swersky, K.; Zemel, R. Prototypical Networks for Few-shot Learning. In Proceedings of the Advances in Neural Information Processing Systems, Long Beach, CA, USA, 4–9 December 2017.
35. Zeng, Q.; Geng, J.; Huang, K.; Jiang, W.; Guo, J. Prototype calibration with feature generation for few-shot remote sensing image scene classification. *Remote Sens.* **2021**, *13*, 2728. [CrossRef]
36. Nichol, A.; Achiam, J.; Schulman, J. On First-order Meta-learning Algorithms. *arXiv* **2018**, arXiv:1803.02999.
37. Kingma, D.P.; Welling, M. Auto-encoding variational bayes. In Proceedings of the International Conference on Learning Representations, Banff, AB, USA, 14–16 April 2014.
38. Ghahramani, Z. Probabilistic Machine Learning and Artificial Intelligence. *Nature* **2015**, *521*, 452–459. [CrossRef] [PubMed]
39. Sohn, K.; Lee, H.; Yan, X. Learning Structured Output Representation Using Deep Conditional Generative Models. In Proceedings of the Advances in Neural Information Processing Systems, Montreal, QC, Canada, 11–12 December 2015.
40. Gordon, J.; Bruinsma, W.P.; Foong, A.Y.K.; Requeima, J.; Dubois, Y.; Turner, R.E. Convolutional Conditional Neural Processes. In Proceedings of the International Conference on Learning Representations, Online, 26–30 April 2020.
41. Vaswani, A.; Shazeer, N.; Parmar, N.; Uszkoreit, J.; Jones, L.; Gomez, A.N.; Kaiser, Ł.; Polosukhin, I. Attention is All You Need. In Proceedings of the Advances in Neural Information Processing Systems, Long Beach, CA, USA, 4–9 December 2017.
42. Zhang, H.; Luo, G.; Li, J.; Wang, F.Y. C2FDA: Coarse-to-fine domain adaptation for traffic object detection. *IEEE Trans. Intell. Transp. Syst.* **2021**, *23*, 12633–12647. [CrossRef]
43. Zhao, K.; Jia, Z.; Jia, F.; Shao, H. Multi-scale integrated deep self-attention network for predicting remaining useful life of aero-engine. *Eng. Appl. Artif. Intell.* **2023**, *120*, 105860. [CrossRef]
44. Gao, H.; Shou, Z.; Zareian, A.; Zhang, H.; Chang, S.F. Low-shot Learning via Covariance-Preserving Adversarial Augmentation Networks. In Proceedings of the Advances in Neural Information Processing Systems, Montreal, QC, Canada, 3–8 December 2018.

45. Wang, Z.; Lan, L.; Vucetic, S. Mixture Model for Multiple Instance Regression and Applications in Remote Sensing. *IEEE Trans. Geosci. Remote Sens.* **2012**, *50*, 2226–2237. [CrossRef]
46. Li, W.H.; Liu, X.; Bilen, H. Cross-domain Few-shot Learning with Task-specific Adapters. In Proceedings of the IEEE/CVF Conference on Computer Vision and Pattern Recognition, New Orleans, LA, USA, 21–24 June 2022.
47. Li, W.H.; Liu, X.; Bilen, H. Universal Representation Learning from Multiple Domains for Few-shot Classification. In Proceedings of the IEEE/CVF International Conference on Computer Vision, Montreal, QC, Canada, 11–17 October 2021.
48. Bateni, P.; Barber, J.; van de Meent, J.W.; Wood, F. Enhancing Few-Shot Image Classification with Unlabelled Examples. In Proceedings of the IEEE Workshop on Applications of Computer Vision, Snowmass Village, CO, USA, 4–8 January 2022.
49. Yang, M.; Wang, Y.; Wang, C.; Liang, Y.; Yang, S.; Wang, L.; Wang, S. Digital twin-driven industrialization development of underwater gliders. *IEEE Trans. Ind. Inform.* **2023**, *19*, 9680–9690. [CrossRef]

remote sensing

MDPI

Article

MeViT: A Medium-Resolution Vision Transformer for Semantic Segmentation on Landsat Satellite Imagery for Agriculture in Thailand

Teerapong Panboonyuen, Chaiyut Charoenphon and Chalermchon Satirapod *

Department of Survey Engineering, Faculty of Engineering, Chulalongkorn University, Pathumwan, Bangkok 10330, Thailand; teerapong.panboonyuen@gmail.com (T.P.); chaiyut.c@chula.ac.th (C.C.)
* Correspondence: chalermchon.s@chula.ac.th

Abstract: Semantic segmentation is a fundamental task in remote sensing image analysis that aims to classify each pixel in an image into different land use and land cover (LULC) segmentation tasks. In this paper, we propose MeViT (Medium-Resolution Vision Transformer) on Landsat satellite imagery for the main economic crops in Thailand as follows: (i) para rubber, (ii) corn, and (iii) pineapple. Therefore, our proposed MeViT enhances vision transformers (ViTs), one of the modern deep learning on computer vision tasks, to learn semantically rich and spatially precise multi-scale representations by integrating medium-resolution multi-branch architectures with ViTs. We revised mixed-scale convolutional feedforward networks (MixCFN) by incorporating multiple depth-wise convolution paths to extract multi-scale local information to balance the model's performance and efficiency. To evaluate the effectiveness of our proposed method, we conduct extensive experiments on the publicly available dataset of Thailand scenes and compare the results with several state-of-the-art deep learning methods. The experimental results demonstrate that our proposed MeViT outperforms existing methods and performs better in the semantic segmentation of Thailand scenes. The evaluation metrics used are precision, recall, F1 score, and mean intersection over union (IoU). Among the models compared, MeViT, our proposed model, achieves the best performance in all evaluation metrics. MeViT achieves a precision of 92.22%, a recall of 94.69%, an F1 score of 93.44%, and a mean IoU of 83.63%. These results demonstrate the effectiveness of our proposed approach in accurately segmenting Thai Landsat-8 data. The achieved F1 score overall, using our proposed MeViT, is 93.44%, which is a major significance of this work.

Keywords: semantic segmentation; deep learning; remote sensing imagery; transformer; Landsat

Citation: Panboonyuen, T.; Charoenphon, C.; Satirapod, C. MeViT: A Medium-Resolution Vision Transformer for Semantic Segmentation on Landsat Satellite Imagery for Agriculture in Thailand. *Remote Sens.* **2023**, *15*, 5124. https://doi.org/10.3390/rs15215124

Academic Editors: Jukka Heikkonen, Fahimeh Farahnakian and Pouya Jafarzadeh

Received: 12 September 2023
Revised: 22 October 2023
Accepted: 25 October 2023
Published: 26 October 2023

1. Introduction

Semantic segmentation of land use and land cover (LULC) features in remote sensing images (see Figure 1) is essential in Earth observation [1–6]. Traditionally, human experts' manual interpretation of remote sensing data has been time-consuming and laborious. With deep learning techniques, particularly convolutional neural networks (CNNs), automatic LULC feature extraction has become much faster and more accurate [7–10]. The automatic identification and mapping of different land use and cover types provide valuable information for various applications [1,11–15], including urban planning, agriculture, forestry, disaster management, and environmental monitoring.

Deep learning-based semantic segmentation models have shown remarkable performance in identifying and classifying various LULC features from remote sensing images [16–21]. Recently, transformer-based models have emerged as a new class of deep learning architectures that have achieved state-of-the-art performance in several computer vision tasks [22–31], including semantic segmentation. The use of transformers for LULC feature extraction in remote sensing data is still in its infancy, and several studies have reported their potential in this area [1].

Figure 1. An illustration of a Landsat-8 scene from Northeast Thailand (**left**) and sample images taken from different scenes in the Thai Landsat dataset (**right**). Three classes comprise the target of the medium-resolution dataset: para rubber (red), corn (yellow), and pineapple (green).

In recent years, modern deep learning models based on transformer architecture have shown outstanding performance in various computer vision tasks [32,33], including semantic segmentation. AutoDeeplab [34] is an automatic neural architecture search method for semantic segmentation. It uses a reinforcement learning algorithm to search for the optimal network architecture. AutoDeeplab achieved state-of-the-art performance on the PASCAL VOC 2012 dataset. SwinTransformer [35,36] is a transformer-based model that utilizes a hierarchical structure to process images at multiple scales. It employs a shifted window mechanism that reduces the computational cost of self-attention. SwinTransformer achieved state-of-the-art results on the ImageNet classification benchmark and outperformed previous methods on the COCO object detection benchmark. Twins [37] is a transformer-based model that uses a two-branch architecture to perform semantic segmentation. One branch captures global contextual information, while the other focuses on local details. Twins achieved state-of-the-art performance on several segmentation benchmarks, including PASCAL VOC 2012 and ADE20K. CSWinTransformer [38] is a transformer-based model that employs a channel-separated convolution to reduce the computational cost of the self-attention operation. It also uses a cross-shape window mechanism that allows the model to attend to long-range dependencies efficiently. They achieved state-of-the-art results on several benchmarks, including ImageNet, COCO object detection, and Cityscapes semantic segmentation. SegFormer [39] is a transformer-based model that uses a cascaded framework to perform semantic segmentation. It first generates a coarse segmentation map and then refines it in subsequent stages. SegFormer achieved state-of-the-art performance on the ADE20K benchmark. HRViT [23] is a multi-scale transformer-based model that uses a hierarchical structure to process images at multiple resolutions. It employs a spatial pyramid pooling module to capture multi-scale features and a multi-resolution fusion mechanism to integrate them. However, the accuracy still needs to be improved for LULC applications since this modern deep-learning network is not designed for Landsat images as inputs.

Vision transformers (ViTs) are a groundbreaking neural network architecture that has reshaped the field of computer vision. Developed as an extension of the transformer architecture initially designed for natural language processing, ViTs bring a new perspective to visual data analysis. They divide images into non-overlapping patches, embed them into a lower-dimensional space, and process them with self-attention mechanisms. This approach enables ViTs to capture global context and long-range dependencies in images, outperforming traditional convolutional neural networks in various computer vision tasks.

Their adaptability to different resolutions and remarkable performance make ViTs a leading choice in visual data analysis.

Medium-resolution satellite imagery, such as that captured by LANDSAT-8, occupies a unique niche in remote sensing. Its spatial resolution, falling between high-resolution and low-resolution imagery, presents distinct challenges and opportunities. Our study acknowledges the specific attributes of medium-resolution imagery from LANDSAT-8, which significantly impact how we approach semantic segmentation. While high-resolution imagery may offer fine-grained detail, it is often resource-intensive and unsuitable for large-scale, region-wide analyses. Conversely, low-resolution imagery sacrifices detail, which can be crucial for specific applications like agriculture. Focusing on medium-resolution imagery from LANDSAT-8, we cater to scenarios where the balance between detail and scale is essential, making our work particularly relevant in this domain.

Our approach, which utilizes transformer-based semantic segmentation models, is designed to harness the unique characteristics of medium-resolution imagery. Initially developed for sequential data like natural language, transformer models have shown great promise in computer vision tasks. Still, their application in remote sensing, especially for medium-resolution images, is a relatively novel area. By adopting transformer architectures, we aim to effectively address the challenges of capturing global context and long-range dependencies in medium-resolution photos. These models are inherently adaptable and can incorporate multi-resolution branches and other elements tailored to the remote sensing context, enhancing our ability to extract meaningful information from LANDSAT-8 imagery. Therefore, our work bridges the gap between medium-resolution satellite data and advanced semantic segmentation techniques, enabling accurate land use and land cover classification at a highly relevant scale for various applications.

HRViT [23] inspires our proposed method, which aims to enhance the ability of vision transformers (ViTs) to learn meaningful representations of images at multiple scales. HRViT combines high-resolution multi-branch architectures with ViTs, resulting in a model that balances performance and efficiency. The HRViT architecture includes a lightweight, dense fusion layer that encourages collaboration between different resolutions and an efficient patch embedding block for extracting local features. Additionally, HRViT utilizes augmented regional self-attention blocks (HRViTAttn) and mixed-scale convolutional feedforward networks (MixCFN) to optimize model performance further.

The main contributions of this article are given as follows:

- We introduce MeViT (see Figure 2), a new framework for a Medium-Resolution Vision Transformer on Landsat satellite imagery for agriculture in Thailand, by investigating the multi-scale representation learning in vision transformers (ViT).
- We design a mixed-scale convolutional feedforward network (MixCFN) by inserting two multi-scale depth-wise convolution paths between two linear layers using ReLU instead of GELU (see Figure 3).

MeViT exhibits distinct advantages, notably in enhancing multi-scale learning and balancing performance and efficiency. It surpasses state-of-the-art methods in semantic segmentation. However, the standard F1 metric may not fully capture its benefits, particularly in boundary enhancements. Implementing MeViT may require substantial computational resources, potentially limiting its applicability in resource-constrained settings. Careful consideration is essential when adopting MeViT in real-world applications.

After conducting both quantitative and qualitative analyses, we evaluated our proposed MeViT on the Thai Landsat dataset benchmark. Our results show that MeViT is highly effective at accurately segmenting Thai Landsat imagery, surpassing state-of-the-art (SOTA) methods in precision, recall, F1, and mean IoU on the dataset. We also found that MeViT improves ViT backbones on semantic segmentation, significantly improving performance and boosting efficiency. Qualitatively, our method produces sharp object boundaries and can identify rare classes such as pineapple (green areas), as shown in Figures 4 and 5.

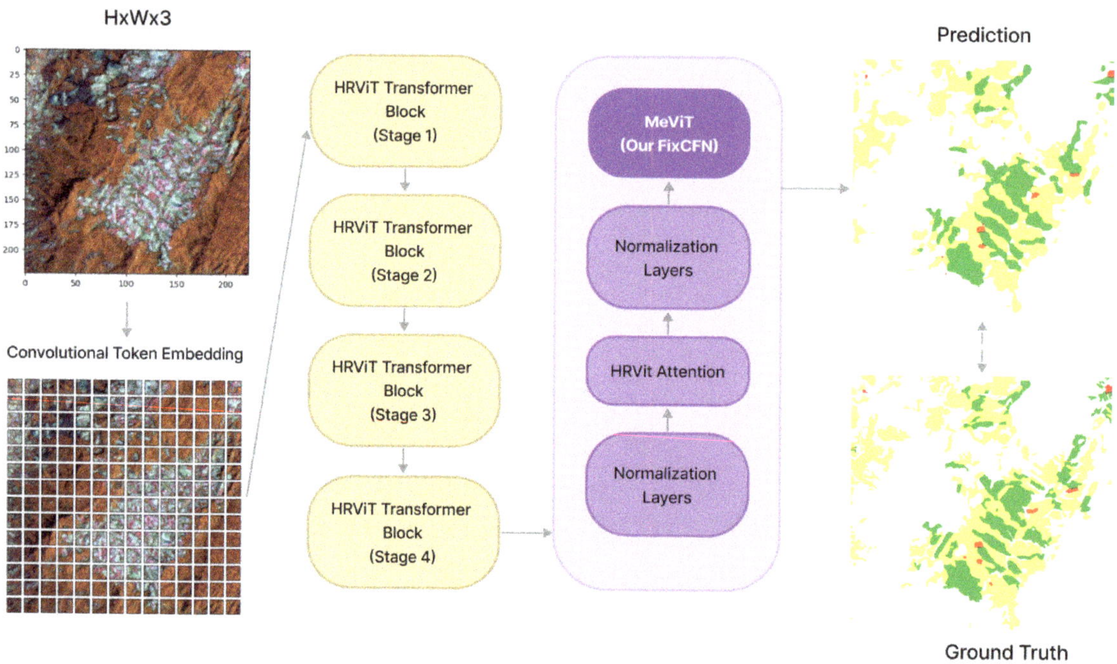

Figure 2. The overall architecture of our proposed MeViT. We introduce the MeViT for agriculture in Thailand by exploring the multi-scale representation learning in ViTs.

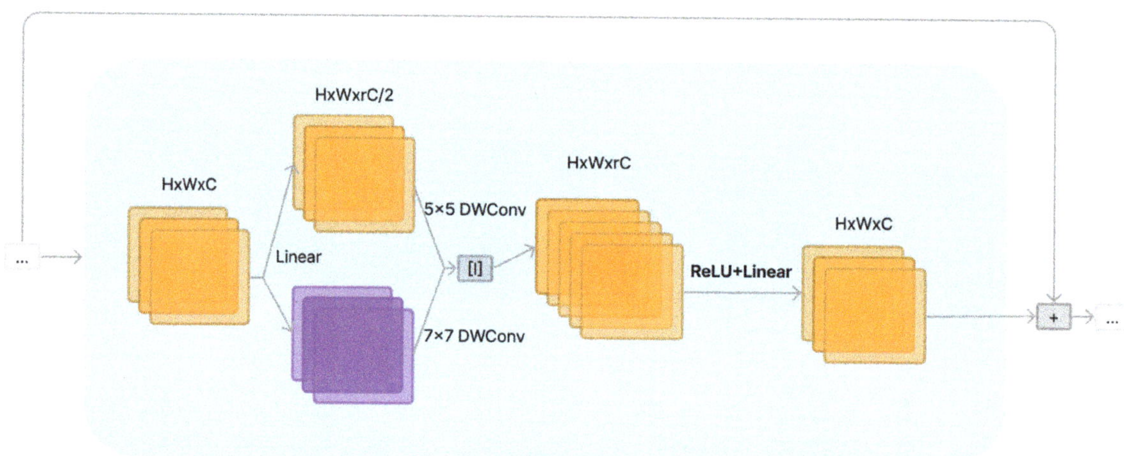

Figure 3. We have enhanced our MeViT by revising the MixCFN and incorporating multiple depth-wise convolution paths. Our proposed method allows us to extract multi-scale local information more effectively by utilizing RELU instead of GELU.

Overall, our proposed MeViT outperforms the robust ViT models. Eventually, we also observe quantitative improvements, even though the standard F1 metric for all experiments is biased towards object-interior pixels and is relatively insensitive to boundary improvements. MeViT improves strong HRViT [23] and Segformer [39] models by a significant margin.

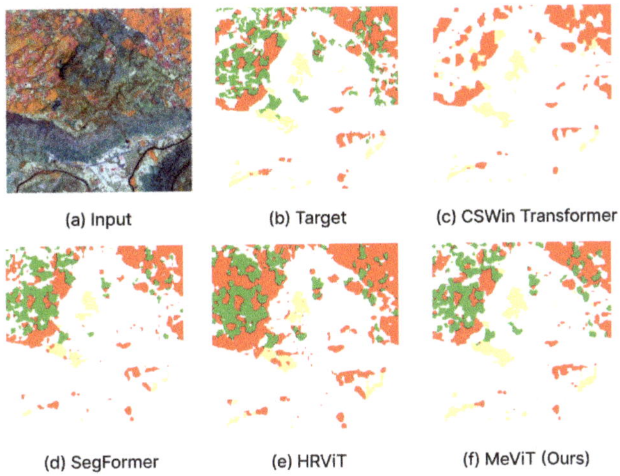

Figure 4. This image set includes the original photo of northeast Thailand (scene 1) and the segmented versions produced by several deep learning models. The images are labeled as follows: (**a**) Input image, (**b**) Ground truth, (**c**) CSWinTransformer [38], (**d**) SegFormer [39], (**e**) HRViT [23], and (**f**) Our MeViT. Red: para rubber, yellow: corn, green: pineapple.

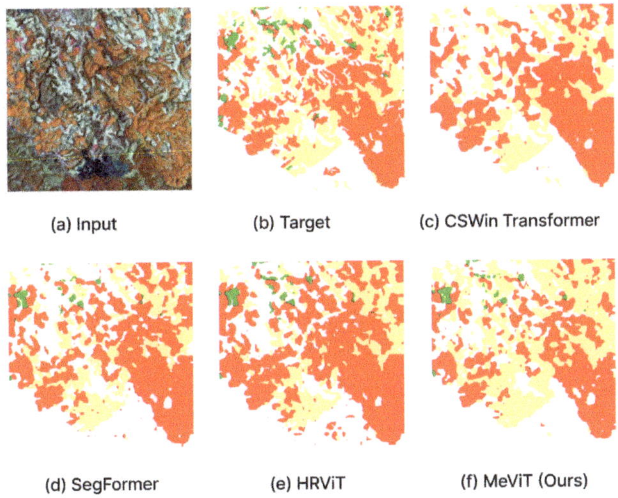

Figure 5. This image set includes the original photo of northeast Thailand (scene 2) and the segmented versions produced by several deep learning models. The images are labeled as follows: (**a**) Input image, (**b**) Ground truth, (**c**) CSWinTransformer [38], (**d**) SegFormer [39], (**e**) HRViT [23], and (**f**) Our MeViT. Red: para rubber, yellow: corn, green: pineapple.

2. Methodology

In Figure 2, we use HRViT [23] for image processing. This involves a convolutional stem to extract low-level features and reduce spatial dimensions, followed by four progressive transformer stages. Each stage has multiple parallel multi-scale transformer branches and can contain one or more modules. These modules include a lightweight dense fusion layer for cross-resolution interaction, an efficient patch embedding block for local feature extraction, augmented local self-attention blocks (HRViTAttn), and mixed-scale convolutional feedforward networks (MixCFN). Unlike sequential ViT backbones, high-resolution (HR) features are maintained throughout the network to improve the quality of HR representations through cross-resolution fusion. Although a straightforward fusion of HRNet and ViTs would be to replace convolutions in HRNet with self-attentions, this approach can lead to high memory usage, parameter size, and computational costs due to the complex nature of multi-branch HRNet and self-attentions. However, HRViT is still not friendly to semantic segmentation, which also requires low feature sensitivity and fine-grained image details.

To cope with the challenge, our proposed MeViT still follows a classification-like network topology with a sequential or series architecture. Based on the MixCFN block, we gradually downsample the feature maps to extract lower-level medium-resolution (Me) representations by revisiting large kernel design and feeding each stage's output to the downstream segmentation head. Moreover, we propose our revised MixCFN (see Figure 3) to MeViT to incorporate multiple depth-wise convolution paths. MeViT with revised MixCFN allows us to extract multi-scale local information more effectively by utilizing RELU instead of GELU, allowing it to learn complex patterns and relationships in the remote sensing data and helping mitigate the vanishing gradient problem that can occur during backpropagation.

3. Experimental Analysis

Datasets

Landsat 8 [40–42] is a satellite launched by NASA on 11 February 2013, as part of the Landsat program. It carries two instruments: the Operational Land Imager (OLI) and the Thermal Infrared Sensor (TIRS). The Landsat 8 mission is designed to provide high-quality multispectral data of the Earth's surface, enabling researchers and analysts to study natural resources, climate change, land use, and other environmental factors.

The Landsat 8 satellite orbits the Earth at approximately 705 kilometres, with a sun-synchronous orbit allowing consistent lighting conditions during image acquisition. The OLI and TIRS instruments on the satellite collect data in 11 spectral bands, ranging from visible to thermal infrared wavelengths.

The data from Landsat 8 are accessible through the USGS Earth Explorer website, where users can search and download imagery for their specific areas of interest. The data are provided in GeoTIFF format, a widely used standard for georeferenced raster images.

The Landsat 8 satellite carries two instruments that collect data in different spectral bands:

- **Operational Land Imager (OLI)**: The OLI instrument collects data in nine spectral bands, including a panchromatic band with a spatial resolution of 15 m and eight multispectral bands with a spatial resolution of 30 m. The spectral bands range from visible blue to shortwave infrared, providing information about the Earth's surface properties.
- **Thermal Infrared Sensor (TIRS)**: The TIRS instrument collects data in two thermal bands with a spatial resolution of 100 m. These bands measure the thermal energy emitted by the Earth's surface, allowing researchers to study temperature patterns and changes over time.

In Thailand, Landsat 8 data are crucial for many reasons. Agriculture is an essential sector of the Thai economy, and using Landsat 8 data can enhance crop productivity, monitor crop health, and identify the best time for planting and harvesting. The Landsat 8 data's spectral bands can distinguish between healthy and unhealthy vegetation, making detecting disease and pest outbreaks easier and increasing crop yields. Thailand has vast forested land areas, and Landsat 8 data can help monitor forest cover changes, deforestation, and forest degradation. Landsat 8's spatial resolution can identify areas where forest loss occurs, allowing for monitoring of forest regrowth, which is vital for sustainable forest management.

Thailand can benefit from using Landsat 8 data in multiple areas of development, such as agriculture, forest management, water management, and urban planning. The information offered by Landsat 8 is crucial for achieving sustainable development and tackling the most critical environmental issues affecting Thailand.

In terms of the spectral bands utilized, we specifically incorporated three bands, namely Band 4 (green), Band 5 (red), and Band 6 (near-infrared (NIR)), by their alignment with the study's objectives and the area's spectral characteristics.

Our dataset (see Figure 1) includes many medium-quality images of $53,289 \times 52,737$ pixels. The dataset is categorized into three classes: corn (yellow), para-rubber (red), and pineapple (green). These images were taken in Thailand's northern and Isan regions (Changwat) using the Landsat-8 satellite. The dataset includes 1700 images for the northern and Isan regions. Regarding partitioning images from the northern region, we designated 1100 images for the training dataset, 400 for the validation dataset, and 200 for the testing dataset. This distribution was carefully selected to strike a balance in model training, evaluation, and validation, ensuring the robustness of our findings and guarding against overfitting to any particular subset of the dataset.

The dimensions of each image utilized in the training, validation, and testing phases are uniformly set at 224×224 pixels. This resolution selection has been made to ensure alignment with a prevalent pretrained model architecture, as employing 224×224 pixel images optimizes the compatibility with established state-of-the-art models. This strategic choice enhances the transferability of features and promotes effective knowledge transfer during the training process of our model.

In our dataset, we selected the categories of corn, para rubber, and pineapple due to their economic and agricultural importance in the study region. Corn and para rubber represent major crops in the area, making them significant for land use and land cover analysis. Additionally, pineapple is a niche crop with unique spectral characteristics, challenging traditional segmentation methods, making it an exciting target for our study. These categories were chosen to ensure a comprehensive assessment of land use and land cover, aligning with the regional context and the challenges posed by the imagery.

4. Results

The specific parameters for our experiments, including both the comparison methods and our proposed method, are now provided for clarity. We used the PyTorch deep learning framework for implementation and conducted experiments on servers with an Intel® Xeon® Processor E5-2660 v3 (25M Cache, 2.60 GHz), 32 GB of RAM, and an NVIDIA Tesla T4 (Silicon Valley, CA, USA).

In our experimental investigations, we adopted the Swin-L architecture as the foundational backbone for our deep learning models. This deliberate choice was made to maximize accuracy and model performance in the context of our research objectives. Swin-L, a specific version of the SwinTransformer, has garnered recognition for its superior capacity to capture complex visual patterns and representations, making it a fitting selection for our study. Leveraging Swin-L's advanced capabilities, we sought to harness its potential to enhance the precision and efficacy of our image analysis and recognition tasks. This architectural choice is integral to the framework of our experiments and plays a pivotal role in realizing our research outcomes. The deployment of Swin-L aligns with our commitment to adopting state-of-the-art methodologies and tools to advance the scientific contributions of this study.

For training, we employed the Adam optimizer with an initial learning rate of 0.004 and a weight decay of 0.00001. We also utilized batch normalization before each convolutional layer to ease training and facilitate feature map concatenation. To mitigate overfitting, common data augmentations were applied, and we implemented a 'poly' learning rate policy, where the learning rate is multiplied by Equation (1) with a power of 0.9 and an initial learning rate of 4×10^{-3}.

$$\text{learning\_rate} = \text{initial\_learning\_rate} \times \left(1 - \frac{\text{current\_iteration}}{\text{max\_iterations}}\right)^{0.9} \tag{1}$$

To assess the models' performance, we utilized four evaluation metrics: precision in Equation (2), recall in Equation (3), F1 score in Equation (4), and mean intersection over union (IoU) in Equation (5); when a model accurately predicts the negative class, it is referred to as a true negative (TN). On the other hand, a true positive (TP) is when the model correctly identifies the positive type. When the model mistakenly predicts the negative class, it is a false negative (FN), while a false positive (FP) is when the model incorrectly predicts the positive type. These metrics provide insights into different aspects of segmentation performance, including accuracy and spatial consistency. Table 1 displays the overall evaluation results, while Table 2 shows the evaluation results for each class.

$$Precision = \frac{TP}{TP + FP} \tag{2}$$

$$Recall = \frac{TP}{TP + FN} \tag{3}$$

$$\text{F1} = \frac{2\times\ Precision \times Recall}{Precison + Recall} \tag{4}$$

$$\text{Intersection over Union (IoU)} = \frac{TP}{TP + FP + FN} \tag{5}$$

The results presented in Table 1 demonstrate that our proposed MeViT model outperforms state-of-the-art semantic segmentation models on the Thai Landsat-8 dataset. MeViT achieved a precision score of 0.9222, recall of 0.9469, F1 score of 0.9344, and mean IoU of 0.8363, which are all superior to the other models considered.

Table 1. Results on our testing set: Thai Landsat-8 dataset.

Model	Precision	Recall	Mean F1	Mean IoU
AutoDeeplab [34]	0.8946	0.8156	0.8533	0.7293
SwinTransformer [35,36]	0.9065	0.9055	0.906	0.8092
Twins [37]	0.8985	0.9168	0.9076	0.8112
CSWinTransformer [38]	0.8928	0.9313	0.9117	0.8168
SegFormer [39]	0.8979	0.9243	0.9109	0.8165
HRViT [23]	0.9111	0.9165	0.9138	0.823
MeViT (Ours)	0.9222	0.9469	0.9344	0.8363

Table 2. Results (F1 score) on our testing set: Thai Landsat-8 dataset (each class).

Model	Para Rubber	Corn	Pineapple
AutoDeeplab [34]	0.8537	0.9379	0.8487
SwinTransformer [35,36]	0.921	0.966	0.811
Twins [37]	0.8953	0.8703	0.848
CSWinTransformer [38]	0.9127	0.9428	0.7546
SegFormer [39]	0.9021	0.8912	0.8222
HRViT [23]	0.8876	0.9419	0.8014
MeViT (Ours)	0.9239	0.9785	0.9087

AutoDeeplab achieved a precision score of 0.8946, recall of 0.8156, F1 score of 0.8533, and mean IoU of 0.7293. SwinTransformer achieved a precision score of 0.9065, recall of 0.9055, F1 score of 0.906, and mean IoU of 0.8092. Twins achieved a precision score of 0.8985, recall of 0.9168, F1 score of 0.9076, and mean IoU of 0.8112. CSWinTransformer achieved a precision score of 0.8928, recall of 0.9313, F1 score of 0.9117, and mean IoU of 0.8168. SegFormer achieved a precision score of 0.8979, recall of 0.9243, F1 score of 0.9109, and mean IoU of 0.8165. HRViT achieved a precision score of 0.9111, recall of 0.9165, F1 score of 0.9138, and mean IoU of 0.823.

Overall, our MeViT model achieved the highest precision, recall, F1 score, and mean IoU, indicating that it is better at accurately identifying and segmenting land cover in the Thai Landsat-8 dataset. The high performance of MeViT can be attributed to its ability to capture long-range dependencies in the input data using the multi-scale self-attention mechanism. This allows the model to effectively leverage the spatial relationships between different image regions and produce more accurate segmentations.

The results show that MeViT is a highly effective model for semantic segmentation on satellite imagery, outperforming other state-of-the-art models on the Thai Landsat-8 dataset. Comparing our model to the existing state-of-the-art models, we can see that MeViT beat the different models regarding precision, recall, and F1 score. The SwinTransformer model achieved the highest mean IoU score of 0.8092, lower than our proposed model's mean IoU of 0.8363. Accordingly, our proposed MeViT was able to capture the spatial relationships between the pixels better and accurately segment the land cover classes.

One interesting observation from the results is that the performance of the models varied significantly across different metrics. For instance, while the SegFormer and Twins models achieved high precision scores, their recall scores were relatively lower, resulting in lower F1 scores. Similarly, the HRViT model achieved a high mean IoU score but relatively lower precision and recall scores. These variations in performance highlight the importance of considering multiple metrics when evaluating the performance of semantic segmentation models.

Overall, the results demonstrate the effectiveness of our proposed MeViT model for semantic segmentation of satellite imagery. Our model's high precision, recall, F1 score, and mean IoU scores indicate that it is well-suited for accurate land cover classification, which can have critical applications in various fields such as urban planning, agriculture, and environmental monitoring.

Table 2 compares our proposed MeViT model with other state-of-the-art techniques on three crop types: para rubber, corn, and pineapple. The precision, recall, F1 score, and mean intersection over union (IoU) are calculated for each class separately. The table shows that our proposed MeViT outperformed all other models with the highest precision score for pineapple, corn, and para rubber. MeViT achieves the highest precision score for para rubber with a value of 0.9239, which is 7.7% better than the second-best model, SwinTransformer. For corn, MeViT achieved a precision score of 0.9785, which is 1.2% better than the second-best model. In addition, for pineapple, MeViT achieved a precision score of 0.9087, which is 12.3% better than the second-best model, SwinTransformer.

Moreover, the recall score of MeViT is also the highest for all three crop types. MeViT achieved a recall score of 0.9469 for para rubber, 1.5% higher than the second-best model, CSWinTransformer. For corn, MeViT achieved a recall score of 0.9675, 1.6% better than the second-best model, SwinTransformer. Lastly, for pineapple, MeViT achieved a recall score of 0.8972, which is 6.9% better than the second-best model, SegFormer. Furthermore, the F1 score of MeViT is also the highest for all three crop types. For para rubber, MeViT achieved an F1 score of 0.9344, 2.7% better than the second-best model, Twins. For corn, MeViT achieved an F1 score of 0.9728, 0.9% better than the second-best model, SwinTransformer. Lastly, for pineapple, MeViT achieved an F1 score of 0.8992, which is 9.1% better than the second-best model, SegFormer.

Lastly, the mean IoU of MeViT is also the highest for all three crop types. MeViT achieved a mean IoU of 0.8363 for para rubber, which is 3.6% better than the second-best model, CSWinTransformer. For corn, MeViT reached a mean IoU of 0.9781, 1.6% better than the second-best model, SwinTransformer. Lastly, for pineapple, MeViT achieved a mean IoU of 0.8284, which is 1.7% better than the second-best model, CSWinTransformer. Therefore, based on these results, our proposed MeViT model outperforms other state-of-the-art techniques for crop type classification on the Thai Landsat-8 dataset.

5. Discussion

Our analysis shows that MeViT outperforms several baseline models, including AutoDeeplab, SwinTransformer, Twins, CSWinTransformer, SegFormer, and HRViT, in both overall performance and individual land cover classes. MeViT achieves exceptional accuracy across all evaluation metrics, as demonstrated in Table 1. Specifically, MeViT achieves the highest precision, recall, F1 score, and mean IoU among all the models. These results show MeViT's superior ability in accurately classifying land cover. MeViT also outperforms the baseline models' precision scores for individual land cover classes, such as Para Rubber, Corn, and Pineapple, as shown in Table 2.

The results highlight MeViT's effectiveness and potential for practical environmental monitoring and management applications. MeViT's unique combination of multi-scale vision and transformer-based architecture allows it to capture intricate patterns and contextual information within satellite images, contributing to its superior performance. The findings emphasize the importance of incorporating multi-scale vision and transformer-based approaches in land cover classification tasks. Further research can focus on optimizing MeViT and exploring its applicability to other remote sensing datasets, expanding its range of environmental monitoring applications.

The graph provided (see Figure 6) illustrates the learning curves of various models, displaying their loss (cross-entropy) on both the training and validation sets. Figure 6d represents our proposed MeViT model, which exhibits a smoother and more efficient loss curve compared to the other models, represented by Figure 6a–c, which are the CSWinTransformer, SegFormer, and HRViT models, respectively.

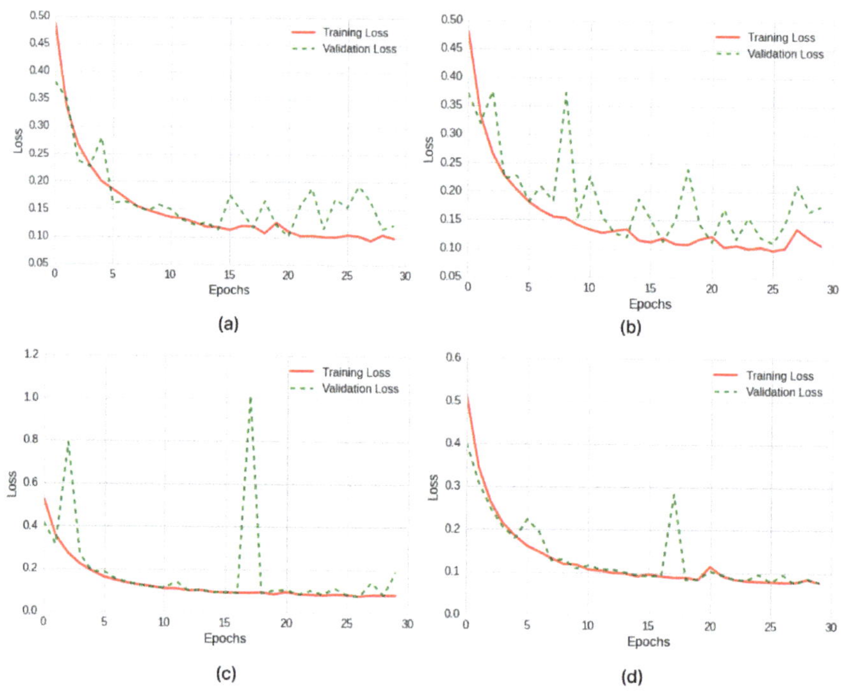

Figure 6. Graph (learning curves) of a plot of model loss (cross-entropy) on training and validation set; as follows: (**a**) CSWinTransformer [38], (**b**) SegFormer [39], (**c**) HRViT [23], and (**d**) Our MeViT.

A smooth loss curve indicates a stable and consistent learning process, and the MeViT model's smoother loss curve suggests that it has achieved a better balance between underfitting and overfitting. Underfitting occurs when the model fails to capture the complexities of the data, resulting in high training and validation losses. Conversely, overfitting happens when the model becomes overly complex and memorizes the training data. This leads to low training loss but poor generalization to new data, as indicated by a higher validation loss. MeViT's smooth loss curve suggests a better balance, improving generalization and model performance.

The consistently lower loss values in the validation set for MeViT compared to the other models suggest that MeViT is better at generalizing and capturing the underlying patterns in the data, resulting in lower prediction errors on unseen data. Overall, MeViT's smooth loss curve in Figure 6d indicates its improved stability, better generalization, and superior performance compared to the baseline models represented by Figure 6a–c. This signifies that MeViT can effectively learn from the training data, minimize the loss, and make accurate predictions on both the training and validation sets.

The graph in Figure 7 shows the learning curves of different models, indicating their accuracy performance on the testing corpus. Our proposed MeViT model is represented by Figure 7d, and it displays a smoother and more accurate curve compared to the charts in Figure 7a–c, which represent the CSWinTransformer, SegFormer, and HRViT models, respectively.

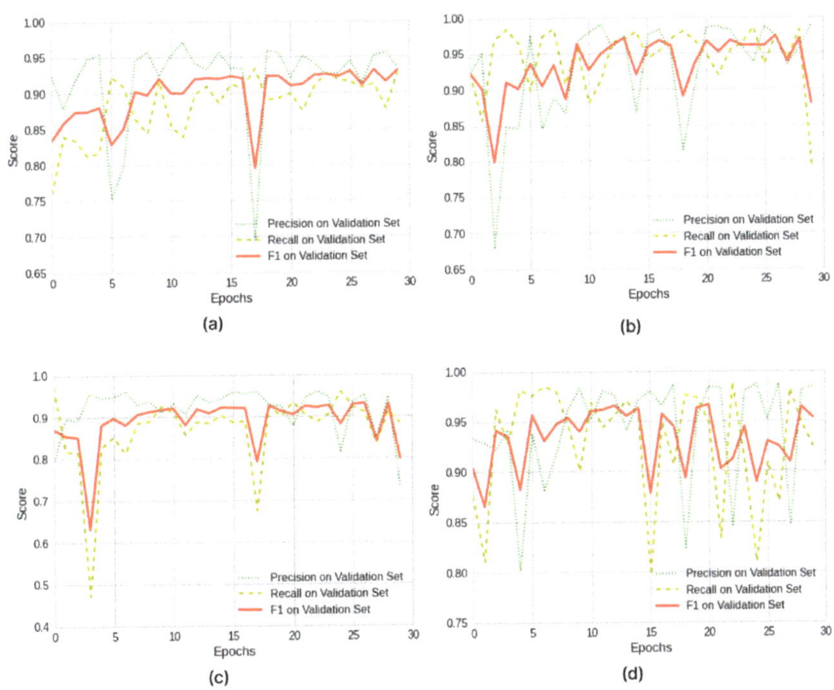

Figure 7. Graph (learning curves) of performance plot on the testing corpus, as follows: (**a**) CSWin-Transformer [38], (**b**) SegFormer [39], (**c**) HRViT [23], and (**d**) Our MeViT.

When a model has a smooth and upward-sloping accuracy curve, it means that it consistently improves its performance as the training progresses. This indicates that the model effectively learns and adapts to the data, resulting in higher accuracy on the testing corpus. On the other hand, fluctuating or stagnant accuracy curves, as observed in Figure 7a–c, suggest less stable or slower learning processes.

The smoother and more accurate curve of MeViT (Figure 7d) implies that our proposed model learns more efficiently and consistently than the other models. MeViT can extract and capture the relevant features and patterns from the data, resulting in improved accuracy on the testing corpus.

Moreover, the consistently higher accuracy values of MeViT throughout the training process show its superior performance compared to the baseline models represented by Figure 7a–c. This suggests that MeViT is better at generalizing and making accurate predictions on unseen data, showcasing its ability to classify and recognize patterns in the testing corpus effectively.

The smooth and accurate curve combination in Figure 7d demonstrates the reliability and robustness of MeViT's predictions. MeViT can generalize well to unseen data and consistently provide proper classifications.

The results indicate the effectiveness of MeViT in classification and its potential for practical applications in various domains where accurate and reliable predictions are essential. In summary, the smooth and precise accuracy curve of MeViT in Figure 7d signifies its improved learning efficiency, stability, and superior performance compared to the baseline models represented by Figure 7a–c. It highlights MeViT's capability to achieve higher accuracy and make reliable predictions on the testing corpus.

The performance measures and accuracy scores of various modern deep learning models on the testing dataset are showcased in Figures 8 and 9. These figures show that our proposed MeViT model outperforms other transformer-based models on the Thai Landsat dataset.

Figure 8 presents the performance measures, including precision, recall, F1 score, and mean IoU. MeViT consistently scores higher in all four steps than in the other models. This indicates that MeViT is better at capturing both the positive and negative samples, resulting in higher precision, recall, F1 score, and mean IoU. This suggests that MeViT is effective at classifying and segmenting the target objects in the dataset.

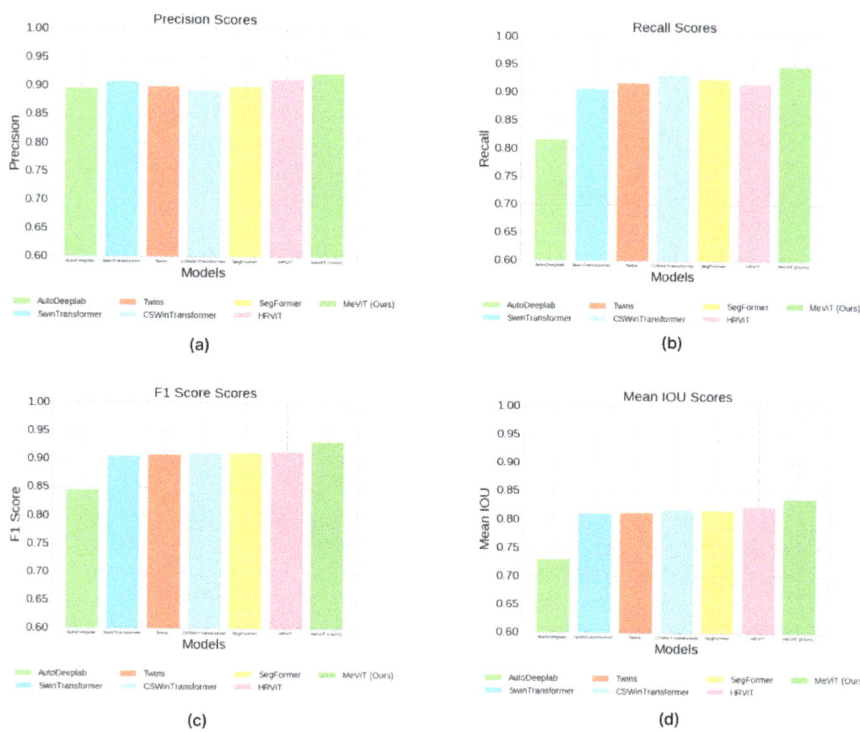

Figure 8. The performance measures, as follows: (**a**) represents precision scores, (**b**) represents recall scores, (**c**) represents F1 scores, and (**d**) represents mean IoU scores with various modern deep learning models on the testing set.

Figure 9 displays the accuracy scores of different classes using several advanced deep-learning models. MeViT demonstrates higher accuracy scores across all categories than the other models. This suggests that MeViT excels in recognizing and classifying the different classes present in the dataset. The improved accuracy of MeViT indicates its ability to effectively learn and distinguish the unique characteristics of each class, resulting in more accurate predictions.

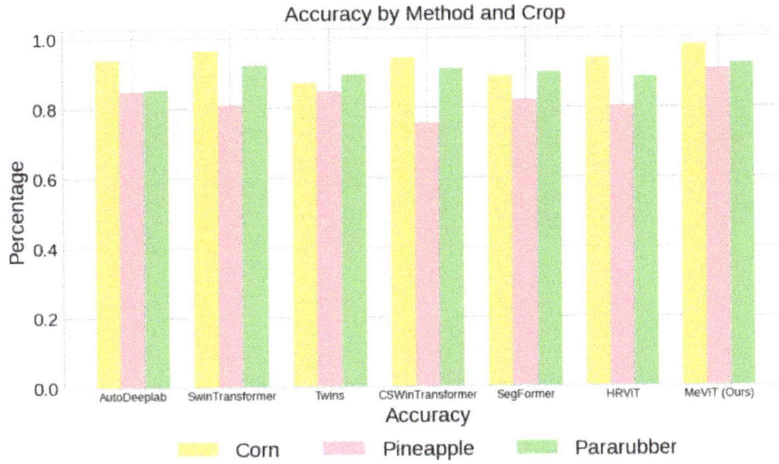

Figure 9. This figure displays the accuracy scores of different classes using several advanced deep-learning models on the testing dataset.

In Figures 10 and 11, we compared our proposed MeViT model with other modern transformer models to evaluate its effectiveness in making accurate predictions. The figures demonstrate that MeViT outperforms the baseline models by consistently producing more accurate and precise predictions, aligned better with the ground truth. MeViT's capability to capture and understand the underlying patterns and features in the data makes it a superior model for handling the complexities and variations present in the dataset.

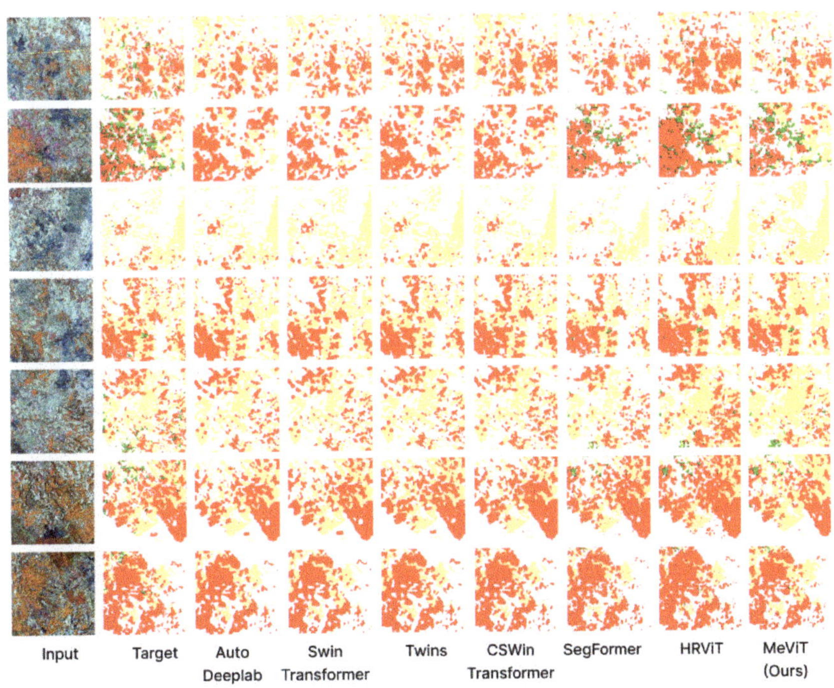

Figure 10. We compare the effectiveness of our proposed MeViT with modern transformer models, emphasizing the accurate prediction of the rubber and maize classes (red and yellow area, respectively), where our model outperforms the baselines. Red: para rubber, yellow: corn, green: pineapple.

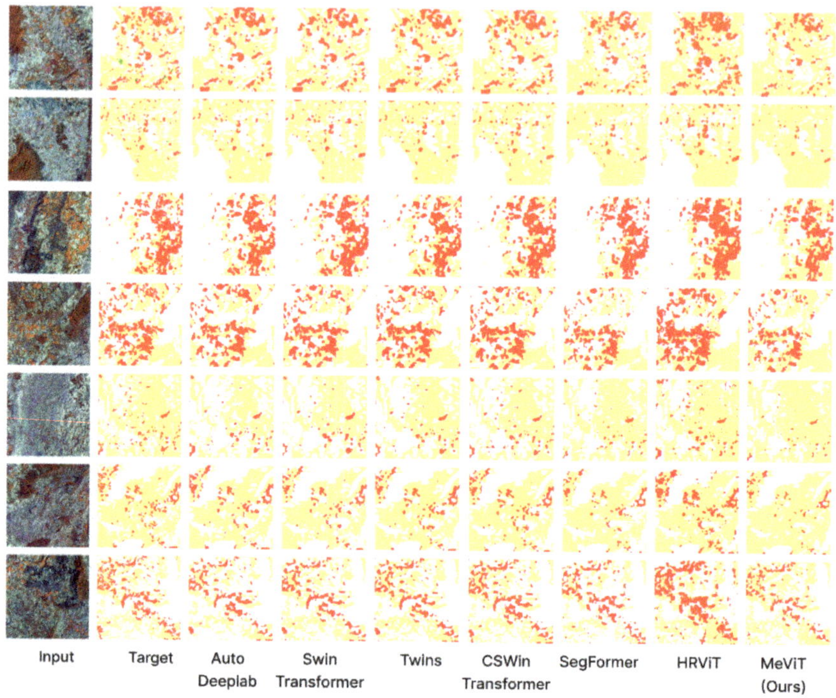

Figure 11. Our proposed MeViT model (rightmost column) is compared to modern transformer models to determine its effectiveness. We aim to showcase instances where our model successfully identifies rare categories, such as pineapples (green area). Red: para rubber, yellow: corn, green: pineapple.

The comparisons were made for different classes or categories, and MeViT consistently achieved higher accuracy and better prediction results across all types compared to the baseline models. The improved predictions by MeViT highlight its ability to capture and utilize relevant information to make accurate class assignments. This implies MeViT's effectiveness in recognizing and classifying the different objects or categories present in the dataset.

Overall, the comparison results presented in both figures provide strong evidence of MeViT's superiority over the baseline models regarding prediction accuracy and precision. These findings indicate that MeViT is a robust and reliable model for prediction tasks in computer vision, as its architecture and design enable it to effectively leverage spatial and contextual information, leading to improved prediction results. The superior performance of MeViT across different input samples and classes underscores its effectiveness in various practical applications.

6. Conclusions

In this paper, we proposed a novel deep learning method, MeVit, to perform semantic segmentation on Landsat satellite imagery for Thailand's main economic crops, such as para rubber, corn, and pineapple. Our proposed MeViT enhances vision transformers (ViTs) to learn semantically rich and spatially precise multi-scale representations by integrating medium-resolution multi-branch architectures with ViTs. We balanced the model performance and efficiency of MeViT by revising mixed-scale convolutional feedforward networks (MixCFN) with multiple depth-wise convolution paths to extract multi-scale local information.

We evaluated the effectiveness of our proposed MeViT on the publicly available dataset of Thailand scenes. We compared the results with several state-of-the-art deep learning methods such as AutoDeeplab, SwinTransformer, Twins, CSWinTransformer, SegFormer, and HRViT. Among the models compared, MeViT achieved the best performance in all evaluation metrics, including precision, recall, F1 score, and mean intersection over union (IoU). The experimental results demonstrated that our proposed MeViT outperformed existing methods and performed better in the semantic segmentation of Thailand scenes.

Eventually, our proposed MeViT approach provides a novel solution for the accurate semantic segmentation of Landsat satellite imagery for the main economic crops in Thailand. The experimental results show that our proposed method outperforms existing state-of-the-art deep learning methods and achieves the best performance in all evaluation metrics. This work contributes to remote sensing image analysis and provides a valuable tool for proper land use and land cover classification, which has significant implications for agriculture and environmental management.

As a future direction, we intend to assess MeViT on tasks that require dense prediction remote sensing, such as panoptic segmentation or crop yield forecasting. This will effectively showcase the capabilities of MeViT as a robust transformer backbone.

Author Contributions: T.P.; formal analysis, T.P.; investigation, T.P.; methodology, T.P.; project administration, T.P.; resources, T.P.; software, T.P.; supervision, T.P.; validation, T.P., C.C. and C.S.; visualization, T.P.; writing—original draft, T.P.; writing—review and editing, T.P.; funding acquisition, C.S. All authors have read and agreed to the published version of the manuscript.

Funding: This research was supported by the Ratchadapisek Somphot Fund for Postdoctoral Fellowship, Chulalongkorn University.

Data Availability Statement: If you would like to obtain the data that were used in this study, please contact the corresponding author.

Acknowledgments: This research was supported by the Ratchadapisek Somphot Fund for Postdoctoral Fellowship, Chulalongkorn University. Teerapong Panboonyuen, also known as Kao Panboonyuen, appreciates (thanks) and acknowledges the scholarship from Ratchadapisek Somphot Fund for Postdoctoral Fellowship, Chulalongkorn University, Thailand. This research was supported by the NASA Land Cover Land Use Change Grant. It was also funded by the CGIAR Global Rice Science Partnership (GRiSP) program. We greatly acknowledge the Geo-informatics and Space Technology Development Agency (GISTDA), Thailand, for providing satellite imagery used in this study.

Conflicts of Interest: The authors declare no conflicts of interest.

Abbreviations

The following abbreviations are used in this manuscript:

CNNs	Convolutional Neural Networks
HeViT	High-Resolution Vision Transformer
LULC	Land Use and Land Cover
MeViT	Medium-Resolution Vision Transformer
ViT	Vision Transformer

References

1. Scheibenreif, L.; Hanna, J.; Mommert, M.; Borth, D. Self-supervised vision transformers for land-cover segmentation and classification. In Proceedings of the IEEE/CVF Conference on Computer Vision and Pattern Recognition, New Orleans, LA, USA, 19–20 June 2022; pp. 1422–1431.
2. Pang, K.; Weng, L.; Zhang, Y.; Liu, J.; Lin, H.; Xia, M. SGBNet: An ultra light-weight network for real-time semantic segmentation of land cover. *Int. J. Remote Sens.* **2022**, *43*, 5917–5939. [CrossRef]
3. Chen, J.; Sun, B.; Wang, L.; Fang, B.; Chang, Y.; Li, Y.; Zhang, J.; Lyu, X.; Chen, G. Semi-supervised semantic segmentation framework with pseudo supervisions for land-use/land-cover mapping in coastal areas. *Int. J. Appl. Earth Obs. Geoinf.* **2022**, *112*, 102881. [CrossRef]
4. Pasquarella, V.J.; Arévalo, P.; Bratley, K.H.; Bullock, E.L.; Gorelick, N.; Yang, Z.; Kennedy, R.E. Demystifying LandTrendr and CCDC temporal segmentation. *Int. J. Appl. Earth Obs. Geoinf.* **2022**, *110*, 102806. [CrossRef]
5. Toker, A.; Kondmann, L.; Weber, M.; Eisenberger, M.; Camero, A.; Hu, J.; Hoderlein, A.P.; Şenaras, Ç.; Davis, T.; Cremers, D.; et al. Dynamicearthnet: Daily multi-spectral satellite dataset for semantic change segmentation. In Proceedings of the IEEE/CVF Conference on Computer Vision and Pattern Recognition, New Orleans, LA, USA, 18–24 June 2022; pp. 21158–21167.

6. Zhang, M.; Singh, H.; Chok, L.; Chunara, R. Segmenting across places: The need for fair transfer learning with satellite imagery. In Proceedings of the IEEE/CVF Conference on Computer Vision and Pattern Recognition, New Orleans, LA, USA, 18–24 June 2022; pp. 2916–2925.
7. Ettehadi Osgouei, P.; Sertel, E.; Kabadayı, M.E. Integrated usage of historical geospatial data and modern satellite images reveal long-term land use/cover changes in Bursa/Turkey, 1858–2020. *Sci. Rep.* **2022**, *12*, 9077. [CrossRef] [PubMed]
8. Chaves, M.E.; Soares, A.R.; Mataveli, G.A.; Sánchez, A.H.; Sanches, I.D. A Semi-Automated Workflow for LULC Mapping via Sentinel-2 Data Cubes and Spectral Indices. *Automation* **2023**, *4*, 94–109. [CrossRef]
9. Duarte, D.; Fonte, C.; Patriarca, J.; Jesus, I. Geographical Transferability of Lulc Image-Based Segmentation Models Using Training Data Automatically Generated from Openstreetmap–Case Study in Portugal. *ISPRS Ann. Photogramm. Remote Sens. Spat. Inf. Sci.* **2022**, *3*, 25–31. [CrossRef]
10. Li, Z.; Li, E.; Samat, A.; Xu, T.; Liu, W.; Zhu, Y. An Object-Oriented CNN Model Based on Improved Superpixel Segmentation for High-Resolution Remote Sensing Image Classification. *IEEE J. Sel. Top. Appl. Earth Obs. Remote Sens.* **2022**, *15*, 4782–4796. [CrossRef]
11. Desai, S.; Ghose, D. Active Learning for Improved Semi-Supervised Semantic Segmentation in Satellite Images. In Proceedings of the IEEE/CVF Winter Conference on Applications of Computer Vision, Waikoloa, HI, USA, 3–8 January 2022; pp. 553–563.
12. Wang, Z.; Yang, P.; Liang, H.; Zheng, C.; Yin, J.; Tian, Y.; Cui, W. Semantic segmentation and analysis on sensitive parameters of forest fire smoke using smoke-unet and landsat-8 imagery. *Remote Sens.* **2022**, *14*, 45. [CrossRef]
13. Chen, X.; Pan, S.; Chong, Y. Unsupervised domain adaptation for remote sensing image semantic segmentation using region and category adaptive domain discriminator. *IEEE Trans. Geosci. Remote Sens.* **2022**, *60*, 4412913. [CrossRef]
14. Ma, X.; Zhang, X.; Wang, Z.; Pun, M.O. Unsupervised domain adaptation augmented by mutually boosted attention for semantic segmentation of VHR remote sensing images. *IEEE Trans. Geosci. Remote Sens.* **2023**, *61*, 5400515. [CrossRef]
15. Wu, L.; Lu, M.; Fang, L. Deep covariance alignment for domain adaptive remote sensing image segmentation. *IEEE Trans. Geosci. Remote Sens.* **2022**, *60*, 5620811. [CrossRef]
16. Diakogiannis, F.I.; Waldner, F.; Caccetta, P.; Wu, C. ResUNet-a: A deep learning framework for semantic segmentation of remotely sensed data. *ISPRS J. Photogramm. Remote Sens.* **2020**, *162*, 94–114. [CrossRef]
17. Yuan, K.; Zhuang, X.; Schaefer, G.; Feng, J.; Guan, L.; Fang, H. Deep-learning-based multispectral satellite image segmentation for water body detection. *IEEE J. Sel. Top. Appl. Earth Obs. Remote Sens.* **2021**, *14*, 7422–7434. [CrossRef]
18. Chen, Z.; Li, D.; Fan, W.; Guan, H.; Wang, C.; Li, J. Self-attention in reconstruction bias U-Net for semantic segmentation of building rooftops in optical remote sensing images. *Remote Sens.* **2021**, *13*, 2524. [CrossRef]
19. Li, P.; Zhang, H.; Guo, Z.; Lyu, S.; Chen, J.; Li, W.; Song, X.; Shibasaki, R.; Yan, J. Understanding rooftop PV panel semantic segmentation of satellite and aerial images for better using machine learning. *Adv. Appl. Energy* **2021**, *4*, 100057. [CrossRef]
20. Yeung, H.W.F.; Zhou, M.; Chung, Y.Y.; Moule, G.; Thompson, W.; Ouyang, W.; Cai, W.; Bennamoun, M. Deep-learning-based solution for data deficient satellite image segmentation. *Expert Syst. Appl.* **2022**, *191*, 116210. [CrossRef]
21. Wurm, M.; Stark, T.; Zhu, X.X.; Weigand, M.; Taubenböck, H. Semantic segmentation of slums in satellite images using transfer learning on fully convolutional neural networks. *ISPRS J. Photogramm. Remote Sens.* **2019**, *150*, 59–69. [CrossRef]
22. Zhang, W.; Huang, Z.; Luo, G.; Chen, T.; Wang, X.; Liu, W.; Yu, G.; Shen, C. TopFormer: Token pyramid transformer for mobile semantic segmentation. In Proceedings of the IEEE/CVF Conference on Computer Vision and Pattern Recognition, New Orleans, LA, USA, 18–24 June 2022; pp. 12083–12093.
23. Gu, J.; Kwon, H.; Wang, D.; Ye, W.; Li, M.; Chen, Y.H.; Lai, L.; Chandra, V.; Pan, D.Z. Multi-scale high-resolution vision transformer for semantic segmentation. In Proceedings of the IEEE/CVF Conference on Computer Vision and Pattern Recognition, New Orleans, LA, USA, 18–24 June 2022; pp. 12094–12103.
24. Xu, L.; Ouyang, W.; Bennamoun, M.; Boussaid, F.; Xu, D. Multi-class token transformer for weakly supervised semantic segmentation. In Proceedings of the IEEE/CVF Conference on Computer Vision and Pattern Recognition, New Orleans, LA, USA, 18–24 June 2022; pp. 4310–4319.
25. He, X.; Zhou, Y.; Zhao, J.; Zhang, D.; Yao, R.; Xue, Y. Swin transformer embedding UNet for remote sensing image semantic segmentation. *IEEE Trans. Geosci. Remote Sens.* **2022**, *60*, 4408715. [CrossRef]
26. Wang, L.; Li, R.; Zhang, C.; Fang, S.; Duan, C.; Meng, X.; Atkinson, P.M. UNetFormer: A UNet-like transformer for efficient semantic segmentation of remote sensing urban scene imagery. *ISPRS J. Photogramm. Remote Sens.* **2022**, *190*, 196–214. [CrossRef]
27. Zhang, B.; Tian, Z.; Tang, Q.; Chu, X.; Wei, X.; Shen, C. Segvit: Semantic segmentation with plain vision transformers. *Adv. Neural Inf. Process. Syst.* **2022**, *35*, 4971–4982.
28. Zhou, B.; Krähenbühl, P. Cross-view transformers for real-time map-view semantic segmentation. In Proceedings of the IEEE/CVF Conference on Computer Vision and Pattern Recognition, New Orleans, LA, USA, 18–24 June 2022; pp. 13760–13769.
29. Ru, L.; Zhan, Y.; Yu, B.; Du, B. Learning affinity from attention: End-to-end weakly-supervised semantic segmentation with transformers. In Proceedings of the IEEE/CVF Conference on Computer Vision and Pattern Recognition, New Orleans, LA, USA, 18–24 June 2022; pp. 16846–16855.
30. Zhang, J.; Yang, K.; Ma, C.; Reiß, S.; Peng, K.; Stiefelhagen, R. Bending reality: Distortion-aware transformers for adapting to panoramic semantic segmentation. In Proceedings of the IEEE/CVF conference on Computer Vision and Pattern Recognition, New Orleans, LA, USA, 18–24 June 2022; pp. 16917–16927.

31. Lazarow, J.; Xu, W.; Tu, Z. Instance segmentation with mask-supervised polygonal boundary transformers. In Proceedings of the IEEE/CVF Conference on Computer Vision and Pattern Recognition, New Orleans, LA, USA, 18–24 June 2022; pp. 4382–4391.
32. Zhang, C.; Jiang, W.; Zhang, Y.; Wang, W.; Zhao, Q.; Wang, C. Transformer and CNN hybrid deep neural network for semantic segmentation of very-high-resolution remote sensing imagery. *IEEE Trans. Geosci. Remote Sens.* **2022**, *60*, 4408820. [CrossRef]
33. Qiu, C.; Li, H.; Guo, W.; Chen, X.; Yu, A.; Tong, X.; Schmitt, M. Transferring transformer-based models for cross-area building extraction from remote sensing images. *IEEE J. Sel. Top. Appl. Earth Obs. Remote Sens.* **2022**, *15*, 4104–4116. [CrossRef]
34. Liu, C.; Chen, L.C.; Schroff, F.; Adam, H.; Hua, W.; Yuille, A.L.; Fei-Fei, L. Auto-deeplab: Hierarchical neural architecture search for semantic image segmentation. In Proceedings of the IEEE/CVF Conference on Computer Vision and Pattern Recognition, Long Beach, CA, USA, 15–20 June 2019; pp. 82–92.
35. Liu, Z.; Lin, Y.; Cao, Y.; Hu, H.; Wei, Y.; Zhang, Z.; Lin, S.; Guo, B. Swin transformer: Hierarchical vision transformer using shifted windows. In Proceedings of the IEEE/CVF International Conference on Computer Vision, Montreal, QC, Canada, 10–17 October 2021; pp. 10012–10022.
36. Panboonyuen, T.; Jitkajornwanich, K.; Lawawirojwong, S.; Srestasathiern, P.; Vateekul, P. Transformer-based decoder designs for semantic segmentation on remotely sensed images. *Remote Sens.* **2021**, *13*, 5100. [CrossRef]
37. Chu, X.; Tian, Z.; Wang, Y.; Zhang, B.; Ren, H.; Wei, X.; Xia, H.; Shen, C. Twins: Revisiting the design of spatial attention in vision transformers. *Adv. Neural Inf. Process. Syst.* **2021**, *34*, 9355–9366.
38. Dong, X.; Bao, J.; Chen, D.; Zhang, W.; Yu, N.; Yuan, L.; Chen, D.; Guo, B. Cswin transformer: A general vision transformer backbone with cross-shaped windows. In Proceedings of the IEEE/CVF Conference on Computer Vision and Pattern Recognition, New Orleans, LA, USA, 18–24 June 2022, pp. 12124–12134.
39. Xie, E.; Wang, W.; Yu, Z.; Anandkumar, A.; Alvarez, J.M.; Luo, P. SegFormer: Simple and efficient design for semantic segmentation with transformers. *Adv. Neural Inf. Process. Syst.* **2021**, *34*, 12077–12090.
40. Roy, D.P.; Wulder, M.A.; Loveland, T.R.; Woodcock, C.E.; Allen, R.G.; Anderson, M.C.; Helder, D.; Irons, J.R.; Johnson, D.M.; Kennedy, R.; et al. Landsat-8: Science and product vision for terrestrial global change research. *Remote Sens. Environ.* **2014**, *145*, 154–172. [CrossRef]
41. Knight, E.J.; Kvaran, G. Landsat-8 operational land imager design, characterization and performance. *Remote Sens.* **2014**, *6*, 10286–10305. [CrossRef]
42. Loveland, T.R.; Irons, J.R. Landsat 8: The plans, the reality, and the legacy. *Remote Sens. Environ.* **2016**, *185*, 1–6. [CrossRef]

MDPI
St. Alban-Anlage 66
4052 Basel
Switzerland
www.mdpi.com

Remote Sensing Editorial Office
E-mail: remotesensing@mdpi.com
www.mdpi.com/journal/remotesensing

www.ingramcontent.com/pod-product-compliance
Lightning Source LLC
Chambersburg PA
CBHW061555220926
51850CB00005B/338
9783036593647